Intelligent Operation and Maintenance for Subsea Production Systems

Baoping Cai · Yiliu Liu · Yonghong Liu ·
Yixin Zhao · Xiaoyan Shao

Intelligent Operation and Maintenance for Subsea Production Systems

中国石油大学出版社
CHINA UNIVERSITY OF PETROLEUM PRESS

Springer

Baoping Cai
College of Mechanical and Electrical Engineering
China University of Petroleum (East China)
Qingdao, Shandong, China

Yonghong Liu
College of Mechanical and Electrical Engineering
China University of Petroleum (East China)
Qingdao, Shandong, China

Xiaoyan Shao
College of Mechanical and Electrical Engineering
China University of Petroleum (East China)
Qingdao, Shandong, China

Yiliu Liu
Department of Mechanical and Industrial Engineering
Norwegian University of Science and Technology
Trondheim, Norway

Yixin Zhao
College of Mechanical and Electrical Engineering
China University of Petroleum (East China)
Qingdao, Shandong, China

ISBN 978-981-96-2676-2 ISBN 978-981-96-2677-9 (eBook)
https://doi.org/10.1007/978-981-96-2677-9

Jointly published with China University of Petroleum Press
The print edition is not for sale in China (Mainland). Customers from China (Mainland) please order the print book from: China University of Petroleum Press.
ISBN of the Co-Publisher's edition: 978-7-5636-8478-6

This work was supported by National Science Fund for Distinguished Young Scholars (52325107).

This Springer imprint is published by the registered company Springer Nature Singapore Pte Ltd.
The registered company address is: 152 Beach Road, #21-01/04 Gateway East, Singapore 189721, Singapore

Foreword by Carlos Guedes Soares

The field of subsea production systems has witnessed tremendous advancements in recent years, driven by the increasing demand for energy and the need to explore more challenging offshore environments. As these systems become more complex and integrated, the challenge of maintaining their reliability grows exponentially. Traditional operation techniques and maintenance strategies are no longer sufficient; what is now needed are intelligent methodologies capable of predicting failures and optimizing maintenance schedules. This book, *Intelligent Operation and Maintenance for Subsea Production Systems*, represents a critical contribution to the growing knowledge of intelligent operation and maintenance schedules.

The authors have structured this book into three interconnected parts: Fault Diagnosis, Fault Prognosis, and Maintenance. This organization reflects the operational lifecycle of subsea systems and reinforces the coherence and continuity of the subject matter. The authors delve into fault diagnosis in the first part, revealing the critical interplay between real-time monitoring and fault identification. Incorporating theoretical insights with practical case studies enriches this part, supporting academic understanding and practical utility. The move to fault prognosis represents an evolution in maintenance strategies, wherein future failures can be anticipated rather than reacted to. This predictive dimension emphasized in the book's second part offers a comprehensive toolkit for researchers and practitioners alike. The punchline of this strict exploration lies in the intelligent maintenance strategies, where already-acquired insights from the prior parts come together to form a proactive approach. Extending the understanding of spare parts management and condition-based monitoring sets the stage for a future where operational and environmental sustainability coexists harmoniously.

This book is a technical manual and a forward-looking vision for future maintenance strategies in subsea operations. In the academic realm, this book will undoubtedly inspire new research directions and offer a rich source of knowledge for graduate students, postdoctoral researchers, and faculty alike. I am confident that this work will inspire further research and development in predictive maintenance, and I strongly recommend it to anyone seeking to deepen their understanding of this critical area.

Carlos Guedes Soares
Distinguished Professor
University of Lisbon, Portugal
Member of the Portuguese National
Academy of Engineering
Lisbon, Portugal

Foreword by Min Xie

In the evolving field of subsea production engineering, we are confronted with the dual challenge of enhancing operational efficiency while ensuring integrity and safety. As educators and researchers, we have the responsibility of developing frameworks and methodologies that bridge the gap between theoretical insights and practical realities. *Intelligent Operation and Maintenance for Subsea Production Systems* is an exemplary endeavor that offers a comprehensive guide that aligns academic inquiry with industrial application.

The three-part structure of this book provides a coherent and systematic exploration of this complex subject matter, encompassing fault diagnosis, prognosis, and maintenance strategies. This book stands out for its rigorous approach to fault diagnosis and prognosis, addressing multiple aspects of system reliability with clarity and depth. The structured discussion of topics such as failure propagation, diagnostic techniques, and remaining useful life prediction provides a framework that is both comprehensive and accessible to researchers, engineers, and professionals. Besides, in the academic area, there has been a growing interest in predictive maintenance, but few resources tackle the subject with the depth and clarity found in this book. It emphasizes the importance of integrated insights from both fault diagnosis and prognosis. One of the key strengths of this work is its interdisciplinary approach—drawing from mechanical engineering, systems theory, and artificial intelligence—creating a fusion of knowledge that is necessary to tackle the complex nature of subsea production systems.

As a professor involved in predictive maintenance and reliability engineering, I find this work to be a timely contribution. The integration of failure dependence models, real-time data analysis, and machine learning-based prediction algorithms offers fresh insights into how maintenance practices can evolve in a more intelligent and data-driven direction. Importantly, the author does not shy away from addressing the limitations of current methods, offering thoughtful discussions on how new techniques such as deep learning and simulation-driven validation can improve predictive accuracy and system robustness.

This book not only enhances our understanding of predictive maintenance but also opens new avenues for research and collaboration between academia and industry. I have no doubt that this book will serve as an invaluable resource and a foundational text for graduate students, researchers, and practitioners alike.

Min Xie
Academician of the European
Academy of Sciences and Arts
Fellow of IEEE
Hong Kong, China

Preface

In the ever-evolving realm of offshore oil and gas exploration and production, the complexity and scale of subsea production systems continue to increase. Failure of subsea production systems could cause serious consequences to the development of offshore oil. These systems, operating in harsh and remote environments, demand a high level of reliability and efficiency. Ensuring their optimal performance is a multifaceted challenge that necessitates advanced methodologies in fault diagnosis, prognosis, and maintenance, each required to meet the growing demands of modern subsea operations. With technological advancements and the integration of intelligent tools, the future of subsea system management calls for innovative solutions.

Intelligent Operation and Maintenance for Subsea Production Systems aims to address this critical need by presenting a comprehensive guide to the latest developments and best practices in the field from both academic and engineering perspectives. This book is structured into three distinct yet interrelated parts: Fault Diagnosis, Fault Prognosis, and Maintenance. Each part is crafted to provide readers with both theoretical foundations and practical applications, enhancing understanding of the basic mechanisms as well as the intelligent technologies that underpin subsea operations and maintenance.

Part I Fault Diagnosis delves into the methods and models used to detect and identify faults in subsea production systems. As these systems operate under extreme conditions, early and accurate fault detection is essential to prevent various failures and minimize downtime. This part explores various diagnostic techniques, from traditional approaches to machine learning algorithms and model-driven methods, offering insights into their implementation and effectiveness.

Part II Fault Prognosis shifts the focus to predicting the remaining useful life of subsea equipment and systems. By predicting potential faults before they occur, operators can make more informed, proactive decisions that prolong the lifespan of their assets, reduce unexpected downtime, and optimize operational efficiency. This part covers a range of classical and intelligent prognostic models, emphasizing the importance of data-driven strategies and other advanced models in enhancing predictive accuracy.

Part III Maintenance brings the discussion full circle by examining the strategies and practices for maintaining subsea production systems. Maintenance is not merely a reactive process but a proactive approach to ensure long-term system reliability and performance. This part addresses various maintenance paradigms, including condition-based maintenance, emergency maintenance, and the integration of multiple considerations to examine maintenance strategies.

As the subsea industry keeps advancing technological innovations, the information and knowledge in this book are set to greatly enhance the field. By bridging the gap between theory and practice, it offers a holistic perspective on the intelligent operation and maintenance of subsea production systems. The interdisciplinary nature of this book makes it a valuable resource for target readers, including students, researchers, practitioners, and engineers in the fields of subsea engineering, operation and maintenance engineering, and artificial intelligence. It is assumed that the readers have basic knowledge of fault diagnosis, fault prognosis, and maintenance, preferably related to subsea production system. We hope readers will find this work both educational and motivating and that it will encourage more research and development in this field.

We extend our heartfelt gratitude to the National Natural Science Foundation of China (No. 52325107, 52171287), the National Key Research and Development Program of China (No. 2022YFC2806102), the Science Foundation for Distinguished Young Scholars of Shandong Province (No. ZR2022JQ25), the Taishan Scholars Project of Shandong Province (No. tsqn201909063), and the Fundamental Research Funds for the Central Universities (No. 24CX10006A) for their generous support and funding, which has been essential in the development of this book.

Some of the research was conducted in collaboration with numerous contributors who have made this book possible. Special thanks go to the academic institutions, particularly Department of Mechanical and Industrial Engineering, Norwegian University of Science and Technology, for their significant contributions to the research and development of the content within this book. We are also deeply grateful to other academic institutions that supported this work, including Ocean University of China, South China University of Technology, Beihang University, Hunan University, Imperial College London, Universiti Teknologi PETRONAS, University of Bologna, China University of Petroleum (Beijing), Texas A&M University, Shenzhen Urban Public Safety Technology Institute, Massachusetts Institute of Technology, National Engineering Research Center of Marine Geophysical Prospecting and Exploration and Development Equipment, National Engineering Laboratory of Offshore Geophysical and Exploration Equipment, China Academy of Safety Science and Technology. Additionally, we extend our gratitude to the companies that provided invaluable industry insights and support, including CNOOC Safety Technology Services Co., Ltd., Shenzhen Branch of CNOOC (China) Co., Ltd., Liuhua Oilfield Operation Area of Shenzhen Branch of CNOOC (China) Co., Ltd., CNOOC EnerTech, Safety and Environmental Protection Branch, CNOOC Offshore Engineering Solutions Co., Ltd., Deepwater Technology Center COOEC Subsea Technology Co. Ltd., BGP Inc., China National Petroleum Corporation, CNPC

Offshore Engineering Company, Ltd., Rongsheng Machinery Manufacture, Ltd., WEFIC Ocean Equipment Manufacturing Co., Ltd., Weichai Power Co., Ltd., etc.

A special acknowledgment to the dedicated Ph.D. students whose research has greatly enriched this work. Their commitment to advancing knowledge in intelligent fault diagnosis, fault prognosis, and maintenance for subsea production systems is commendable. We would like to specifically mention Xiaoyan Shao, Xiangdi Kong, Chao Yang, Yuandong Wang, Yanping Zhang, Xuelin Liu, Rongkang Li, Rui Zhang, Haojie Li, Hongyan Fan, etc., whose efforts have been crucial in making this book comprehensive.

Finally, we extend our appreciation to the editors of Springer and China University of Petroleum Press, as well as all those who have supported the creation of this book. Your expertise and dedication have been invaluable in bringing this book to fruition.

Qingdao, China — Baoping Cai
Trondheim, Norway — Yiliu Liu
Qingdao, China — Yonghong Liu
Qingdao, China — Yixin Zhao
Qingdao, China — Xiaoyan Shao

Introduction

The subsea production industry operates in some of the most challenging environments on Earth, where extreme pressure, remote locations, and harsh conditions present constant threats to system integrity. Subsea production systems, which form the backbone of offshore oil and gas extraction, consist of highly complex and interdependent components such as subsea Christmas trees, pipelines, valves, and hydraulic control systems. The reliability and availability of these systems are significant, as failures can lead to costly downtime, environmental disasters, and significant operational hazards. The traditional approach to maintenance, which relies on corrective or scheduled preventive strategies, is no longer sufficient in the face of growing operational demands and environmental pressures. As industries move towards a more intelligent and data-driven future, predictive maintenance strategies have emerged as indispensable solutions for improving system availability, minimizing downtime, and reducing the cost of operations and maintenance. This book, *Intelligent Operation and Maintenance for Subsea Production Systems*, provides a comprehensive framework for addressing these challenges through exploration of the advanced methodologies and tools necessary for fault diagnosis, fault prognosis, and maintenance strategies. The primary aim of this book is to offer both theoretical insights and practical approaches for the intelligent operation and maintenance of subsea production systems, by integrating academic research with industrial applications.

This book is structured into three parts, each addressing a critical aspect of the operation and maintenance lifecycle of subsea production systems: Fault Diagnosis, Fault Prognosis, and Maintenance.

- Part I focuses on fault diagnosis, which involves identifying and understanding faults as they occur in real-time or retrospectively. As subsea systems become increasingly reliant on a network of sensors and monitoring technologies, the ability to accurately diagnose issues in components such as valves, control systems, and hydraulic systems is essential. This part introduces digital twin techniques for identifying composite and minor faults, digital twin-assisted methods, and optimizing sensor placement to enhance diagnosis accuracy. Chapters also cover the fault diagnosis methods for subsea control systems, electro-hydraulic systems, and the full-stage stable method for subsea production system. By combining theoretical principles with practical cases, this part equips readers with the necessary tools to handle various fault scenarios in subsea production systems.

- Part II of the book moves beyond fault diagnosis to fault prognosis, where the focus shifts to predicting future failures before they occur. Predictive maintenance is a key element in reducing unplanned downtime and ensuring the operational continuity of subsea production systems. In this part, various models are examined for predicting the Remaining Useful Life (RUL) of critical components, which is a cornerstone of predictive maintenance strategies. Seven chapters cover a range of topics, including RUL prediction based on both small and large datasets, multi-cause failure prediction, hybrid models for multi-stage degradation, and the role of failure dependence in failure prediction. Special emphasis is placed on how industrial incomplete information, numerical simulations, and artificial intelligence models can be integrated to create robust predictive frameworks that predict faults in subsea production systems.
- Part III, final part of the book, focuses on intelligent maintenance strategies, which bring together the insights gained from fault diagnosis and prognosis to form a proactive approach to maintenance. This part includes six chapters that discuss topics such as spare parts management based on condition monitoring, maintenance delays, and discrete-state monitoring approaches. It also explores strategies for managing heterogeneous failure dependences, joint optimization of emergency maintenance with Condition-based Maintenance (CBM), and sustainable maintenance strategies. This part highlights how intelligent optimization can guide maintenance decisions, reduce maintenance costs, and contribute to the overall sustainability of subsea operations.

Through these three interconnected parts, this book provides a comprehensive and practical guide to the intelligent operation and maintenance of subsea production systems. By combining fault diagnosis, fault prognosis, and advanced maintenance strategies, operators can significantly enhance the reliability, safety, and efficiency of subsea production systems. The theoretical models presented throughout the book are complemented by real-world case studies, demonstrating the practical application of these methodologies in various subsea environments. Whether the reader is an academic researcher, an engineer, or an industry professional, this book offers valuable insights into the cutting-edge methods for managing subsea production systems in a rapidly evolving industry.

Contents

Part I
Fault Diagnosis

In subsea production systems, fault diagnosis is an essential aspect of maintaining system integrity and ensuring continuous operation. These systems operate in complex and often extreme environments, where even minor faults can escalate into critical failures if not addressed promptly. Fault diagnosis in such systems requires advanced methodologies that account for the composite nature of faults, the vast array of sensors, and the intricate interplay between hydraulic, electric, and mechanical subsystems. Part I *Fault Diagnosis* explores key techniques and approaches to address these challenges effectively.

This part begins with a focus on diagnosing composite and minor faults, where the development of digital twin models and enhanced validation methods offer robust solutions for detecting and understanding fault behaviors in real time. The unique Bidirectional Data Consistency Evaluation framework is another critical contribution, which has proven particularly useful in measuring the difference between virtual and real data in subsea equipment. An innovative approach to optimal sensor placement is presented to maximize fault diagnosis efficiency, ensuring that sensors are strategically positioned to capture critical data. Additionally, the diagnosis of faults in subsea control systems is explored, highlighting the complexity of hydraulic control systems and redundant closed-loop feedback control systems. The issue of concurrent faults is addressed by introducing models capable of diagnosing multiple simultaneous failures in electric-hydraulic systems, which is especially pertinent to complex operations like subsea blowout preventer (BOP) systems. Finally, intelligent full-stage stable fault diagnosis method for subsea production systems is examined, combining model-based and data-driven fault diagnosis models. By integrating theory with practical case studies and experiments, this part provides a comprehensive framework for advancing fault diagnosis in subsea production systems.

Chapter 1
Fault Diagnosis for Composite Faults and Minor Faults

Abstract To deal with the problem of weak fault characteristics and misdiagnosis in one-time fault diagnosis, a cross-validation enhanced digital twin-driven fault diagnosis methodology for is proposed. The models for fault diagnosis using Bayesian networks are constructed, integrating both virtual and real data, specifically designed for addressing composite faults and minor faults. A multiple diagnosis mechanism of diagnosis, verification and re-diagnosis is established through data exchange between the digital twin model and the fault diagnosis model. This fault diagnosis method is applied to a subsea production system in the South China Sea. The results indicate that compared to a one-time fault diagnosis, this method can identify minor faults and composite faults in the subsea production system accurately.

1.1 Introduction

Monitoring the status of subsea production systems is a guarantee for safe production. Minor and composite failures are common faults. Minor faults are the early forms of faults in production systems. The characteristic of it is weak signal features, and the system changes caused by faults are not obvious. Usually, major faults and performance degradation typically evolve from minor faults. Obtaining information of minor faults can ensure the normal operation of subsea production systems. Influenced by the high cost of handling faults, faults that do not affect the system are usually not handled. This would lead to the coexistence of multiple faults and frequent occurrence of composite faults. Obtaining the location of faults is an important prerequisite for intelligent operation and maintenance of subsea production system.

The current fault diagnosis method applied in field is the threshold-based method. When the sensor exceeds the threshold, an alarm is triggered, and the abnormal location is manually determined based on the alarm. This method is difficult to diagnosis minor faults because minor faults are difficult to detect large system changes. In addition, inference algorithms such as neural networks, deep learning, and Bayesian networks are studied and used for fault diagnosis of subsea production systems. They accelerate the diagnosis speed and improve the intelligence level of diagnosis.

B. Cai et al., *Intelligent Operation and Maintenance for Subsea Production Systems*,
https://doi.org/10.1007/978-981-96-2677-9_1

More importantly, they can provide the location of the fault automatically, speeding up the process of obtaining fault information. These methods can be called once diagnosis, which involves obtaining sensor data and inputting it into a diagnostic model to obtain diagnostic results. This method is prone to misjudgment and missed diagnosis, resulting in a high misdiagnosis rate.

A cross-validation enhanced digital twin-driven fault diagnosis methodology for minor faults and composite faults is proposed. Control, loss and fault parameters are presented and used for building a digital twin model. Bayesian Network is used to establish a fault diagnosis model. Single and cumulative errors are used to measure the difference between digital twin models and physical systems. A verification feedback method is used to check the diagnosis results. A composite fault diagnosis mechanism is proposed for fault diagnosis of composite faults. Data of four days of a subsea production system in the South Sea of China is used to demonstrate the proposed methodology. The results show that the method can improve the diagnosis accuracy for minor faults and composite faults effectively.

1.2 Digital Twin-Driven Fault Diagnosis Model for Composite Faults

The digital twin-driven fault diagnosis method that integrates both virtual and real data is shown in Fig. 1.1. During a progress of fault diagnosis, the digital twin-driven fault diagnosis goes through three different stages in which various data are obtained and used in separate models. When a fault occurs, the output of the physical system deviates from normal value under its influence. In a subsea production control system, the output is in the form of the pressure sensor signal and the monitor signal of the electronic system, such as current and voltage. At this moment, the digital twin model is still running to simulate the normal state of physical system because the diagnosis results have not been obtained. At the same time, the error between the output of digital twin model and the physical system is enormous. Based on these errors, it is easy to judge whether a fault in the system exists. Therefore, a judgment method should be used at the first stage. When the system is considered to have a fault, the data of the physical system are collected and input into the fault diagnosis model in the second stage. When only one obvious fault occurs, the existing fault diagnosis methods can achieve a good diagnostic accuracy (Feng and Zhao 2021). However, when minor faults occur, the effect of fault diagnosis is poor. Therefore, it is necessary to verify the diagnostic result with the help of the digital twin model in the third stage. The fault diagnosis results can also be input into the digital twin model so that the digital twin model can simulate an abnormal system.

Through the above stages, a dynamic digital twin driven fault diagnosis of the subsea production control process can be realized. However, according to the on-site feedback, during the operation of the subsea production, some existing faults cannot be removed in time. Thus, a fault diagnosis model for composite faults that combines

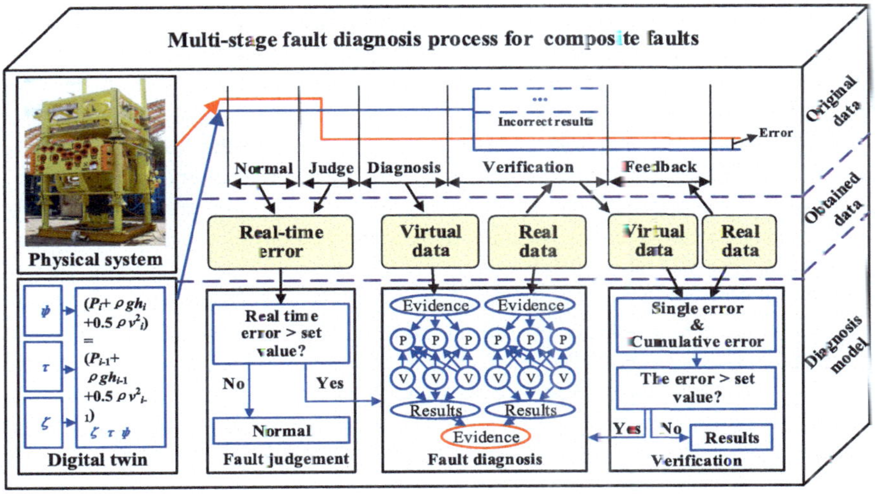

Fig. 1.1 Digital twin-based fault diagnosis method

the real data from physical system and the virtual data from digital twin model is needed. The proposed method encompasses three key models, a dynamically updated digital twin model, a fault diagnosis model for composite faults combining virtual and real data, and a dynamic error calculation model, all of which can be readily obtained.

1.2.1 Dynamically Updated Digital Twin Model

The hydraulic control system regulates fluid flow through valves and pipelines to achieve specific functions. Thus, the description state of the fluid is the core of the digital twin modelling of the subsea hydraulic control system. Bernoulli Equation is a form of energy conservation in fluid mechanics, which is usually used to describe fluid state (Nisar et al. 2022).

In a subsea production control system, three pieces of information are essential: pressure loss, control, and fault information. The parameters are usually updated with the valve as a node to ensure accurate pressure and flow information. For a certain valve, the digital twin model before and after the valve is shown below.

$$p_i + \rho g h_i + \frac{1}{2}\rho v_i^2 = \left(p_{i-1} + \rho g h_{i-1} + \frac{1}{2}\rho v_{i-1}^2\right)\zeta_{i-1}\tau_{i-1}\varphi_{i-1} \tag{1.1}$$

where p_i is the pressure of the i th valve, ρ is the density of fluid, g is acceleration of gravity, v_i is flow rate of fluid near the ith valve. ζ, τ, φ are loss, control, and fault parameters, respectively.

From Eq. (1.1), three parameters are added to the traditional Bernoulli Equation, that is ζ, τ, φ, ζ is the pressure loss. The local pressure loss and the pressure loss along the way are difficult to obtain in the subsea production control system, making it difficult to calculate the pressure loss according to the actual fluid Bernoulli. Data in field show that a pressure loss parameter can replace it. τ is the control parameter, and the calculation method is shown in Eq. (1.2). The control parameter is necessary to the valve closure, which affects the fluid state directly before and after the valve. This information can be easily obtained by reading the control signal of the system. φ is the state of the local circuit. Running to simulate the normal state of the system is a part of the functions of a digital twin model. More importantly, running to simulate an abnormal state is another function. Unlike the simulation model, the digital twin model simulates the physical system's abnormal state just because the physical system is abnormal. For this, the φ should be updated promptly and its value is decided according to the diagnosis result. Thus, a feedback algorithm is developed to assign a value to this parameter, introduced in next section.

$$\tau = \begin{cases} 0.9 + \varepsilon & \textit{the valve is opened} \\ 0.2 - \sigma & \textit{the valve is closed} \end{cases} \tag{1.2}$$

where ε and σ are adjusting parameters used to fine-tune the control parameters under different conditions. In most hydraulic systems, the closed valve means that no fluid flows through the valve. Similarly, the open valve means that the fluid can pass through completely. However, the function loss is inevitable with the use of the valve. A small control parameter such as 0.9 for an opening valve and 0.2 for a closing valve is appropriate. In fact, the performance of most valves is better than the set parameters. Thus, two adjustment parameters are set to ensure the accuracy of the digital twin model. Their change will not exceed 0.05, which is very small compared with the control parameters.

1.2.2 Fault Diagnosis Models for Composite Faults

BN is an effective tool for fault diagnosis. The feature of BN is that the output results are expressed in the form of probability. When BN is applied to fault diagnosis, the probability approximates the degree of component failure. In this study, it is necessary to receive the failure degree of a component so that the local circuit φ state can be obtained. Usually, establishing a fault diagnosis model includes three steps: structure modeling, parameter modeling, and fault reasoning.

The structure of a fault diagnosis model is shown in Fig. 1.2. In a subsea production control system, the pressure transmitters (PT, P in Fig. 1.2) and valves (V in Fig. 1.2) are the main components. Normally, the failure of valves can cause changes in the indication of PT. Thus, the valves should be the parent node of the PT node. In this model, each node contains two states: normal (N) and abnormal (A). In the beginning, the data can be obtained from the data of PT. Thus, the input of the diagnosis model is P_1–P_m. The results can be obtained by diagnosing with original data. The verifying information will be obtained after the above results are verified by digital twin. Hence, a diagnosis model combining the original data and verifying results should be established to optimize the results. Figure 1.2 shows that in the process of fault diagnosis with cross-validation, some states of the parent node are input as known information. This information is obtained from the feedback of the digital twin model.

With a cross-validation fault diagnosis model, signal fault can be diagnosed easily and has satisfactory accuracy. However, when composite faults exit, the verification model can cause problems. One fault is obtained and the other is ignored. Thus, a re-diagnosis model is needed as shown in Fig. 1.2. At this stage, the digital twin model is running in a state with certain faults, and the error between digital twin model and

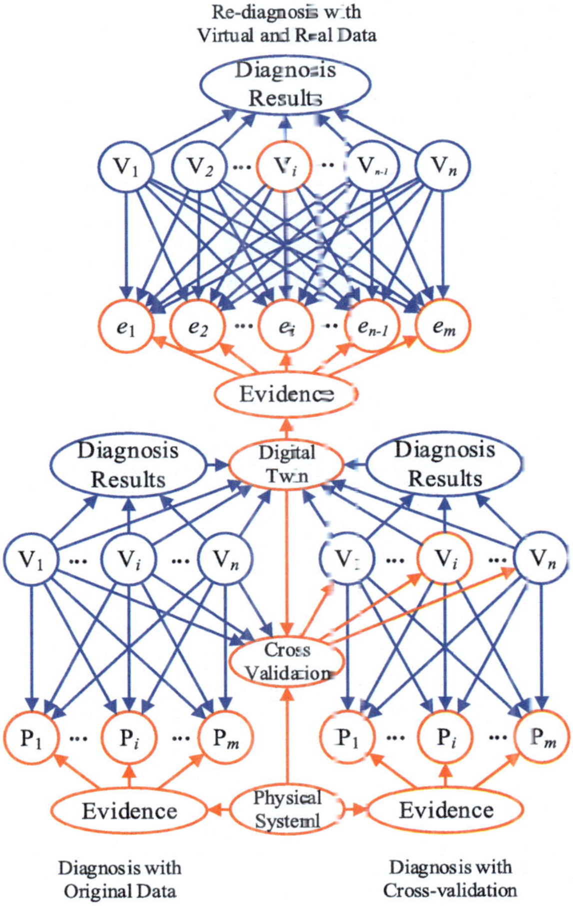

Fig. 1.2 Fault diagnosis model for composite faults

physical system is affected by the fault that has not been diagnosed. Therefore, the parent node of this model is the valve, and the child node is the deviation between the virtual and actual data.

Once the structure of the diagnosis model has been determined, the prior probabilities and the condition probabilities are required to be specified. Two methods can be used to obtain them. The first is through experience, and the second is through training. In this study, the SCM data are challenging, especially fault data. The logic of the hydraulic system is straightforward and does not need to be explored if a schematic diagram is obtained. Thus, prior and condition probabilities are determined by experience. In a fault diagnosis model built using BN, the prior probabilities are not critical factors of all the child nodes, and conditional probabilities are given to obtain a posterior probability that characterizes the possibility of a fault. For valves, the probability of failure is far lower than expected. Referring to existing research (Tang et al. 2022), the prior probability of failure is 4%, and the normal probability is 96%. Conditional probability is obtained according to the relationship between hydraulic circuit components. For example, when a valve fails, the probability of an associated abnormal PT is 95%. In the process of re-diagnosis, when a valve fails, the probability of an associated abnormal error is 93%. In this way, a conditional probability table is obtained.

After the diagnosis model is constructed, it can be used for fault diagnosis. For a node X with parent nodes U and children nodes Y, the conditional probability from U to X is defined as $\pi_X(U)$, and the conditional probability from Y to X is defined as $\lambda_Y(X)$. For some known evidence E, the joint marginal for the family of X with the parent nodes U_i and children nodes Y_i is given by

$$P(XU, E) = \lambda_E(X)\Theta_{X|U}\prod_i \pi_X(U_i)\prod_j \lambda_{Y_j}(X) \tag{1.3}$$

where $\lambda_E(X)$ is evidence indicators and $\lambda_e(X) = 1$ if $x \epsilon X$ is consistent with evidence e $\epsilon\, E$ and zero otherwise. The causal and diagnostic massage can be defined as follows:

$$\lambda_X(U_i) = \sum_{XU\setminus\{U_i\}} \lambda_E(X)\Theta_{X|U}\prod_{k\neq i} \pi_X(U_k)\prod_j \lambda_{Y_j}(X) \tag{1.4}$$

$$\pi_{Y_j}(X) = \sum_U \lambda_E(X)\Theta_{X|U}\prod_i \pi_X(U_i)\prod_{k\neq j} \lambda_{Y_k}(X) \tag{1.5}$$

A node can be sent to a neighbor only after it has received messages from all other neighbors. When a node has a signal neighbor, it can immediately send a message to that neighbor, which includes the life node X with a single parent U, for which

$$\lambda_X(U) = \sum_X \lambda_E(X)\Theta_{X|U} \tag{1.6}$$

It also includes a root node X with a single child Y, for which

$$\pi_Y(X) = \lambda_E(X)\Theta_X \tag{1.7}$$

During fault diagnosis, the data of PT are input into the fault diagnosis model, and posterior probability of fault nodes is updated using the belief propagation algorithm above. The new data should be mapped to the adjacent state space by multiplying by the load matrix with adjacent frequency and modulation index. Generally, the larger the posterior probability of fault, the higher the possibility of occurrence of the corresponding fault. The fault diagnosis methodology can offer the probability of a fault, but not draw a definite diagnostic result. Instead, a threshold needs to be set. When the failure probability of a component exceeds this threshold, the component is considered as a failure. According to engineering experience, it is usually set to 78%.

When a posterior probability is obtained, it is assigned to φ as

$$\varphi = \begin{cases} 1 - P(V) & P(V) \geq 0.65 \\ 0.9 + \kappa & P(V) < 0.65 \end{cases} \tag{1.8}$$

where κ is the adjusted parameter for fine-tuning the digital twin model, and $P(V)$ is the probability of the corresponding valve failure.

1.2.3 Dynamic Error Calculation

In this study, errors are used to measure the difference between the digital twin model and the physical system. Two errors are used in this study. The calculation method is shown in Eq. (1.9) and Eq. (1.10):

$$E_s = \frac{|P_r - P_{DT}|}{P_r} * 100\% \tag{1.9}$$

$$E_p = Sum(E_s) \tag{1.10}$$

where E_s is the single error. $E_s = [e_{s1}, e_{s2}, \ldots, e_{sn}]$, e_{si} is the i-th single error. P_r is data from the physical system, and P_{DT} is data from the digital twin model. $P_r = [P_{r1}, P_{r2}, \ldots, P_{rn}]$, $P_{DT} = [P_{DT1}, P_{DT2}, \ldots, P_{DTn}]$. And P_{ri} is the data from the i-th PT in a physical system. E_p is a cumulative error.

In fact, the error exists all the time. When a fault occurs, the error will become larger. Hence, an error threshold is necessary. In this study, they are set through actual data.

With the above algorithms, the digital twin-driven fault diagnosis method combining virtual and real data for composite faults is shown in Fig. 1.3. First, the error between digital twin model and physical system is calculated. If the error exceeds a set threshold, one or more faults are in the system. The data from physical system is used for fault diagnosis of a single fault. When the results of fault diagnosis are obtained, the fault parameter φ is updated so that the digital twin model can run to simulate a fault state. Currently, the error is recalculated to determine whether the fault diagnosis result is correct. If the error still exceeds the threshold value, the current diagnosis result is wrong, and this information will be fed back to the diagnosis model for cross-validation diagnosis. If composite faults occur, all errors in this process will exceed the threshold. The fault that causes the minimum error will be considered as the one that has occurred and can be introduced into the digital twin model. New errors are calculated and input into a re-diagnosis model with the combination of virtual and real data. In this way, the single fault and composite faults of the subsea production control system can be diagnosed.

1.3 Cross-Validation Enhanced Fault Diagnosis Model for Minor Faults

The cross-validation enhanced digital twin-driven fault diagnosis methodology is shown in Fig. 1.4. In this method, four parts are included: physical system, fault diagnosis model, digital twin model, and error discrimination model. The physical system is the monitored object and many Sensors (*S* in the Fig. 1.4) are installed in it. Thus, the data used for establishing digital twin and fault diagnosis models are mainly obtained from physical system. When the control information and system parameters are received, the digital twin model can simulate the normal state of the physical system. At the same time, data from sensors are obtained by the fault diagnosis model for judging the physical system's state. The fault parameters in the digital twin model are assigned when a fault message is received after reasoning by the fault diagnosis model. If a fault of a component occurs, the digital twin model would run to simulate the fault state of the physical system. Getting completely correct results all the time is a challenge. Therefore, a verification method is necessary. The error discrimination model is used to calculate the single error and cumulative error. When the error exceeds the limit value, the result of fault diagnosis is considered wrong, which would be input into the diagnosis model as information for re-diagnosis to obtain a new result. It is noted that the process of diagnosis verification and re-diagnosis is continuous until the error between the digital twin model and the physical system is less than the threshold. In a subsea production control system, the input of the digital twin model is the inlet pressure, and the output is the value of Power Transmitter (PT) mounted at different locations. This framework relies on three key algorithms: a digital twin modeling method integrating control, loss, and fault parameters; fault diagnosis algorithms combined feedback data from digital twin; and an interactive promotion algorithm between digital twin and fault diagnosis models.

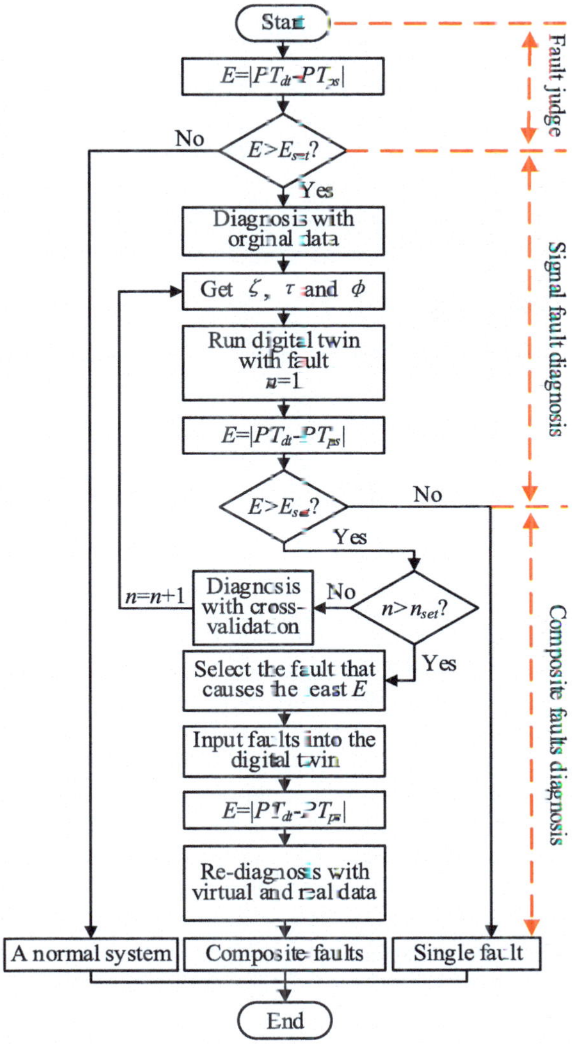

Fig. 1.3 Process of digital twin-driven fault diagnosis combining virtual and real data

1.3.1 Digital Twin Model Integrating Control, Loss and Fault Parameters

Generalized digital twin refers to the simulation process to fully use physical models, sensors, operating history, and other data to integrate multiple disciplinary and multiple scales. As a mirror image of the physical product in the virtual space, the digital twins reflect the full life cycle process of the corresponding physical product (Lv et al. 2023). When the digital twin is applied to the subsea production systems, a 4d model of digital twin is required. In this model, four modules are included, that

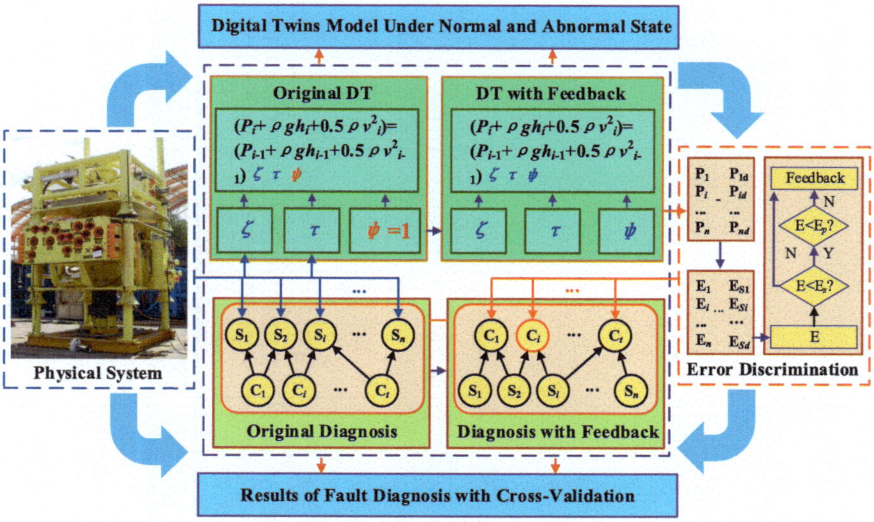

Fig. 1.4 The cross-validation enhanced digital twin driven fault diagnosis methodology

is, geometric model, production state model, health management model and environmental model. The geometric model is used to display the appearance of subsea equipment. In addition, it is used to display production states and health states. In this model, some techniques such as flat display, 3D display, and virtual reality can be used. The production state model is used to describe the current production states. In this model, sensor signals, control signals, and fault signals are required. The health management model usually includes fault diagnosis, fault prognosis and maintenance algorithm. They are used to judge the state of system, calculate the RUL and provide a good maintenance plan. The environmental model includes lots of parameters that affect the subsea production system. The production state model directly determines the performance of fault diagnosis. Because almost all the data used for fault diagnosis are from production state model. The geometric model is just used to display the results of fault diagnosis, and its quality does not affect the accuracy of diagnosis. In this way, the production state model is the focus of the research.

1.3.1.1 Bernoulli Equation and Pressure Loss of Real Fluid

Under the control command of valves and the guidance of pipelines, the fluid in the hydraulic control system flows achieve corresponding functions. In a subsea production control system, the description state of the fluid is the core of the digital twin modeling of the subsea hydraulic control system. Bernoulli Equation is a form of energy conservation in fluid mechanics, which is usually used to describe fluid state (Nisar et al. 2022).

A flow section of the tube is taken to get the actual Bernoulli Equation of fluid. The areas at both ends of the flow tube are A_1 and A_2, respectively. And then, a tiny stream is taken out of the flow tube, and the cross-sectional areas are dA_1 and dA_2. Besides, the pressure, flow rate, and height are p_1, u_1, z_1 and p_2, u_2, z_2. Then, Eq. (1.11) can be obtained after integrating the flow tube separately.

$$\int_{A_1}\left(\frac{p_1}{\rho g}+z_1\right)u_1 dA_1+\int_{A_1}\frac{u_1^2}{2g}u_1 dA_1=\int_{A_2}\left(\frac{p_2}{\rho g}+z_2\right)u_2 dA_2+\int_{A_2}\frac{u_2^2}{2g}u_2 dA_2+\int_q h_w' dq \tag{1.11}$$

where ρ is the fluid density in the flow tube, g is the gravitational acceleration, h' w is the energy loss, and q is the tube flow. The calculation method of q refers to Eq. (1.12).

$$dq = u_1 dA_1 = u_2 dA_2 \tag{1.12}$$

In Eq. (1.11), the average flow velocity v is used to replace the uneven flow velocities u at points A_1 and A_2. Besides, an error correction coefficient α is the difference between the fluid kinetic energy calculated by the average flow rate and the actual kinetic energy. The calculation method is as

$$\alpha=\frac{\int_A \rho \frac{u^2}{2} u dA}{\frac{1}{2}\rho A v v^2}=\frac{\int_A u^3 dA}{v^3 A} \tag{1.13}$$

In addition, the energy loss h' w is replaced by the average energy loss h_w as

$$h_w=\frac{\int_q h_w \prime dq}{q} \tag{1.14}$$

Equation (1.15) can be obtained by bringing Eqs. (1.12)–(1.14) into Eq. (1.11).

$$\frac{p_1}{\rho g}+z_1+\frac{\alpha_1 v_1^2}{2g}=\frac{p_2}{\rho g}+z_2+\frac{\alpha_2 v_2^2}{2g}+h_w \tag{1.15}$$

where α_1 and α_2 are the error correction coefficient of A_1 and A_2.

After simplification, the Bernoulli Equation of the actual fluid is shown in (1.16)

$$p_1+\rho g h_1+\frac{1}{2}\rho\alpha_1 v_1^2=p_2+\rho g h_2+\frac{1}{2}\rho\alpha_2 v_2^2+\triangle p_w \tag{1.16}$$

where h_1 and h_2 are the height of fluid and $\triangle p_w$ is the pressure loss.

Pressure loss is common in a fluid system, and the calculation method of the generator is shown in Eq. (1.17)

$$\sum\triangle p_w=\sum\triangle p_\lambda+\sum\triangle p_\eta \tag{1.17}$$

where Δp_{λ} is the pressure loss along the way, friction within the fluid and between the fluid and the pipe wall is the cause. Δp_{η} is the local pressure loss. Their calculation method is shown below.

$$\Delta p_{\lambda} = \frac{32ulv}{d^2} \tag{1.18}$$

$$\Delta p_{\eta} = \eta \frac{\rho v^2}{2} \tag{1.19}$$

where l is the length of the pipe, d represents the diameter of the distance, and η is the local resistance coefficient.

1.3.1.2 Digital Twin Model for Subsea Production Control System

In a hydraulic control system, three pieces of information are essential: pressure loss information, control information, and fault information. The parameters are usually updated through the valve. For a certain valve, the digital twin model before and after the valve is shown below.

$$p_i + \rho dh_i + \frac{1}{2}\rho v_i^2 = (p_{i-1} + \rho dh_{i-1} + \frac{1}{2}\rho v_{i-1}^2)\zeta_{i-1}\tau_{i-1}\varphi_{i-1} \tag{1.20}$$

From Eq. (1.20), three parameters are added to Bernoulli Equation, that is ζ, τ, φ. ζ is the pressure loss, and it different from the loss in Bernoulli Equation. Since the local pressure loss and the pressure loss along the way are difficult to obtain in the subsea production control system, however, from the data it is found that a pressure loss parameter can replace it. τ is the control parameter, and the calculation method is shown in Eq. (1.21). The control parameter is necessary to description the state of a valve. When it is closed, the pressure would change. This information can be easily obtained by reading the control signal of the system. φ is the state of the local circuit. Running to simulate the normal state of the system is one of the functions of a digital twin model. More importantly, running to simulate an abnormal state is another function. Unlike the simulation model, the digital twin model runs into an abnormal state because the physical is abnormal. For this, the φ should be timely updated, and its value is decided according to the diagnosis result. Thus, a feedback algorithm is developed to assign a value to this parameter, introduced in next section.

$$\tau = \begin{cases} 0.9 + \varepsilon & \textit{the valve is opened} \\ 0.2 - \sigma & \textit{the valve is closed} \end{cases} \tag{1.21}$$

where ε and σ are adjusting parameters used to fine-tune the control parameters under different conditions.

1.3.2 Fault Diagnosis Model Combined Feedback Data from Digital Twin

BN is an effective tool for fault diagnosis. The feature of BN is that the output results are expressed in the form of probability. When BN is applied to fault diagnosis, the probability approximates the degree of component failure. In this work, it is necessary to receive the failure degree of a component so that the local circuit φ state should be obtained. Usually, establishing a fault diagnosis model includes three steps: structure modeling, parameter modeling, and fault reasoning.

The structure of a fault diagnosis model is shown in Fig. 1.5. In a control system, the Sensors (S in Fig. 1.5) and Components (C in Fig. 1.5) are the main module of it. Usually, the failure of components can cause changes in indication of sensors. Thus, the components should be the parent node of the sensor node. Then each node contains two states: normal (N) and abnormal (A). In the beginning, the data can be obtained from the physical system. In this model, the input of the diagnosis model is S_1–S_m. The result was obtained by diagnosing with original data. After the result is verified, the feedback information can be input into the fault diagnosis model, and the result is obtained by diagnosing with feedback data. At this time, some parent nodes become the input node. It should be noticed that the same data of sensor is also input into the feedback diagnosis model.

Once the structure of the diagnosis model is determined. The prior probabilities and the condition probabilities are required. Two methods can be used to get them. The first is obtaining by experience, and the other is training. In this study, the data in

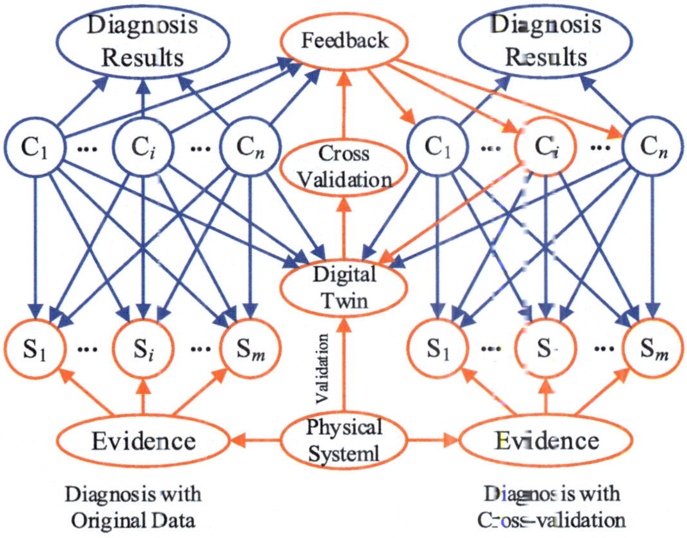

Fig. 1.5 The fault diagnosis model combined feedback data from the digital twin

SCM is challenging, especially the fault data. Besides, the principle of the hydraulic system is straightforward and does not need to be explored if a schematic diagram is obtained. Thus, prior probabilities and condition probabilities can be determined by experience. In a fault diagnosis model built by BN, the prior probabilities are not critical because all the child nodes, and conditional probabilities are given to obtain a posterior probability that characterizes the possibility of a fault. For valves, the probability of failure is far lower than expected. Referring to existing research (Tang et al. 2022), the prior probability of failure is 4%, and the normal probability is 96%. Conditional probability is obtained according to the relationship between hydraulic circuit components. For example, when a valve fails, the probability of an associated abnormal PT is 95%. In this way, a conditional probability table is obtained.

After the diagnosis model is constructed, it is used for fault diagnosis. For a node X with parent nodes U and children nodes Y, the conditional probability from U to X is defined as $\pi_X(U)$, and the conditional probability from Y to X is defined as $\lambda_Y(X)$. For some known evidence E, the joint marginal for the family of X with the parent nodes U_i and children nodes Y_i is given by

$$P(XU, E) = \lambda_E(X)\Theta_{X|U} \prod_i \pi_X(U_i) \prod_j \lambda_{Y_j}(X) \tag{1.22}$$

where, $\lambda_E(X)$ is evidence indicators and $\lambda_e(X) = 1$ if $x \in X$ is consistent with evidence $e \in E$ and zero otherwise. The causal and diagnostic massage can be defined as follows:

$$\lambda_X(U_i) = \sum_{XU \backslash \{U_i\}} \lambda_E(X)\Theta_{X|U} \prod_{k \neq i} \pi_X(U_k) \prod_j \lambda_{Y_j}(X) \tag{1.23}$$

$$\pi_{Y_j}(X) = \sum_U \lambda_E(X)\Theta_{X|U} \prod_i \pi_X(U_i) \prod_{k \neq j} \lambda_{Y_k}(X) \tag{1.24}$$

A node can be sent to a neighbor only after it has received messages from all other neighbors. When a node has a signal neighbor, it can immediately send a message to that neighbor, which includes the life node X with a single parent U, for which

$$\lambda_X(U) = \sum_X \lambda_E(X)\Theta_{X|U} \tag{1.25}$$

It also includes a root node X with a single child Y, for which

$$\pi_Y(X) = \lambda_E(X)\Theta_X \tag{1.26}$$

During fault diagnosis, the data of PT are input into the fault diagnosis model, and posterior probability of fault nodes is updated using the belief propagation algorithm above. The new data should be mapped to the adjacent state space by multiplying by the load matrix with adjacent frequency and modulation index. Generally, the larger the posterior probability of fault, the higher the possibility of occurrence of the corresponding fault. The fault diagnosis methodology can offer the probability of a fault, but not draw a definite diagnostic result. Instead, a threshold needs to be set. When the failure probability of a component exceeds this threshold, the component is considered as a failure (Cai et al. 2014). According to engineering experience, it is usually set to 78%.

When a posterior probability is obtained, it is assigned to φ as

$$\varphi = \begin{cases} 1 - P(V) & P(V) \geq 0.65 \\ 0.9 + \kappa & P(V) < 0.65 \end{cases} \tag{1.27}$$

where, κ is the adjusted parameter for fine-tuning the digital twin model, and $P(V)$ is the probability of the corresponding valve failure.

1.3.3 Interactive Promotion Algorithm Between Digital Twin and Fault Diagnosis Models

For verification, two kinds of errors are necessary: single error and cumulative error. They are essential for ensuring diagnostic accuracy and increasing degree of fault diagnosis.

By calculating the single error, the state of the specified valve can be corrected. In a digital twin model, pressure loss, control, and fault information can influence it by changing parameters. It can easily be seen that the pressure loss information and control information obtained from the control information. Thus, if a single error is too large, the fault parameter here is wrong. The calculation method for a single error is shown in Eq. (1.28).

$$E_s = \frac{|P_r - P_{DT}|}{P_r} * 100\% \tag{1.28}$$

where E_s is the single error. $E_s = [e_{s1}, e_{s2}, \ldots, e_{sn}]$, e_{si} is the i-th single error. P_r is data from the physical system, and P_{DT} is data from the digital twin model. $P_r = [P_{r1}, P_{r2}, \ldots, P_{rn}]$, $P_{DT} = [P_{DT1}, P_{DT2}, \ldots, P_{DTn}]$. And P_{ri} is the data from the i-th PT in a physical system.

A single error is very effective in correcting an error diagnosis. However, incorrect fault diagnosis usually shows that the fault data is considered normal when the fault feature is fragile. That is, the fault is not distinguished. Under these circumstances, a single error is usually not enough. So, a cumulative error is also necessary because it can make up for this misdiagnosis. The calculation method of cumulative error is shown in Eq. (1.29).

$$E_p = Sum(E_s) \tag{1.29}$$

where E_p is a cumulative error.

Table 1.1 The cycle validation diagnostic method of fault diagnosis with digital twin

1	**Get** ζ, τ;	12	**Calculate** $\Phi = [\varphi_1, \varphi_2, \ldots, \varphi_n]$;	23	**Get** ζ, τ;
2	**Set** $\varphi_{all=1}$;	13	**Get** ζ, τ;	24	**Get** $P_r = [P_{r1}, P_{r2}, \ldots, P_{rn}]$;
3	**Get** $P_r = [P_{r1}, P_{r2}, \ldots, P_{rn}]$;	14	**Get** $P_r = [P_{r1}, P_{r2}, \ldots, P_{rn}]$;	25	**Run** Digital twin model;
4	**Run** Fault diagnosis model;	15	**Run** Digital twin model;	26	**Calculate** $E_s = [e_{s1}, \ldots, e_{sn}]$, E_p;
5	**Calculate** $\Phi = [\varphi_1, \varphi_2, \ldots, \varphi_n]$;	16	**Calculate** $E_s = [e_{s1}, \ldots, e_{sn}]$, E_p;	27	**Calculate** $j = j + 1$
6	**Run** Fault diagnosis model;	17	**end While**	28	**end While**
7	**Calculate** $E_s = [e_{s1}, e_{s2}, \ldots, e_{sn}]$, E_p;	18	**Set** $j = 1$	29	**Run** Digital twin model;
8	**While** $E_s > E_{ss} \parallel E_p > E_{sp}$	19	**While** $E_p > = E_{sp}$ && $j < j_{max}$	30	**Calculate** $E_s = [e_{s1}, \ldots, e_{sn}]$, E_p;
9	**While** $E_s > = E_{ss}$	20	**Set** V_j = *abnoral* in fault diagnosis model;	31	**end while**
10	**Set** V_i = *abnoral* in fault diagnosis model;	21	**Run** Fault diagnosis model;	32	**Output** $P = [P(V_1), \ldots, P(V_n)]$;
11	**Run** Fault diagnosis model;	22	**Calculate** $\Phi = [\varphi_1, \varphi_2, \ldots, \varphi_n]$;	33	**End**

With both types of errors, the cycle validation diagnostic method with the digital twin model is shown in Table 1.1.

1.4 Cases of Subsea Production System

1.4.1 Introduction of Subsea Production System in South China Sea

A subsea production system in the South China Sea with composite faults and minor faults are shown in Figs. 1.6 and 1.7. The system is divided into the offshore and the subsea parts. The main function of the offshore part is to control the actions of the subsea system and provide power for related actions. As shown in the figure, the hydraulic power unit and electronic power unit are the main modules of offshore part. The hydraulic power unit is the main hydraulic power supply module. It charges the subsea accumulator to ensure the pressure of the subsea accumulator and the normal function of the subsea hydraulic system. The electronic power unit has two functions. One is to provide sufficient power supply for subsea equipment, and the other is to

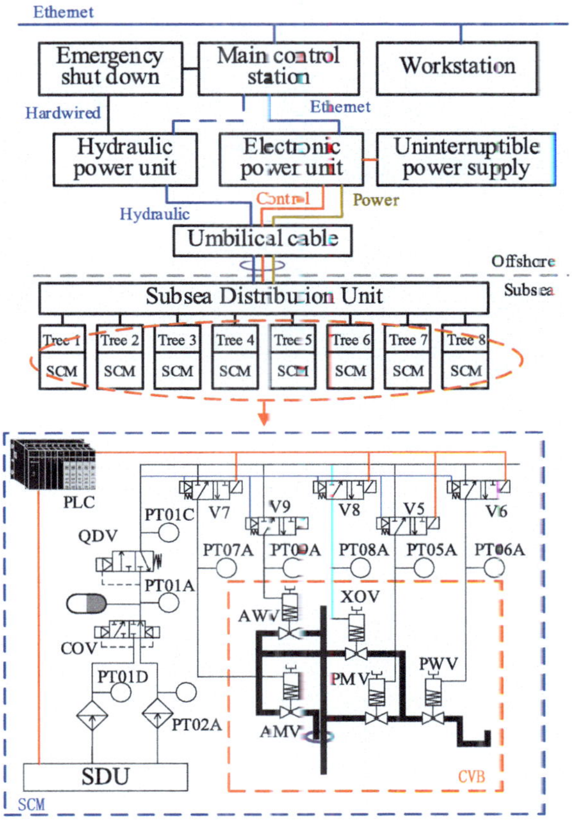

Fig. 1.6 A subsea production system in South China Sea (for composite faults)

convert the signals from main control station into signals that can be transmitted to subsea through the umbilical cable. In addition, an emergency shutdown module is used to shut down the whole system in case of an emergency. AN uninterruptible power supply module can provide power for the system in case of power failure to ensure the normal operation of the system. When the offshore signal is transmitted to the subsea, the subsea distribution system is used to distribute electro-hydraulic control and power signals to the eight subsea Christmas trees.

1.4.2 Composite Faults

The Subsea Control Module (SCM) is the core for each subsea Christmas tree. Its main function is to control the production valve mounted on the body of a subsea Christmas tree. Figure 1.6 shows that the hydraulic and electronic control signals are all received by SCM from SDU, and that electronic control signal is received by PLC to control valves. A Choice Operation Valve (COV) is used to select the

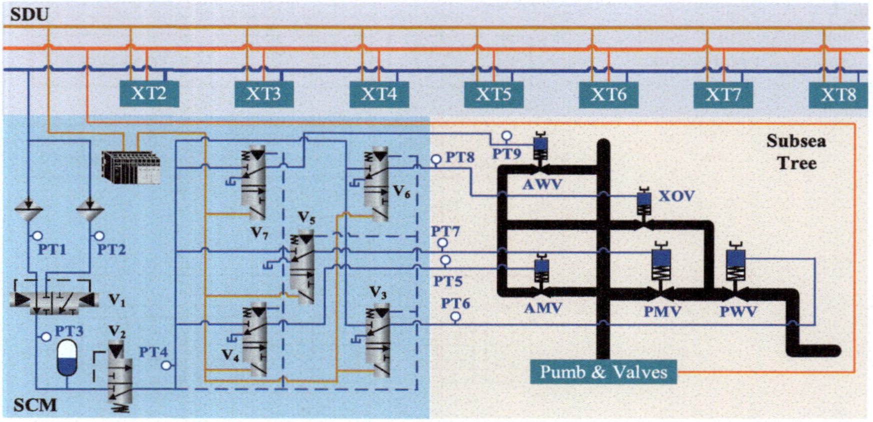

Fig. 1.7 A subsea production system in South China Sea (for minor faults)

hydraulic supply circuit. Then, a Quick Down Valve (QDV) is used to shut-off the SCM in an emergency. In addition, five valves are the main components of the subsea Christmas tree, namely, the Production Main Valve (PMV), Production Wave Valve (PWV), Annual Main Valve (AMV), Annual Wave Valve (AWV), and Translator Valve (XOV). During the production, the pump lifts underground oil, the PMV and PWV are opened to collect oil in the central manifold and transported to the platform. When the annular space pressure is higher than normal, the AMV and XOV will be opened for pressure relief. AAV is used mainly for workover and chemical input. From the description above, the PMV, PWV, and AMV are opened, and the XOV, AWV are closed when oil and gas production is in progress. These main valves are controlled by hydraulic circuits in SCM. Five valves (V5–V9 in Fig. 1.6) are used to control the flow of hydraulic oil to these valves. The command sending from the driller's room can be transmitted to subsea through the umbilical cable as an electrical signal. SDU transmits the signal to the corresponding subsea Christmas tree, the valves in SCM begin to work after receiving the electrical signal. Data for one month in a row is collected for this experiment and 24 data are collected every day. Before collecting data, a large maintenance program was completed and thus, there are fault data. The faulty data used in this experiment are divided into two types. One is real fault data. They were stored when a fault occurs. These fault characteristics are obvious because stopping the system for maintenance only occurs when the fault is very serious. However, this will lead to a sudden stop of production, which can be expensive because an unreasonable stop planning especially sudden shutdown can lead to waste of a large costs of supporting facilities. The diagnosis method is required to be able to identify smaller degrees of faults. The other data are obtained by simulation based on existing fault data. The whole control circuit is divided with the valve as the dividing point. When a fault occurs in a circuit, the pressure of related circuit should be 5–75% of the normal pressure. Values in this range are taken at certain intervals to ensure that all levels of faults are taken. In

Table 1.2 Data composition used in experiment

Faulty valve (s)	Number of date	
	Data from filed	Data by simulation
Normal	348	36
COV	8	36
QDV	10	36
CV5	18	36
CV6	14	36
CV8	16	36
CV5 and CV6	1	36
CV5 and CV8	2	36
QDV and CV5	1	36
QDV and CV8	1	36

this way, the possible working conditions of the whole system are included in the experiment. The specific data composition is shown in the Table 1.2. Because the actual fault data are difficult to obtain, all the fault data that can be obtained are used in the experiment. Their numbers are listed in the second column of the table and the data obtained through simulation are listed in the third column.

1.4.2.1 Output of Digital Twin Model

The output of the physical system is shown in Fig. 1.8. For a subsea production control system, the main output that can be obtained is the data of PT. In this system, the data of PT02A, PT01A, PT01C, PT05A, PT06A, PT07A, PT08A and PT09A in a month are obtained. The data show that the pressure fluctuates, but the fluctuation is small. This is because the subsea production system is affected by a complex production environment. In this environment, minor pressure loss commonly occurs. In this system, other valves are opened except for departure V8 and V9. The left ordinate is used for the PT of the open valve. The pressure fluctuates around 234 psi, and the tendency of change is the same. When a valve is opened, the output pressure is determined by the input pressure. In this system, the input is PT02A, and thus, the changes in the other valves depend on the fluctuations of PT02A. Their pressure difference is small, and the maximum pressure difference does not exceed 5psi, which is negligible relative to the normal value of 234psi. When V8 and V9 are closed, their outputs are very small. The vertical coordinate on the right as shown in Fig. 1.8 is their value. These valves hardly fluctuate because their output depends on the reaction force of the controlled object. In this system, the force mainly comes from the spring in the controlled valve.

The output of the digital twin model is shown in Fig. 1.9. Similar to Fig. 1.8, PT08A and PT09A use the right ordinate, and the rest use the left ordinate. In the built model, PT02A from the field is the only input, and the other PTs are output by

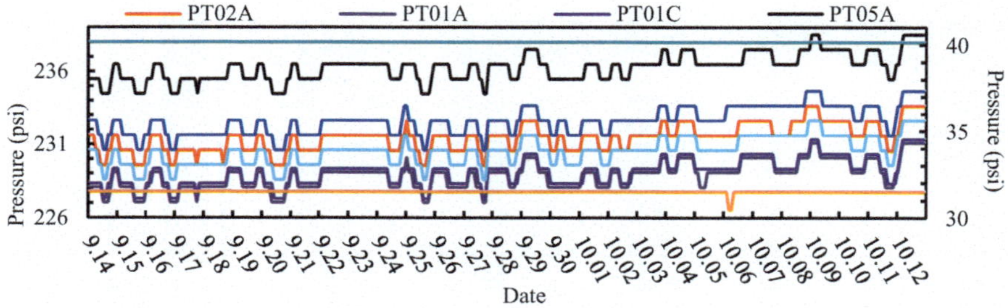

Fig. 1.8 Field data of subsea production system

the digital twin model. Compared with the output of the physical system, the digital twin model has a similar output trend, but the specific values are different. This is because the control parameters and loss parameters of each valve are obtained based on the average value, which may fluctuate in a single data. The output fluctuation of PT08A and PT09A differs from the physical system because their data are calculated according to PT02A, which affects their change trend. In fact, this error is small enough to be ignored. Figures 1.8 and 1.9 show an error between the output of the digital twin model and the actual system output. The reasons for these errors vary, such as sensor data fluctuation and digital twin model calculation error but the trends are consistent. After calculation, they are less than 1%, which can be ignored.

The average error between the digital twin model and the physical system is calculated as shown in Fig. 1.10. The errors are low, with the largest error being 0.00428 psi, which can be ignored because the normal working pressure is 234 psi. In addition, the errors of PT08A and PT09A are large because, in the digital twin model, the output of these two valves is calculated through the system input pressure and control and loss parameters. This method is different from the physical system, resulting in relatively large errors. At the same time, the percentages are large because their value is small. In the opening valve, the errors of PT05A, PT06A and PT07A are larger than those of PT01A and PT01C because the cumulative parameters are used in the digital twin model. For example, when the output of QDV is calculated, two

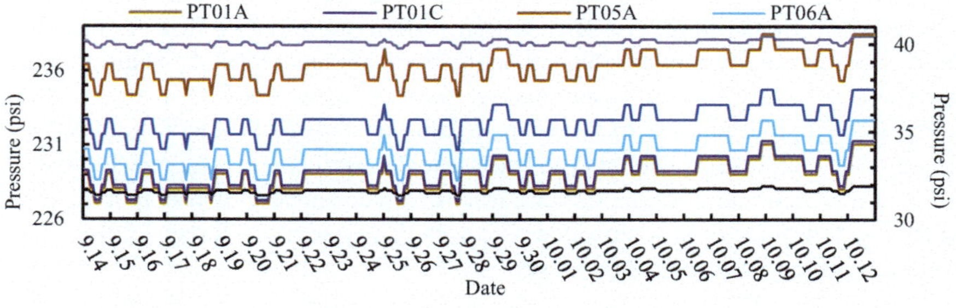

Fig. 1.9 Output of digital twin model when the physical system is normal

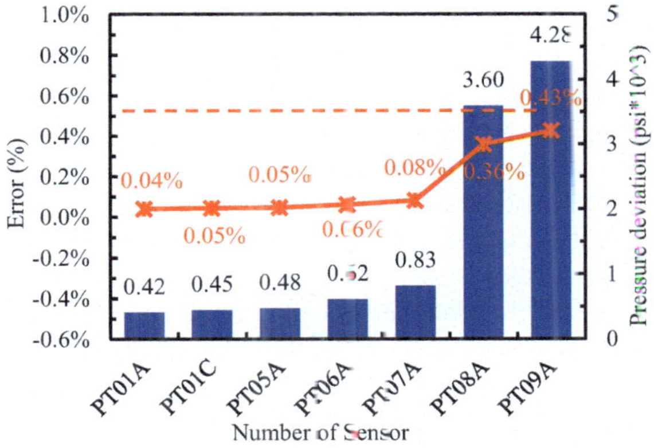

Fig. 1.10 Error between digital twin and physical system

control parameters and two loss parameters are used. However, when the output of V7 is calculated, three are used. Errors of more parameters are accumulated, which may lead to a larger error. In terms of percentage, these errors are all lower than 1%, which meets the requirements of the field.

1.4.2.2 Fault Diagnostic Results

Nine faults are set in the database according to historical data and field investigation shown in Table 1.2. These faults are divided into three categories: single fault, composite faults that do not affect each other, and composite faults that affect each other. Single faults are the most likely faults. Thus, the faults of each valve are set with different degrees from slight to obvious. When the fault is very slight, the component can be considered as having no fault. Therefore, 36 normal data are from the simulation. Some valves cannot influence each other, such as CV5 and CV3 and CV5 and CV6. They are different from the valves that influence each other and thus, their faults are set. They represent composite faults in two situations, concluding an opening valve and a closing valve, or two closing valves. In the composite faults that affect each other, the fault of COV is not set. According to a site investigation, the fault COV installed in an accumulator can be easy to distinguish, and its failure has little impact on the fault identification of other valves.

The diagnosis results with the set faults are shown in Fig. 1.11. The figure shows that the diagnostic accuracy for the single fault remains good regardless of whether the digital twin model is used for fault diagnosis. Combining the digital twin model yields limited improvement in the accuracy of fault diagnosis on single fault. In fact, the performance of fault diagnosis on obvious faults is good when a diagnosis model without digital twin model is used. However, the data used in this study contain many minor faults that cannot be easily distinguished. A slight improvement occurs when the digital twin model is used in fault diagnosis. The condition of composite faults that do not affect each other and a single fault diagnosis is the same. However, when composite faults that affect each other are diagnosed, the diagnostic accuracy without digital twin model is lower than 80%. When the digital twin model is used, the accuracy is higher than 90%. The digital twin model improves the fault diagnosis through the following aspects. First, the digital twin model provides cross-validation in the stage of fault diagnosis of a single fault. In this way, the diagnostic accuracy of minor faults can be improved, which is the basis of fault diagnosis of composite faults. Second, virtual data are used to diagnose composite faults, which results in higher accuracy.

Dynamic update is one of the important characteristics of a digital twin model and is also an important feature that distinguishes it from traditional simulation. Specifically, when the physical system is in a normal state, the digital twin model should run in a normal state and be consistent with the physical system. When one or several faults occur, the information should be timely, and the digital model should run in an abnormal state. A time series is used to explore the dynamic performance of the established digital twin model as shown in Fig. 1.12a. One-month data is used to establish these time series, and twelve hours of data are used every day, from 2:00, 4:00, 8:00, to 24:00. Four faults containing three types of faults mentioned before are included. The specific fault type and occurrence time are shown in Table 1.3. The data of PT05A and PT08A are obtained. When setting data are input into the model, the results are obtained on time, and the change process of the digital twin model cannot be known. Therefore, three sampling points are set during each data set process, as shown in the sub abscissa of Fig. 1.12b–e.

Figure 1.12a shows that the digital twin model can quickly reflect when three types of faults occur. The delay is almost indistinguishable at the beginning and ending of the fault. The error is mainly reflected in the judgment of the fault degree. More precise sampling of when a single fault occurs is shown in Fig. 1.12b, c. The figure shows that there are some delays in the digital twin model regardless of the diagnosis process of CV5 or CV8.

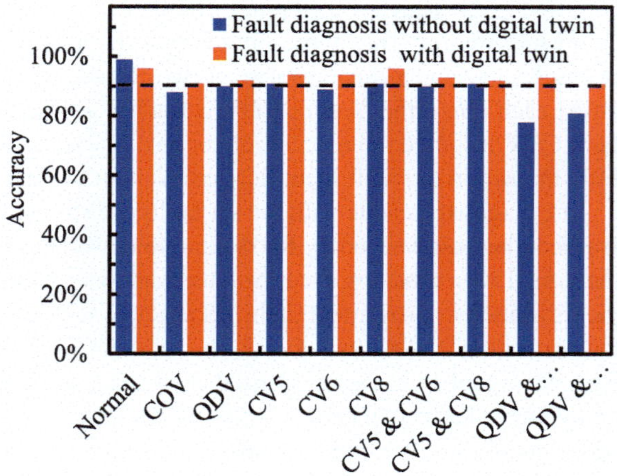

Fig. 1.11 Accuracy of fault diagnosis

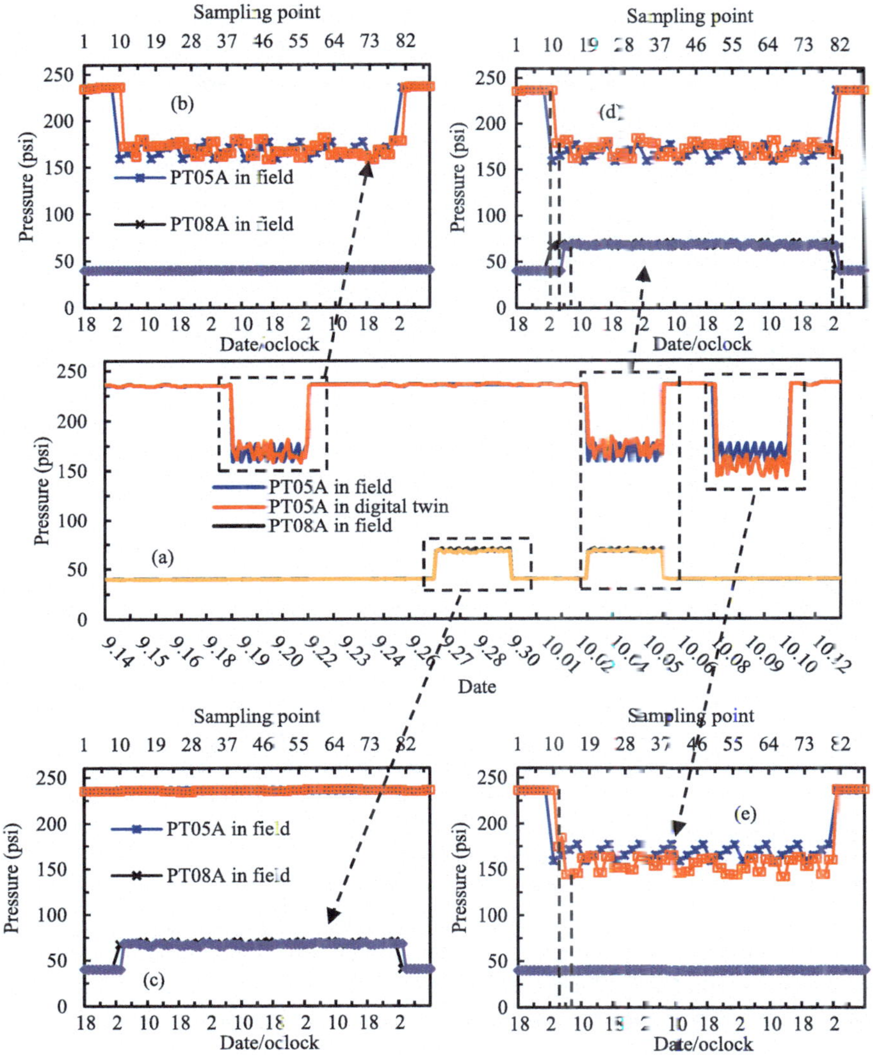

Fig. 1.12 Dynamic performance of digital twin model

Table 1.3 Data composition used in the experiment

Date	Faults	Fault type
9.19–9.21	CV5	Single fault
9.27–9.29	CV8	
10.03–10.05	CV5 and CV8	Composite faults that do not affect each other
10.08–10.10	QDV and CV5	Composite faults that affect each other

Moreover, the delay of the fault occurrence process is greater than that of the normal recovery process. In fact, cross-validation is usually not required when the system recovers from a failure state. As a result, the digital twin model takes less time to recover from the fault state, thereby allowing it to respond faster.

A more precise sampling of the occurrence of composite faults that do not affect is shown in Fig. 1.12d. The change of the digital twin model is divided into two stages. In the first stage, the CV5 fault is diagnosed, and in the second stage, CV8 is diagnosed. After detailed study, the pressure change caused by the failure of the opening valve is larger. Therefore, the opening valve is usually diagnosed first when composite faults occur. When the system returns to normal, the digital twin model also recovers after a short delay, similar to the delay time when a single fault occurs.

Figure 1.12e shows a more precise sampling of the occurrence of composite faults that affect each other. The change in the digital twin model is also divided into two stages. In the first stage, the fault of QDV is diagnosed while in the second stage, CV5 is diagnosed. In this fault, the error caused by the QDV fault is much larger than that caused by the CV5 fault because the QDV fault will affect all subsequent valve outputs. Thus, the QDV fault can be diagnosed more easily. When the system returns to normal, the digital twin model recovers after a short delay, similar to the delay time when a single fault occurs. The above description clearly shows that the proposed digital twin model can accurately and quickly follow the dynamic changes of the subsea production control system in the occurrence or disappearance of a fault.

The response characteristics of the digital twin model are explored further by calculating the average time spent in different diagnosis stages of the four faults as shown in Fig. 1.13. The results show that for the fault diagnosis of the single fault, majority of the time is spent on the fault judgment and fault diagnosis cross-validation stage while less time is spent on the fault diagnosis of composite faults. The reason may be because the characteristics of some faults are not obvious, leading to wrong diagnosis. When composite faults are diagnosed, the time spent on the cross-validation stage and fault diagnosis with virtual and real data increases. The results show that the longest time spent is not more than 20 ms, which is acceptable for a subsea production control system. Generally, minor faults in the subsea production system can be difficult to address and are often not handled due to safety concerns. Minor faults can provide a basis for the next major overhaul and long-term planning to ensure production safety.

1.4.3 Minor Faults

After the electronic control signal, high pressure control signals, and low-pressure control signal are transferred to the subsea production system. And then, these three signals are received by a Subsea Distribution Unit (SDU) and distributed to eight subsea Christmas trees. SCM is the core of a subsea Christmas tree. Its main function is to control the production valve mounted on the body of a subsea Christmas tree. As shown in Fig. 1.7, the hydraulic and electronic control signals are all received

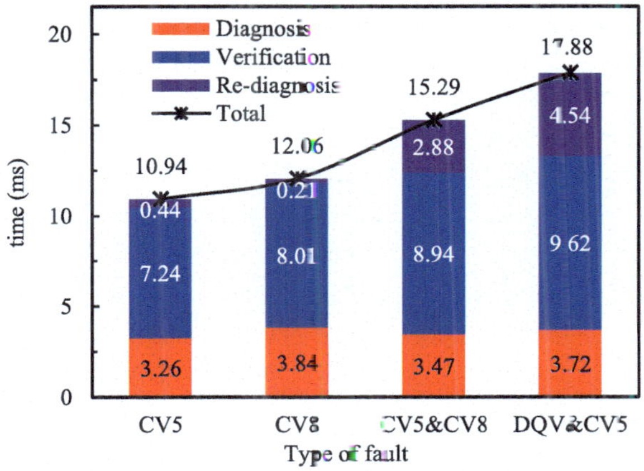

Fig. 1.13 Response time for impassable faults

by SCM from SDU, electronic control signal is received by FLC to control valves. SCM receives two hydraulic pressure supplies to ensure reliable control. A hydraulic selector valve (V1 in Fig. 1.7) is used to select the hydraulic supplies circuit. The subsea environment is complex, and the emergency shutdown function is an essential guarantee for oil and gas production safety which is realized by an emergency shut off valve (V2 in Fig. 1.7).

Five valves are the main components of a subsea Christmas tree, that is, the Production Main Valve (PMV), Production Wave Valve (PWV), Annual Main Valve (AMV), Annual Wave Valve (AWV), and Translator Valve (XOV). During production, the pump lifts oil from underground, the PMV and PWV are opened to collect oil in the central manifold and then transported it to the platform. When the annular space pressure is higher than normal, the AMV and XOV would be opened for pressure relief. AAV is mainly used for workovers and input of chemical substances. From the description above, the PMV, PWV, XOV, and AWV are opened, and the AMV is closed when oil and gas production is in progress. These valves are controlled by hydraulic in SCM. So, five valves (V5–V9 in Fig. 1.7) are used to control the flow of hydraulic oil to these valves. When the driller's room sends out the command, they would be transmitted to subsea through an umbilical cable as an electrical signal. SDU transmits the signal to the corresponding subsea Christmas tree. The valve in SCM receives the electrical signal to action and drive the valve at the body of the subsea Christmas tree. The data from four days are used for research. According to the field, the failure frequency of the subsea production system is low, so the data with failure is collected during long-term operation. And it is not collected from data in the four days. Some fault data are obtained through fault simulation, and the specific information on data composition to obtain sufficient fault data is shown in the Table 1.4. The experimental data is exported as an Excel file, and the algorithm is run in MATLAB 2019b. The computer server with an NVIDIA RTX1660s and R7-5800 CPU to execute proposed approach.

Table 1.4 Data from subsea production system

Valve	Normal state	ζ	τ	Fault abbreviation	Number of fault data	Number of normal data
V_1	Open	1.000	0.998	FD-V1	96	68
V_2	Open	1.000	0.988	FD-V2	96	
V_3	Close	0.110	0.989	FD-V3	96	
V_4	Open	1.006	0.994	FD-V4	96	
V_5	Close	0.117	0.997	FD-V5	96	
V_6	Close	0.120	0.999	FD-V6	96	
V7	Close	0.124	0.986	FD-V7	96	

1.4.3.1 Output of Digital Twin Model Under Different Conditions

(1) Output of digital twin model when the physical system is normal

The digital twin model is established based on the method mentioned previously. In this method, the selection of ζ and τ is the core. The average of all data is calculated and used to determine the values of the two parameters. Their specific values are shown in Table 1.4.

The data of pressure transmitters PT2, PT3, PT4, PT5, PT6, PT7, PT8, and PT9 is obtained from the field. The four-day monitoring data is shown as Fig. 1.14. The pressure fluctuates slightly because the subsea production control system is affected by the outside environment, and the pressure loss is variable. It can also be seen in Fig. 1.14a that, with an opening valve, the fluctuating pressure fluctuates around 234 psi, and the tendency of change is the same. Interestingly, PT5 pressure is the lowest because PT5 is the farthest from the pressure input end and has experienced more pressure loss. The difference between PT5 and other PT is less than 5%, and these losses are low enough to be negligible. Figure 1.14b shows that the pressure would become low and doesn't change when the valve is closed. That's because when the valve is closed, the pressure in the pipe depends on the load outside. In this study, the load is generated by the internal spring force of the controlled valve, which is small and relatively stable.

The output of the digital twin model is shown as Fig. 1.15. In the model, PT2 from the field is the only input, and the other PTs are output by the digital twin model. The valve of ζ and τ is shown in Table 1.4. The mean value of each PT is used to calculate these two parameters. Figure 1.15a shows that the trend of PT in a digital twin is similar to that of field data. Contrary to Fig. 1.14a, the minimum value is not the PT5, because the valve of PT5 is influenced by several control parameters, fault parameters and causes errors in the data due to their overlapping with each other. Figure 1.15b shows that the value of PTs fluctuates, which is different from Fig. 1.14b. But the value of PT is determined by the load when the valve is closed. However, in the digital twin model, they are calculated by Bernoulli's Equation in

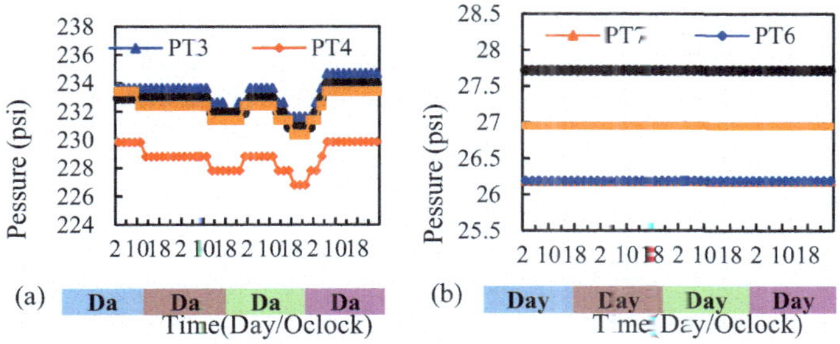

Fig. 1.14 Field data of subsea production system. **a** High pressure, **b** low pressure

the digital twin model. Still, the value is low, and the difference is so low that it is ignored. Through field experiments, this error meets the needs of the field.

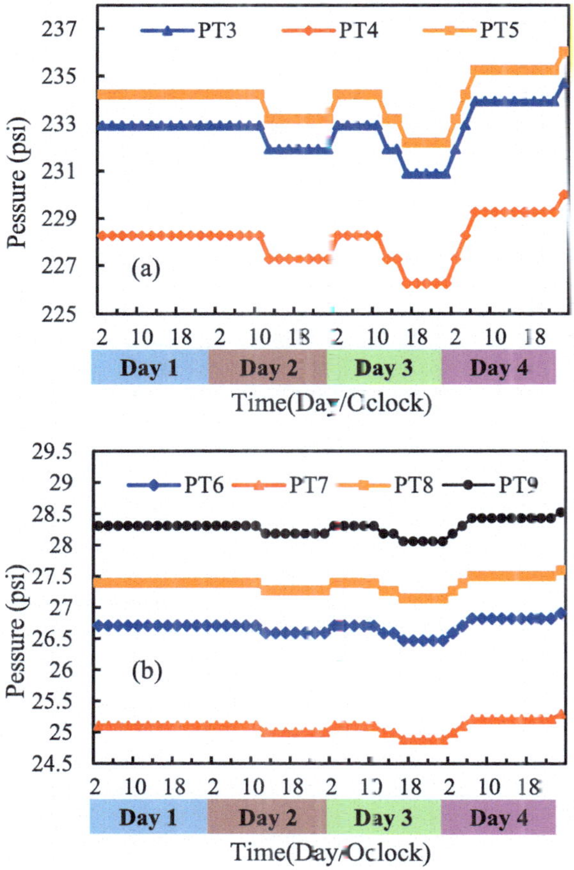

Fig. 1.15 Output of digital twin model when the physical is normal. **a** High pressure, **b** low pressure

To further study the performance of the digital twin model, the real-time error is calculated and shown in Fig. 1.16. The error is calculated by the percentage of the difference between the output of the digital twin model and the field data. Figure 1.16 shows that the error is changing, but the value is lower than 6%. Obviously, the error of PT5 is the largest. An interesting phenomenon is that the error of PT6-PT9 is higher than PT3-PT5, Because the valve in front of that valve is closed, and its value is small. Thus, a small error will result in a large percentage.

The average error of each PT is shown as Fig. 1.17. Most errors are less than 3%, and the average error of PT7 is 4.4%. Figure 1.14 shows that the value of PT7 is about 26.5 psi, and thus the error of it is about 1.17 psi. This value is minimal compared to the system with a normal pressure of 234 psi. The phenomenon that the error of PT6-PT9 is higher than PT3-PT5 still exists. There are some small disturbances during the data collection process. Usually, typical noise reduction methods such as wavelet transform can effectively handle them (Cai et al. 2016). However, in this research, they are not reduced. There are two reasons for this. Firstly, Due to the

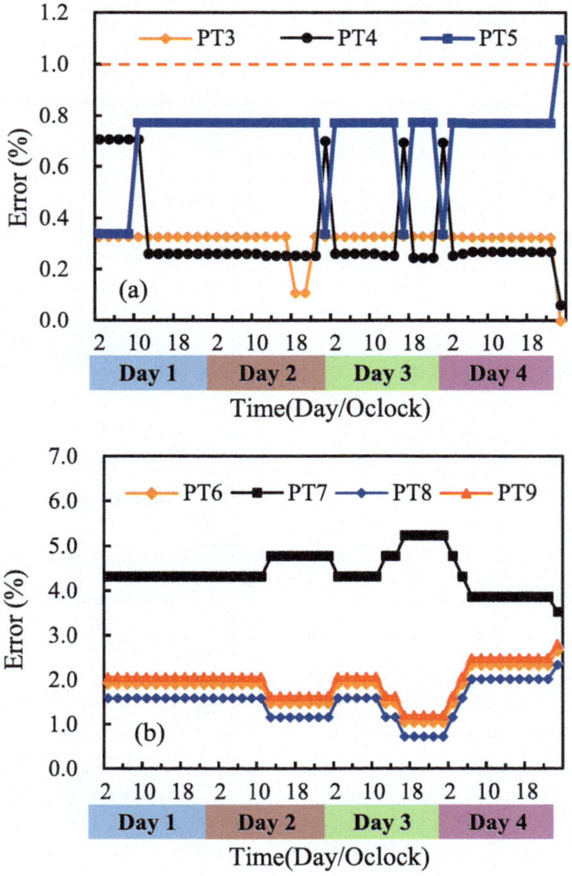

Fig. 1.16 Real time error. **a** High pressure PTs, **b** low pressure PTs

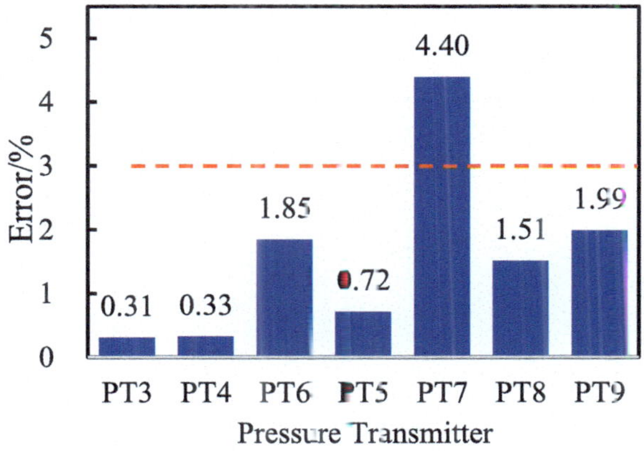

Fig. 1.17 Average error between digital twin and physical system

stability of the subsea production system, the sampling frequency is low. This results in some noise being unable to be collected, and the noise collected is minimal. Their impact on fault diagnosis is minimal. It was found that these disturbances exhibit a trend of jumping up and down centered around theoretical values. This has little impact on the error of the digital twin model. Their amplitude is low, which makes them have almost no impact on the effect of fault diagnosis.

(2) Output of digital twin model when a fault occurs

The changes of PT with a failure component are shown in Fig. 1.18. Two types of valves are included in a subsea production control system: opened valves and closed valves. From the field, the performance is different when there is a failure.

For opening valves, the main failure mode is leakage. Since it allows oil to flow through for a long time, the valve experiences vibration and liquid friction. When a leakage happens, the pressure would be lost, which can cause the value of PT to decrease, as shown in Fig. 1.18a. When a leakage occurs in V1, the value of PT2 is normal, and the value of PT3-PT5 decreases compared with the normal value. In addition, the value of PT6-PT9 is unchanged. The leakage of V1 has little influence on PT2 cause PT2 is in front of V1, and SDU influences the value of PT2. However the other PTs are behind V1. Thus, the leakage wastes a part of the pressure, and the value of the other PTs are low. Similarly, the value of high PT obviously drops when the leakage occurs on the related valve.

For daily closed valves, the main failure mode is internal leakage. The valve cannot block the hydraulic oil due to the failure of the inner sealing material of the valve under long-term pressure. The common phenomenon of this failure is that the value of the PT behind the valve increases, as shown in Fig. 1.18b. Only the value of the PT behind the failure valve is affected since the several valves closed daily directly control the subsea Christmas tree function valve. For example, when V3 fails, the value of PT6 increases and the other PTs are not influenced.

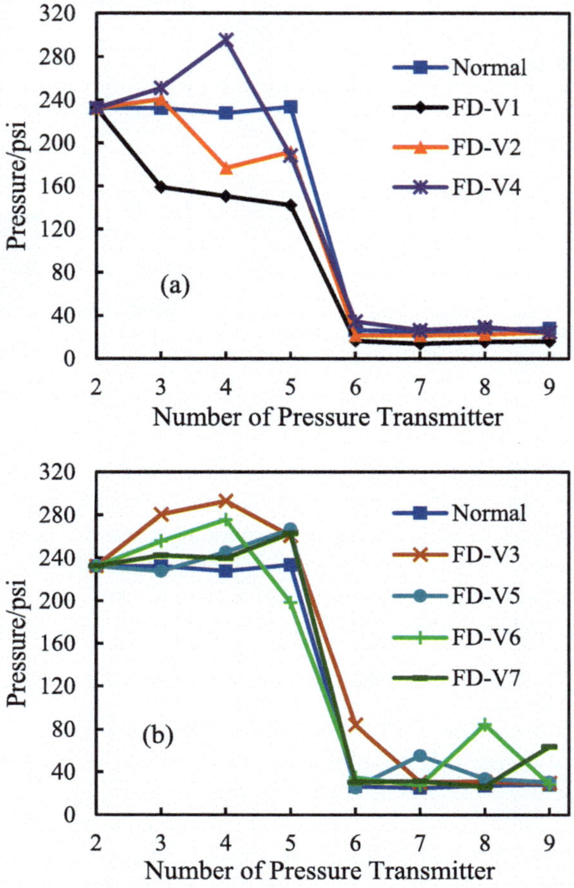

Fig. 1.18 Data of fault. **a** Opened valve failure, **b** closed valve failure

It should be noted that there are few data with fault. In Fig. 1.18, some data are a single fault data, and some are the mean values of several data. They can represent the trend of PTs when a component fails. In fact, the actual situation is more complex.

The same with the digital twin model in a normal state of the subsea production control system, the data of PT2 is input into the digital twin model, and the fault parameter is set to 0.35–0.75. This range of fault parameters is obtained through field data. It should be explained that the range of it is not confirmed strictly. Because the propose is to observe the system's response after the fault exists in the digital twin model. Faults within this range of parameters have occurred in field, so it can be used to study changes in the digital twin model. The set fault parameters are different because the fault degree is not diagnosed. The results are shown in Fig. 1.19. From Fig. 1.19a, the trends of PT3-PT5 are similar to the filed data when an opened valve fails. Compared with the filed data, the fluctuation of the result of the digital twin model is small. Because constant fault parameters influence the output of the digital twin model without updating the fault diagnosis model. Moreover, many unimportant factors are not taken by the digital twin model. The result of the digital twin model with a failure closed valve is also shown in Fig. 1.19b.

From the results, the digital twin model can well describe the trend of PTs values after a valve failure. However, there are differences in details, even large errors. Because a consistency model has not been established in this study. A consistency maintenance model is mainly used to adjust the digital twin model to make it consistent with reality. The function of the digital twin model is to improve the diagnostic accuracy of the fault diagnosis model. Thus, a dynamic trend simulation is sufficient. Besides, fault parameters are not updated dynamically, which is also a fundamental reason for this phenomenon.

The errors with different faults are shown in Fig. 1.20. It shows that the errors are less than 25%, and most of them are less than 15%. When a dynamic update model is used, they are less than 9%. All the errors are larger than the error without faults. Because the digital twin model is built based on normal data. It is worth noting that

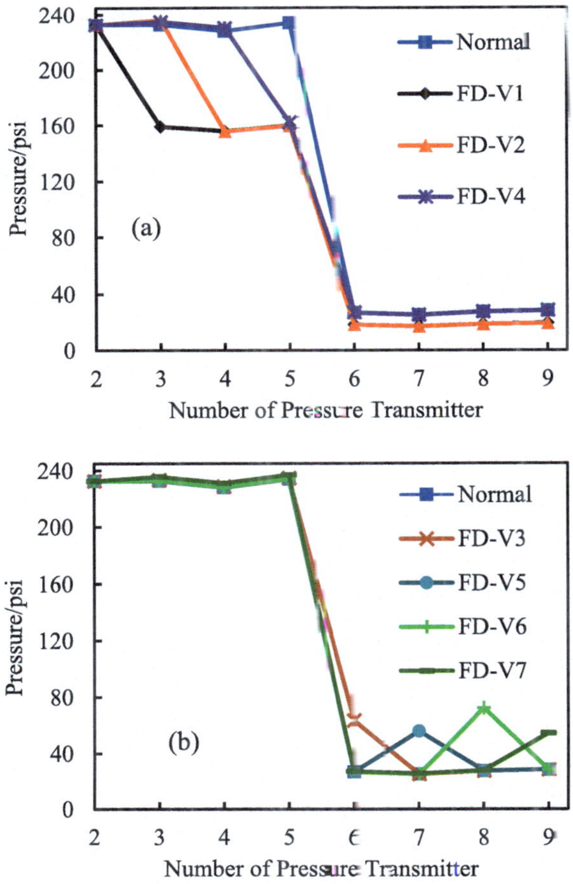

Fig. 1.19 Output of digital twin model when a fault occurs. **a** Opened valve failure, **b** closed valve failure

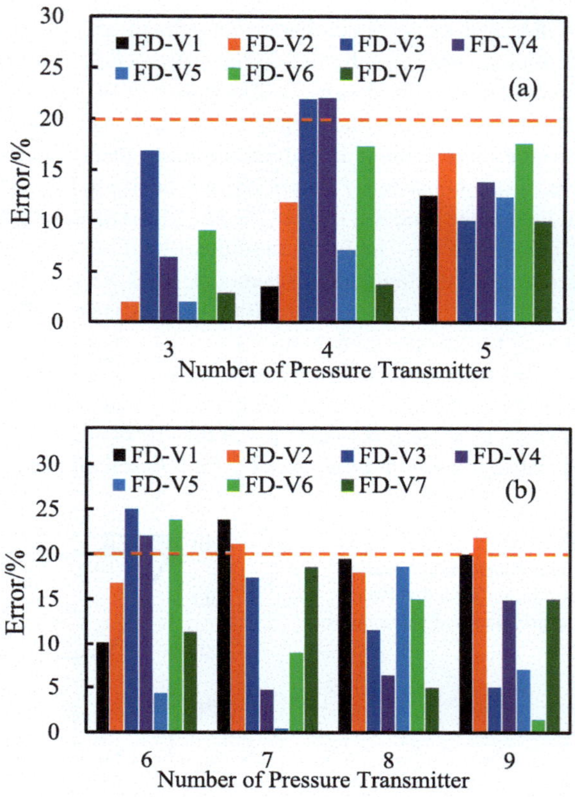

Fig. 1.20 Error of digital twin model when a fault occurs. **a** High pressure PTs, **b** low pressure PTs

the error of PT3 is small. Because, except for the fault in V1, other faults have little impact on PT3. In addition, the error of high PT is similar to that of low PT, which is different from the previous. Indeed, the fault parameters have little effect on the low PTs whose values are relatively small.

(3) Output of digital twin model when a valve is opened or closed

The changes of pressure when the state of the valve changes is shown in Fig. 1.21. In fact, field data is not collected. The results of the digital twin model are only analyzed through the principle of the hydraulic circuit. The changes of pressure after the valve closed are shown in Fig. 1.21a. When the V2 is closed, the oil is cut off. Thus, the value of the PTs behind V2 decreases. The same occurs when V4 is closed, and they conform to the principle of the hydraulic system. When a valve is opened, the changes of pressure are shown in Fig. 1.21b. The value of PT increases significantly.

1.4.3.2 Effect of Proposed Methods

(1) The faults used in experiment

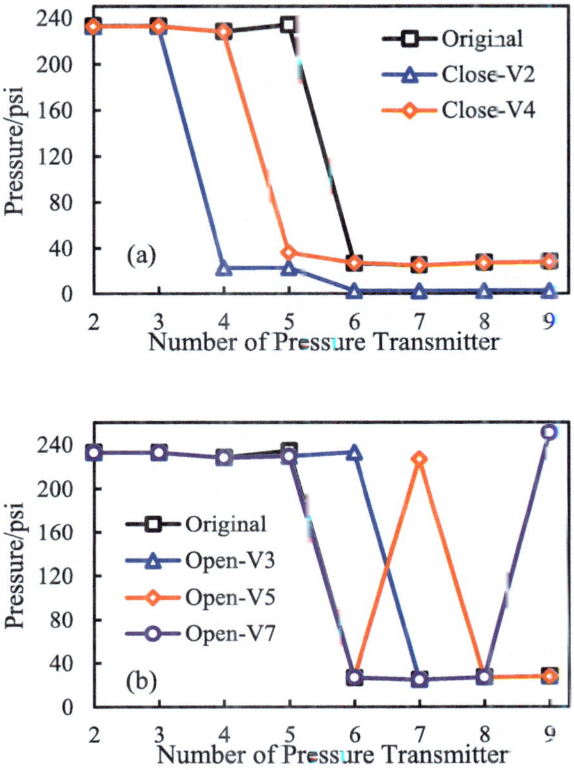

Fig. 1.21 The value of PT when **a** close the valve **b** open the valve

Data for four days in a row are collected for this experiment. Before collecting data, a large maintenance program was completed and thus, there are fault data. The faulty data used in this experiment are divided into two types. One is real fault data. They were stored when a fault occurs. These fault characteristics are obvious because stopping the system for maintenance only occurs when the fault is very serious. However, this will lead to a sudden stop of production, which can be expensive because an unreasonable stop planning especially sudden shutdown can lead to waste of a large costs of supporting facilities. The diagnosis method is required to be able to identify smaller degrees of faults. The other data are obtained by simulation based on existing fault data. The whole control circuit is divided with the valve as the dividing point. When a fault occurs in a circuit, the pressure of related circuit should be 0–90% of the normal pressure. Values in this range are taken at certain intervals to ensure that all levels of faults are taken. The degree of failure and threshold are set by combining field data and some experiments. Because there are few faults in the field, sufficient data is a prerequisite to ensure the accuracy of the model.

According to the feedback of the platform, it is rare for a component to fail. It is the same; a component with minor faults is only found in regular maintenance. Therefore, partially failed components are common. Their quantity distribution should meet the normal distribution, as shown in Fig. 1.22. In this experiment, failure is defined as the percentage of the valve and its front and rear hydraulic lines that can perform their normal functions. In fact, other faults exist. All the faults in this study are the faults of valves. There are two reasons for this. Firstly, the historical data shows that faults in a subsea production system are mostly concentrated in valve connections, including internal and external leaks. In addition, some data comes from simulations due to the limited amount of data containing faults. It is difficult to accurately simulate faults in valves or pipelines. In the field, locating the faults of a relevant circuit can meet the requirements of rapid maintenance. Based on these considerations, the failures of valves can be considered as failures of the hydraulic circuit where the valve is located. According to the field investigation, when the failure parameter is greater than 0.55, the system can operate normally, and minor faults are considered to have occurred. When it is less than 0.55, an obvious fault occurs. Of course, when the fault parameter is greater than 0.9, the component is considered normal. In this way, 740 data points are used to study the performance of the fault diagnosis model, including 68 normal data, 224 minor fault data, and 448 apparent faults. The number of faults at each level is the same in many studies (Cai et al. 2021). In this case, accuracy is difficult to reflect the diagnostic performance of the method in actual situations. In the production process, there are many faults with a failure degree of around 0.55, and there are few faults with complete or almost no failure. The accuracy that is calculated in a database with a reasonable distribution of faults can reflect the effectiveness of the method when applied in the field.

(2) Fault diagnostic accuracy of the proposed method

The fault diagnosis model is shown in Fig. 1.23. It can be seen from the Fig. 1.6 that the V_1 and V_2 can influence all the PT. So, they are the parent nodes of all the PTs (P in Fig. 1.23). In addition, the V_3–V_7 influences the P_5–P_9. Thus, they are the parent node of corresponding PT. The verification algorithm and feedback algorithm are introduced in Sect. 1.3.3.

The accuracy of the fault diagnosis model is shown in Fig. 1.24. In Fig. 1.24, the abscissa is the fault number, the fault number in V1 is 1, the fault number in V2

Fig. 1.22 Fault simulation parameters and distribution

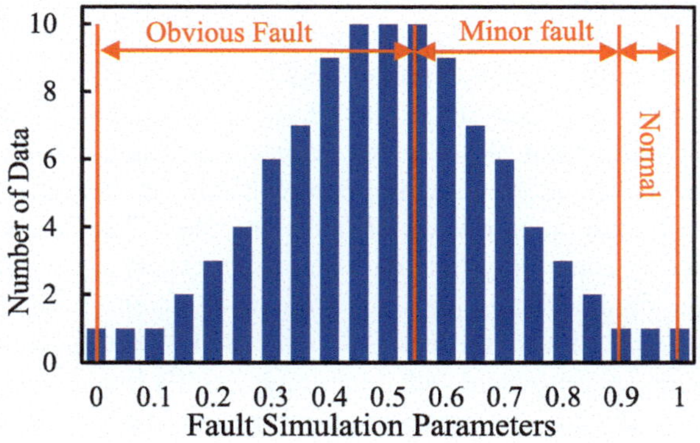

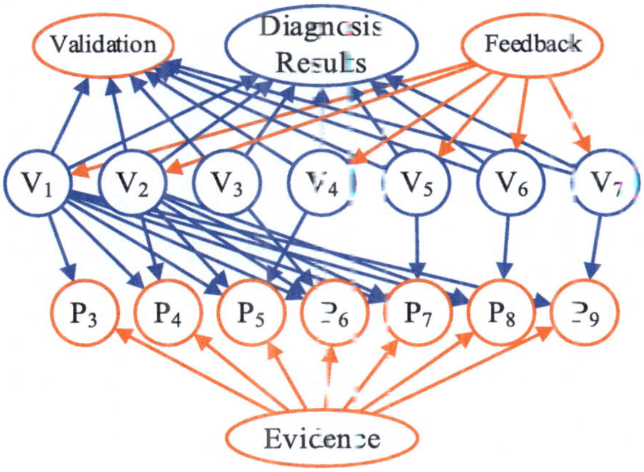

Fig. 1.23 Fault diagnosis model for the subsea production control system

is 2, and so on. What's more, 0 represents normal. The accuracy of fault diagnosis with a digital twin is a little higher than that without a digital twin model slightly. It also can be seen that the improvement is not apparent. Such a small promotion is dispensable. A notable phenomenon is that the accuracy for normal data, combined with the digital twin model, decreases. The reason may be that the accumulated error of the digital twin model exceeds the set value during the simulation. To study the indeed reason of it, the obvious faults and the minor faults are studied separately, and the result is shown in Fig. 1.25. Besides, the accuracy for obvious faults is almost the same, and they are all higher than 90%. However, the accuracy for minor faults is extremely different, some are even lower than 80%. They are different when the digital twin model is used for verification with the fault diagnosis model. Most of them are over 90%. Through various research, the accuracy of fault diagnosis for the obvious and mixed fault is achieved; however, the minor fault is difficult. The number of obvious fault data throughout the data is almost twice that of minor fault data. This is why the accuracy of fault diagnosis with the digital twin model is not improved significantly in Fig. 1.24.

(3) Influence digital twin model on fault diagnosis model

To research the factors that influence the fault diagnosis, the more detailed results are shown in the matrix in Figs. 1.26, 1.27, 1.28 and 1.29. In each matrix, the abscissa represents the actual fault of the input data, and N is the data in a normal state. The ordinate represents the fault type identified by the fault diagnosis model. It is easy to see that the larger the number on the diagonal, the better the diagnostic performance. It is notable that the sum of each vertical line number is not the total number of data because some data are identified as two or more faults. At the same time, the color depth also reflects the capitalization of numbers in the table. The darker the color, the larger the number.

Figure 1.26 shows that the fault diagnosis model can almost identify all the fault data, and most of wrong faults are that one is identified as another. The results are

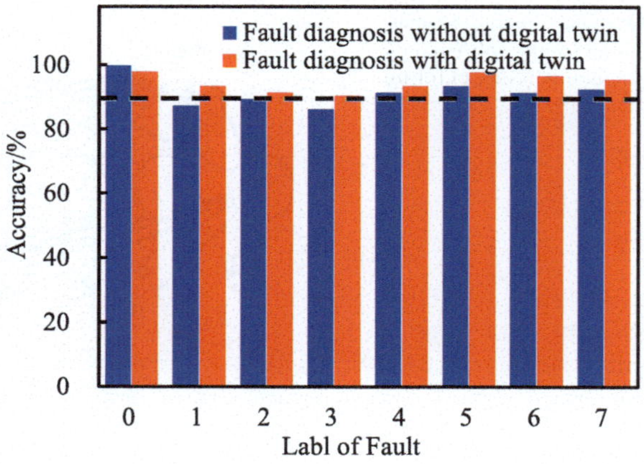

Fig. 1.24 Accuracy of fault diagnosis for all data

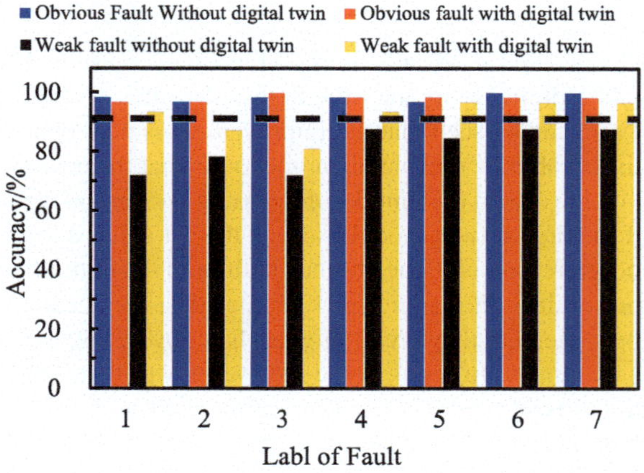

Fig. 1.25 Accuracy of obvious and weak faults

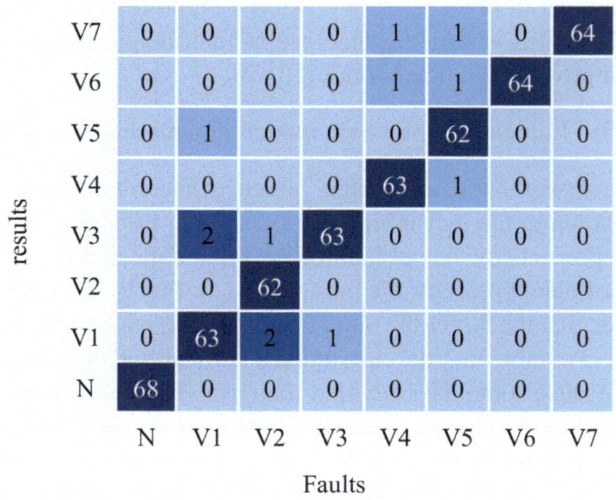

Fig. 1.26 Result of fault diagnosis without digital twin model for obvious faults

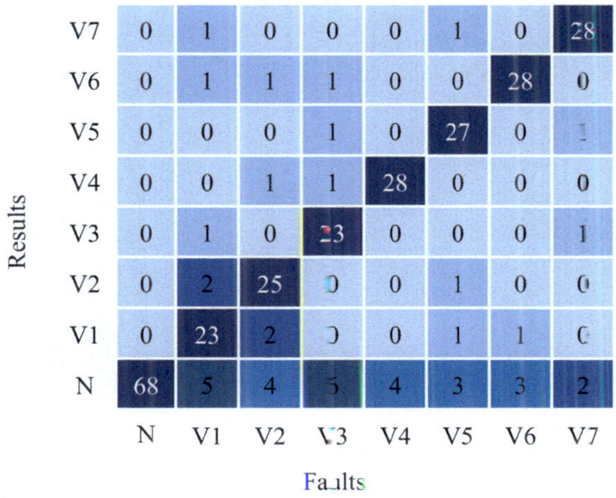

Fig. 1.27 Result of fault diagnosis model without digital twin model for minor faults

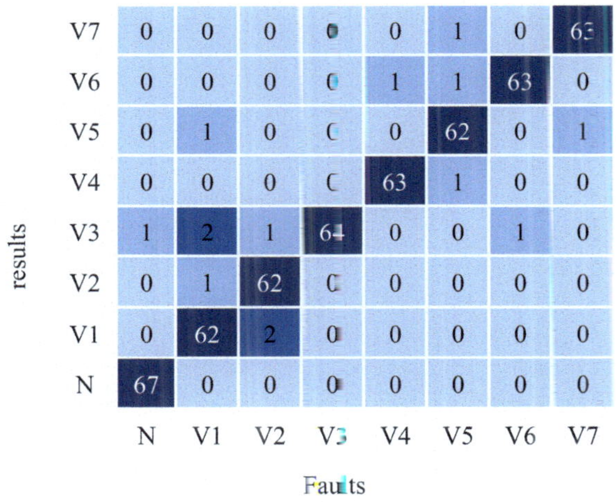

Fig. 1.28 Result of fault diagnosis model with digital twin model for obvious faults

different when minor faults are involved, as shown in Fig. 1.27. Due to the minor fault characteristics, the fault diagnosis model cannot identify all faults, leading to the failure data not being recognized in most wrong diagnoses. When the fault diagnosis model combined with the digital twin model is used for diagnosis, the effect of fault diagnosis for obvious faults is similar to the original shown in Fig. 1.28. Because the original diagnosis effect is good enough. Obviously, the performance of fault diagnosis for minor fault as shown in Fig. 1.29. Most unknown faults have been identified, and good diagnostic accuracy is obtained. It can be seen from the comparison that the digital twin model has a significant effect on improving the diagnosis accuracy for minor faults.

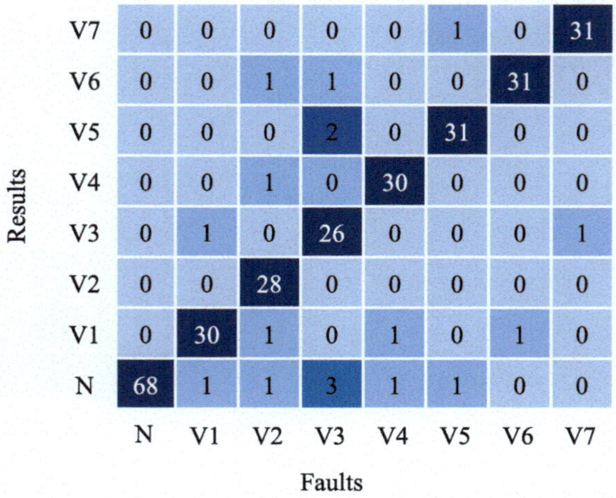

Fig. 1.29 Result of fault diagnosis model with digital twin model for minor faults

Chapter 2
Digital Twin-Assisted Intelligent Fault Diagnosis

Abstract As engineering systems grow increasingly complex, traditional fault detection methods struggle to maintain high accuracy and reliability. This chapter introduces a novel fault diagnosis framework for hydraulic systems, leveraging DT technology. The framework combines a virtual model, built with Modelica, with real-time system data through an innovative bidirectional data consistency evaluation mechanism. To further enhance the data's reliability, a two-dimensional signal warping algorithm is applied. The optimized data is then used to train a multi-channel one-dimensional convolutional neural network-gated recurrent unit model, which captures both spatial and temporal features, improving fault detection accuracy. The method is validated using a subsea blowout preventer in laboratory settings, achieving an accuracy rate of 95.62%. This approach outperforms existing methods, offering a robust and scalable solution for predictive maintenance in complex hydraulic systems by integrating DT technology, data optimization, and advanced deep learning techniques.

2.1 Introduction

As the global subsea equipment industry embraces intelligent technology, there is a significant shift towards automation and smart systems. This transformation is primarily driven by the need to improve operational efficiency and minimize downtime in subsea environments. Fault detection and diagnosis in subsea equipment play a critical role in ensuring the reliable operation of these complex systems, which operate in challenging conditions and are subject to high risks.

Traditional fault diagnosis techniques in subsea systems often depend on expert knowledge and manual analysis, which are labor-intensive and prone to human error. These conventional approaches, while useful, are limited by the availability of skilled technicians and the complexity of fault scenarios in deep-sea operations. In contrast, intelligent fault diagnosis (IFD), which leverages machine learning algorithms has emerged as a promising alternative. IFD can automate fault detection and classification, significantly reducing diagnostic time and improving accuracy. However, traditional IFD methods often rely on manually crafted features and require large amounts

B. Cai et al., *Intelligent Operation and Maintenance for Subsea Production Systems*,
https://doi.org/10.1007/978-981-96-2677-9_2

of labeled data, which are difficult to obtain, especially in subsea applications where data is sparse and expensive to collect.

To overcome these challenges, this chapter proposes an innovative DT-assisted IFD framework that integrates real and virtual data for enhanced fault diagnosis. By employing a bidirectional data consistency evaluation mechanism, this framework aims to optimize the alignment of real-time sensor data with simulated data from a digital replica of the subsea system. The framework is designed to address the issue of limited fault data, which is common in subsea environments, by improving data reliability and diagnostic performance. The approach combines physical entities, virtual models, and data analytics into a comprehensive system, offering a more robust solution for fault detection and predictive maintenance in subsea hydraulic systems.

2.2 Proposed IFD Method

DT technology is fundamentally composed of five dimensions: physical entities, virtual models, connections, data, and services. In this chapter, the DT system achieves the integration of the five dimensions in the following ways. The system relies on subsea equipment (such as BOP) as physical entities and comprehensively collects its operating status, environmental conditions and fault information, ensuring that the virtual model can accurately reflect the actual operating conditions (Longoo et al. 2023). The Modelica language is used to build a high-fidelity virtual model that covers all subsystems of the hydraulic control system and can accurately simulate different fault scenarios. The virtual model can be dynamically adjusted and calibrated through data interaction with the physical entity (Carlin et al. 2024). The system realizes the data flow between physical entities and virtual models through a two-way data transmission mechanism, which ensures that the virtual model can reflect the state changes of the real system (Qi et al. 2021). The system combines real-time data of physical entities and simulated data generated by virtual models, and continuously optimizes data consistency through the bidirectional data consistency evaluation model, ensuring high data quality and reliability (Jiang et al. 2021). The DT system ultimately transforms data analysis results into decision support in actual operations through services such as fault diagnosis and optimized control. These services can respond to changes in system status in real time and provide operators with timely and accurate suggestions.

An intelligent diagnostic DT-assisted framework for hydraulic system faults is proposed in this chapter. The framework describes a complex three-stage diagnosis method, as shown in Fig. 2.1. Each stage targets a different function: The function of the first stage is to generate real data and virtual data. The function of the second stage is to verify the consistency of the data and fuse real data and virtual data into twin data. The function of the third stage is to train the diagnostic model using twin data.

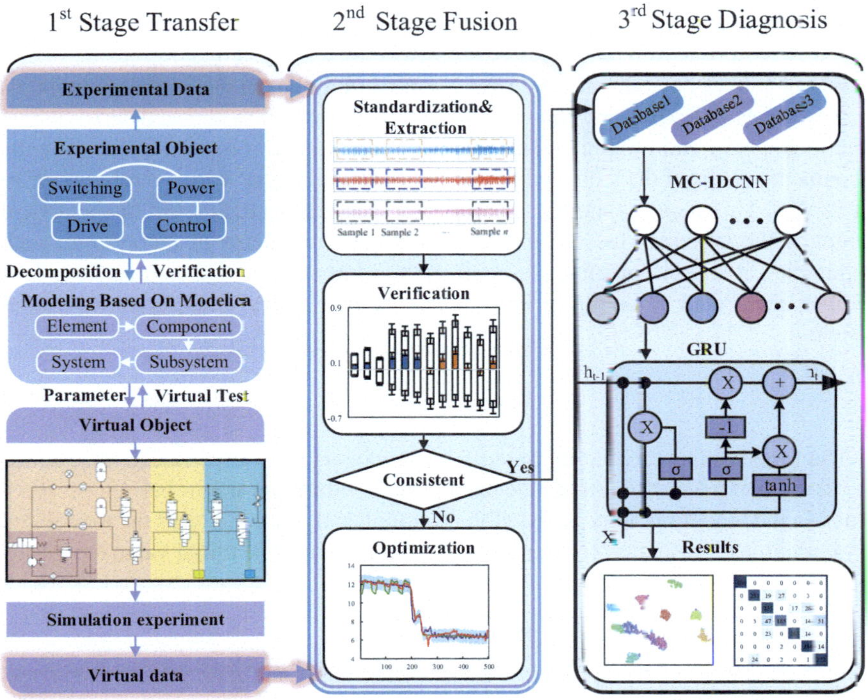

Fig. 2.1 Framework of the proposed IFD method

2.2.1 *Virtual Model Parameter Transfer Based on Modelica*

Engineering systems with high functional density (such as subsea equipment, aircraft engines or intelligent manufacturing systems) usually involve the coupling of multiple physical phenomena (such as thermal, mechanical, electromagnetic, etc.). The complexity and diversity of the equipment make it difficult for traditional fault diagnosis methods to fully cover all possible failure modes.

Modelica, an object-oriented high-level modeling language is used to model and simulate physical systems. Modelica language modeling has the advantages of openness, physical consistency, and scalability. These advantages make it very suitable for creating multi-domain virtual models of complex systems, such as hydraulic control systems integrated with electromechanical fluids. This chapter uses the Modelica language to build a virtual model. The element-component-subsystem-system construction method simplifies the complexity of the model. This is also conducive to system expansion. Furthermore, components or subsystems can be reused in different systems, thus improving the efficiency and reliability of the entire system.

Taking the hydraulic control system shown in Fig. 2.2 as an example. The system can be decomposed into power subsystem, switching subsystem, control subsystem, and drive subsystem. The driving subsystem can be decomposed into components such as valves and cylinders; The valve components can be decomposed into dynamic functions, ports, and other elements; The dynamic function is used to describe the dynamic behavior of the valve, which relates the travel signal y_s to the relative valve travel y_{Rel}. This requires the use of Modelica language to define the mathematical model of the system. This includes the physical components of the system, such as pumps, valves, and pipelines, as well as their interactions and connections. This involves integrated modeling of fluid dynamics, mechanics, and electronic control systems.

$$y_{Rel} = f(y_s, t) \tag{2.1}$$

The time related calculation process of the port is illustrated in Fig. 2.3, which plays a crucial role in facilitating the interaction of information flow between components. In the modeling process, the global symbol analysis transforms the model into a computable representation, simplifying its complexity. The initial value represents

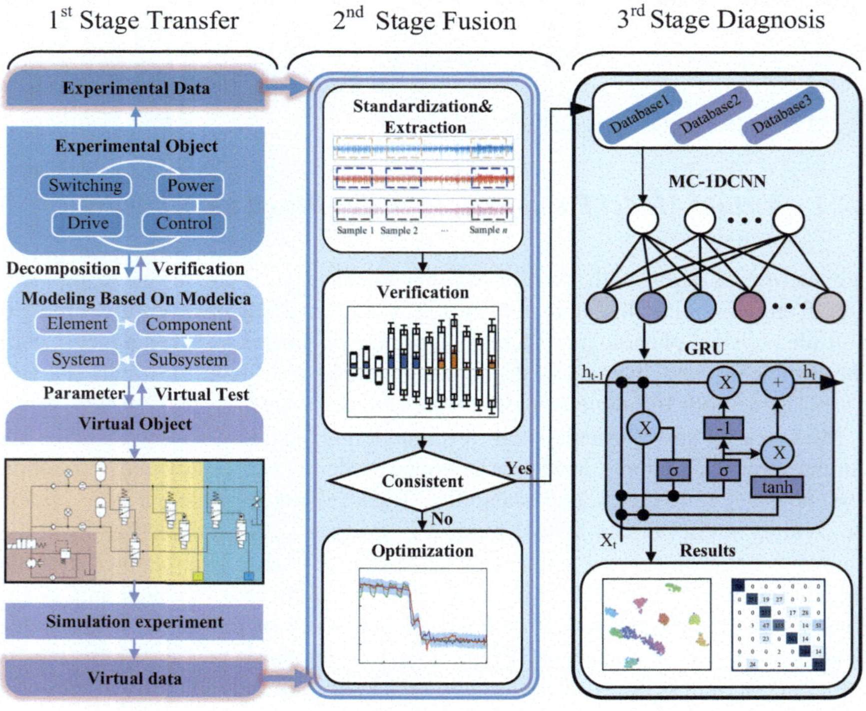

Fig. 2.2 The modeling of hydraulic control system using Modelica

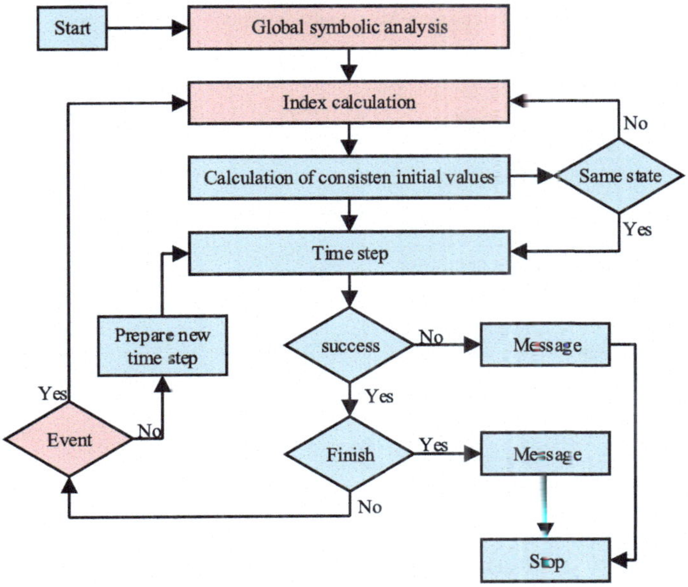

Fig. 2.3 The interaction process of information flow between components

the state of the model at the beginning of the calculation. An event refers to a specific moment in time when an equation becomes discontinuous or when there is a change in the equation's structure during the simulation operation. This process involves converting the virtual model into a computable form. The model is then represented in the internal format of the solver. A consistent initial value and start time-domain simulation are determined. After each successful time step, the solver checks whether an event has occurred. If an event is detected, the solver determines the exact time of the event as accurately as possible, halts the continuous integration process, and reinitializes the model. Virtual models can generate simulated data through mathematical model calculations.

After building the system model under normal working conditions, the next step is to add fault conditions and generate virtual data. This process involves the following steps, as shown in Table 2.1. By following these steps, the model can accurately simulate the operating state of the hydraulic system. High-fidelity simulation data is generated for training and validation of the fault diagnosis model.

Table 2.1 Detailed process and steps for adding fault conditions

Step	Do
1	**Set** input system parameters $\Phi = [\varphi_1, \varphi_2, \varphi_3, \ldots, \varphi_n] = [P, Q, V, \ldots, T]$ **Simulate** fault state
2	**Set** Time step $t_s = t_2 - t_1 = t_3 - t_2 = \ldots t_n - t_{n-1}, T = [t_1, t_2, t_3, \ldots, t_n]$ **Calculate** state changes
3	**Process** fault conditions **Set** conditions $C_{set} = [c_1, c_2, c_3, \ldots c_n]$
4	**Run** model
5	**If** $t < t_n$ **Continue** updating **Calculate** $\Phi_t = [\varphi_{1t}, \varphi_{2t}, \ldots, \varphi_{nt}]$ **If** $C_k = C_{set}$ Record output signals $S = [s_1, s_2, \ldots, s_m]$ **End** **else** **Feedback and return to Step 1** **End** **else** Record output signals $S = [s_1, s_2, \ldots, s_m]$ **Feedback and return to Step 4** **End**
6	**Record** output signals $S = [s_1, s_2, \ldots, s_m]$ to form virtual data
7	**Optimize** virtual data using 2DSW for alignment with real data

2.2.2 *Virtual Real Data Fusion Using Bidirectional Data Consistency Evaluation Framework*

The 2DSW algorithm enhances fault data quality in Modelica by aligning or matching two-dimensional signals, ensuring spatial and shape consistency (Fu et al. 2008). The 2DSW algorithm adapts the template to the waveform by minimizing the distance between them. A two-dimensional grid of warping points is overlaid onto the signal template, as depicted in Fig. 2.4. Each intersection represents a warping point, forming warping regions. Warping points are active if their surrounding areas contain parts of the template; otherwise, they are passive. The template is adjusted by shifting active warping points.

The shift of P_i minimizes the cost function, using the normalized Euclidean distance, between the adapted waveform and the reference waveform. The normalized Euclidean distance $d^{(2n)}$ between vectors $y^{(1)}$, $y^{(2)}$ is given by the following equation under the condition $X^{(1)} = X^{(2)}$.

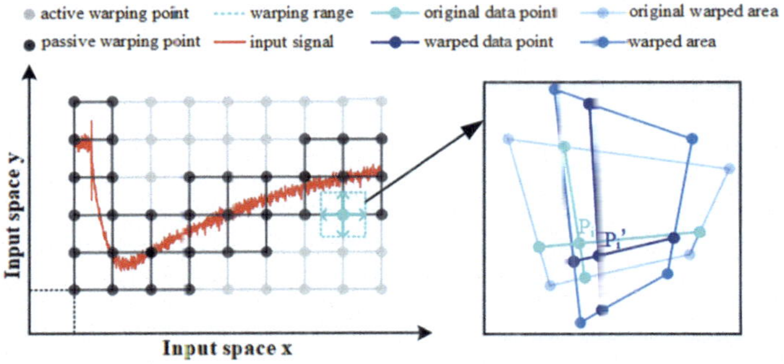

Fig. 2.4 The warping process of signal points on a template covered by a 2D network

$$d^{(2n)}(Y^{(1)}, Y^{(2)}) = \frac{\sqrt{\sum_{i=1}^{|\gamma|} (y^{(1)} - y^{(2)})^2}}{|\gamma|} \tag{2.2}$$

where $Y^{(1)}$ and $Y^{(2)}$ represent the y-values of vectors $y^{(1)}$ and $y^{(2)}$ Compared with Euclidean distance, its standardized version describes the range of comparison between two sets. The condition $X^{(1)} = X^{(2)}$ guarantees the same domain of $X(\bullet)$.

In optimization, virtual data is the source signal and real data is the target signal. The BA consistency method evaluates if the optimized virtual data meets preset standards, enhancing alignment with real data. The 2DSW method can be expressed as the following mathematical model:

$$\min_{W} \sum_{i=1}^{m} \sum_{j=1}^{n} d_{i,j} w_{i,j} + \lambda \sum_{i=1}^{m} \sum_{j=1}^{n} |w_{i+1,j} - w_{i,j}| + |w_{i,j-1} - w_{i,j}| \tag{2.3}$$

where W represents the twisted path, $d_{i,j}$ represents the distance between the source signal and the target signal at their positions(i,j), and λ is a parameter that controls the regularization term. In the recursive process, it is necessary to calculate each $w_{i,j}$ corresponding minimum distance $D_{i,j}$ and path point $P_{i,j}$, so that after calculating the complete matrix D, the optimal path can be obtained through backtracking.

The Dynamic programming algorithm is used to solve the above optimization problems. The specific process is as follows: Initialize $D_{1,1} = d_{1,1}$, $D_{i,1} = +\infty$, $D_{1,J} = +\infty$, $P_{i,1} = (1,1)$; For $i = 2 \to m$, $j = 2 \to n$; calculate $D_{i,j}$ and $P_{i,j}$. Backtracking the optimal path to obtain distorted paths W; For each $w_{i,j} \in W$, update $B'_{i,j} = B_{i,j} + w_{ij}$.

$$D_{i,j} = \min \begin{cases} D_{i-1,j} \\ D_{i,j-1} \\ D_{i-1,j-1} \end{cases} + d_{i,j} \tag{2.4}$$

$$P_{i,j} = \begin{cases} (i-1,j) & (D_{i-1,j} \leq D_{i,j-1} \text{ and } D_{i-1,j} \leq D_{i-1,j-1}) \\ (i,j-1) & (D_{i,j-1} \leq D_{i-1,j} \text{ and } D_{i,j-1} \leq D_{i-1,j-1}) \\ (i-1,j-1) & (D_{i-1,j-1} \leq D_{i-1,j} \text{ and } D_{i-1,j-1} \leq D_{i,j-1}) \end{cases} \quad (2.5)$$

For engineering equipment, by comparing and analyzing virtual data with real data, determine whether the operating status of the hydraulic control system meets expectations, and then simulate faults. To input real and virtual data into the 2DSW algorithm using mathematical expressions, A_{obs}, B_{vir} and B'_{vir} are defined. The observation matrix A_{obs} describes the state of the hydraulic control system under specific operating conditions C (such as control commands, blowout events, and environmental changes).

$$A_{obs} = \begin{bmatrix} a_{11} & a_{12} & \cdots & a_{1m} \\ a_{21} & a_{22} & \cdots & a_{2m} \\ \vdots & \vdots & a_{ij} & \vdots \\ a_{n1} & a_{n2} & \cdots & a_{nm} \end{bmatrix} \quad B_{vir} = \begin{bmatrix} b_{11} & b_{12} & \cdots & b_{1m} \\ b_{21} & b_{22} & \cdots & b_{2m} \\ \vdots & \vdots & b_{ij} & \vdots \\ b_{n1} & b_{n2} & \cdots & b_{nm} \end{bmatrix} \quad (2.6)$$

In the equation, $\alpha_{i,j}$ is the measured value of the observed parameter α_i, and this time is t_i, n and m is the subscript number. C represents the state of the system over a period of time (set to $t_i - t_j$) and $A(t_j) = (\alpha_{1j}, \alpha_{2j}, \ldots, \alpha_{nj})$ represents the time series of A_i. Evaluation matrices B_{vir} and B'_{vir} represent the simulated output of the virtual model under the same operating conditions. B_{vir} is the mapping of A_{obs} in the virtual model. B'_{vir} is the optimized matrix of B_{vir} in the virtual model. Record $B_i = (b_{i1}, b_{i2}, \ldots, b_{im})$ as the state of the virtual model over a period of time, and record $(B_j) = (b_{1j}, b_{2j}, \ldots, b_{nj})$ as the time series of B.

Based on the above theory, establish bidirectional data consistency evaluation framework as shown in Fig. 2.5. The optimized object needs to initialize the twisted path W and iteratively optimize the twisted path W until similar requirements are met. Twist matrix $B'_{m \times n}$ is obtained based on the twisted path W. The consistency check of BA chart requires calculating the average and standard deviation of data A and data B.

Taking this research object as an example, A represents real data, B represents virtual data, and B' represents optimized virtual data. First, the input virtual data and real data are preprocessed, including denoising, standardization, and feature extraction, to ensure that the data are compared on the same scale. Then, the virtual data and real data are preliminarily aligned in time using the dynamic time warping method. Based on the preliminary alignment, the 2DSW algorithm is applied to further bend and adjust the signal. The alignment of the signal is gradually optimized by minimizing the weighted sum of the time difference and the amplitude difference. During this process, the algorithm continuously adjusts the position of the signal on the time axis and amplitude until the optimal matching solution is found. Next, the mean square error is used to quantify the error between the virtual data and the real data. Based on the results of the error evaluation, the virtual data is iteratively calibrated.

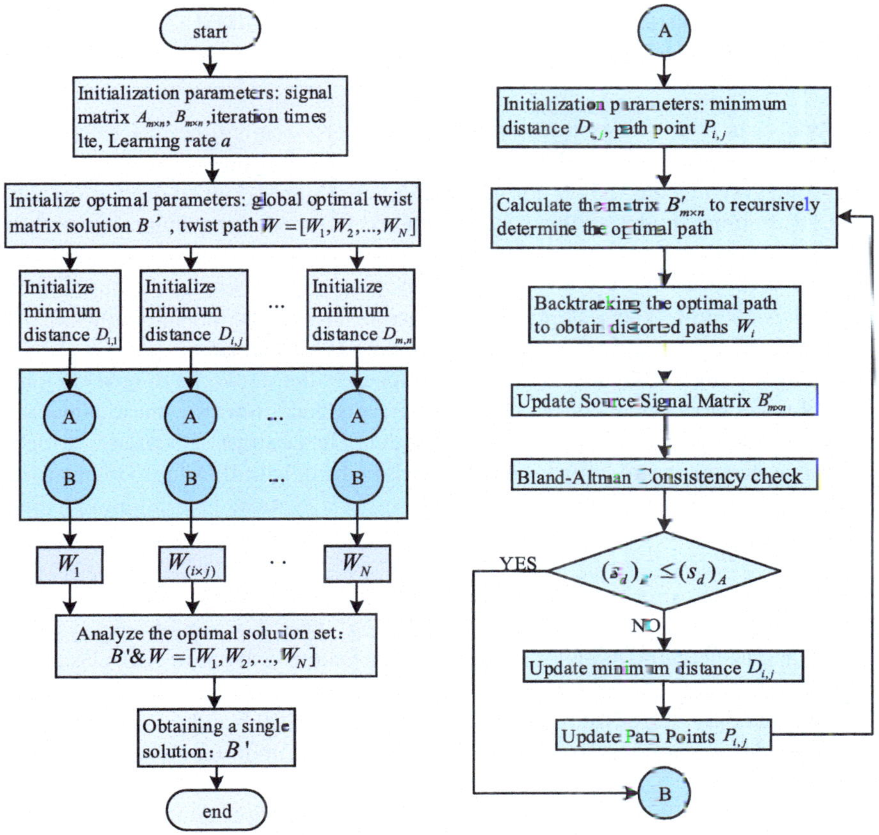

Fig. 2.5 The bidirectional data consistency evaluation framework process of combining virtual and real data

This includes adjusting the model parameters and optimizing the simulation algorithm to gradually reduce the difference. Finally, the optimized virtual data and real data are evaluated again. When the difference in time and amplitude drops to the minimum threshold, it is confirmed that the simulated data has met the requirements for being used as twin data.

The virtual model is regularly calibrated with real data to enable the model to simulate the real system. Although the two-way communication link between the real entity and the virtual model is not real-time, it is very reliable. The adjusted virtual data is defined as twin DT data if it meets the consistency criteria. The bidirectional data consistency evaluation framework ensures the accuracy and reliability of the data. Although existing studies have used DT technology to perform IFD on various devices and systems, this study is the first to propose combining DT technology with the 2DSW optimization algorithm to calibrate the virtual model. Make its output data closer to the actual measurement data.

2.2.3 Construction of Fault Diagnosis Model Based on MC1DCNN-GRU

DL models can process classification problems by processing time series data. In a hydraulic system, pressure and flow signals can be used as inputs to a DL training model, and system status labels corresponding to the signals can be used as outputs of the DL training model. In this way, the trained DL model can diagnose the fault type of the hydraulic system through the input pressure and flow signals. Among them, the sources of the training input signals are varied, including data obtained from a physical prototype system, data obtained from a virtual model built using the Modelica language, and data optimized using the 2DWS algorithm.

When it comes to IFD tasks, the ability of deep belief networks to provide joint probability distributions for input data is acknowledged. However, these networks face challenges when analyzing real value data. In contrast, CNNs are widely regarded as the most commonly used DL models in the field of IFD due to their strong supervised feature learning capability. RNNs, especially those incorporating gating units like LSTM and GRU, have demonstrated significant advantages in sequence modeling.

To leverage the strengths of both CNNs and RNNs, this chapter proposes a joint model called MC1DCNN-GRU, as shown in Fig. 2.6. This model employs two distinct neural network modules to accomplish feature extraction and classification tasks. The MC1DCNN module is responsible for extracting features from the temporal data through convolution operations, while the GRU module captures long-term dependencies in the sequence data, facilitating accurate fault type prediction and classification.

The 1DCNN module is responsible for extracting features from temporal data through convolution operations. The key formulas related to this process are provided below:

$$Z_{i,j} = \sum_{k=0}^{K-1} X_{i+k,j} W_k \tag{2.7}$$

where $X \in R^{T \times C}$ represents the one-dimensional temporal data input for the one-dimensional convolution operation, T represents the time step size, and C represents the number of channels; $Z \in R^{T' \times F}$ represents the output of convolution operations, T' represents the time step after convolution, and F represents the number of convolution kernels; $W \in R^{K \times C}$ represents a convolutional kernel, K indicating the size of the convolutional kernel.

$$h_{i,j} = \max_{k=0}^{K-1} Z_{i+k,j} \tag{2.8}$$

where $h \in R^{T' \times F}$ represents the output of one-dimensional pooling operation and T' represents the time step after pooling.

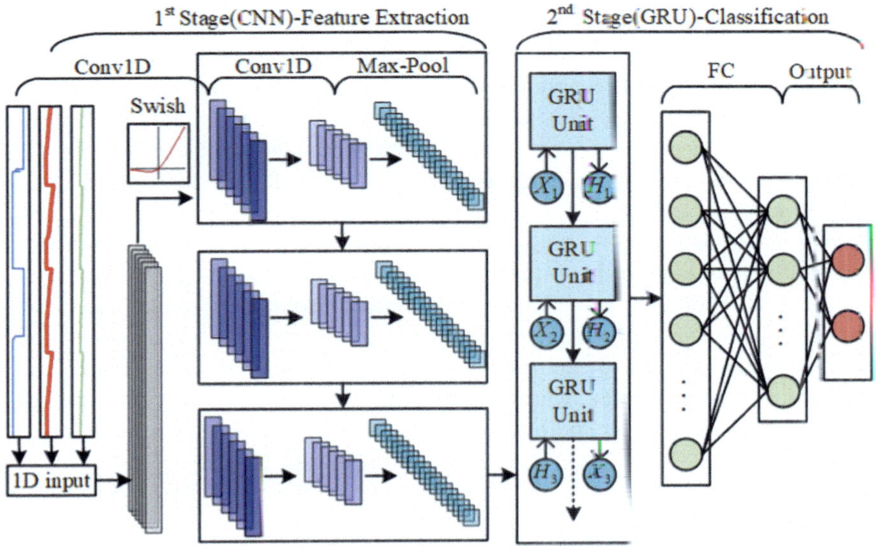

Fig. 2.6 The process of the MC1DCNN-GRU joint model for input signals

In this module, the activation function used is Swish. While ReLU is commonly used in deep neural network design, its limitation lies in producing zero output for negative inputs, which can restrict its learning ability. Swish is a newly developed activation feature that overcomes this limitation. It possesses characteristics such as being lower bounded, upper unbounded, smooth, and non-monotonic. The formula for the Swish activation function is as follows:

$$f(x) = x \cdot \sigma(\beta x) = \frac{x}{1 + e^{-\beta x}} \tag{2.9}$$

where σ represents the sigmoid function, which β is a learnable parameter.

The GRU module accepts one-dimensional feature vectors h as input and completes the classification task by recursively updating the hidden state. Specifically, let the hidden state of the GRU module be $h_t \in R^D$, where D represents the dimension of the hidden state. The calculation formula for the gated recursive unit is as follows:

$$r_t = \sigma(W_r x_t + U_r h_{t-1} + b_r) \tag{2.10}$$

$$z_t = \sigma(W_z x_t + U_z h_{t-1} + b_z) \tag{2.11}$$

$$\tilde{h}_t = \tanh(Whx_t + r_t \odot U_h h_{t-1} + b_h) \tag{2.12}$$

$$h_t = (1 - z_t) \odot h_{t-1} + z_t \odot \tilde{h}t \tag{2.13}$$

where $x_t \in R^{F'}$ represents the output features of the 1D-CNN module, r_t and z_t represents the gating vector, $\hbar_t$ represents the candidate hidden state, σ and Θ represents the product of the sigmoid function and the vector element by element, respectively.

The classification of one-dimensional time series data from hydraulic systems can be achieved by integrating the previously mentioned modules into the MC1DCNN-GRU model. Enhancements in model performance can be attained through the adjustment of model parameters and optimization of the network structure. Fault diagnosis methods in deep learning technologies, such as GAN and data augmentation, usually require a large amount of high-quality data to train reliable models. However, it is very difficult to collect a large amount of data when a complex system fails. The DT method is conducive to creating multi-domain virtual models of complex systems and accurately simulating the dynamics within the system. Simulation generates virtual data and integrates real data, which enhances the diversity and capacity of the data set. Rich data sets can train more accurate and reliable fault diagnosis models.

2.3 Case Study of Subsea BOP System

This section is divided into subsections to systematically present the proposed fault diagnosis method and its validation. It begins with the description of the hydraulic control system (Sect. 2.3.1), followed by the verification of the virtual model (Sect. 2.3.2), the process of collecting and fusing real and virtual data (Sect. 2.3.3), the optimization of data consistency using the 2DSW algorithm (Sect. 2.3.4), and the performance optimization of the MC1DCNN-GRU model (Sect. 2.3.5). These subsections are structured to provide a step-by-step explanation, from system setup to model evaluation.

2.3.1 Description of Hydraulic Control System for Subsea BOP

The subsea BOP system is of utmost importance in ensuring safe working conditions for deep-water drilling operations. At the heart of the subsea BOPs lies the BOP control system, which is a complex electromechanical hydraulic composite system. In order to improve working reliability, the BOP system generally adopts a full-process redundant design. However, this design also leads to an increase in data volume and difficulty in identifying fault characteristics, which brings challenges to fault diagnosis. The research conducted on fault diagnosis of subsea BOP control systems can also provide valuable insights for diagnosing faults in other complex control systems with redundant designs.

One of the key components of the BOP system is its control system, which includes an electronic control system, a supporting hydraulic control system, and a sensor system. The main circuit of the hydraulic control system is controlled by the main control pump and the main control solenoid valve. This circuit is further divided into a high-pressure circuit and a low-pressure circuit using a diversity valve. The high-pressure circuit obtains hydraulic power from the hydraulic pump and accumulator group, enabling the closing and opening actions of the ram BOP. This high-pressure circuit is responsible for sealing and opening the well. The low-pressure circuit controls the locking and opening of the locking device. The main control pressure for both circuits is regulated by a pressure reducing valve. The entire system is equipped with a dual redundant control system, which consists of a blue box and a yellow box. Each control box independently performs all the necessary functions of the subsea BOP group.

For large-scale engineering equipment such as BOPs, due to the fact that most of the equipment is in normal operation, the actual fault data in the system is incomplete and unbalanced. This makes building high fidelity virtual models a challenge. To address this challenge, an experimental platform was established based on the HYSY981 drilling platform BOP control system as a prototype (refer to Fig. 2.7). The purpose is to obtain complete data in advance based on known schematics and signals in the engineering system. In this chapter, six common fault states and one normal state were simulated, and a total of seven operational data were collected as shown in Table 2.2. Fault simulation is completed through some manual valves and replacement of faulty components. The experiments are conducted on a workstation equipped with an Intel Core i7-10700K processor (3.8 GHz), 32 GB DDR4 memory, and an NVIDIA RTX 3080 GPU. This hardware configuration ensures computational efficiency in complex simulations and deep learning training.

The complete data workflow is summarized in Fig. 2.8. Firstly, the behavior of the virtual model under normal working conditions and fault conditions is simulated. Secondly, the data collected from the experimental device under the corresponding conditions are used to verify the accuracy of the virtual model. Then, the virtual data is optimized by bidirectional data consistency evaluation (BDCE) framework using the real data as a template. Finally, the optimized virtual data and experimental data are combined as twin data to enhance the diversity and reliability of the data set. The data set will be used to train and verify the fault diagnosis model. In future work, it is planned to collect data from actual subsea deployed equipment in a few years. These real data with maintenance records and actual fault occurrences will be used to further validate the fault detection model. This step is crucial to ensure the accuracy and reliability of the model in real-world applications.

2.3.2 *Verification of Virtual Models*

This section describes the verification process of the correctness of the underlying components in the model. In order to make the virtual model as close as possible to the actual system, the model of the original physical entity, including its structure, parameters, and behavior, is reconstructed numerically, as shown in Fig. 2.9.

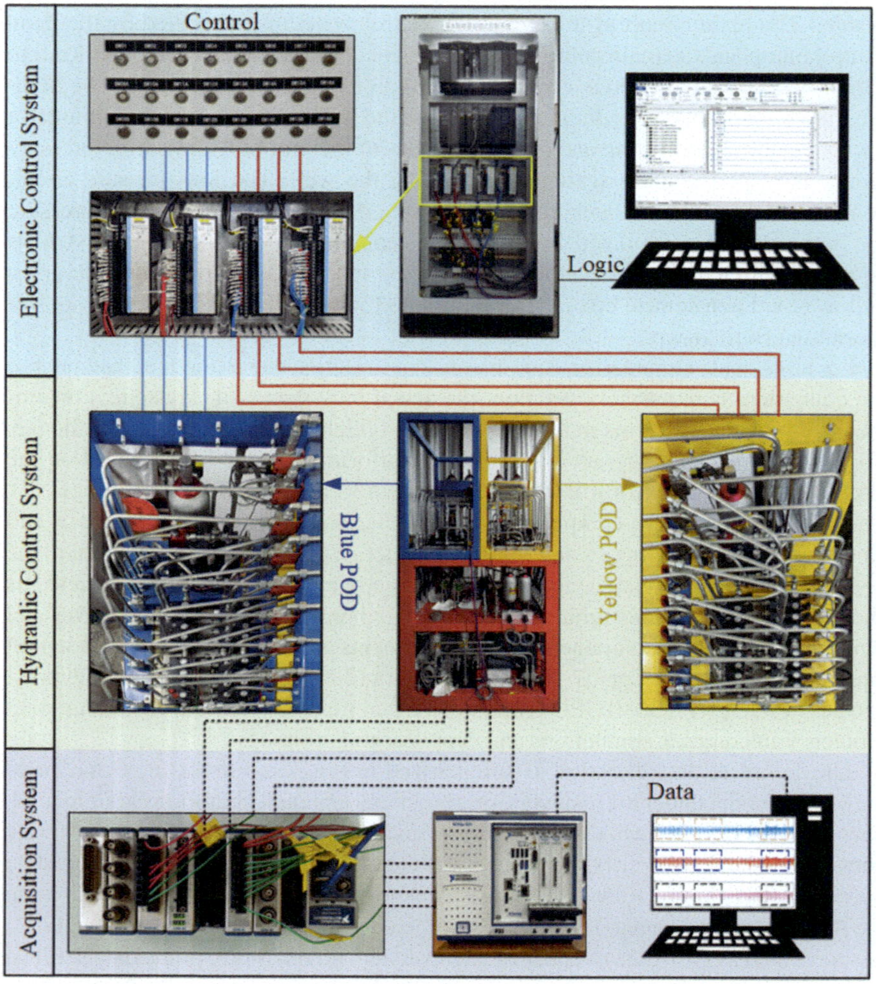

Fig. 2.7 The subsea BOP control system experimental platform

The hydraulic control system of the subsea BOP functions in parallel with multiple sets of pilot operated solenoid valves. A set of execution subsystems is selected as a representative model for verification because it has a simple structure and is also expandable. In the Drive Unit Model in Fig. 2.9, the quantitative pump supplies oil, and the parameters of the hydraulic resistance, proportional relief valve, and proportional speed regulating valve are initially set to constant values. The cycle of the reversing valve is set to 3 s. A damping mechanism composed of a spring and a mass block is used to replace the load. The interface variables (pressure, flow) between the hydraulic cylinder and the reversing valve and the key parameters of

Table 2.2 Description of each fault state and source and quantity of dataset

The working states of the hydraulic control system	Total number of data dataset			Label
	Real dataset A	Virtual dataset B	Twin dataset C	
Normal	1200	1000	1000	1
Accumulator hydraulic leakage	1200	1000	1000	2
Pump leakage	1200	1000	1000	3
Proportional valve core stuck	1200	1000	1000	4
Leakage of bypass solenoid valve	1200	1000	1000	5
Main circuit solenoid valve leakage	1200	1000	1000	6
The shuttle valve core is stuck	1200	1000	1000	7

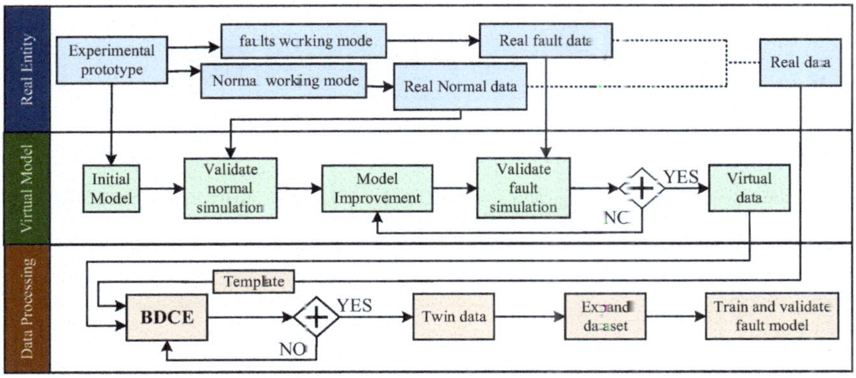

Fig. 2.8 The complete data workflow

the hydraulic cylinder (speed, acceleration) are selected to analyze and evaluate the simulation results, as shown in Fig. 2.10.

The simulation results are analyzed to ensure the validity of the model. The pressure fluctuations of the reversing valve A and B ports are consistent with the pressure fluctuations of the hydraulic cylinder A and B ports. The pressure value of the reversing valve T port is always 0, which is consistent with the actual situation. When the hydraulic rod is extended, the reversing valve A port is the oil flow outlet, and the hydraulic cylinder A port is the oil flow inlet. The flow rates of the two ports are equal in size and opposite in direction, which is consistent with the simulation curve in the figure. The pressure value curve of the reversing valve A and B ports is consistent with the cycle of the reversing valve. The periodic pulsation curve of the pressure value of the reversing valve A and B ports is also consistent with the periodic pulsation curve of the pressure value of the hydraulic cylinder A and B ports. The velocity and acceleration curves of the mass block correspond one by one to the actual motion state of the hydraulic rod in the hydraulic control system of the blowout preventer. The correctness of the model established is proved.

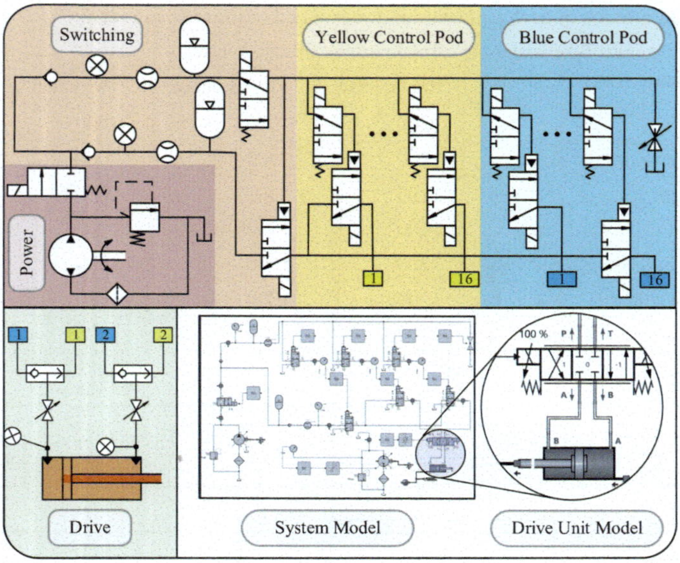

Fig. 2.9 The process of establishing representative models selected from hydraulic control systems

Fig. 2.10 The simulation results of several key interface variables in the virtual model

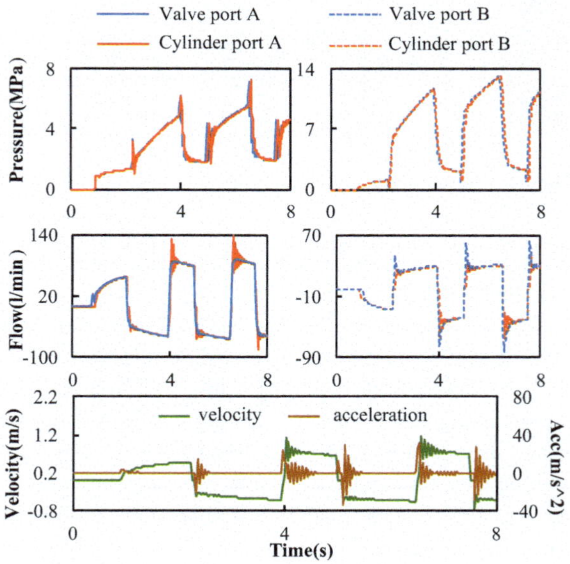

The verification confirmed the accuracy of the basic components in the model. It also laid the foundation for the subsequent work of building a system model and collecting data sets. When the requirements are met, the virtual model can simulate multiple scenarios, including behavior analysis under extreme conditions and abnormal operating conditions. The simulation not only includes static structures, but also covers dynamic operating processes. This enables each subsystem and component of the system to be finely analyzed and optimized in a virtual environment. In this case, new virtual data can be provided for engineering equipment. In the results and discussion section of the proposed method, rich data support is needed. However, data from the engineering system is not enough to support such a discussion. Therefore, this chapter focuses on building a virtual model of the subsea blowout preventer hydraulic control system experimental prototype, rather than modeling the entire engineering system.

2.3.3 Collection of Virtual and Real Data

The sensor acquisition system consists of four hydraulic sensors and two flow sensors, which are evenly divided into two parts and placed in blue and yellow boxes respectively, as well as related hardware and software. It is placed at the corresponding nodes in the hydraulic control circuit. Each of these six faults has 1200 sets of data, and with normal data, there are a total of 8400 sets. Each state corresponds to two hydraulic sensors and one flow sensor. The sampling frequency is 1024 Hz. The sample duration is 8 s. The final data volume is 8400 * 3 * 1024 * 8. The collected sensor data can be used as real data for future research on fault diagnosis methods.

At the same time, the virtual data of the hydraulic control system under all states can be obtained by adjusting the parameters of the virtual model. The core of this process is obtaining fault virtual data, which includes collecting the output pressure and flow data of the virtual sensor. The simulation duration for each run is 8 s, and each sample consists of 1024 * 8 * 3 data points. By combining the real system fault states mentioned above and selecting appropriate nodes to collect virtual data of the same type, the three-channel one-dimensional signals corresponding to the six common fault states shown in Fig. 2.11 are obtained. The (A) and (B) in the Fig. 2.10 represent the real Dataset A (dark) and the simulated Dataset B (light), respectively. (1)–(6) represent 6 types of fault labels, 2–7.

This chapter simulated seven common states, and each state is divided into three datasets. The collected dataset was labeled as Dataset A, Dataset B, and Dataset C, as shown in Table 2.2. Dataset A represented the original real data of a real entity. Real world data inherently contains noise and anomalies that may not exist in simulated data. Dataset B referred to the virtual data of the virtual model. In the virtual model, these faults were parameterized as workspace variables and configured using Modelica language. Dataset C represented twin data generated by 2DSW optimization of Dataset A and Dataset B. Among them, the faults simulated using virtual models correspond to those simulated in experiments. The virtual data also

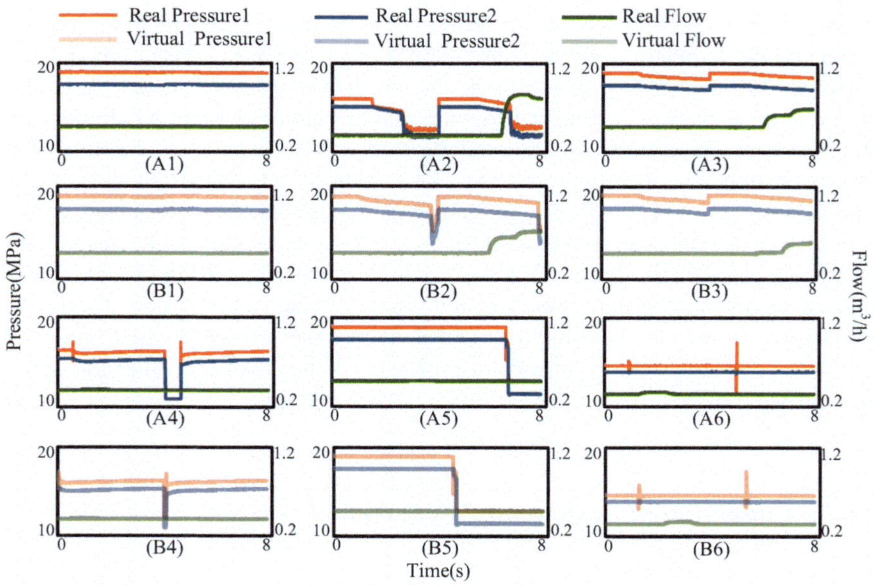

Fig. 2.11 Time-domain waveforms of the raw output flow: (A)–(B) represent dataset A–dataset B; (1)–(6) represent label 2 –label 7

corresponds to the real data. The new fault that differs from the experiment was not simulated on the virtual model.

The three (A, B, C) datasets are reasonably divided into six types according to different proportions to facilitate better analysis and performance evaluation. Their numbers and proportions are shown in Table 2.3. Type1 and Type2 are datasets of real data and twin data in different proportions, respectively. Type3 is a dataset of real and virtual data with the same proportion. Type4 and Type5 are datasets containing the same number of real data and virtual data, respectively. Type6 is a dataset that contains a larger amount of real data.

Table 2.3 Sources and numbers of the dataset in each type

Type	Source of dataset	Number of dataset
Type1	A:C	1000:4000
Type2	A:C	2500:2500
Type3	A:B	2500:2500
Type4	A	5000
Type5	B	5000
Type6	A	6000

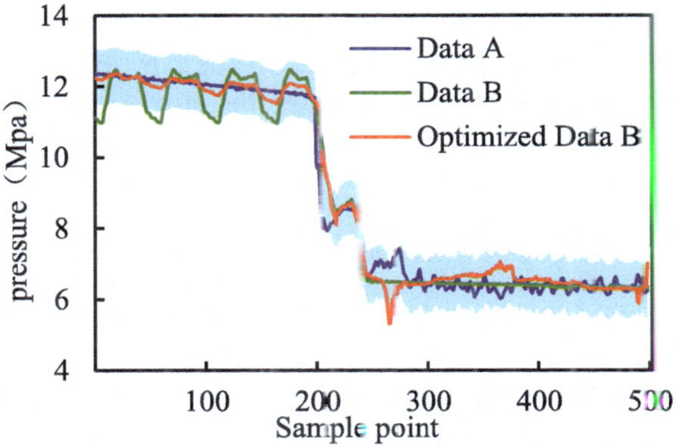

Fig. 2.12 Optimization results of 2DSW algorithm for virtual data

2.3.4 *Ragu2DSW Algorithm Optimization Process for Virtual Data*

The 2DSW optimization algorithm is the core of the bidirectional data consistency evaluation framework. It provides technical support for data consistency evaluation. This consistency is the basis for ensuring that the virtual model accurately reflects the real system state. By evaluating and optimizing data consistency, the reliability of simulation data can be greatly improved. This enables it to serve as a trusted substitute for real data, thereby effectively supporting fault diagnosis in digital twin systems.

The virtual data is optimized using the 2DSW algorithm to obtain the optimized data, as shown in Fig. 2.12. The optimization process involves 40 initialization iterations with a learning rate of 0.1. The search range of the 2DSW algorithm is set to [− 0.1, 0.1] to ensure reasonable alignment of time and amplitude. The maximum number of iterations of the algorithm is set to 1000 to balance the computational accuracy and time overhead. When the loss function change between two consecutive iterations is less than 10^{-6}, the algorithm stops iterating and is considered to have converged. It can be seen from the figure that the accuracy of the sequence alignment has been improved. The virtual data has been successfully optimized.

2.3.5 *Optimization and Hyperparameter Adjustment of MC1DCNN-GRU Model*

The optimization process of the MC1DCNN-GRU model mainly focuses on the optimization of the feature extraction layer, the optimization of the recurrent unit, and the optimization of the model architecture. First, the MC1DCNN module was optimized, and the number and size of the convolution kernels were determined through experiments, so that the features extracted by the model at different scales can capture the temporal changes of the fault signal to the greatest extent. At the same time, the configuration of the pooling layer was experimented, and by adjusting the size and step size of the pooling window, the information loss was reduced while retaining important features. Secondly, in the GRU module, the number of hidden layers and the dimension of the hidden units were experimented many times to determine the optimal configuration. We found that appropriately increasing the dimension of the hidden unit can better capture the long-term and short-term dependencies in the time series data, but at the same time, we also noticed the increase in computational overhead, so we balanced performance and computational cost. Finally, the model architecture was adjusted through cross-validation, including increasing and decreasing the number of convolutional layers and GRU layers, and finally an architecture that can ensure moderate model complexity and high accuracy was selected.

In terms of hyperparameter tuning, a combination of grid search and random search was adopted, and the effects of different hyperparameter combinations were evaluated through five-fold cross validation. Specifically, for the learning rate, multiple learning rates were tried from 0.0001 to 0.01, and 0.001 was finally selected as the optimal learning rate, which not only ensured the convergence speed but also avoided the model from falling into the local optimum. For batch size, different batch sizes of 16 to 128 were tested in the experiment, and the batch size of 64 was finally selected. This is a compromise choice, which not only ensures the stability of model training but also optimizes the utilization of computing resources. For regularization parameters, L2 regularization was introduced to prevent overfitting. Through experiments, 0.0005 was determined to be the optimal regularization parameter value, which effectively reduced the risk of model overfitting. For activation function, the effects of ReLU and Swish activation functions were compared, and it was found that Swish function performed better on complex data sets. Finally, the Swish activation function is used in the model to improve the expressiveness of nonlinear features. In addition, to avoid overfitting, the early stopping strategy is introduced in the training process. When the performance on the validation set does not improve within 10 consecutive epochs, the training is stopped, and finally the model with the best performance on the validation set is selected.

Through the above optimization and hyperparameter adjustment, the MC1DCNN-GRU model showed significant advantages in complex fault diagnosis tasks. The final accuracy, precision, recall and F1 score of the model were better than the baseline model before optimization. In addition, the model showed good generalization ability when processing different types of fault data.

2.4 Results and Discussion

This section is divided into subsections to present the outcomes of each stage of the proposed method. The subsections follow the flow of the case study, starting with an analysis of the real and virtual data results (Sect. 2.5.1), followed by the performance evaluation of the DT-assisted fault diagnosis model (Sect. 2.5.2), a discussion on the impact of different virtual and real data ratios (Sect. 2.5.3), and concluding with a comparison of the proposed model with other deep learning methods (Sect. 2.5.4) These subsections are organized to provide a comprehensive understanding of the model's effectiveness across various evaluation metrics and scenarios.

2.4.1 Results of Virtual and Real Data

The collected virtual data needs to be analyzed to verify whether it truly reflects the expected fault behavior. This involves comparing with actual system or known fault case data. Verifying the accuracy of simulated data is a crucial step in ensuring the reliability of fault diagnosis models. Each dataset contains 1000 sets of data. BA consistency verification was conducted on a random set of data from Dataset A and a random set of data from Dataset B, as shown in Fig. 2.11d–f. Similarly, BA consistency verification was conducted on a randomly selected set of data A and a set of data C generated through data B, as shown in Fig. 2.11a–c. Each group of data contains three channels, namely two hydraulic data channels and one flow data channel. Therefore, six visualization results were obtained, as shown in Fig. 2.13.

By comparing these results, it can be observed that the unoptimized virtual data exhibits significant differences in mean and square deviation compared to the real data. The 95% confidence interval shows a high degree of overlap, indicating a low level of data consistency. This highlights the need for further optimization to improve the quality of the virtual data.

Randomly select 200 samples from Datasets A, B, and C respectively. The samples in Datasets A and Datasets B, Datasets A and Datasets C are combined in pairs,

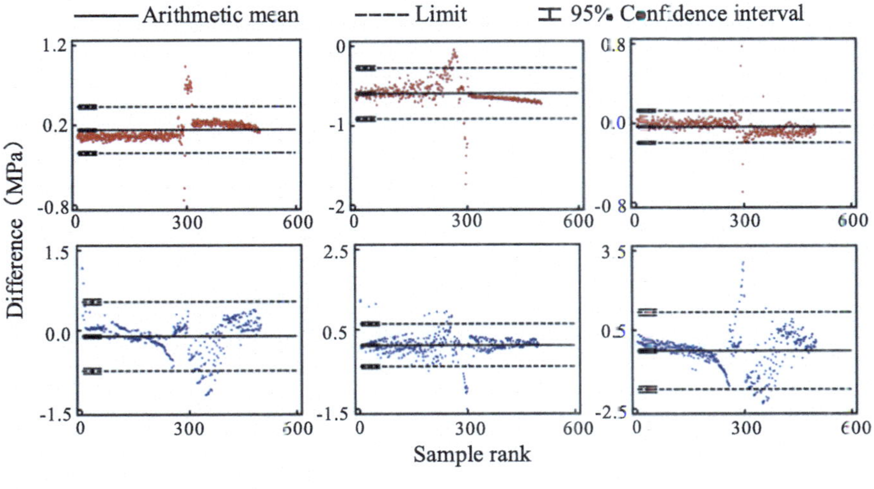

Fig. 2.13 Consistency verification results of BA on any data

Fig. 2.14 MSE results of the two sets of data

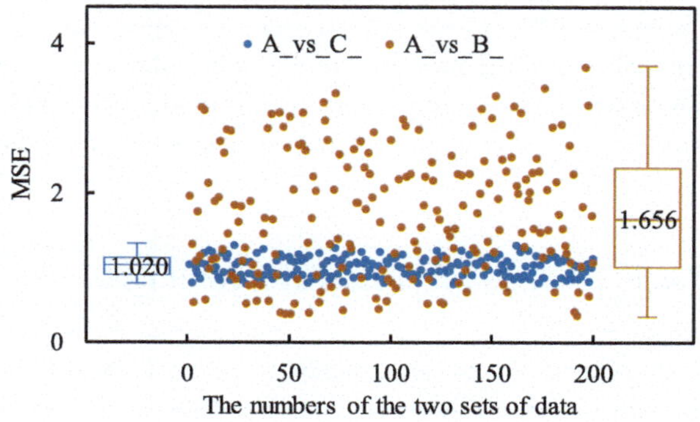

and then the mean square error (MSE) is calculated respectively, as shown in the Fig. 2.14. The calculation results show that the mean square error of the optimized virtual data using the 2DSW technology is reduced by about 38.41% compared with the traditional method.

To further validate the consistency of the optimized data, twenty sets of data were randomly selected from the six types of fault signals in Dataset A and Dataset C for BA consistency evaluation. The average mean and average consistency limit are displayed in Fig. 2.15, where tags 1–6 represent the six fault types of Dataset A, and tags 7–12 represent the six fault types of Dataset C. The results clearly indicate that the optimized virtual data exhibits higher consistency and reliability when compared to the original data.

In summary, when establishing the virtual model layout, the data consistency assessment model can evaluate the correlation between different data sources to ensure the accuracy and reliability of the virtual model. The 2DSW optimization algorithm can adjust the virtual data in time and amplitude to greatly improve its matching degree with the real data. The dynamic programming algorithm is used to solve the optimal twisted path to generate high-quality twin data. This data fusion

Fig. 2.15 The results of average and average consistency limit

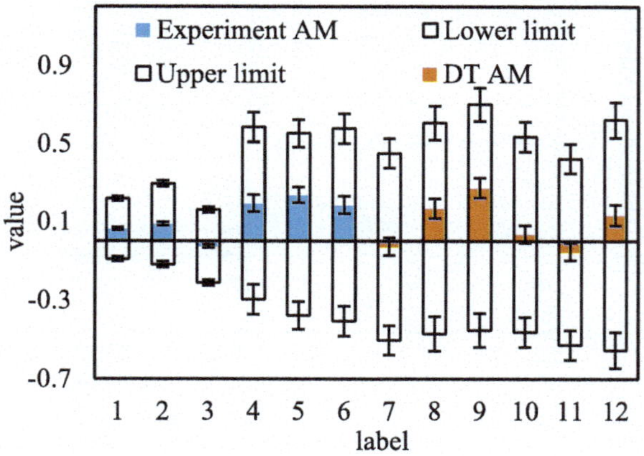

technology not only improves the diversity and reliability of model training data, but also provides effective simulation data when data is scarce, solving the common problem of insufficient data in fault diagnosis.

2.4.2 The Performance of DT

This section primarily focuses on evaluating the effectiveness of the training data assisted by the DT method. Three types of datasets, Type4, Type2 and Type6, as shown in Tables 2.2 and 2.3, are compared as training data inputs. Compared with Type4, Type2 uses more twin data. Compared to Type4, Type6 uses more real data. The test data consists of raw signals and all types are diagnosed using the MC1DCNN-GRU diagnostic model. The classification performance of the proposed model is evaluated using five parameters: sensitivity, specificity, accuracy, Hamming loss, and precision. In the evaluation, TP represents the true positive count, TN represents the true negative count, FP represents the false positive count, and FN represents the false negative count.

Considering the computational complexity issue, redundant neurons and layers are removed in the deep learning model. This ensures that model complexity and computational requirements are reduced without significantly reducing model performance. To further simplify the complexity, DT technology, 2DSW optimization and deep learning models are used for parallel computing. Multi-threading and GPU acceleration technology are used to increase computing speed so that the system can run more efficiently in a single hardware environment. To reduce the impact of randomness on the results, all methods were run 5 times, with each method running 5 independent runs using random data. Figure 2.16 illustrates the average test diagnostic results of the DL method trained with three types of training data over the five runs.

By observing Type4 and Type6, the MC1DCNN GRU diagnostic model performs well in a wide range of real-world data. As the number of iterations increases, both types of accuracy tend to reach their maximum values, but larger capacity real datasets help train more robust and reliable fault diagnosis models. It shows that larger capacity real datasets provide a more comprehensive overview of potential fault conditions and system behavior. By observing Type2 and Type4, real datasets inherently contain noise and anomalies that may not exist in simulated data, so the training results have a certain degree of variability and unpredictability. As more twin data is input into the algorithm, the calculated threshold becomes more accurate, leading to an improvement in the accuracy of the algorithm. The results converge towards the maximum value, indicating that the hydraulic control system fault diagnosis model based on DT exhibits higher accuracy and effectiveness compared to the model based on the original data.

This section emphasizes the potential of DT assisted IFD framework to address the limitations of traditional DL based fault diagnosis methods by leveraging the synergy

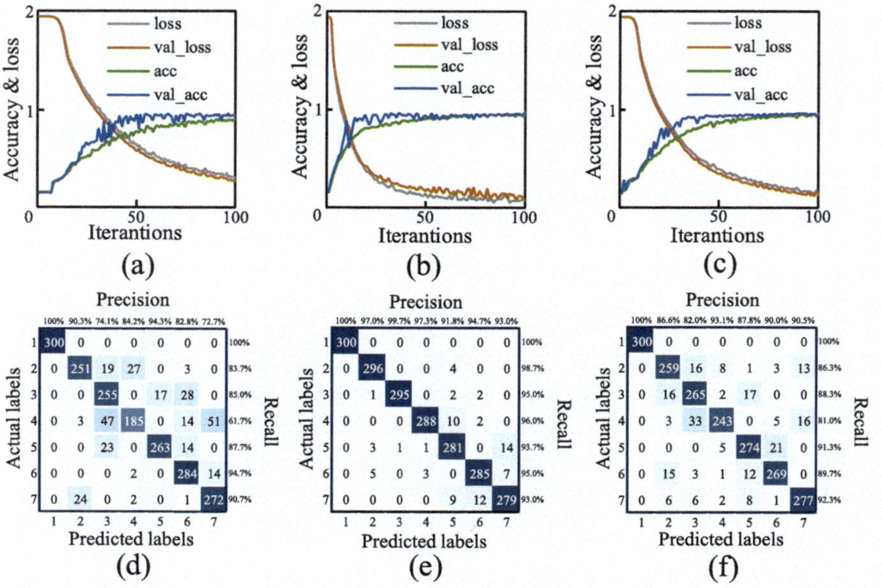

Fig. 2.16 Average test diagnostic results of DL method trained with training data: **a**, **d** represent type 4; **b**, **e** represent type 2; **c**, **f** represent type 6

between virtual and real data. By combining the strengths of both data types, the DT-assisted approach can significantly enhance the robustness and reliability of fault diagnosis models.

2.4.3 The Impact of the Ratio of Input Virtual to Real Data on the Model

This section primarily focuses on evaluating the impact of the ratio of input virtual to real data on the model. The training data inputs for comparison consist of six types of datasets, namely type1, type2, type3, type4, type5 and type6, as shown in Table 2.3. The test data used is the original signal, and all datasets are diagnosed using the MC1DCNN-GRU diagnostic model. The diagnostic results are compared using five parameters: sensitivity, specificity, accuracy, Hamming loss, and Precision. The comparison results are presented in Fig. 2.17 and Table 2.4.

The performance of the DT-assisted fault diagnosis model is influenced by the proportion of virtual and real data. When using Type4 as the training data, the model becomes too sensitive to the data. Due to fluctuations in the data, the fault diagnosis model struggles to accurately determine the trend of signal changes, resulting in a high misdiagnosis rate (SPE = 86.22%) and missed diagnosis rate (SEN = 85.33%).

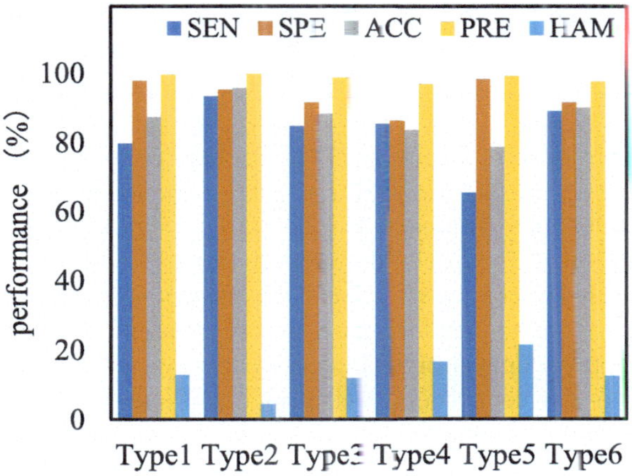

Fig. 2.17 The comparison of five diagnostic parameters for six data types

Table 2.4 Comparison of diagnostic results with five parameters

	SEN (%)	SPE (%)	ACC (%)	PRE (%)	HAM (%)
Type1	79.78	97.78	87.17	99.59	12.83
Type2	93.33	95.11	95.62	99.76	4.38
Type3	84.67	91.48	88.11	98.66	11.89
Type4	85.33	86.22	83.44	96.89	16.56
Type5	65.38	98.23	78.54	99.27	21.46
Type6	88.98	91.55	89.86	97.67	12.47

In contrast, Type6 combines richer real-world datasets, resulting in improved accuracy and reliability of fault diagnosis (ACC = 89.86%) compared to Type4. It ensures the balance of the training dataset and the reflection of the real operating environment as a training set. The sensitivity has decreased, but the misdiagnosis rate and missed diagnosis rate are still relatively high.

On the other hand, when using Type1 and Type5 as the training data, the fault diagnosis model becomes relatively lower, which effectively avoids misdiagnosis (SPE = 97.78%, 98.23%). However, the missed diagnosis rate remains high (SEN = 79.78%, 65.38%). Therefore, it is necessary to strike an appropriate balance between specificity (SPE) and sensitivity (SEN).

When Type2 is used as the training data, the overall performance is better than that of Type3. The fault diagnosis model effectively avoids the interference of data fluctuations and detects anomalies in a timely manner (SEN = 93.33%). Furthermore, the DT technology also effectively avoids misdiagnosis (SPE = 95.11%). In the comparison between Type2 and Type4, it is shown that the accuracy of the MC1DCNN-GRU model trained with 2DSW improved data is improved by 12.5%.

Both Type1 and Type2 are datasets that mix twin data with real data, with relatively few real data. When Type1 and Type2 are used as datasets to train the model, the evaluation results indicate that the diagnostic model shows good diagnostic performance. Therefore, twin data can help repair the lack of faulty data.

In terms of diagnosis, DT technology, combined with deep learning methods, can extract key fault features from a large amount of twin data and build a more accurate fault diagnosis model. This data-driven approach significantly improves diagnostic accuracy. The results show that by comprehensively integrating the five dimensions of physical entity, virtual model, connection, data and service, the performance of the DT system in fault diagnosis has been significantly improved. Specifically, the data support of physical entities ensures high-precision simulation of virtual models. The two-way connection mechanism enables the virtual system to respond to changes according to the real system. The consistency optimization of data improves the reliability of the diagnostic model. Ultimately, the fault diagnosis results provided by the service provide effective decision support for the operator.

2.4.4 The Performance of the MC1DCNN-GRU Diagnostic Model

This section focuses on testing the diagnostic performance of the developed MC1DCNN-GRU using raw data. The performance of this method is compared with three different DL methods, namely MC1DCNN, LSTM, and GRU. From the Dataset A, 4900 sets of data are randomly selected as training data, while the remaining 2100 sets of data are used as test data. To reduce the impact of randomness, both the MC1DCNN-GRU and the comparison methods are run 10 times. Figure 2.18 and Table 2.5 present the average test diagnostic results of the 5 DL methods over the 10 runs.

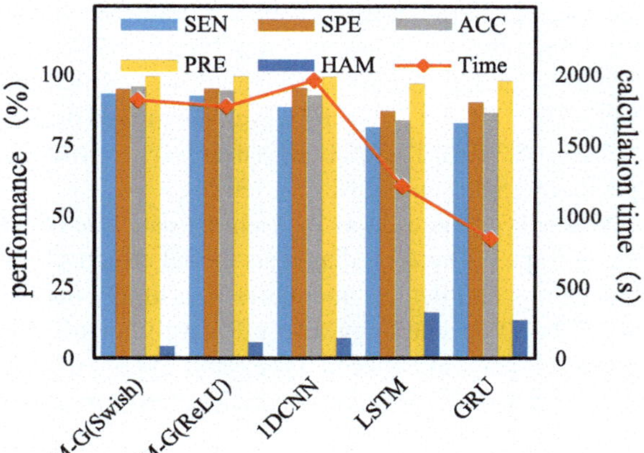

Fig. 2.18 The comparison of six diagnostic parameters for five diagnostic models

Table 2.5 Comparison of diagnostic results with six parameters (M-G refers to MC1DCNN-GRU)

	SEN (%)	SPE (%)	ACC (%)	PRE (%)	HAM (%)	Time (s)
M-G(Swish)	93.33	95.11	95.62	99.76	4.38	1821
M-G(ReLU)	92.60	95.24	94.23	99.69	5.77	1775
1DCNN	88.67	95.43	92.63	99.59	7.37	1961
LSTM	81.59	87.36	83.71	97.07	16.29	1214
GRU	83.09	90.49	86.40	98.23	13.60	841

The average accuracy of the MC1DCNN-GRU method for 2100 sets of test data is 95.62%, while the average accuracy of the comparison methods is 94.23%, 92.63%, 83.71%, and 86.4% respectively. Although LSTM and GRU have faster computational times, their accuracy significantly lags behind the MC1DCNN-GRU method. The performance of M-G (Swish) and M-G (ReLU) results is compared. It can be observed that the Swish activation function shows a greater improvement in model performance compared to the ReLU activation function when the calculation time is not significantly different. Therefore, it is verified that the MC1DCNN-GRU constructed using the Swish activation function achieves the highest diagnostic accuracy and is more effective than other popular DL methods.

In addition, compared to other fault diagnosis methods, the DT-driven MC1DCNN-GRU method (when A:B = 1:1) demonstrates better performance. The advantage of the DT method lies in its deep integration of virtual and real data. By ensuring consistency between the real and virtual systems of the BOP, the constructed DT can effectively fuse signals from different sources such as simulation data, historical data, and real data for diagnosis. The integration of more knowledge and information leads to more effective and reliable diagnosis.

On the other hand, data-driven methods heavily rely on historical data. In harsh subsea environments, the accuracy of data-driven methods can be easily affected due to the susceptibility of signals to unknown factors. These same conclusions can be observed from Fig. 2.16. In the unpredictable subsea environment, the real data of complex hydraulic systems will change with the changes in the environment. The training of traditional data-driven fault diagnosis models only relies on a large amount of historical data input. When data changes, the model trained through historical data experiences misdiagnosis and missed diagnosis.

Conversely, twin data is based on real data, which is constantly updated with changes in real data. It can help fault diagnosis models continuously respond to the challenges brought by new environments. This ability to adapt to data changes ensures that the DT-assisted MC1DCNN-GRU method remains robust and reliable even under varying environmental conditions.

Chapter 3
Optimal Sensor Placement for Fault Diagnosis

Abstract The optimal placement of a limited number of sensors in subsea hydraulic systems is crucial for enhancing the efficiency of fault diagnosis. This involves determining the most reasonable positions for these sensors. To get good diagnostic results with fewer sensors, a sensor placement method optimized by discrete particle swarm algorithm is proposed. Response time model and propagation path model are established to measure sensor performance. A discrete particle swarm optimization algorithm is used to calculate the optimal solution for the optimal placement of sensors. The method is used in the hydraulic control system of the subsea blowout preventer. The result shows that the proposed method reduces the number of sensors and data redundancy effectively.

3.1 Introduction

Sensors are used for signal extraction in status monitoring and fault diagnosis of subsea control systems. Effective placement of sensors is essential to improve diagnosis efficiency and accuracy. Ideally, a sensor network should be configured to facilitate fault detection and maximize diagnostic performance. However, the number of sensors is constrained by space resources and costs. Additionally, the redundant data generated by multiple sensors pose challenges for high-speed storage and effective analysis of information. Therefore, arranging a limited number of sensors in the most reasonable positions within the structure—known as optimal sensor placement (OSP)—is a critical issue in fault diagnosis.

The primary purpose of solving the OSP problem encompasses two main aspects. First, it aims to achieve better diagnostic performance, including enhanced diagnostic accuracy and increased diagnostic speed. Second, maximizing the diagnosis range while minimizing the spatial variation of communication distances among sensors is also a crucial goal. By strategically placing sensors, one can reduce redundancy and enhance the coverage of critical areas, thereby improving the system's ability to detect and diagnose faults accurately and promptly. Despite advancements in this field, there is still room for improvement in developing algorithms and methodologies

B. Cai et al., *Intelligent Operation and Maintenance for Subsea Production Systems*,
https://doi.org/10.1007/978-981-96-2677-9_3

that can quickly and accurately solve discrete sensor placement problems, especially in complex and dynamic environments.

Compared to common mechanical and electrical structures, faults in hydraulic systems within engineering equipment are often more concealed and less apparent. The sealed nature of hydraulic control systems necessitates the pre-installation of sensors within the hydraulic circuit for effective fault diagnosis. Ensuring that sensors are optimally placed can significantly enhance the accuracy and reliability of fault detection. Strategic sensor placement can help identify potential issues early on, leading to timely maintenance and reduced downtime. Therefore, developing robust optimization methods that account for the unique characteristics of hydraulic systems is essential for achieving efficient and accurate fault diagnosis.

To improve the performance of fault diagnosis of the hydraulic control system, this chapter proposed a sensor placement method optimized by discrete particle swarm algorithm. A discrete particle swarm optimization algorithm is used to calculate the optimal solution for the optimal placement of sensors. The simulation case of typical multi-circuit hydraulic control systems proves that the proposed method has fast convergence speed and optimization results. A real case of a subsea BOP control system proves the efficiency of the proposed method in reducing the number of sensors and data redundancy.

3.2 Sensor Placement Methodology

A typical multi-circuit hydraulic control system is shown in Fig. 3.1. The system consists of m hydraulic control loops C_1, C_2, ..., C_m, and each loop includes n_1, n_2, ..., n_m group control circuits. The flow of the discretization optimization methodology of hydraulic control system is shown in Fig. 3.1. First, the system's fault propagation and sensor response time models are established. Then, the influence of the rules of sensor placement on diagnosis efficiency from these models are determined. After that, an optimization model of sensor placement is established. The optimization objects of sensor placement include the number distribution and placement of sensors. The optimization constraint condition is the optimization objective function that corresponds to time and the maximum sensor interval. Finally, particle swarm optimization algorithm is used to obtain the optimal solution for the optimal placement of sensor. In the optimization process, Monte Carlo simulation optimization method is used for the optimization of particles. The Monte Carlo simulation model is established based on the sensor response time model.

3.2.1 Fault Propagation and Sensor Response Time Model

The relationship between fault propagation and sensor response time is shown in Fig. 3.2. Due to the viscosity of fluid, it takes time $T(d)$ from the circuit to leak until

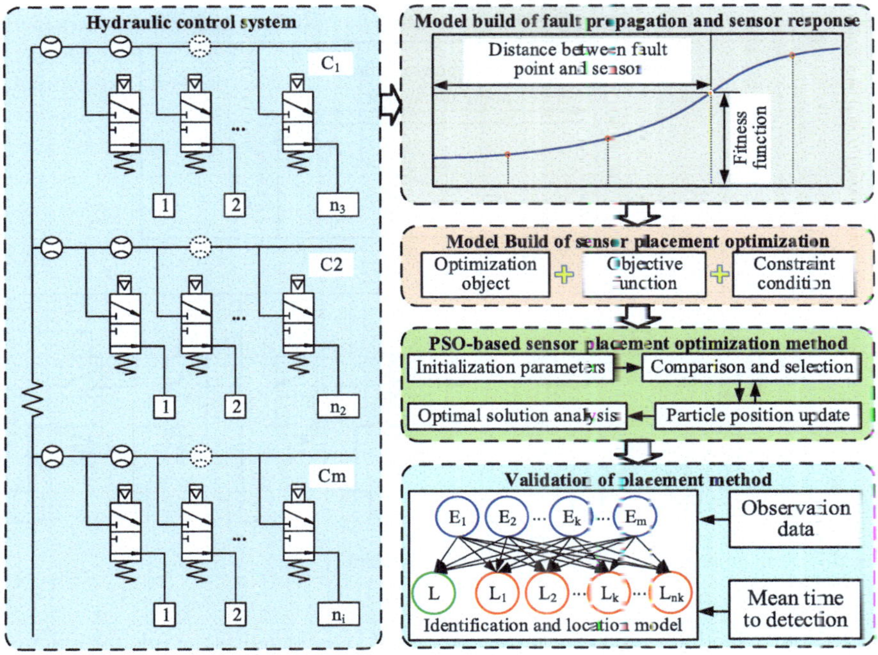

Fig. 3.1 Discrete optimization methodology flow of hydraulic control system

flowmeter detects fault. T is related to the position d from flowmeter to circuit where the fault occurs. The sensor detection time function is defined as follows:

$$f = T(d) \tag{3.1}$$

As shown in the Fig. 3.2, it is assumed that the distance of each circuit is equal, which is d0. When No. 3 circuit leaks, the time when the fault is detected at positions f3, f4, f5, and f6 is $T(d0)$, $T(2d0)$, $T(3d0)$, $T(4d0)$, respectively.

In this work, $T(d)$ is evaluated by a simulation model built based on Simulink of MATLAB. The structure of hydraulic system is built with the SimHydraulics tool of Simulink. The operating pressure of the system, the diameter of the pipeline, the response curve of valve and other parameters are used as the input parameters of the simulation model.

Sensors are built through independent mathematical models to be closer to the real situation. The sensor reading model is as follows.

$$y = Ku + f + e \tag{3.2}$$

where K is the process gain; u is the sensor input; f is the sensor bias; y is the sensor reading; e is the noise at sensor which follows a normal distribution.

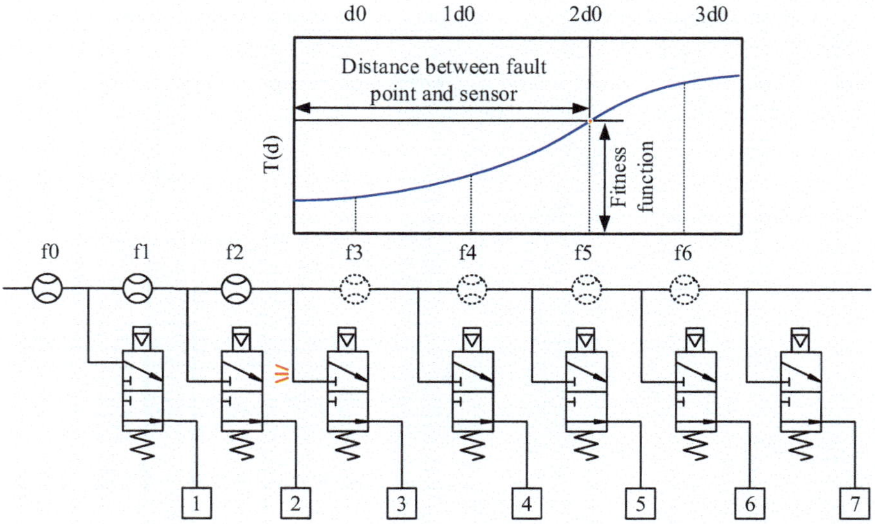

Fig. 3.2 Fault propagation and sensor response time model

To reduce the interference of noise, a threshold is selected as the rule for judging leakage. When the sensor reading is below 2% of normal, the sensor is considered to have identified a leak. At this time, the distance to the event of leakage in the simulation is taken as the value of T(d). Furthermore, the distribution of $T(d)$ can be also obtained through experiments, history date, or expert experience.

3.2.2 Sensor Placement Optimization Model

The placement optimization model of sensor includes the optimization objects, optimization targets and constraints. The sensor placement optimization object models S and X are shown in the formula:

$$\begin{cases} S = [s_1, s_2, \ldots, s_m](\forall 1 \le k \le m, s_k \le n_k) \\ X = \begin{bmatrix} x_1^1 & x_1^2 & \ldots & x_1^{s1} & O_{n1-s1} \\ x_2^1 & x_2^2 & \ldots & x_2^{s2} & O_{n2-s2} \\ \ldots & \ldots & \ldots & \ldots & \ldots \\ x_m^1 & x_m^2 & \ldots & x_m^{sm} & O_{ni-sm} \end{bmatrix} (\forall 1 \le k \le m, x_k^{sk} = n_k) \end{cases} \tag{3.3}$$

where S is the vector of number of sensors, s_1, s_2, …, s_m are the number of sensors allocated to circuits C_1, C_2, …, C_m respectively. X is sensor position placement matrix, and x_{ij} represents the position coordinates of the j-th sensor of the i-th loop in system.

The two major goals to solve the problem of sensor placement optimization are the cost of the number of sensors and the efficiency of monitoring. Assuming that system has q failures during operation, the objective function of mathematical model for optimization of sensor placement can be defined by following formula:

$$\begin{cases} \min S_{sum} = s_1 + s_2 + \cdots + s_i \\ \min T_{sum} = \sum_1^p T_k \end{cases} \tag{3.4}$$

where S_{sum} is the total number of system sensors, T_k is the minimum detection time of the k-th fault, and T_{sum} is the sum of the minimum fault detection time in life cycle of system.

The objective function described in the above formula needs to meet the maximum fault detection time constraint, that is, when a fault occurs and is not handled in time, it will cause potential production risks and higher intervention costs. Therefore, sensor placement should be satisfactory in that the detection time of all faults does not exceed the tolerable maximum detection time T_{tol}.

$$\forall k \in (1, p), T_k \le T_{\text{tol}} \tag{3.5}$$

3.2.3 Sensor Placement Optimization Based on Discrete PSO

To optimize placement of sensors, reasonable optimization criteria that can meet the design requirements must be determined, and simple and efficient optimization methods should be selected. The PSO algorithm was originally a swarm intelligence-based evolutionary computing technology inspired by the predation behavior of birds. In this section, discrete PSO is used to optimize the above model. During the iterative process, the particle's position and velocity update model is built. A correction algorithm is used to solve the problem of particles exceeding the limit.

Assuming that current is the t-th generation, the i-th particle in D-dimensional search space can be expressed as follows:

$$X_i^t = (x_{i1}^t, x_{i2}^t, \ldots, x_{id}^t) \tag{3.6}$$

$$V_i^t = (v_{i1}^t, v_{i2}^t, \ldots, v_{id}^t) \tag{3.7}$$

$$P_i^t = (p_{i1}^t, p_{i2}^t, \ldots, p_{id}^t) \tag{3.8}$$

where $X_i{}^t$ is the particle position of the i-th particle in the t-th generation, $V_i{}^t$ is the particle velocity of the i-th particle in the t-th generation, and $P_i{}^t$ is the particle individual optimal currently of the i-th particle in the t-th generation.

By evaluating fitness value of each particle, the current individual optimal and current population optimal of each particle in the t-th generation are determined, and then the speed and position of each particle are updated as follows.

$$V_{id}^{t+1} = \omega V_{id}^{t} + c_1 r_1 (p_{id}^{t} - x_{id}^{t}) + c_2 r_2 (p_{gd} - x_{id}^{t}) \tag{3.9}$$

$$x_{id}^{t+1} = x_{id}^{t} + v_{id}^{t+1} \tag{3.10}$$

where V_{id}^{t+1} is the velocity of the i-th particle in the d-th dimension during the t-th iteration; $i = 1, 2, \ldots, q$, which is the i-th individual, q is population size; $t = 1, 2, \ldots, t_{max}$ represents the number of iterations; $d = 1, 2, \ldots, D$ represents the dimension of optimization space; w is called the inertia weight, and its function is to control the influence of current speed on back; c_1, c_2 are acceleration coefficients, and its function is to make particles have self and social cognition ability, usually positive constant; r_1, r_2 are random numbers uniformly distributed on [0,1], used to maintain the diversity of group; ${x_{id}}^{t+1}$ is the i-th particle in the t-th time The current position of the d-th dimension during iteration; p_{id}^{t+1} is the position of the d-th individual extreme point of the i-th particle during the t-th iteration; P_{gd} is the best global extreme searched by entire particle swarm so far in the d-th dimension location of the value point.

To prevent particles from moving away from the search space, the particle velocity v in each dimension is usually limited to the range of $[-v_{max}, v_{max}]$. The choice of v_{max} will affect the global and local search capabilities of the algorithm. If its value is too large, then the particles will fly from the optimal solution. If deviation from the optimal solution is too small, then it will fall into a local optimal. Therefore, v_{max} is generally set as the variation range of each dimension variable without the need for fine selection and adjustment. When the example exceeds the maximum position limit X_{limit} during the update process, a correction algorithm is used to determine and correct the particle position. The specific procedure of the algorithm is shown in Table 3.1.

The sensor placement optimization method based on particle swarm algorithm is shown in Fig. 3.3. The specific steps include initialization parameter: space dimension d, population particle number N, maximum iteration number ger, position parameter limit X_{limit}, and speed parameter limit V_{limit}. Appropriate optimal parameters are selected: inertia weight ω, self-learning factor c_1, global learning factor c_2. Moreover the global optimal solution $pbest$, historical optimal solution $Gbest = [gbest_1, gbest_2, \ldots, gbest_N]$ is initialized. Before the application of the PSO approaches, an analysis has been performed to determine the best parameter selection. The size of the particle matrix and the number of generations for each of the networks are chosen after several tests. The particle size refers to the number of rows in the initialization matrix used at the beginning of the search in the algorithm. This matrix is full of possible random solutions for sensor placement. First, a particle size is selected to run the test to select the appropriate number of generations. Then, the tests are repeated by varying the particle size and keeping the number of generations according to the previous result.

Table 3.1 Particle position boundary constraint correction algorithm

No.	Algorithm
1	**Input**: Particle position x(1 to d)
2	**Initialization**: Space dimension d, X_{limit}
3	x(n,:) = sort(x(n,:));
4	for m = 1:d
5	if(x(n,m) < 1)
6	x(n,m) = 1;
7	end
8	if(x(n,m) > (X_{limit} -d + m))
9	x(n,m) = X_{limit} -d + m;
10	end
11	if(m < d)
12	if(x(n,m) = = x(n,m + 1))
13	x(n,m + 1) = x(n,m + 1) + 1;
14	end
15	end
17	**Output**: x(1 to d)

In the iterative process, a fitness function is used to evaluate the quality of the particles in the iterative process. In this work, the fitness function model is built based on the Monte Carlo simulation algorithm combined with sensor response time model. In the Monte Carlo simulation, the failure rate of the directional valve of each circuit can be built based on gamma process and is affected by factors such as frequency of use, pressure, and external environment. The comparison results are used to update the particle optimal solution and the global optimal solution particle speed and particle position. The final global optimal solution and local optimal solution is output when the number of iterations meets the requirements. Finally, a single optimal sensor placement solution is obtained by analyzing the optimal solution set.

3.2.3.1 Validation of Proposed Methodology

Usually, a series of solution sets is obtained through the above-mentioned algorithm. To further select a single optimal solution from the solution set while verifying the effectiveness of the proposed algorithm, a leakage identification and location model of the hydraulic control system was developed on the basis of the Bayesian network. Each hydraulic control system model corresponds to an independent recognition and positioning reasoning network. This model can be used for sensor data mining and fault identification.

First, the observation matrix of a sensor is defined as follows.

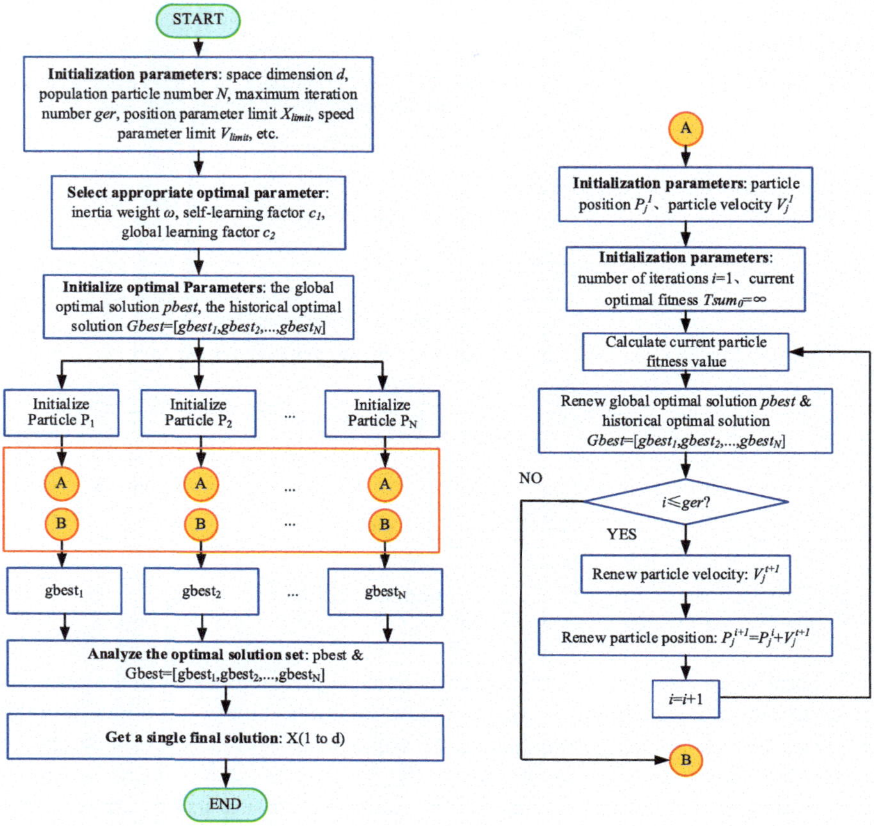

Fig. 3.3 Sensor placement optimization method based on particle swarm algorithm

$$R = [r_1, r_2, \ldots, r_m] \tag{3.11}$$

where r_1, r_2, ..., r_m is the readings from a sensor group that consists of m sensors. In this study, the observations of the flowmeter group are converted into the sensor's observation matrix.

The vector composed of the value of the parent node for fault identification and location is defined as E. The sensor observation matrix R is further transformed into the fault symptom matrix E as follows.

$$E = [e_1, e_2, \ldots, e_m], e_k = \frac{|R(1:k)|}{k}, 1 \le k \le m \tag{3.12}$$

The established Bayesian network model of the hydraulic control system for leakage identification and location is shown in Fig. 3.4. The model is composed of fault symptom node $\{E_1, E_2, \ldots, E_m\}$, leak identification node L and leak location

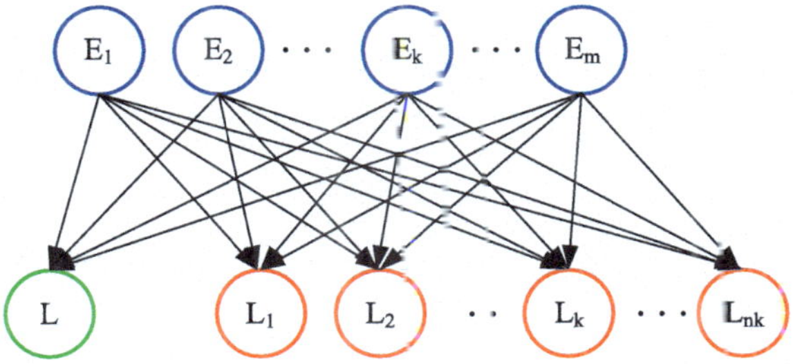

Fig. 3.4 Leakage identification and location model of the hydraulic control system

node $\{L_1, L_2, \ldots., L_{nk}\}$. The failure symptom node $\{E_1, E_2, \ldots, E_m\}$ represents the value of a continuous variable $\{e_1, e_2, \ldots, e_m\}$. The elements of the vector E must be discretized as the input information for identification and positioning. Leak identification node L and leak location node $\{L_1, L_2, \ldots., L_{nk}\}$ are all Boolean variables. Leak identification node L indicates whether a fault exists in the loop. When $L = 1$, leakage exists in the system. When $L = 0$, the circuit is operating normally. The leak location node $\{L_1, L_2, \ldots., Lm\}$ corresponds to the health status of each hydraulic component of the loop respectively. When $L_k = 0$, the k-th component works normally. When $L_k = 1$, it means that the k-th component has leaked. The mean time to detection (MTTD) is selected as the quantitative evaluation index of the sensor placement solution. MTTD represents the average time required for the leak identification model to identify the leak from the moment the leak occurs. In addition, the positioning accuracy of the leakage identification and location model can also be used as an auxiliary evaluation index.

3.3 Case of Common Multi-circuit Hydraulic Control Systems

3.3.1 Description of the System

This section selects a hydraulic control system with 30 control circuits as the research object. In this system, the 30 control circuits correspond to different control functions of the system. The distribution of the number of sensors and the optimal placement plan of different distribution forms were studied separately, and the optimization results were discussed and analyzed.

In the discrete particle swarm optimization algorithm, the inertia weight is $\omega = 0.6$, the self-learning factor $c1 = 0.6$, and the global learning factor $c2 = 0.4$. The maximum number of iterations is 500, and the population size is $N = 20$. In the Monte Carlo simulation of comparison and selection, the failure rate of directional valve of each circuit obeys a gamma process and is affected by factors such as frequency of use, pressure, and external environment. The failure rate of different components varies. The simulation time is set to 10 years. To ensure the stability of the optimization results, 30 simulations are performed for each optimization, the average value is calculated, and the average minimum fault detection time is used as the fitness value of the particles.

3.3.2 Multi-circuit Hydraulic Control System with Single Loop

The iterative optimization process of the optimal solution under different numbers of sensors in the hydraulic control system with single circuit case is shown in Fig. 3.5, where *d* is the number of additional sensors in the circuit. The optimization reaches the optimal solution interval when the iteration reaches 80 and converges to the optimal solution after the iteration reaches 100–300 times. In particle swarm optimization algorithms, more iterations usually lead to better results. To avoid the situation where the result converges to the local optimal solution, a random factor is set in the iterative process. Iteration is also an uncertain process. During our attempts, cases where the optimal solution comes later were noted. Therefore, we set the number of iterations to 500 to ensure that the optimization results can converge with the optimal solution during the iteration process. The particle swarm algorithm with current parameters has high-efficiency convergence, and the proposed methodology can provide an effective way to solve the problem of optimal placement of sensors. The final optimization results obtained by this methodology are shown in Table 3.2.

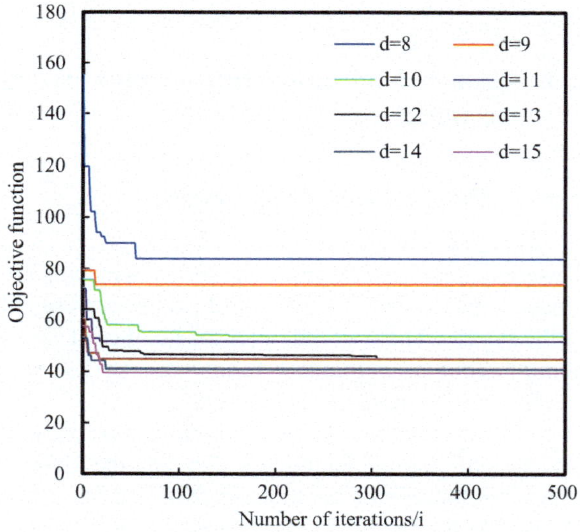

Fig. 3.5 Iterative optimization process of optimal solution under different number of sensors

Table 3.2 Optimal solution of hydraulic control system with single loop

Number of sensors	Optimal sensor placement
8	4, 5, 8, 9, 10, 15, 20, 25
9	4, 8, 11, 14, 15, 18, 21, 25, 28
10	1, 3, 6, 8, 11, 13, 14, 15, 20, 25
11	1, 2, 4, 6, 8, 10, 13, 17, 20, 24, 27
12	1, 3, 4, 6, 8, 11, 14, 16, 18, 19, 21, 25
13	1, 2, 4, 5, 7, 8, 10, 12, 15, 18, 19, 20, 25
14	1, 4, 5, 6, 7, 8, 10, 12, 14, 15, 19, 20, 23, 26
15	1, 3, 4, 6, 8, 9, 10, 13, 14, 15, 15, 18, 20, 21, 24

3.3.3 *Multi-circuit Hydraulic Control System with Double Parallel Loops*

The objective function of the optimal solution for sensor placement in the control system case is shown in Fig. 3.6. According to space and functional requirements, the hydraulic control system may also be designed as a structure of two parallel loops. This section discusses the optimization results of sensor placement for double parallel loops multi-circuit hydraulic control system. Each loop includes 15 control circuits. The number of sensors for each hydraulic branch loop is 4, 5, 6, 7, and 8, and cross-simulation optimization is performed under different sensor quantity allocations.

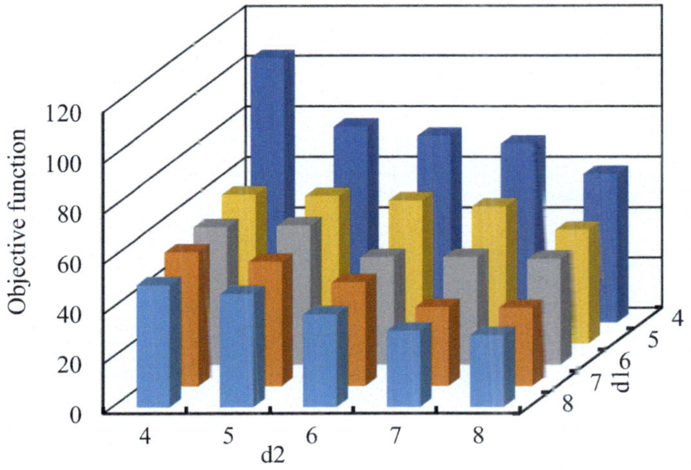

Fig. 3.6 Optimal solution of hydraulic control system with double parallel loops under different sensor quantity allocations

3.3.4 Results and Discussions

In this section, the MTTD is used as the evaluation index of the placement strategy. The relationship between the MTTD of hydraulic control system with single circuit and two parallel circuits and the number of sensors is shown in Fig. 3.7. As the number of sensors increases, MTTD continues to decrease. For the hydraulic control system with single circuit, when the number of sensors is greater than 12, the increase in the number of sensors will not improve the diagnostic efficiency significantly. When the MTTD requirement is 60 min, 10 sensors are needed to meet the requirement. For the system with two parallel circuits, 11 sensors are needed to meet the requirement of MTTS in 60 min. When the number of sensors is greater than 12, the increase in the number of sensors will improve diagnostic efficiency. When 16 sensors are used the minimum fault detection time is less than 30 min. When the number of sensors is low, the optimal solution has better diagnostic efficiency in the single circuit hydraulic control system. When the number of sensors is greater than 8, the optimal solution has better diagnostic efficiency in the two parallel circuits of a hydraulic control system. When the number of sensors is greater than 12, the increase in sensors will result in a very limited improvement in the diagnostic efficiency of the system with single circuit. Nevertheless, good diagnostic benefits can still be obtained in the two parallel circuits of a hydraulic control system.

The number of occurrences of all sensor position numbers in the optimal solution is calculated by analyzing all the optimal sensor placement schemes of the single-loop hydraulic control system and double-circuit hydraulic control system. The frequency of the sensor position number in the optimal solution can be obtained, as shown in Fig. 3.8. The optimal placement of different conditions and the number of sensors has certain common features. Safety–critical locations, such as 1, 4, 8, and 14, are frequently faulty. Therefore, these positions have a higher frequency of occurrence in the optimal solutions. At the same time, the sensor placement of some positions also has a correlation, such as [2, 3], [5, 6, 7], [20, 21], most of the optimal solutions

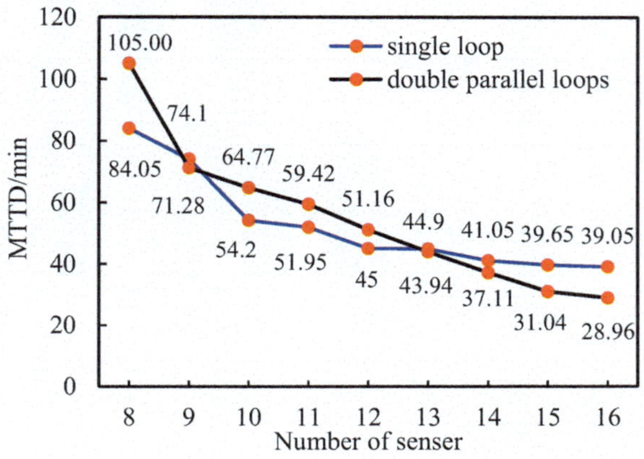

Fig. 3.7 Relationship between MTTD and number of sensors

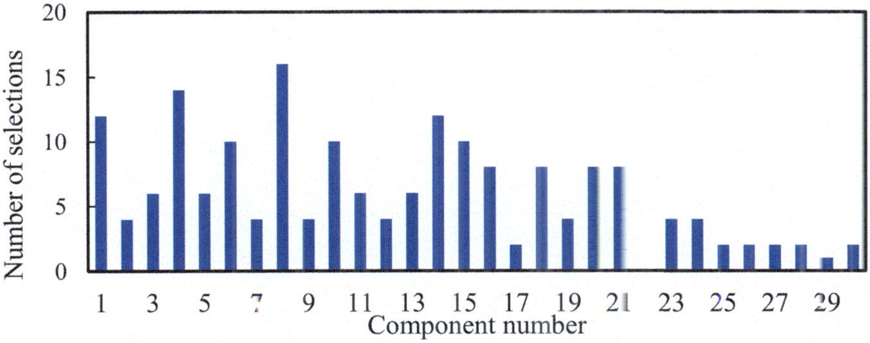

Fig. 3.8 Frequency of the sensor position number in the optimal solution

must install sensors in these combinations to obtain positional continuity and stable diagnosis efficiency. These mechanisms ensure the safe and reliable operation of the system.

3.4 Case of Actual Subsea BOP Hydraulic Control System

3.4.1 Description of Hydraulic Control System of Subsea BOP

A BOP system plays an extremely important role in providing safe working conditions for drilling activities in ultra-deep-water regions (Casillas et al. 2013). The safe operation of the hydraulic system is the key to the reliability of the BOP system. This chapter takes the hydraulic control circuit of the NOV 3000 m water depth BOP carried by the HYSY-981 as the research object. To ensure high reliability, the subsea BOP control system is structurally divided into a yellow pod, a blue pod and an emergency hydraulic backup system (EHBS). The structure of the hydraulic control circuit of the subsea BOP is shown in Fig. 3.9. The yellow and blue pod control units are identical and redundant. Each control pod includes a set of pressure supply and storage modules and hydraulic control circuits. During operation, only one pod is in operation, and the other pod is used as redundancy. In addition, the subsea BOP control system is also equipped with a set of EHBS, which is used for the emergency backup control of the two major functions of ram BOP closing and shearing BOP closing. EHBS contains an independent pressure storage system and two control circuits that correspond to the control functions. EHBS can ensure the normal operation of the key functions of subsea BOP in emergency situations. Leakage is one of the most important reasons for the failure of hydraulic systems (Casillas, Garza-Castañón et al. 2015). In the current design of the hydraulic control

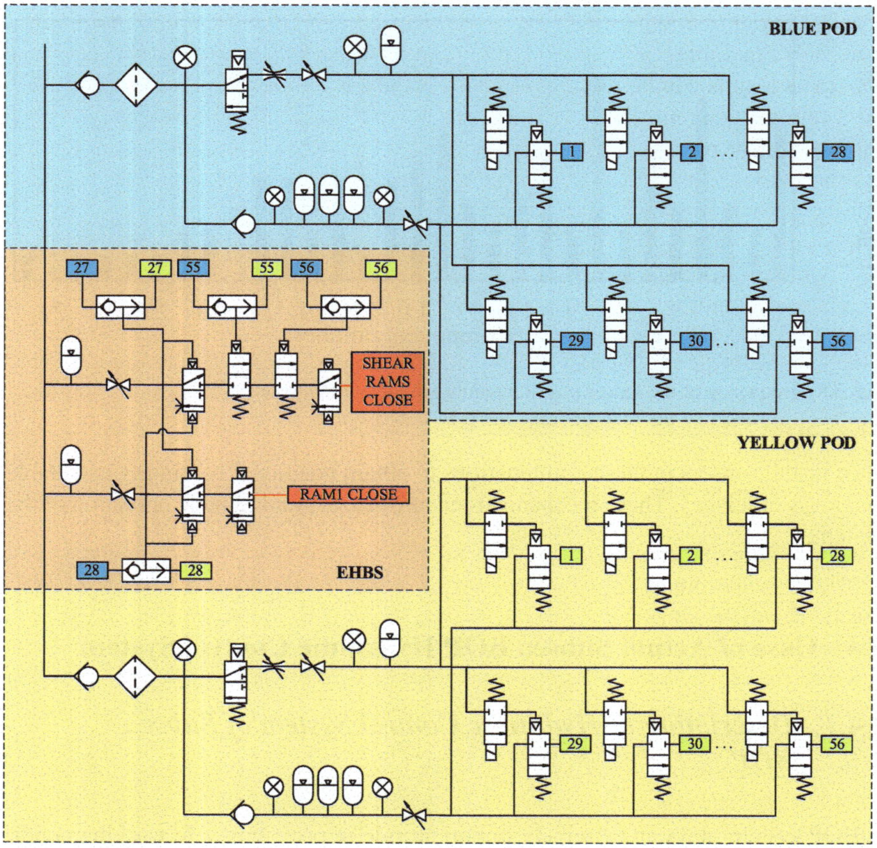

Fig. 3.9 Hydraulic control circuit of subsea BOP

circuit of subsea BOP, the placement method of evenly distributing the sensors in the system is used. The problem with this method is that it does not consider the difference in the working pressure and frequency of use of hydraulic components.

3.4.2 Sensor Placement Optimization for Hydraulic Control System of Subsea BOP

Before using the proposed methodology to optimize the sensor placement, the redundant logic and circuit structure of the hydraulic control must be analyzed first. According to the common energy and system structure of the hydraulic control circuit of the subsea BOP, the yellow and blue pod control units are divided into four control circuits. Based on the two control functions, the EHBS is divided into two control circuits. The hydraulic components that correspond to each control circuit are shown in Table 3.3. Each control circuit is optimized separately. The same optimization results can be used to simplify the optimization process because the blue and yellow pod control units are mutually redundant.

The sensor placement of the hydraulic control of the subsea BOP is optimized by the proposed method. The number of sensors for circuit 1–8 is 8–14, the number of sensors for circuit 9 is 1–3, and the number of sensors for circuit 10 is 1. The sensor placement of 10 hydraulic circuits is optimized. In the optimization process, the number of population particles is 8, the maximum number of iterations is 500, the inertia weight $w = 0.4$; the self-learning factor $c_1 = 0.6$, and the group learning factor $c_2 = 0.4$. In the Monte Carlo simulation of comparison and selection, the failure rate of the directional valve of each circuit obeys a gamma process, and is affected by factors such as frequency of use, pressure, and external environment. The simulation time is set to 10 years. To ensure the stability of the optimization results, 30 simulations are performed for each optimization, the average value is calculated, and the average minimum fault detection time is used as the fitness value of the particles.

According to the optimization results, the optimal solutions of each circuit under different sensor numbers are obtained. The MTTD and the number of sensors for each circuit are shown in Fig. 3.10. For the two low-voltage circuits of circuit 1/5 and circuit 2/6, under the same number of sensors, it has a higher MTTD than circuit 3/7 and circuit 4/8. The reason is that the relative error of the low-pressure circuit sensor is larger than that of the high-pressure circuit. With the increase in the number of sensors, the positioning efficiency of the low-voltage circuit is greatly increased. In contrast, after the number of sensors exceeds 11, the marginal benefit of the diagnostic efficiency of the high-pressure circuit is very limited as the number of sensors increases.

Table 3.3 Hydraulic components corresponding to each control circuit

No.	Hydraulic components
Circuit 1	B_CHE1, B_RV, B_TV, B_GV, B_Acc, B_CCSV1-28
Circuit 2	B_CHE1, B_RV, B_Acc1 to 3, B_SPM1-28
Circuit 3	B_CCSV1-28
Circuit 4	B_SPM1-28
Circuit 5	Y_CHE1, Y_RV, Y_TV, Y_GV, Y_Acc, Y_CCSV1-28
Circuit 6	Y_CHE1, Y_RV, Y_Acc1 to 3, Y_SPM1-28
Circuit 7	Y_CCSV1-28
Circuit 8	Y_SPM1-28
Circuit 9	EHBS_SRC1-4
Circuit 10	EHBS_RC1-2

Fig. 3.10 Relationship between the MTTD of each circuit of the BOP and number of sensors

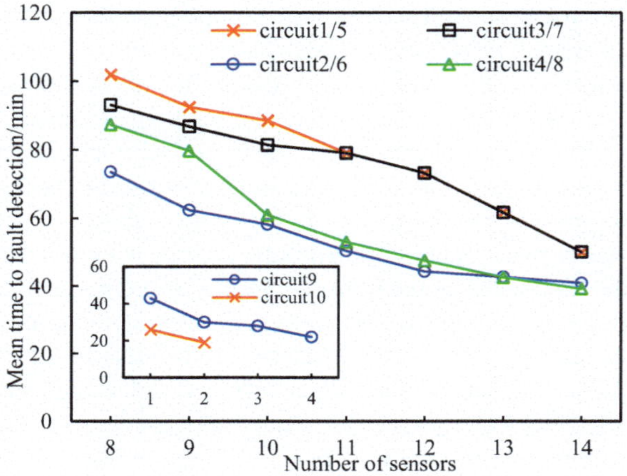

3.4.3 Results and Discussions

Providing an optimal combination of the optimal solutions of each circuit is necessary for the optimal solution of the subsea BOP system. The 7203 kinds of combination with 34–60 sensors are simulated, and the optimal solution is finally obtained using different sensor numbers. Taking a widely used evenly distributed sensor placement scheme as a comparison, the relationship between the MTTD of each circuit and the number of sensors is shown in Fig. 3.11. The OSP scheme obtained by this methodology can increase diagnostic efficiency by more than 10%. The acceptable MTTD is 450 min. In this methodology, 45 sensors can meet the requirements, whereas the existing methodology requires at least 51 sensors to meet the requirements. Considering that this methodology fully considers the difference in the degradation speed of the hydraulic components in the circuit, it can give full play to the detection and diagnosis efficiency of the sensor and provide an effective way to solve the problem of discrete optimal placement of the sensor.

Fig. 3.11 Relationship between MTTD of each circuit and number of sensors

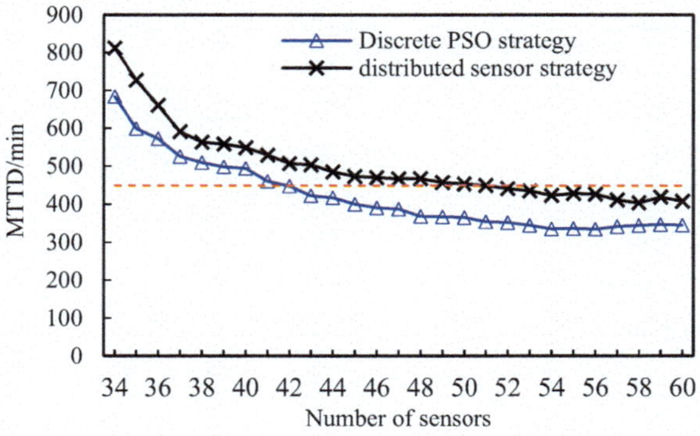

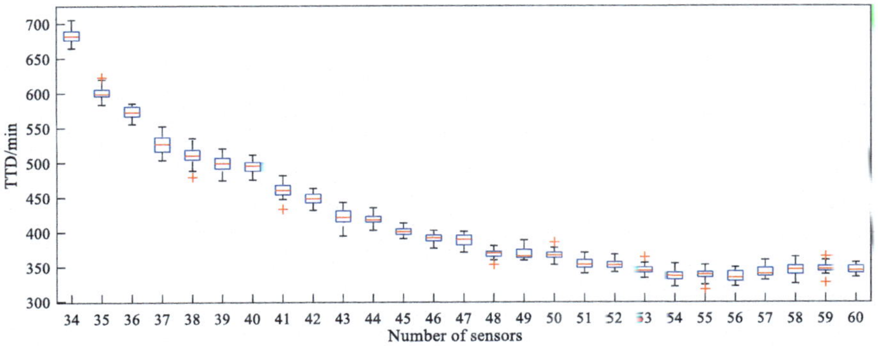

Fig. 3.12 TTD distribution under discrete PSO strategy

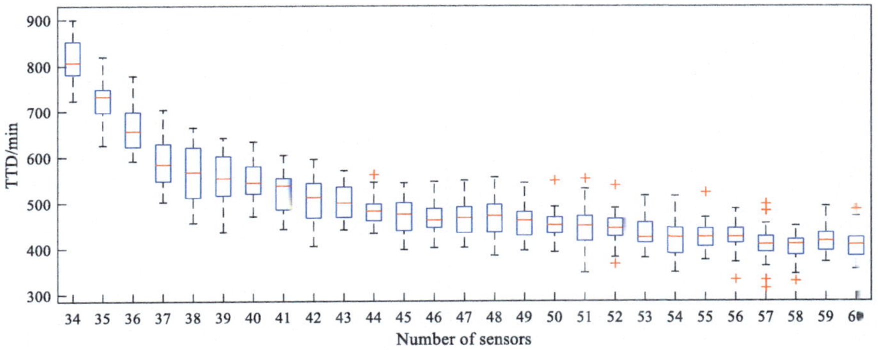

Fig. 3.13 TTD distribution under distributed sensor strategy

To verify the robustness of the optimal solution of the proposed algorithm, data analysis is performed on the time to detection (TTD) of all simulation cases. The box plots of TTD under discrete PSO strategy and distributed sensor strategy are shown in Figs. 3.12 and 3.13, respectively. Under the discrete PSO strategy, the robustness of the leak detection of the system has significant improvement. In terms of singular values, the discrete PSO strategy has a total of 9 singular values, and all singular values do not exceed 10% of the average value. In contrast, 14 singular values were under the uniform distribution strategy. Some singular values still exist in many sensors. Some singular values are higher than the average value by more than 20%, thereby causing potential safety risks for the hydraulic control system.

Chapter 4
Fault Diagnosis for Subsea Control System

Abstract This chapter introduces two fault diagnosis models for two types of subsea control systems. To address the limitations posed by a finite number of sensors and the concealment of faults, a three-model-driven fault diagnosis model with regard to energy, fluid and information is proposed for complex hydraulic control system. The diagnostic model is finalized through the establishment of a fault reasoning model using a Bayesian network. To solve the problem of unclear fault characteristics in multi module redundant systems. A model-driven methodology is proposed to the fault diagnosis of complete closed-loop control systems configured by multiple modular redundant technology. Multi monitoring information together with the basic information is used for fault symptom input. Inference probability is converted into diagnostic results. A dual redundant control system for subsea blowout preventers is used to study the performance of the proposed methods. The results indicate that the methods are effective for fault diagnosis of complex hydraulic control system and redundant closed-loop control systems.

4.1 Introduction

The hydraulic control system is one of the most critical subsea control systems, whose components mainly include pumps, pressure control valves, direction control valves, actuator, and storage vessels (Jia et al. 2010; Coelho and Cunha 2011; Azzam et al. 2023). The hydraulic control system may generate various faults due to factors such as its internal pressure, external pressure, fatigue, and load impact. Hydraulic systems have the characteristics of strong fault concealment, significant sensor delay, and a complex signal transmission mechanism. Therefore, the current security guarantee of the hydraulic system is mainly completed by the offline test. Online fault diagnosis for hydraulic control systems is an important problem that needs to be solved.

Due to the strong correlation of the hydraulic system, the model-driven method has always been an effective approach to solve the problem of fault diagnosis of the hydraulic control system. There are two main aspects of challenges using model-driven methods for fault diagnosis. On the one hand, the strong interaction between

B. Cai et al., *Intelligent Operation and Maintenance for Subsea Production Systems*,
https://doi.org/10.1007/978-981-96-2677-9_4

components is very complicated. The coupling of these interactions brings challenges to model establishment. On the other hand, the type and quantity of the sensors in hydraulic control system are limited due to the closure characteristics of the system (Kelley and Hagan 2020). Generally, only a limited number of pressure sensors and flow sensors are placed in a hydraulic control system (Luo 2022). In this work, multi-model collaboration methods are crucially used as a key to solving the above problems. Three models established through energy theory, thermodynamics, fluid mechanics, information conversion theory and signal processing interaction for seeking solutions to solve the challenges brought by the dynamic complexity of hydraulic system and the lack of its fault mechanism. Motivated by the considerations, a three-model-driven fault diagnosis method is proposed for complex hydraulic control system. Three hydraulic control system models around energy, fluid and information are proposed and established. A redundant control system for subsea BOP is used as a case to demonstrate the proposed method.

In addition to that, complex subsea control systems typically consist of several closed-loop control circuits. The functional relationships between different variables, and even the operating characteristics of the entire system, change due to the closed-loop feedback. The fault diagnosis and fault location of closed-loop feedback system is a challenging problem. The inclusion of feedback links enhances the robustness of the overall system. Consequently, when a fault is minor or in its initial stage, there may be slight disturbances observed in the corresponding residual signal. These disturbances are mitigated by the closed-loop system, resulting in a reduced likelihood of successfully detecting such faults. Moreover, faults can propagate through feedback loops within the closed-loop system, ultimately leading to widespread alarms throughout the entire system and thereby complicating fault localization and isolation (Zou et al. 2018, Chen et al. 2019).

To meet the requirements of high reliability and robustness, a modular redundancy technique is often used to design system hardware structure in offshore oil exploration and equipment development. The modular redundancy technology includes controller redundancy, bus redundancy, network redundancy, server redundancy and power supply redundancy (Yang et al. 2019). The redundant control system can analyze some monitoring data and output results according to the principle of majority voting by the voter to ensure the functional stability of the control system under component failure and signal interference. The use of redundant systems complicates the control system and simultaneously poses a new challenge for fault diagnosis of multi-redundant systems. However, it leads to a high computational burden and an interference to data analysis and diagnosis. In addition, the operating environment of some control systems of offshore oil equipment is complicated, and historical data have limited contribution to fault diagnosis. Therefore, a model-driven methodology is proposed to the fault diagnosis of complete closed-loop control systems configured by multiple modular redundant technology. The performance of control systems is dynamically assessed based on continuous sensor data. A double modular redundant control system for subsea BOP is used as a case to demonstrate the proposed method.

4.2 Fault Diagnosis for Complex Hydraulic Control System

4.2.1 Energy Flow Model

The energy undergoes various transformations in the working process of the hydraulic control system. The forms of energy in the system mainly include electrical energy, hydraulic energy and mechanical energy. The interconversion of energy in these forms has accomplished the controlling function. Faults in the system can cause abnormalities in the efficiency and speed of energy transfer during these conversions. Therefore, the establishment of the energy flow model in the working process is an important means of diagnosing the failure of the hydraulic control system.

The energy flow model of hydraulic control system is given in Fig. 4.1. In the hydraulic control system, almost all the energy comes from the electric energy input. The task of converting electrical energy into hydraulic energy is performed by a pump Hydraulic energy is accumulated in an accumulator. When the control command is received, hydraulic energy is distributed to the control target through various directional control valves.

The input power of the motor P' can be calculated as follows

Fig. 4.1 The energy flow model of hydraulic control system

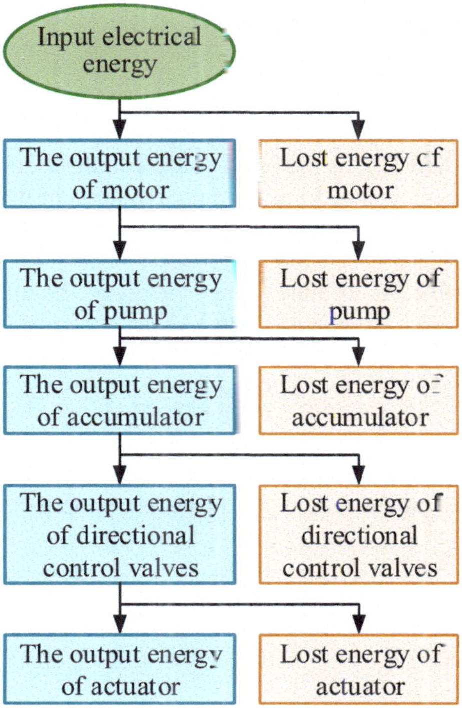

$$P' = \sqrt{3} \times U \times I \tag{4.1}$$

where U is the input line voltage of the motor, I is the input line electric current of the motor. The U and I can be measured by a sensor attached to the power supply.

The input energy of the motor W_m^{in} can be calculated as follows.

$$W_m^{in} = P' \times t_m \tag{4.2}$$

where t_m is the working time of the motor.

The output power of the motor P can be calculated based on the output torque between the motor and the pump. After the output torque T of the motor is measured, the output power of the motor can be calculated as follows.

$$P_m^{out} = \frac{\pi Tn}{30} \tag{4.3}$$

where n is the motor speed.

During the working process of the accumulator, the hydraulic energy conversion follows the ideal gas law. That is, when the state of the accumulator changes, there is no heat energy exchange with the outside world. The energy stored in the accumulator is calculated by numerating the work done on the airbag during the stamping process, which is shown as follows.

$$E_{acc} = P_0 V_0 \ln \frac{P_{acc}}{P_0} \tag{4.4}$$

where E_{acc} is the energy stored in the accumulator, P_0 is the initial pressure of the accumulator, V_0 is the initial volume of the storage device, P_{acc} is the pressure of the accumulator.

During the working process of the direction control valve and pressure control valve, the absolute value of energy is difficult to be directly calculated. In the proposed model, the relative energy of the input and output can be calculated as follows.

$$\frac{E_{in}}{q_{in} P_{in}} = \frac{E_{out}}{q_{out} P_{out}} \tag{4.5}$$

where E_{in} is the input energy, q_{in} is the input flow, P_{in} is the input pressure, E_{out} is the output energy, q_{out} is the output flow, P_{out} is the output pressure.

4.2.2 Fluid Flow Model

The fluid flow model mainly includes the circulation process of the oil between the oil tank, pump group, accumulator, valve, actuator. The core of fluid flow model is mass-flow-velocity model. According to the monitoring information of the sensor, a mass-flow-velocity model is established. Fluid flow model for hydraulic pilot directional valve is given as Fig. 4.2.

A widely used model to determine mass flow of a fluid through a known orifice that is commonly utilized in modeling valves is given as follows.

$$\dot{m} = \begin{cases} kap_u & \text{if} \frac{p_d}{p_u} < r_{th} \\ kap_u\left(\frac{p_d}{p_u}\right)^{\frac{1}{\gamma}}\sqrt{1-\left(\frac{p_d}{p_u}\right)^{\frac{\gamma-1}{\gamma}}} & \text{otherwise} \end{cases} \tag{4.6}$$

where p_u is the upstream absolute pressure, p_d is the downstream absolute pressure. m is the mass flow through the valve. γ is the ratio of the specific heat at constant pressure to the specific heat at constant volume of the fluid. In hydraulic control system of industrial system, the values of γ is approximately 1.4. The terms of a, b, k, and r_{th} are given as follows.

$$a = \sqrt{\frac{\gamma}{R}\left(\frac{2}{\gamma+1}\right)^{\frac{\gamma+1}{\gamma-1}}} \tag{4.7}$$

$$b = \sqrt{\frac{2\gamma}{R(\gamma-1)}} \tag{4.8}$$

$$r_{th} = \left(\frac{2}{\gamma+1}\right)^{\frac{\gamma}{\gamma-1}} \tag{4.9}$$

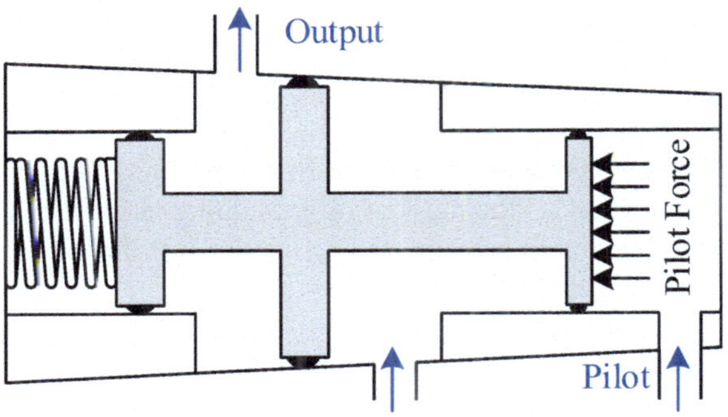

Fig. 4.2 Fluid flow model for hydraulic pilot directional valve

$$k = \frac{A_v C_f}{\sqrt{T}} \tag{4.10}$$

where R is the oil constant. The value of R is approximately 287 J/(kg K). The threshold r_{th} marks the change in dynamics between choked and unchoked flow regimes. T is the fluid at the orifice, A_v is the effective orifice area through which fluid is flowing, and C_f is the experimentally determined discharge coefficient.

The working medium of the exit of the direction control valve will flow to the energy storage element, such as accumulators, the controlled object converting hydraulic energy into mechanical energy, or hydraulic motors and hydraulic cylinders. In fluid flow model, energy storage elements and controlled objects are modeled as closed chambers, which can output external energy. In the established model, the quality flow of the direction control valve output needs to be converted to the volume change of chambers. Before the model is established, some assumptions should be given as follows:

(1) The transform process is isothermal.
(2) The temperatures of the flow already in the chambers are approximately equal to the temperatures of the flow flowing through the valve.
(3) The pressure at the actuator chamber and lines can be approximated by the pressure at the valve outlet.

Under the condition that the chambers have no leakage, the change of volume is given as follows.

$$p_d V_d = mRT \tag{4.11}$$

In the established model, the above formulas are transformed as follows.

$$V_d = RT\frac{m_0 + \Delta m}{p_d} \tag{4.12}$$

The change rate of volume is as follows.

$$\dot{V}_d = RT\frac{\dot{m}p_d + \Delta m\dot{p}_d - m_0\dot{p}_d}{p_d^2} \tag{4.13}$$

where p_d is the pressure at the downstream port of the valve, V_d represents the variable volume of the downstream system (consisting of lines and one of the chambers of the actuator), m_0 is the mass of oil presenting in the downstream system at the moment of activation, $\dot{m}$ is the mass flowing through the valve, Δm is the cumulative integral of $\dot{m}$, which is shown as follows.

$$\Delta m = \int_0^t \dot{m}(\tau)d\tau \tag{4.14}$$

4.2.3 Information Flow Model

The correct transmission of information is an important factor for the control system to complete its mission. The information transmission form of the hydraulic control system respectively includes electronic control signal, low-pressure oil signal and high-pressure oil signal. Due to the spread between multi-components, the information flow model is an important tool for interactive interaction between reaction components. In this work, the characteristics obtained by the information flow model mainly include two aspects: time characteristics and content characteristics. The information transmission form of the hydraulic control system respectively includes electronic control signal, low-pressure oil signal and high-pressure oil signal. In this section, the information flow model, which includes control signal transmission and execution of the control system, is established as shown in Fig. 4.3. During the work of the hydraulic control system, the control signal is first issued in the form of electrical signal. After receiving the control signal, the electromagnetic direction valve begins to work, and the low-pressure oil pressure rises from p_{li} to p_{lw}. When the pressure of the low-pressure hydraulic oil reaches the pilot pressure p_{pil}, the hydraulic pilot valve begins to work. High-pressure oil pressure rises from phi to p_{hw}. Parameters p_{li}, p_{lw}, p_{pil}, p_{hi}, p_{hw} are considered to be the time feature of the information flow model obtained in this process can be obtained through the pressure signal measured by the sensor.

In the information flow model, in addition to the pressure and flow characteristics measured by the sensor, the response time is also an important node. The response time (Δt1, Δt2, Δt3, Δt4, Δt5 in Fig. 4.3) can be described as the time from the signal of each link between sending and the downstream responding and are considered to be the time feature of the information flow model obtained in this process. The interval between each conversion of the signal can be obtained through the pressure signal collected by the electronic control feedback and the sensor.

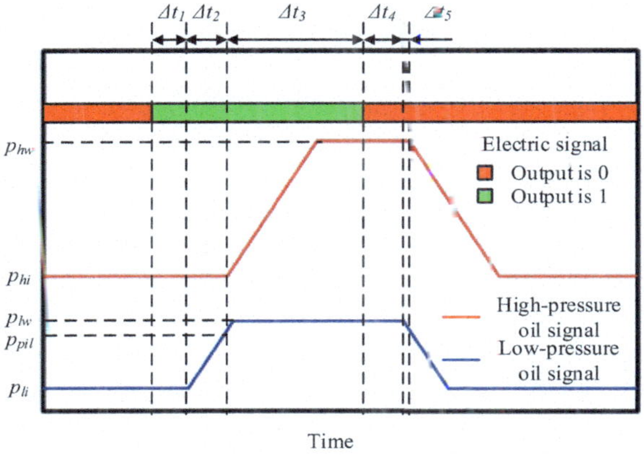

Fig. 4.3 The information flow model of hydraulic control system

4.2.4 Three-Mode-Driven Hydraulic System Fault Diagnosis Framework

The proposed three-mode-driven hydraulic system fault diagnosis framework is given as Fig. 4.4. Firstly, the proposed energy model, fluid flow model and information flow model are established respectively. Afterwards, a hydraulic system fault diagnostic model with a multi-model is established based on the BN. Fault feature extraction algorithm and diagnostic reasoning algorithm are given by the established BN model. Then, fault identification rules based on probability is given to gain the result of fault diagnosis. Finally, verification and validation are carried out to confirm the correctness and performance of the model. The establishment of the model has been given previously. Generally, the model parameters are input according to the main features of the system. The features include geometric size, system structure, workflow and fluid characteristics. When the system information cannot be fully obtained, parameter recognition is used to improve the model. In addition, in some non-linear links, the correction of parameters is used to improve the accuracy of diagnosis. Other steps are taken in the remainder of this section.

The process of the establishment of the BNs model mainly includes two steps: structural modeling and parameter modeling. Structural modeling includes definition nodes and mapping the relationship between nodes. In this work, network structure is established based on causal relationship between nodes. BNs parameter model consists of prior probability of root nodes and conditional probability of leaf nodes. The purpose of parameter modeling is to obtain the conditional probability table of the model. The parameter structure is established using Noisy-OR and noisy-MAX models. The established fault diagnostic models based on BN are shown in Fig. 4.5. The model includes the model node layer, the intermediate node layer, the fault signals layer and the fault reasoning layer. The model node layer is a key parameter. The nodes of energy model include the energy efficiency of components $\eta_1, \eta_2, \ldots, \eta_a$ and the energy value of keys $E_1, E_2, \ldots, E_a$. The nodes of the flow model include the energy efficiency of components $m_1, m_2, \ldots, m_a$ and the energy value of keys $L_1, L_2, \ldots, L_a$. The nodes of information model include the energy efficiency of components $\Delta t_1, \Delta t_2, \ldots, \Delta t_a$ and the energy value of keys $p_1, p_2, \ldots, p_a$. During the inference, the node status of the model node layer and observation nodes are used as the input of the BN model. The status of nodes in the model node layer are obtained by the established model.

The middle nodes of the hydraulic control system include observation pressure nodes ($p_s^1, p_s^2, \ldots, p_s^i$), predictive pressure nodes ($p_m^1, p_m^2, \ldots, p_m^i$), observation flow nodes ($L_s^1, L_s^2, \ldots, L_s^j$) and forecast flow nodes ($L_m^1, L_m^2, \ldots, L_m^j$). Observation values are collected by the system's sensors that are independent of models. Forecast values are calculated based on the model established. The differences between the predictive value and the observation value ($\Delta p_m^1, \Delta p_m^2, \ldots, \Delta p_m^i, \Delta L_m^1, \Delta L_m^2, \ldots, \Delta L_m^j$) are calculated as the failure characteristics of the extraction. The output of the model is the fault probability of the faults nodes ($F_1, F_2, \ldots, E_{i+j}$) obtained by the BN inference.

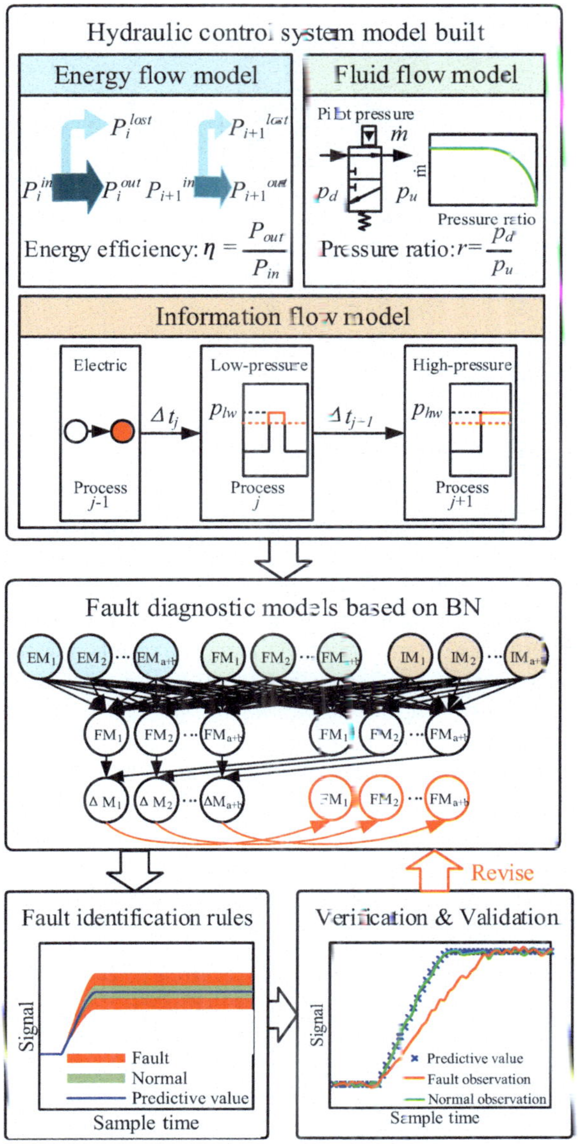

Fig. 4.4 Flow chart of the three-model-driven fault diagnosis method

BNs parameter model consists of prior probability of root nodes and conditional probability of leaf nodes. The purpose of parameter modeling is to obtain the conditional probability table of the model. The parameter structure is established using Noisy-OR and noisy-MAX models.

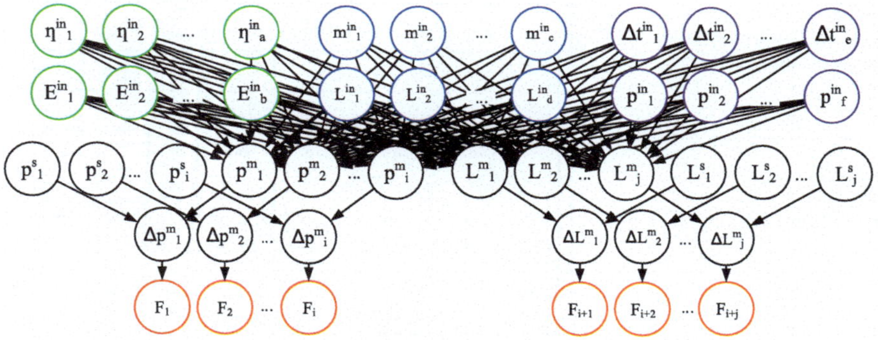

Fig. 4.5 Fault diagnostic models based on BN

4.2.5 Fault Identification Rules

Due to the fault inference model established based on BN, the obtained fault diagnosis results are presented in the form of probability. Overall, corresponding faults have higher probability of occurrence with increasing posterior probability. To increase accuracy and robustness of fault diagnosis, a series of fault identification rules were reported for determination of diagnostic results. The threshold with higher diagnostic accuracy is selected as the fault identification rules. The nodes in the fault layer mainly include two states: 'Normal' and 'Fault'. Fault identification is implemented based on the deviation of the observation value and predictive value of the pressure and flow signal. The observation value of pressure and traffic signals comes from data collected by the sensor. The predicted value comes from the results of the model interaction and operation based on the analysis of historical data. Through several rounds of selection, testing, and optimization, the optimal threshold with higher diagnostic accuracy is selected as the fault identification rules. This work has formulated the following three rules for fault judgment:

a. The fault is determined has occurred when the deviation of the observation value and predictive value is larger than 15%;
b. The fault is determined may occur when the deviation of the observation value and predictive value is larger than 5% and smaller than 15%;
c. The fault is determined that has not occurred when the deviation of the observation value and predictive value is smaller than 5%

4.2.6 Verification and Validation

Model verification and validation are significant aspects of fault diagnosis because they provide reasonable confidence in diagnostic results. Verification is defined as a process of determining how accurately a computer program solves equations of mathematical models. That is whether the system built above meets the performance specifications stated. Validation is defined as a process of determining degree to which a model accurately represents the real world from the perspective of intended users. That is whether the system model is close enough to the physical system, and whether the performance specifications and system constraints are correct. In this study, model verification is carried out by data collected when the system is normal without fault. Data samples of faults are used to validate models.

4.3 Fault Diagnosis for Closed-Loop Feedback Control System

4.3.1 Proposed Fault Diagnosis Methodology

As shown in Fig. 4.6, a closed-loop control system has four components, namely controller, actuator, process, and sensors. In the practical application of control systems, some algorithms are often used to adjust closed-loop feedback control to achieve production goals and system stability, robustness and other requirements, such as PID control, robust control, optimal control, etc. Control loop performance can be affected by any problem with the components. The system shown in Fig. 4.6 is an LTI system. Normally, the system is subject to uncertainties derived from the LTI characteristics. Uncertainties or nonlinearities challenge the inference and diagnosis of faults in the linear system. As one of the most effective tools for dealing with uncertain problems, BNs can effectively overcome difficulties caused by many uncertain factors in fault diagnosis. In this work, BNs are used to the system performance evaluation and fault diagnosis inference model built.

The general electro-hydraulic composite control system can often be described as typical second-order inertia plus pure delay link. The transfer function $G(s)$ if the system is shown as follows.

$$G(s) = \frac{Ke^{-\tau s}}{(T_1 s + 1)(T_2 s + 1)} = \frac{\omega_n^2}{s^2 + 2\xi\omega_n s + \omega_n^2} \tag{4.15}$$

where K, τ, T_1, T_2 are constants, ξ is the damping ratio of the system, ω_n is the natural frequency of the system without damping.

Considering feedback link $H(s)$, the transfer function of the system can be described as follows.

Fig. 4.6 Block diagram of a closed-loop control system

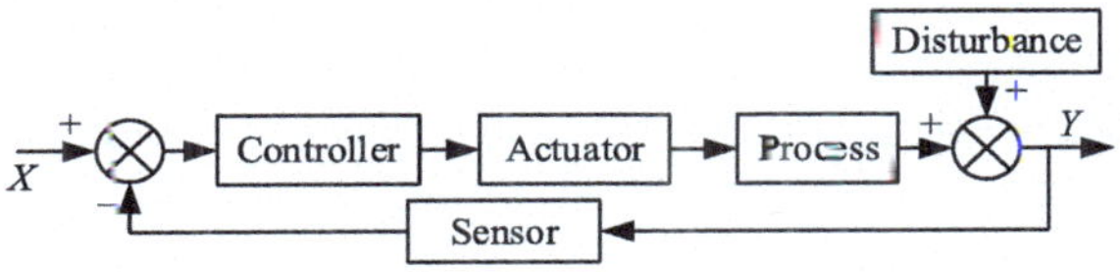

$$\frac{Y(s)}{X(s)} = \frac{G(s)}{1 + G(s)H(s)} \tag{4.16}$$

where $Y(s)$ is the output of the system, $X(s)$ is the input of the system.

There are three main types of control system faults, namely, sensor faults, electrical component faults, and mechanical or hydraulic faults. Sensors are the core of closed-loop control systems. Usually, sensors faults can occur such as drift and gain change during use. Reading of sensors is the ground of a control system. Therefore, the performance and monitoring of the system are affected directly by sensor faults. Electrical components mainly include controllers and redundant input/output modules. Due to the modular redundancy design of these components, faults of electrical components rarely affect control performance. Meanwhile, the root cause of failure becomes difficult to find at the same time. Actuator mechanical or hydraulic failures include motor drive faults, proportional valve sticking, or leakage of other hydraulic circuits and components. The performance of the control system would be affected by these faults. These effects may be similar to the effects caused by sensor faults and electrical components. Therefore, the use of monitoring information such as flow as supporting evidence is necessary when determining the fault.

The flow chart of the proposed fault diagnosis method of multiple modular redundant closed-loop control system is shown in Fig. 4.7. Firstly, the main parameters of the system are analyzed to obtain the forward channel transfer function $G(s)$ and feedback link transfer function $H(s)$. The system information including control system structure, redundancy mode, transfer function and main parameter settings can be obtained based on the system's software and hardware structure. Afterwards, a fault diagnosis model is built based on Dynamic Bayesian Network (DBN), including structural model and parameter model. Each time slice of the proposed DBN model includes sensor layer nodes ($K1, K2, U1, U2, f1, f2, e1, e2, S1, S2, S$), performance layer nodes ($P1, P2, \ldots, Pn$), monitors layer nodes ($M1, M2, \ldots, Mn$) and fault layer nodes ($F1, F2, \ldots, Fn$). Finally, the formulation of fault identification criteria, and the model verification and optimization are implemented.

4.3.2 Structure Modeling of DBNs

The DBNs model for redundant closed-loop control system diagnosis is shown in Fig. 4.8. Each slice of the model mainly includes four layers, that is, sensors, performances, monitors, and faults. Sensors are the key to a closed-loop control system, and sensor reading is affected by the sensor's operating characteristics and noise. The characteristic nodes of the sensor mainly include process gain (K), sensor bias (f), sensor input (u) and sensor noise (e). Redundantly configured sensors are used to improve the robustness and reliability of the control system. In the proposed model, the readings of mutually redundant sensors are defined as S1 and S2 respectively. Safety–critical systems often use redundant voting results of 1002 and 2003 as the final sensor reading (S). Nodes K, f, u and e are given an initial value and

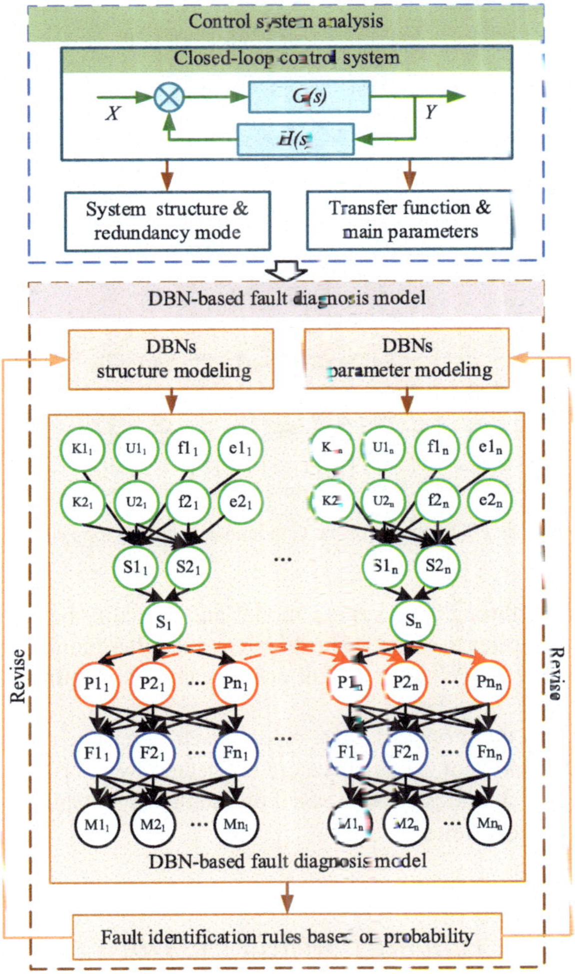

Fig. 4.7 Flow chart of multi-redundancy closed-loop control system fault diagnosis methodology

analyzed according to the system parameters and updated with time. S1, S2 and S updated in real time with the reading of the control system. System performance of the control system mainly includes maximum dynamic deviation (*d*), adjustment time (*ts*), control state (*cs*), residual error (*Re*), steady-state error (*ess*), etc. These performance indicators are calculated from changes in sensor readings. Performance indicators may change with system operation. Faults of the control system have a significant impact on performance changes. There is also a correlation between different performance indicators. Monitoring information mainly includes the pressure and flow of some control links in the system and alarm information of some electrical and hydraulic components. This information serves as an auxiliary part to investigate the system performance and is important evidence for analyzing the root cause of faults. The fault layer calculates and infers the probability of each

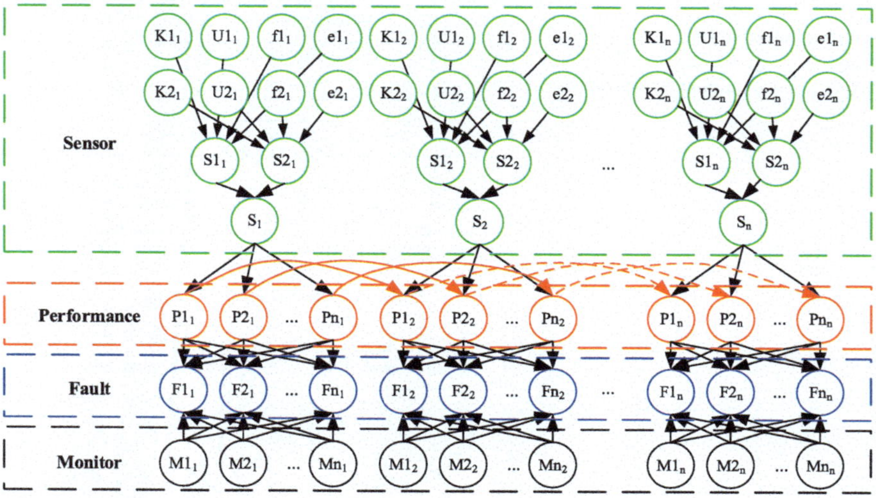

Fig. 4.8 DBNs model of multi-redundancy closed-loop control system fault diagnosis

fault through the system model and causality between faults and symptoms. The performance of the control system and other monitoring information is the basis for inference. Whether a fault occurs can be determined by established fault judgment rules.

In the proposed DBN model, each time slice corresponds to the system state at that moment. Time slices are connected by inter-slice arcs. The structure model of inter-chip arcs is built based on the iterative relationship of system performance at different times.

4.3.3 Parameter Modeling of DBNs

4.3.3.1 Sensors Layer

Obtaining data from sensors is the beginning of the fault diagnosis process. The accuracy of reading directly affects the performance of diagnosis. The process of reading from the sensor is affected by such factors as change of gain, change of input signal, sensor bias, and change of variance in the measurement. Sensor reading is modeled by

$$y = Ku + f + e \tag{4.17}$$

where K is the process gain; u is sensor input; f is sensor bias; y is sensor reading; e is the noise at sensor, noise e has the distribution shown as follows.

$$e \sim N(o, \sigma^2) \tag{4.18}$$

where σ is the standard deviation of the normal distribution

The initial performance parameters of sensor are determined by prior knowledge. Faults on sensors can change the actual parameter values. In a redundant control system, sensor reading results used for feedback usually pass a voting link to obtain a stable control function.

4.3.3.2 Performances Layer

The indicators for performance evaluation of electro-hydraulic control system are shown in Fig. 4.9. Indicators of control system performance usually include rapidity, accuracy and stability. PID parameters, actuators and feedback links are optimized in the design of a closed-loop control system to meet the requirements of these three indicators. However, the performance of a control system can be affected by its faults. A typical DBNs model based on dynamic performance evaluation control system is shown in Fig. 4.10.

The residual error *Re*(t) refers to the difference between control variable and input control target at time t. *Re*(t) can be defined by

$$Re(t) = y(t) - O \tag{4.19}$$

where O is the input control target.

System state includes two states, that is, steady-state and regulation. Due to the presence of noise, the steady-state is defined as the state when the *Re*(t) continues to fluctuate within 2% of the input target value.

Maximum dynamic deviation d is the magnitude of deviation at the moment when the deviation *Re* reaches its largest value within an adjustment period *Td*. The

Fig. 4.9 Dynamic performance indexes of closed-loop control system

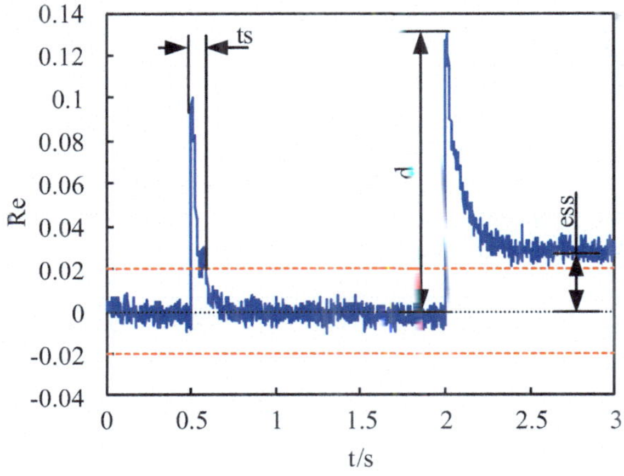

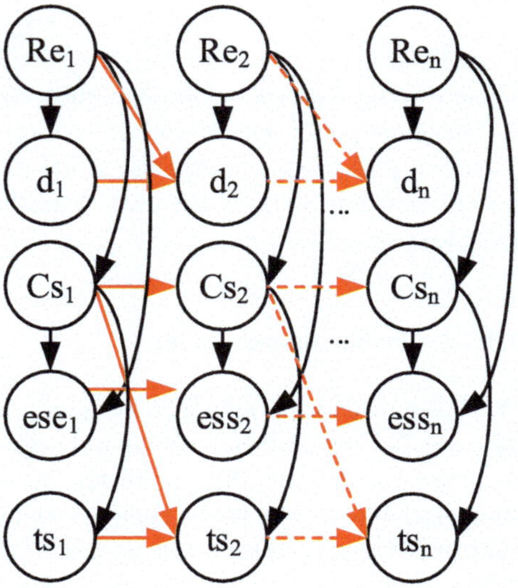

Fig. 4.10 DBNs model of control system performance layer

definition of d is as follows.

$$d = MAX\{|Re(t)|, t \in T_d\} \tag{4.20}$$

The steady-state error *ess* is the difference between average fluctuation value and control target value after system is adjusted to steady-state. When the system is in steady-state, $ess(t)$ can be calculated as follows.

$$ess(t) = y_{0.02\%} - O \tag{4.21}$$

where $y_{0.02\%}$ is the steady state value of the system, which is taken from the mean value when the amplitude is less than 0.02%.

The adjustment time *ts* is the time for the system to readjust from steady-state interruption to another steady-state. In general, *ts* can be obtained by the analytical method under a normal system. The analytic method is solved by the formula to obtain adjustment time *ts* of system. The analytic method algorithm, as shown in Table 4.1, can be used in this case.

In the evaluation of dynamic characteristics and fault diagnosis and reasoning, indicators are set as dimensionless to increase the scope of application of model and accuracy of diagnosis. θ can be defined by

$$\theta = \frac{\tau}{t} \tag{4.22}$$

The dimensionless adjustment time is shown as follows.

Table 4.1 Algorithm for *ts* evaluation of closed-loop control system

No.	Algorithm
1	**Input:** Transfer Function G(s), Input Function X(s)
2	**Initialization:** d, n
3	Y(s)=G(s)·X(s)
4	y(t)=ilaplace(Y(s))
5	[yupper, ylower] = envelope (*y*)
6	for(t=0:t:t++)
7	{
8	if(abs(yupper)<0.02||abs(ylower)<0.02)
9	ts=t
10	end if
11	break
12	}
13	**Output:** Transient Time ts

$$T_s = \frac{ts}{\theta} \tag{4.23}$$

4.3.3.3 Monitors Layer

The parameters d and ts are important to control indicators for judging faults. Due to the feedback link of a closed-loop control system, component failures cause cascading changes in other system performance, and the presence of different faults may cause similar system changes. Therefore, monitoring information is used as an additional basis for fault identification. Monitoring information usually includes information such as flow, current, voltage, and voter status.

The faults layer is mainly used to identify and determine the probability of fault. Generally, fault node mainly includes two states, that is, Normal and Fault. Some fault nodes may include Normal, High, and Low states, such as sensor drift. High and Low states respectively indicate two fault states of sensor positive drift and negative drift. The conditional probability of fault node is obtained from Noisy-OR and Noisy-MAX models.

When a node is Boolean type namely, the node has two states: normal and abnormal. The node status depends on data detected by sensors. Conditional probability table of a node can be simplified with noisy-OR model. Specifically, supposing that several causes $X_1, X_2, \ldots, X_n$ lead to an effect Y. All nodes are Boolean and conditional probability table can be generated with only n parameters, and q_1, q_2, …, q_n are obtained by

$$P(Y = 1|X_1, X_2, ..., X_n) = 1 - \prod_{i=1}^{n} q_i \quad (4.24)$$

where q_i is probability for each parent node. q_i can be expressed by

$$P(Y \leq y|X) = \prod_{i=1,xi\neq 0}^{n} \sum_{y'=0}^{y} q_{i,y'}^{xi} \quad (4.25)$$

where false Y when X_i is true, and all other parents are false. y' is the iteration variable in the accumulation process, $y' \leq$ Y.

Noisy-MAX is the general form of Noisy-OR in the field of non-Boolean type variables. A node has three states when non-Boolean, that is, higher, lower, and middle, with which the noisy-OR model cannot work. Suppose that non-Boolean variables X_1, X_2, ..., X_n lead to the effect variable Y, conditional probability table can then be established by using the noisy-MAX model which is shown as follows.

$$P(Y = y|Xi = xi, Xj = 0[\forall j, j = i]) = q_{i,y}^{xi} \quad i = 1, \ldots, n \quad (4.26)$$

$$P(Y = y|X) = \begin{cases} P(Y \leq 0|X) & y = 0 \\ P(Y \leq y|X) - P(Y \leq y - 1|X) & y > 0 \end{cases} \quad (4.27)$$

where X represents parent node configurations of Y; $X = x_1, x_2, \ldots, x_n$, and P $(Y = 0|X_1 = 0,..., X_n = 0) = 1$.

All parametric model building is done offline. The conditional probability table is obtained through the establishment of the parametric model offline, which can be directly used for online fault diagnosis and inference. The online diagnostic inference process can be completed within 0.1 s.

4.3.3.4 Fault Identification Rules

Due to the fault inference model established based on BN, the obtained fault diagnosis results are presented in the form of probability. Overall, corresponding faults have higher probability of occurrence with increasing posterior probability. To increase accuracy and robustness of fault diagnosis, a series of fault identification rules were reported for determination of diagnostic results (Cai et al. 2017). The threshold with higher diagnostic accuracy is selected as the fault identification rules. The nodes in the fault layer mainly include two states: 'Normal' and 'Fault'. Fault identification is implemented based on posterior probabilities of fault nodes based on provided evidence. Based on the analysis of historical data, this work has formulated the following three rules for fault judgment:

a. The fault is determined has occurred when the probability that the status of nodes in fault layer is 'fault' larger than 80%;
b. The fault is determined may occur when the probability that the status of nodes in fault layer is 'fault' larger than 30% and smaller than 80%;
c. The fault is determined has not occurred when the probability that the status of nodes in fault layer is 'fault' smaller than 30%.

4.4 Cases of Subsea BOP Control System

4.4.1 Description of Subsea BOP Control Systems

Subsea BOP is a safety barrier to prevent blowout accidents during ocean oil exploration and drilling. Subsea BOP system plays an extremely important role in providing safe working conditions for drilling in deep-water. Faults of the subsea BOP could cause a catastrophic accident. The subsea BOP control system is a complex electro-hydraulic integrated system. The working environment of BOP is harsh, and the safety and the reliability requirements are extremely high. As a result, BOP system adopts a full-process redundant design and a large number of feedback links to improve working reliability. Taking the HYSY981 drilling platform as the prototype, a BOP hydraulic control system prototype was built in the laboratory as shown in Figs. 4.11 and 4.12.

In Fig. 4.11, the built system includes hydraulic power unit, blue box, yellow box and pressure regulating unit. The hydraulic power unit is used to provide pressure supply to the accumulator. The blue pod and the yellow pod are redundant hydraulic control pods that work in the subsea. Each hydraulic control box includes blue/yellow box switching system, energy storage system, pilot pressure control circuit, work pressure regulator circuit and work pressure control circuit. The pressure regulating unit is used to regulate the pressure flowing to the controlled object. The blue pod and the yellow pod can implement the redundant hydraulic control function of 16 routes. The subsea BOP control system used in the case covers all links that may be used by hydraulic systems such as pressure supply, pressure storage, pressure regulation, direction control, etc. The research on the fault diagnosis of the subsea BOP control system can also provide a reference for the fault diagnosis of almost all kinds of hydraulic system.

In this subsea BOP control system, there are two stress specifications that are used as pilot accumulator and working accumulator device. The normal working pressure of the pilot accumulator is set to 5.5 MPa to provide low-pressure oil as a control solution. The normal working pressure of the working accumulator is set to 10 MPa for high-pressure oils that drive BOP. Compensated chamber solenoid valves (CCSVs) are used as electromagnetic direction valve in the system. CCSV converts the electrical signal into a pilot oil signal. Sub plate mounted (SPM) control valves are used as hydraulic direction valve in the system. SPM control valve is used to

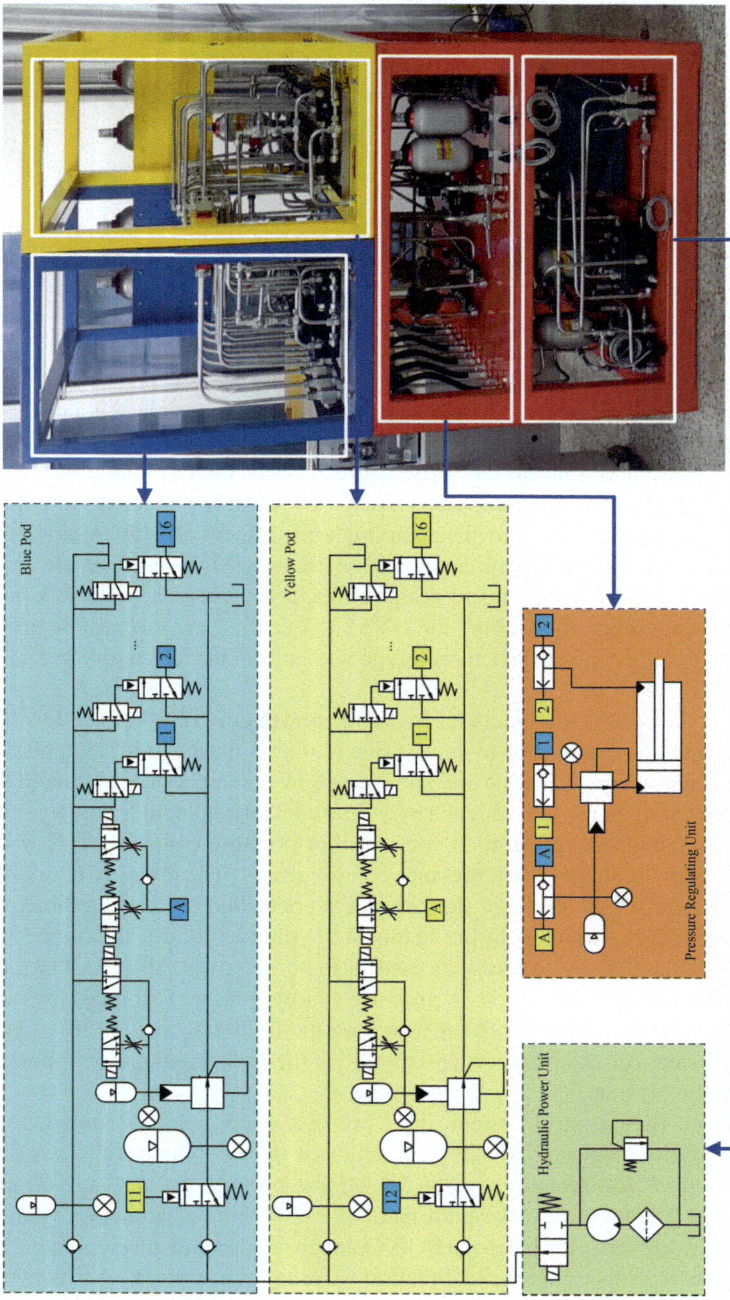

Fig. 4.11 Subsea BOP control systems prototype

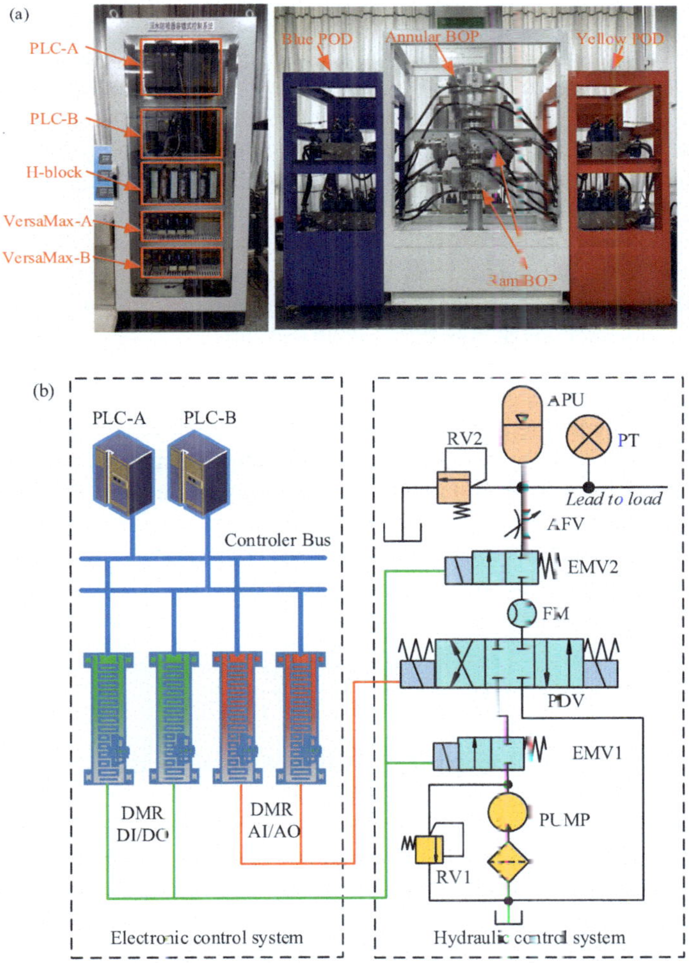

Fig. 4.12 Subsea BOP control systems **a** External view. **b** Schematic diagram

receive CCSV's pilot oil signals and uses oil with working pressure to complete the BOP control. There is a total of two kinds of sensors used in the laboratory. Pressure transmitters are used to collect the pressure signal. The collected pressure signals include the pressure of pilot accumulator, work accumulator, regulating accumulator and each control circuit. Flow sensors are used to collect the flow signal. The collected flow signals include the flow of the oil provided by the HPU to the pilot accumulator and the work accumulator. The measurement range of the used pressure transmitters is 0–10 MPa. The measurement range of the flow sensor used is 0-20m3/h. NI9203 current input module is used for data collection system and is installed on the CDAQ-9188 chassis. The single frequency is set to 200KS/s.

In this work, we built an experimental platform as shown in Fig. 4.12a with the BOP control system of the HYSY981 drilling platform as a prototype. The system includes an electronic control system, supporting hydraulic control system and sensor system. A dual modular redundant pressure control loop system as shown in Fig. 4.12b was selected as a case study. The electronic control system is based on dual redundant GE Fanuc 90–70 PLC. PLC-A and PLC-B are redundant PLC mainframes, which include CPU module, Ethernet communication module and GBC communication module respectively. The redundant voting of the digital output is implemented by the I/O module located in the H-block. VersaMax-1 and VersaMax-2 are redundant remote I/O modules which are located underwater in actual work and mainly perform functions such as analog input and output and high-speed counting. A 1oo2 voting rule is used for dual redundant I/O interface Discrete PID links are used in CPU in the form of algorithms to improve performance. The hydraulic control part includes a hydraulic pump that provides pressure, a proportional directional valve as the actuator of the system, and a control object is an accumulator group of hydraulic power units (HPU). In addition to monitoring sensors, system flow rate and the status monitoring information of electronic modules are also used for the diagnosis and location of system faults.

The control flow of the system of the subsea BOP is shown in Fig. 4.13. The transfer function of the proportional directional valve is shown as follows.

$$G(s) = \frac{523,500}{s^2 + 87.35s + 10,470} \tag{4.28}$$

The voting rule of 1002 adopts the principle of average value. That is, the output result of voting is the average of the input values.

The sensor reading and flow rate change of a control cycle of this control system are shown in Fig. 4.14. Under normal operation, when factors such as environmental or pressure release led to the pressure change, the sensors transmit this change to the CPU of PLC through the dual redundant AI channel. CPU obtains the control signal according to input target pressure and PID regulator. Afterwards, the proportional directional valve as an actuator pair converts control electrical signals into the corresponding action to inject or release pressure in the accumulator. The flow of injected or released fluid can be monitored by a flow meter.

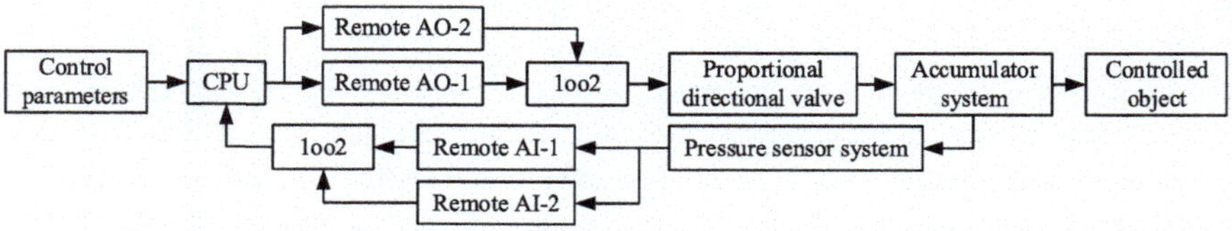

Fig. 4.13 Flow of control system of subsea BOP

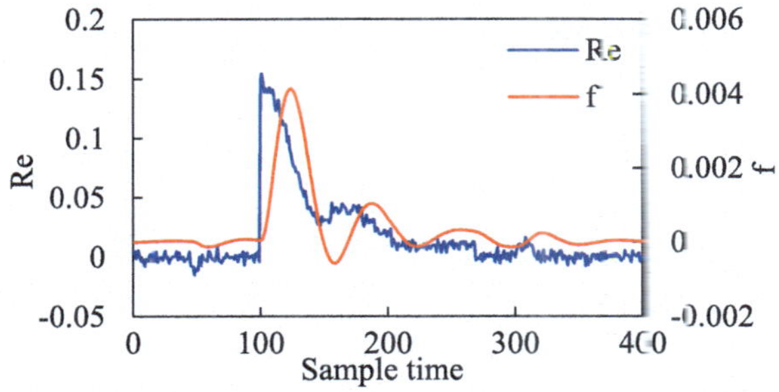

Fig. 4.14 Accumulator pressure control process in subsea BOP system

4.4.2 *Case of Complex Subsea Hydraulic Control System*

4.4.2.1 Model Established of Subsea BOP Control System

Energy flow model of the subsea BOP control systems is given in Fig. 4.15 The established model consists of 5 modules, including the hydraulic power unit, the blue/yellow box switch unit, the blue box control unit, the yellow control unit and the BOP stack. The components in hydraulic power unit mainly include motors and pumps. The input electrical energy is converted into hydraulic energy in this part. The functions of hydraulic switching units mainly control the direction of energy transfer. In normal failure work, the energy loss of this link is ignored. The channel 11–12 of the hydraulic power unit and the blue/yellow box control unit are coordinated, and the energy loss of its directional valve is discussed in the blue/yellow pod control unit. The structure of the blue/yellow pod control unit with redundant working module is completely consistent. The blue/yellow box control unit consists of a pilot hydraulic circuit and a working hydraulic circuit, respectively. The pilot hydraulic circuit mainly includes the pilot storage and the electromagnetic directional valve CCSV. The working hydraulic circuit mainly includes working storage capacity and hydraulic replacement valve SPM control valve. The BOP STACK section is mainly used to convert SPM control valve output hydraulic energy to mechanical energy.

Fluid flow model is built based on the design parameters of the valve hydraulic components and the geometric structure of the oil circuit block. The relationship between the pressure ratio of the oil flow of the valve to the input and output of the valve hole is given by Fig. 4.16. The fault status of the valve is described by calculating the flow characteristic curve of different valves at different valves. The curve of the normal working valve is a working characteristic curve under the opening area of 100% *Av*. The relationship between the pressure of the accumulators and time during pressure supply process is given by Fig. 4.17. The pressure changes of the pressure of different valves under the two types of accumulators are calculated separately. The flow characteristic curve of the valve and the pressure characteristic curve of the accumulator are used to iterate over time during the working process of the hydraulic

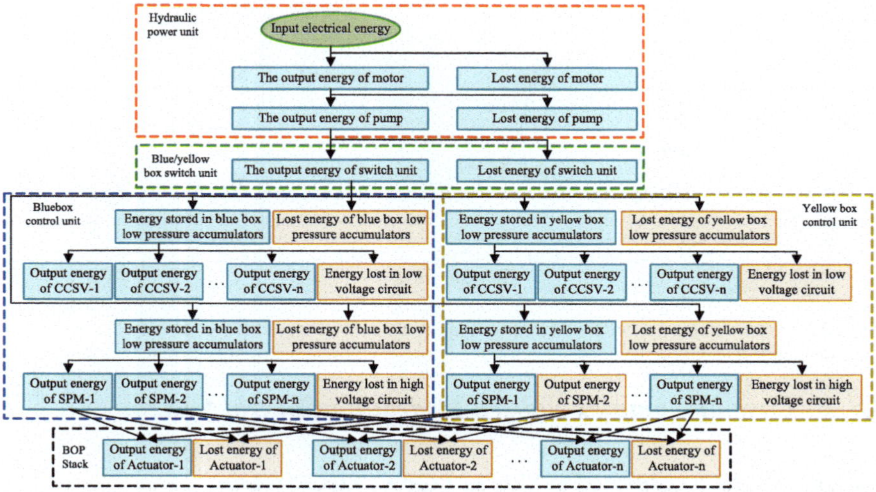

Fig. 4.15 Energy model of the subsea BOP control systems

control system. In this study, the step length of the iteration process is selected to 0.1 s.

The information flow model is established according to the design parameters of the valve and the space location information of the component. In this study, the working pressure of the pilot accumulator is set to 5.5 MPa and the working pressure of the working accumulator is set to 10 MPa. The adjustment of the pressure first is completed by the overflow valve set at HPU. The adjustment of the work pressure is completed by the electromagnetic direction valve of the pressure regulation circuit. The trigger pilot pressure SPM control valve is 5 MPa. That is, when the receiving CCSV's first control pressure reaches 5MPA, the SPM control valve is turned on to start working. Due to the viscosity of fluids, the response time of CCSV and SPM

Fig. 4.16 The flow characteristic curve of valve calculated by the fluid flow model

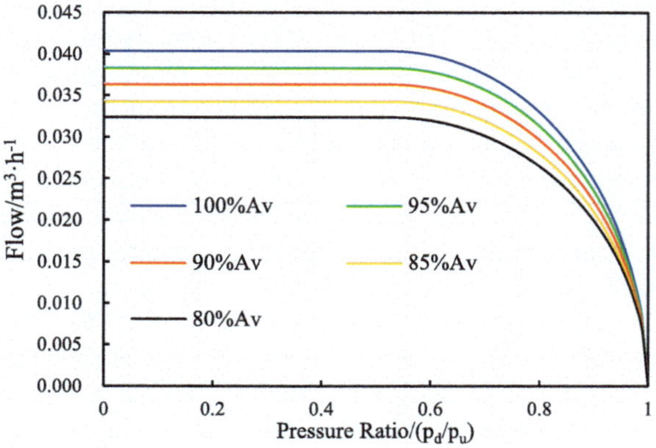

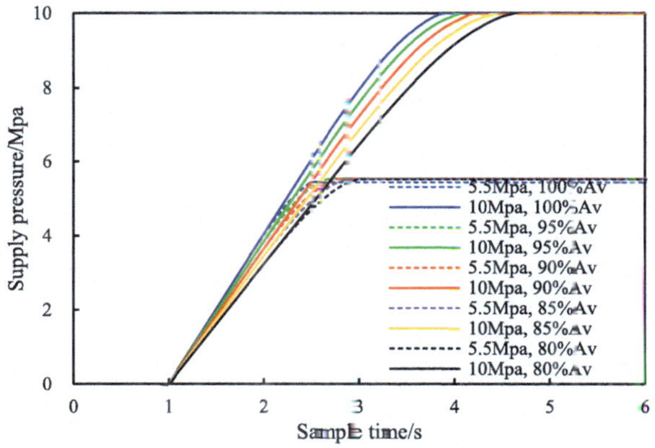

Fig. 4.17 The characteristic pressure curve of the accumulator calculated by the fluid flow model

control valve of different channels is affected by the distance between the valve and the accumulator. The response time is evaluated by a simulation model built based on Simulink of MATLAB.

4.4.2.2 Verification of the Proposed Model

In this chapter, the sensor signals of the three processes include the pilot accumulator pressurize, the work accumulator pressurize and the blue/yellow pod switching signals are collected in the experiment. Normal signals are used to verify the model proposed. In the collected signals, p1, p2, p3 and p4 represent the signals collected by the pressure transmitters T1, T2, T3 and T4 respectively. And f1, f2, f3 and f4 represent the signals collected by the flow sensors F1, F2, F3 and F4 respectively.

Using hydraulic power unit to press the pilot accumulator is a necessary preparation link for the control system of BOP. In this experiment, the pilot pressure is set to 5.5 MPa. The pressure signals during the process and the changes in prediction pressure obtained by the model are given in Fig. 4.18. It can be seen from the figure that the predicted value of the model is well fit the pressure changes to the pilot accumulator press process.

Using hydraulic power unit to press the work accumulator is an important guarantee for the reliable work of the control system of BOP. The working accumulator is the direct power source of the core functions such as opening, closing, locking, and unlocking of BOP. Taking the working accumulator process of the blue pod control unit as an example, the changes in pilot pressure, work pressure, and flow signal during the process are shown in Fig. 4.19.

The redundant blue box control unit and the yellow box control unit are the core subsystems for reliability control of the BOP. During the work of the BOP, the blue box/yellow box switch is achieved by pressing the work storage of the blue and yellow box. Channel 11 and 12 of the hydraulic system are used to achieve the choice of blue and yellow boxes, respectively. In PLC programming, the control

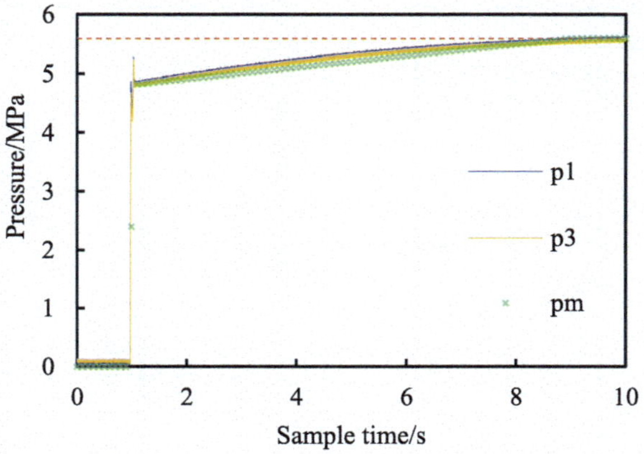

Fig. 4.18 Signal in pilot accumulator pressurize process

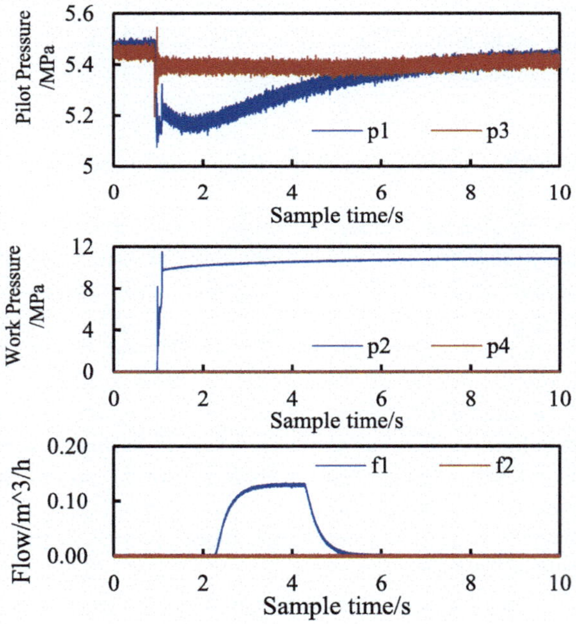

Fig. 4.19 Signal in work accumulator pressurize process

coils of CCSV11 and CCSV12 are set to the interoperability relationship to achieve switching. The changes in pilot pressure, work pressure, and flow signal during the process are shown in Fig. 4.20.

Fig. 4.20 Signal in blue and yellow pod switching process

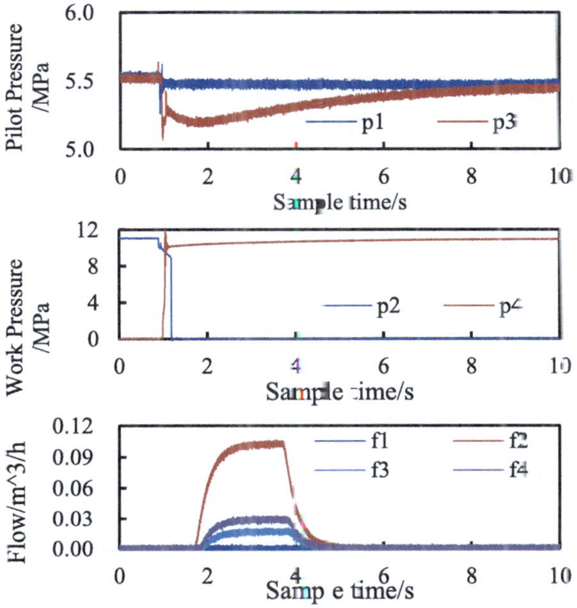

4.4.2.3 Validation of the Proposed Model

In this part, the samples collected by the experiment are used for the model. Because each component has different missions when controlling different working processes. The work process of three typical subsea BOP control system is simulated in the experiment, including pilot accumulator pressurize, work accumulator pressurizes and blue and yellow pod switching. There are 23 kinds of possible faults in the 3 processes simulated. A total of 100 sample signals of each fault in each process were collected. Table 4.2 summarizes the main faults. It can be seen from the table that the BOP control system is a complex system with multi-component coordination. All parametric model building is done offline. The online diagnostic inference process can be completed within 0.1 s.

Pumps, switching valve, pipeline and pilot accumulator are used in pilot accumulator pressurize. The accuracy of the fault diagnosis during the pilot accumulator pressurize process is shown in Fig. 4.21. The total accuracy of the fault diagnosis in this process is 95.80%. The accuracy of all fault diagnosis is higher than 90%. The minimum diagnostic accuracy appears in the branch line leakage fault of the pipeline. The possible reason is that branch line's geometric structure is relatively complicated. Therefore, branch line's modeling error is relatively large.

Pumps, switching valve, work accumulator, redundant accumulator and pipeline is used in work accumulator pressurize. The accuracy of the fault diagnosis during the work accumulator pressurize process is shown in Fig. 4.22. The total accuracy of the fault diagnosis in this process is 93.00%. The accuracy of all fault diagnosis

Table 4.2 Main faults simulated in the experiment

Component	Fault	Mark
Pump	Electrical Power supply fault	PUMP_EP
	Leakage	PUMP_L
Switching valve	Leakage	SV_L
	Valve core wear	SV_CW
	Blocked	SV_B
Pipeline	Main line leakage	PIPE_ML
	Branch line leakage	PIPE_BL
Redundant accumulator	Airbag leakage	RA_AL
	Oilbag leakage	RA_OL
Pilot accumulator	Airbag leakage	PA_AL
	Oilbag leakage	PA_OL
Work accumulator	Airbag leakage	WA_AL
	Oilbag leakage	WA_OL
CCSV	Leakage	CCTV_L
	Valve core wear	CCSV_CW
	Blocked	CCSV_B
SPM control valve	Leakage	SPM_L
	Valve core wear	SPM _CW
	Blocked	SPM _B
Supply pipeline	Main line leakage	SP_ML
	Branch line leakage	SP_BL
Discharge pipeline	Main line leakage	DP_ML
	Branch line leakage	DP_BL

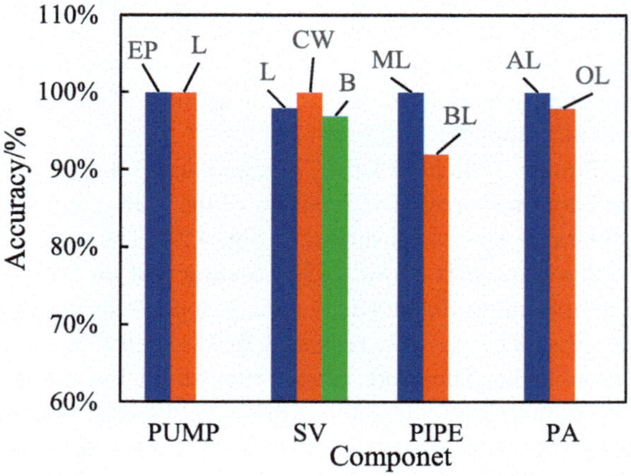

Fig. 4.21 The accuracy of the fault diagnosis during the pilot accumulator pressurize process

is higher than 90%. Except for the branch line leakage fault which is similar to the accuracy of the pilot accumulator pressurize process. The accuracy of pump leakage faults is relatively low. The analysis of the sample shows that the faults are relatively misdiagnosed with the switching valve leakage failure. The reason for this phenomenon is that the work pressure is high, and the differences in faulty characteristics are not obvious.

Pilot accumulator, work accumulator, CCSV, SPM control valve, supply pipeline and discharge pipeline are used in blue and yellow pod switching. The accuracy of the fault diagnosis during the blue and yellow pod switching process is shown in Fig. 4.23. The total accuracy of the fault diagnosis in this process is 96.86%. The accuracy of all fault diagnosis is higher than 85%.

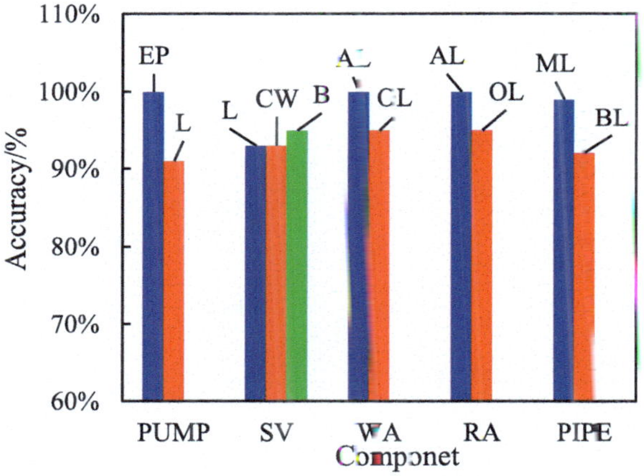

Fig. 4.22 The accuracy of the fault diagnosis during work accumulator pressurize process

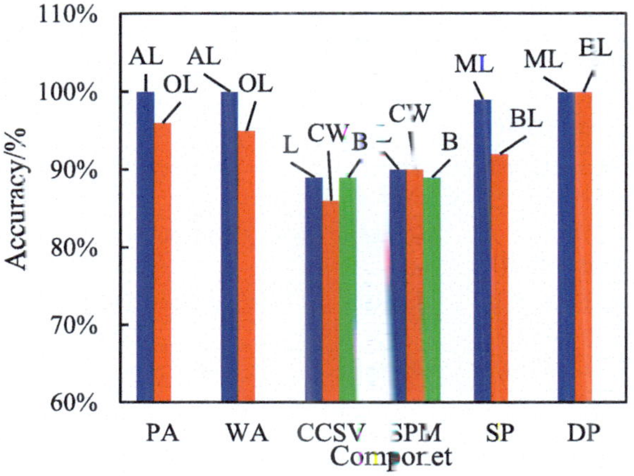

Fig. 4.23 The accuracy of the fault diagnosis during blue and yellow pod switching process

4.4.3 *Case of Subsea Closed-Loop Feedback Control System*

4.4.3.1 Fault Diagnosis Modeling

The causal relationship between the main faults and fault symptoms of the control system are shown in Table 4.3. Adjust Time (*ts*) History Flow (*FH*) Regulated Flow (*FR*) AI Alarm (*AIA*) are used as fault symptoms. *ts* obtained from the results of performance evaluation. *FH*, *FR* and *AIA* are the monitoring information of the

Table 4.3 Causality between faults and symptoms

Fault	Adjust Time (*ts*)	History Flow (FH)	Regulated Flow (FR)	AI Alarm (AIA)
Pump Leakage (PL)	High	Normal	Normal	Normal
Accumulator Airbag Leakage (AAL)	High	High	High	Normal
Proportional Valve spool stuck (PVS)	High	Normal	High	Normal
Accumulator Hydraulic Leakage (AHL)	High	High	High	Normal
Electromagnetic Valve Leakage (EVL)	High	Normal	High	Normal
Sensor Drift High (SD_H)	Normal	Low	Normal	Normal
Sensor Drift Low (SD_L)	Normal	High	Normal	Normal
Sensor Gain High (SG_H)	Normal or High	Extremely Low	Normal or Low	Normal
Sensor Gain Low (SG_L)	Normal	Extremely High	Normal or High	Normal
AI Drift High(AD_H)	Normal or High	Extremely Low	Normal or Low	Alarm
AI Drift Low(AD_L)	Normal	Extremely High	Normal or High	Alarm
AI Gain High (AG_H)	Normal	Low	Normal	Alarm
AI Gain Low (AG_L)	Normal	High	Normal	Alarm
AI Gain Low (AG_L)	Normal	High	Normal	Alarm

system. *ts*, *FH* and *FR* is a continuous monitoring quantity, which needs to be discretized into qualitative state indicators during the inference process. Standard values are set based on a set of normal signals. When the value is within $\pm$ 5% of the standard value, it is defined as being in the "Normal" state. When the value is greater than 5% and less than 10% of the standard value, it is defined as "extremely high", defined as "high" when the value is greater than 10% of the standard value. When the value is less than $-$ 5% and greater than $-$ 10% of the standard value, it is defined as "extremely low", defined as "low" when the value is less than -10% of the standard value. The *AIA* state comes from the alarm information of the PLC system. Normally, AIA is in "Normal" state. When the PLC detects that the voltage or current of the AI channel is abnormal, it will trigger an alarm. Currently, the AIA is defined as the "alarm" state. The faults of hydraulic system mainly include PL, AAL, PVS, AHL and EVL. Among them, the leakage of the accumulator airbag and blockage of proportional valve spool may cause the system's failure to return to a steady-state or a large steady-state error of the system Faults of sensor system mainly include sensor drift and gain. These faults affect sensor readings, and lead to a steady-state monitoring error is output as Normal. At the same time, because the system reacts to the sensor's erroneous readings, the output of *ts* is wrong as normal. Flow monitoring can detect these misjudgments. Some sensor faults and hydraulic component fault signals are shown in Fig. 4.24.

The I/O channel fault signal of the redundant electronic control system is shown in Fig. 4.25. AI channel drift, gain faults and sensor faults affect the system by sensor readings. The information on this kind of fault is very similar to sensor faults. The AI reading difference alarm on the voting system monitor can isolate the AI channel fault and sensor fault.

At the same time, faults of electrical components such as Pow, CPU and DIO are also considered. The causality of these faults is shown in Table 4.4. CPU Alarm (*CA*), DI/DO Alarm (*DA*) and *AIA* are used as fault symptoms of electrical faults. CA and DA come from the alarm information of PLC CPU and DI/DO channel respectively which is similar to *AIA*. According to the proposed method, a DBNs model for performance evaluation and fault diagnosis of the subsea BOP control system is established as shown in Fig. 4.26.

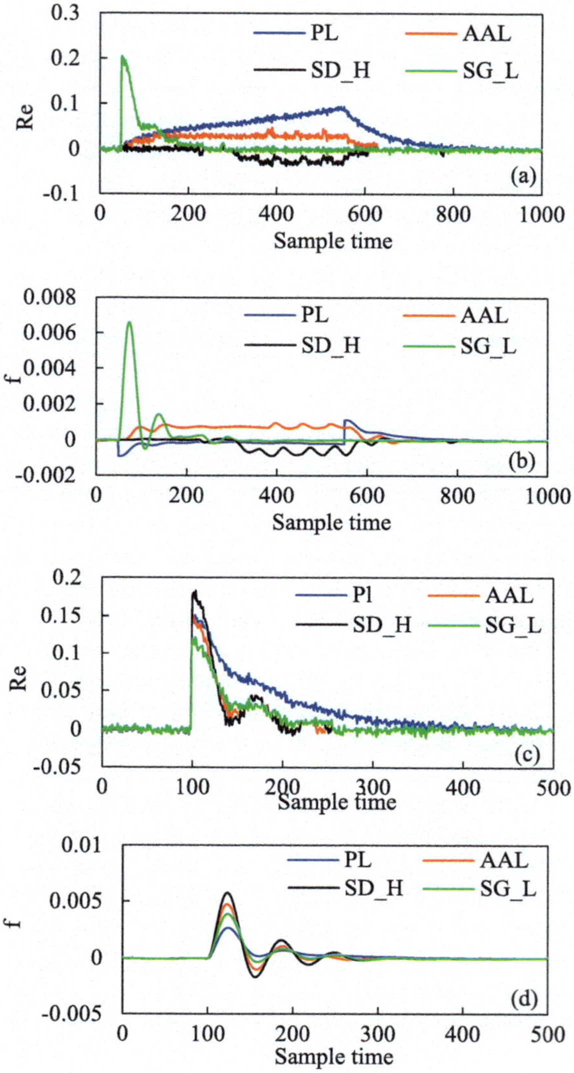

Fig. 4.24 Hydraulic system faults and sensor faults. **a** Re at fault happening. **b** Flow at fault happening. **c** Re in fault. **d** Flow in fault

4.4.3.2 Results and Discussions

(1) Fault diagnosis results

A total of 1800 sets of pressure and flow monitoring samples, including 200 sets of normal and 100 sets of each fault, are used for verification. The sampling frequency of the system is 100 Hz. Table 4.5 summarizes the results of fault diagnosis for all samples. The diagnostic accuracy of all samples reaches 95.94%. Among them, the

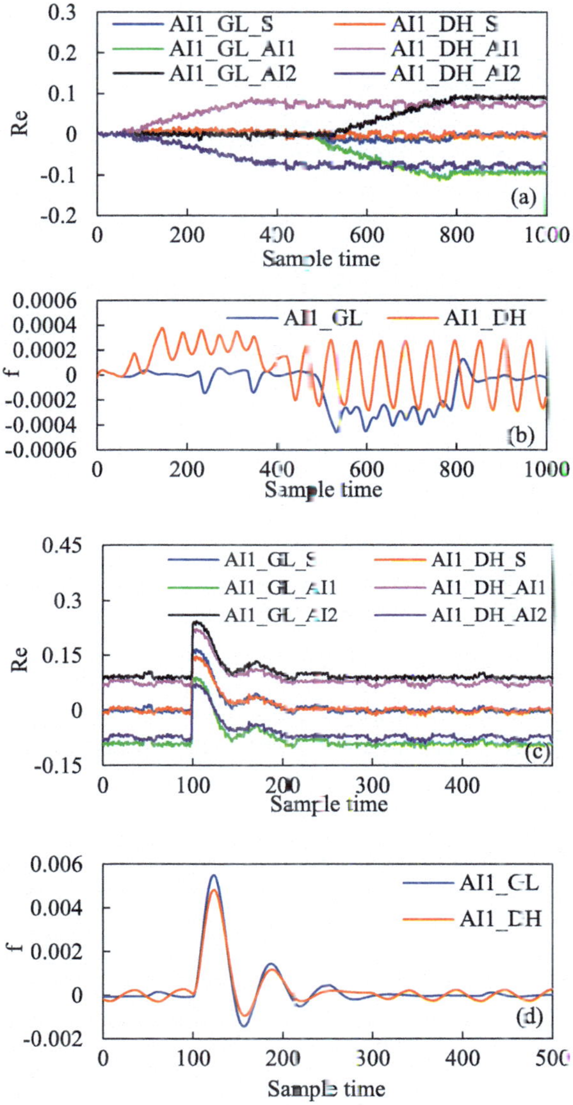

Fig. 4.25 Redundant electronic system faults. **a** Re at fault happening. **b** Re in fault. **c** Flow at fault. Happening. **d** Flow in fault

Table 4.4 Conditional probability table

Fault	CPU Alarm (CA)	DI/DO Alarm(DA)	AI Alarm (AIA)
Power Supply Failure (Pow)	Alarm	Alarm	Alarm
CPU Failure (CPU)	Alarm	Normal	Normal
DI/DO Channel Failure (DIO)	Normal	Alarm	Normal

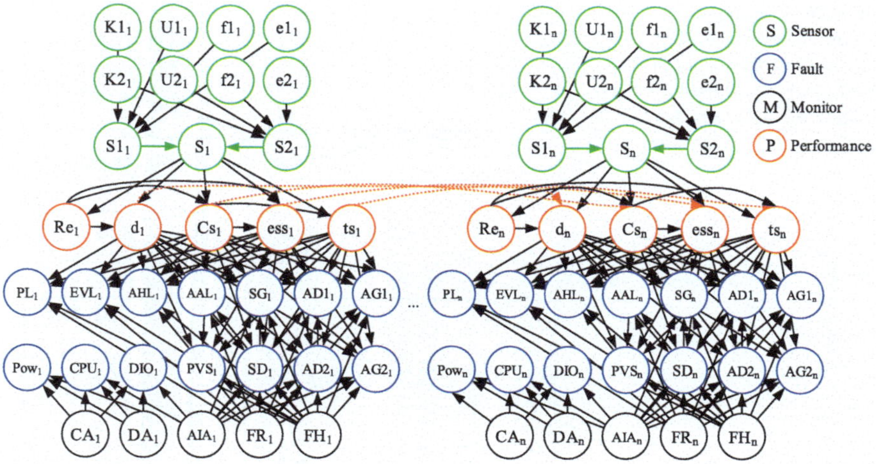

Fig. 4.26 DBNs and fault diagnosis of subsea BOP control system

hydraulic system faults, sensor faults and electrical faults diagnosis accuracy rates are 95.80%, 93.00% and 96.86% respectively.

(2) Influences of sampling frequency on fault diagnosis results

In this system, the same number of samples are obtained again by setting different sampling frequencies, and the number of misdiagnosed samples is shown in Fig. 4.27.

Table 4.5 Results of fault diagnosis of subsea BOP

Faults		Misdiagnosis	Misdiagnosis of other faults
Hydraulic system faults	PL	2	0
	AAL	0	0
	PVS	5	0
	AHL	7	7(From EVL)
	EVL	7	5(3 from AHL, 2 from normal)
Sensor faults	SD_H	11	5(From SG_H)
	SD_L	8	11(4 from SG_L, 4 from AHL, 3 from mormal)
	SG_H	5	0
	SG_L	4	0
	AG_H	4	0
	AG_L	3	0

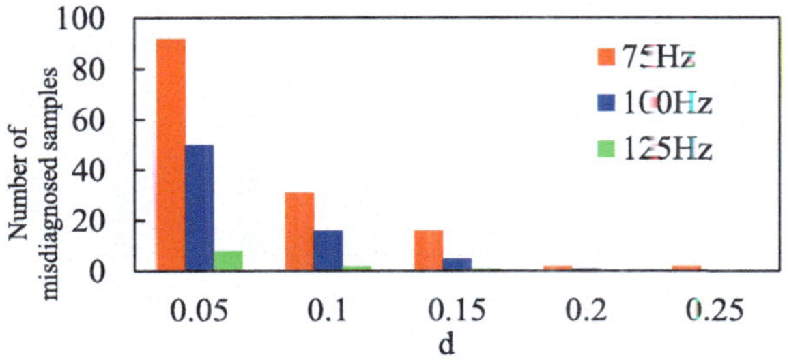

Fig. 4.27 Influences of sampling frequency

When the disturbance is relatively minor, the fault misdiagnosis rate increases significantly because the extraction of dynamic performance is greatly affected by noise interference. The sampling frequency can be increased to reduce the diagnostic accuracy of noise on sample information. The cost of increasing the sampling frequency is an exponential increase in the amount of calculation. The upper limit of the sampling frequency is restricted by the hardware parameters of sensor and PLC. At present the sampling frequency of sensors and PLC hardware used in mainstream subsea BOPs can provide the sampling frequency below 120 Hz.

(3) Influences of noise on fault diagnosis results

The performance of information monitoring and fault diagnosis is limited by the complicated working environment of subsea BOP. Input and output channels and sensor readings are usually disturbed by noise due to the electronic equipment interference and changes in the physical environment. In the previous results, according to the historical records, the energy of the noise distribution at the input and output channels of the control system is given as follow: $\sigma_{in} = 5 \times 10^{-6}$, $\sigma_{out} = 5 \times 10^{-6}$, and the energy of the noise distribution of sensor collection point $\sigma_s = 1 \times 10^{-6}$. These values are taken from historical data obtained from continuous monitoring of the system. The BOP control system works underwater, whose complex environment makes noise unstable. Therefore, studying the impact of the transformation between noise at different positions on the accuracy of diagnosis is necessary. The influence of the input channel, the output channel and the sensor noise in the diagnosis results are shown in Figs. 4.28, 4.29 and 4.30.

Fig. 4.28 Influences of noise at input channel

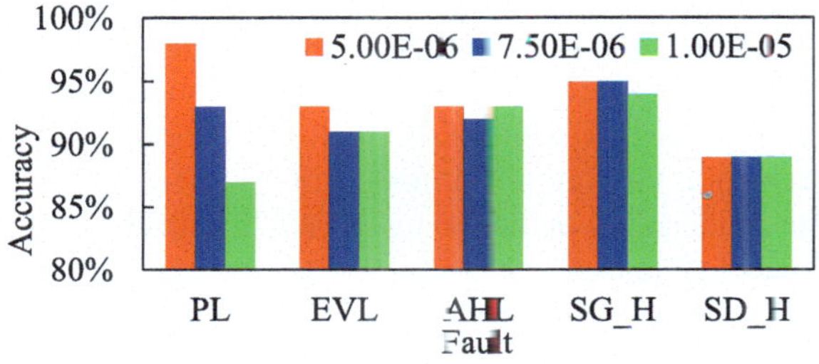

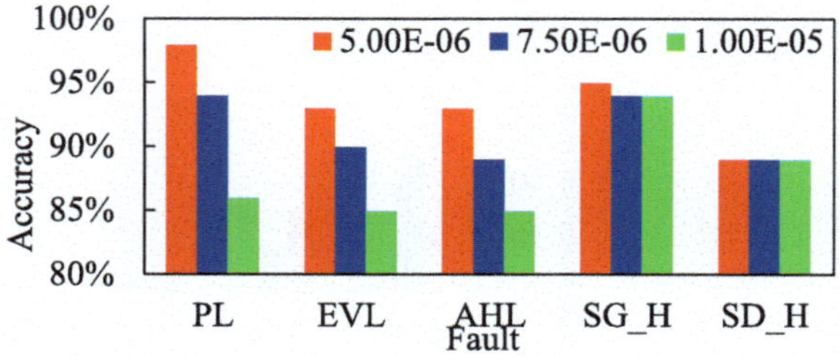

Fig. 4.29 Influences of noise at output channel

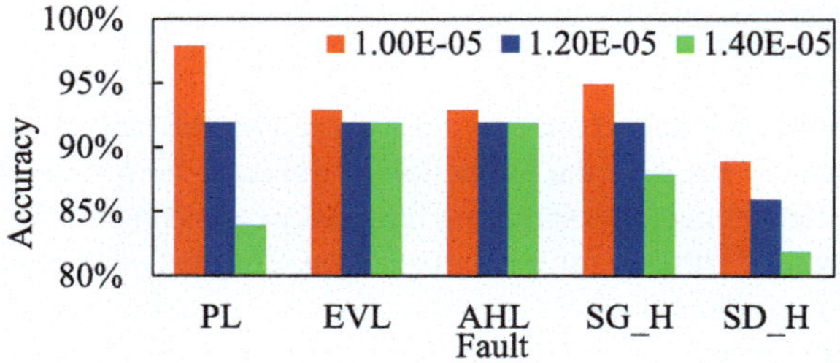

Fig. 4.30 Influences of noise at sensor

The diagnostic performance of the PL fault is significantly affected by all three types of noise. This is because pump leakage is the only fault that relies on adjustment time as a single isolated indicator. Compared to PL fault, EVL and AHL are more significantly affected by output channel noise. Because the noise at output channel caused oscillations in the flow monitoring, the accuracy of the identification of the two types of faults decreased. Sensor drift and gain are only affected by sensor noise. The noise at sensor has a certain impact on the stability of the entire system, which interferes with the identification of system performance. When the fault is relatively minor, the increase in noise also has an impact on the identification between drift and gain.

(4) Influences of redundancy level on fault diagnosis results

The simulation model used is a dual modular redundant control system (DMR), which has been widely used in mainstream BOP products. Due to the requirement of high reliability, some advanced redundancy methods have been proposed. For example, a BOP control system with a triple modular redundant controller as the core is designed (Cai et al. 2012). This method has higher reliability and is used in the 3000 m water depth BOP products of Rongsheng Machinery Manufacture Corp., Ltd., Huabei oilfield, Hebei, China. At the same time, Cai et al. also proposed a dual redundant control system (DDMR) model, established its reliability model and verified that this method has a higher reliability than that of TMR (Cai et al. 2012). In this work, the simulation models of TMR and DDMR are also established to verify the applicability of the proposed fault diagnosis method in these advanced redundant control systems.

The simulation models of TMR and DDMR are established, and four faults including SG_H, SD_H, AG_H and AD_H, which are closely related to the redundant

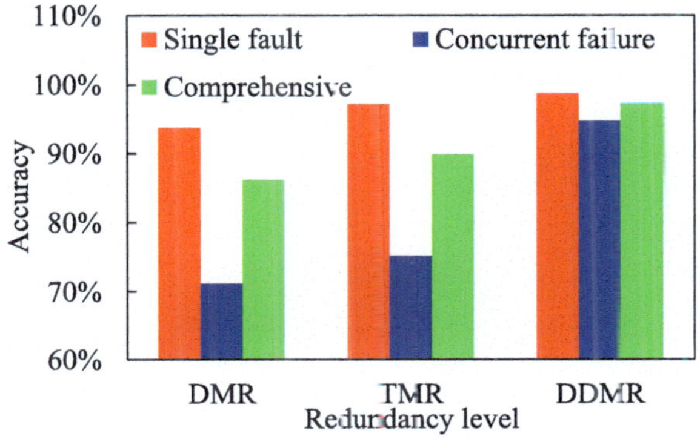

Fig. 4.31 Influences of redundancy level

system, are selected. At the same time, considering the concurrent faults caused by common cause faults of electrical components during system operation, the simulation obtained 400 sets of two-point concurrent fault samples. The diagnostic accuracy of these three systems is shown in Fig. 4.31.

The proposed fault diagnosis method has a fault diagnosis accuracy rate of more than 85% for three systems, including the concurrent faults. Specifically, for single faults, the accuracy of the diagnostic method for all three systems is above 90%, and the accuracy of TMR and DDMR systems can be increased to above 97%. For two-point concurrent faults, the diagnostic accuracy of the DMR system is only 71.25%. TMR and DDMR systems can effectively improve the diagnostic accuracy of the concurrent faults. There are many misjudgments which affect voting results and system's diagnostic results when two-point concurrent I/O channel drift and gain faults of DMR and TMR are diagnosed. The DDMR system is able to solve such problems effectively with its more advanced redundancy method. It not only maintains better stability of control functions when concurrent faults occur, but also improves the diagnostic accuracy of the concurrent faults to 94.75%.

Chapter 5
Concurrent Fault Diagnosis for Electric-Hydraulic System

Abstract Effective fault diagnosis of electrical control systems enhances both economic benefits and operator safety. However, the complexity of electro-hydraulic control systems, which consist of multiple components, poses a significant challenge for concurrent fault diagnosis. To identify faults with similar features accurately, A fault diagnosis method based on multi model fusion is proposed. OOBNs (object-oriented Bayesian networks) is used to establish the sub-models. The D-S evidence theory is used to establish the concurrent fault diagnosis and reasoning model. A hydraulic control system with multiple branches for subsea blowout preventers is used to study this method. The results shows that the method is effective in the fault diagnosis of faults with similar features.

5.1 Introduction

Due to the diverse types and numerous components of the electro-hydraulic system, there are complex interactive relationships among components. Influenced by mechanisms such as cascading failure and common-cause failure, compared with some overall systems, such as rotating equipment and reciprocating equipment, electro-hydraulic system faults may not appear in the form of a single fault. The phenomenon that faults have occurred simultaneously in two or more points is called a concurrent fault. The diagnostic model for a single fault is unsuitable for diagnosing of concurrent faults.

The fault characteristics of complex faults have a nonlinear coupling relationship (Amin et al. 2021), which means that when two or more faults occur at the same time, the system response generated by their interaction is not simply equal to the superposition of each single fault effect. For example, in an electro-hydraulic control system, the failure of one hydraulic component may cause pressure fluctuations, which in turn affect the working state of other related components; At the same time, the electrical control part may also cause misoperation due to the influence of abnormal signals, thus forming a complex fault scene with multiple factors interweaving and mutual cause and effect. In this case, the traditional diagnostic model based on single

B. Cai et al., *Intelligent Operation and Maintenance for Subsea Production Systems*,
https://doi.org/10.1007/978-981-96-2677-9_5

fault analysis may not be able to accurately identify and analyze such complex fault phenomena. Therefore, the single fault diagnosis model is not used to directly diagnose complex faults. The complex mechanism of concurrent faults has always made the diagnostic performance of concurrent fault improvement an important task.

The diagnostic strategy of concurrent fault is usually divided into two categories. One of these strategies is to convert the problem of concurrent fault into a dual classification problem of parallel (Ning et al. 2022). This strategy has limited performance when the complicated system is used as a diagnostic object. Another strategy is to treat concurrent faults as a new state of a node in reasoning (Xu et al. 2020). This strategy has greatly improved the number of space dimensions, leading to the difficulty of diagnosis time to meet the online diagnosis requirements.

In this chapter, the diagnosis problem is divided into multiple fault diagnostic sub-models based on analyzing the system structure and work process. The sub-model is established based on OOBNs (object-oriented Bayesian networks) for preliminary diagnosis and reasoning. The concurrent fault diagnosis and reasoning model is established based on the D-S evidence theory. The evaluated fault probability and diagnostic belief degree of each fault are combined. Fault identification rules combined with fault probability and diagnostic belief are established. A control system for subsea BOP is used as a case to demonstrate the proposed method.

5.2 Concurrent Fault Diagnosis Model

In this work, concurrent fault diagnosis problem is classified as non-independent parallel dual classification. The flowchart of the proposed concurrent fault diagnosis method of electric-hydraulic control system is shown in Fig. 5.1. Firstly, the diagnosis problem is divided into multiple fault diagnostic sub-models based on analyzing of the system structure and work process. The sub-model is established based on OOBNs for preliminary diagnosis and reasoning. The classification results are further processed to arrange combinations using possible fault combinations. Possible fault combinations are used to arrange combinations using possible fault combinations. The concurrent fault diagnosis and reasoning model is established based on D-S evidence theory. Evaluate the fault probability and diagnostic belief degree of each fault combination. Fault identification rules combined with fault probability and diagnostic belief are established.

5.2.1 Preliminary Diagnosis and Reasoning Sub-models Based on OOBNs

BNs is an important probabilistic graphical model that can effectively deal with various uncertainty problems based on probabilistic information representation and

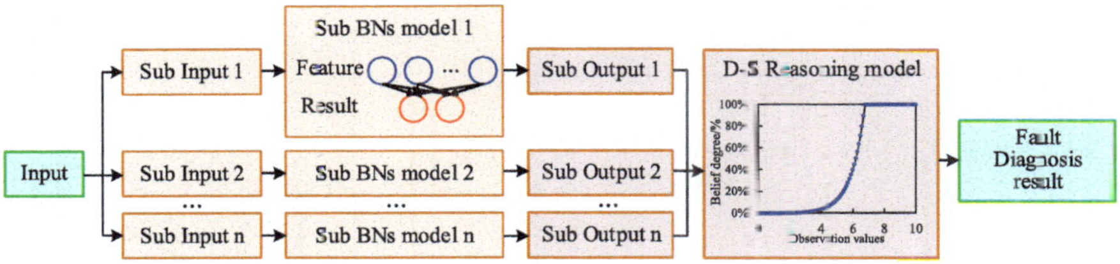

Fig. 5.1 The proposed concurrent fault diagnosis method

inference (Cai et al. 2017). Hence, this tool is deeply researched and widely used in fault diagnosis. Complex BN models often include copies of almost-identical network fragments. The occurrence of such repetitive structures is exploited during model construction. Similar to computer programming, a generic network fragment is called a class, and each network fragment produced by instantiating the class is called an object. The object-oriented approach performs excellent performance in some characteristics, including encapsulation, inheritance, polymorphism, and modularity. This approach is introduced into BNs and called OOBNs. Instead of specifying the same network fragment multiple times, a generic network fragment that instantiates the required number of times is constructed.

The proposed OOBN structure model is shown in Fig. 5.2 Generally, the fault reasoning of the same type of diagnostic object in the system is completed by the same diagnostic reasoning sub-model. Repeated inference BN model is encapsulated as an object (labeled *inst.*). Each sub-model consists of an input signal layer, a fault symptoms layer, a fault reasoning layer and a result output layer. The input signal layer is the input of the fault diagnostic sub-model. The information on the signal layer comes from the monitoring data of the electrical system and the signal of sensors in the hydraulic system. The monitoring data of the electrical system includes the status of the control point, system variable value, system alarm information and system monitoring information. Hydraulic systems' sensor signals usually include pressure signals and flow signals. Symptoms of faults are the quantitative expression of phenomenon caused by faults. These symptoms are the basis of fault reasoning. The node of the fault signal layer is obtained according to the input signal. Using extracted fault symptoms input to the model to perform reverse Bayesian reasoning. The node of the fault layer is a fault probability obtained by reasoning.

In the OOBNs, the operator can renew input signal layer based on the real-life situations in the on-line phase or automatically obtain it by an established communication interface. The prior probabilities of root nodes in each layer have to be specified. The primary and prior probabilities of faults symptoms nodes are obtained from expert knowledge in the off-line phase. Then, they can be updated by analyzing the input signal layer. The conditional probabilities of fault nodes are obtained from expert knowledge and historical sensor data in the off-line phase. In general, in fault diagnosis, the prior probabilities of all fault nodes are identical to emphasize posterior probabilities with new observations. The conditional probability of the fault node is obtained from the Noisy-OR and Noisy-MAX models.

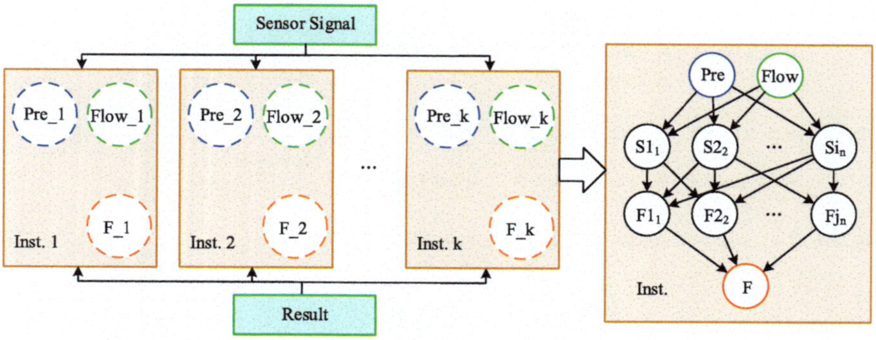

Fig. 5.2 The proposed OOBNs preliminary diagnosis model

When a node is Boolean type namely, the node has two states: normal and abnormal. The node status depends on data detected by sensors. The conditional probability table of a node can be simplified with a noisy-OR model. Specifically, supposing that several causes $X_1, X_2, \ldots, X_n$ lead to an effect Y. All nodes are Boolean and conditional probability table can be generated with only n parameters, and q_1, q_2, …, q_n are obtained by

$$P(Y = 1|X_1, X_2, \ldots, X_n) = 1 - \prod_{i=1}^{n} q_i \tag{5.1}$$

where q_i is the probability for each parent node, q_i can be expressed by

$$P(Y \leq y, X) = \prod_{i=1,x_{i=0},0}^{n} \sum_{y'=0}^{y} q_{i,y'}^{x_i} \tag{5.2}$$

where false Y when X_i is true, and all other parents are false. y' is the iteration variable in the accumulation process, $y' \leq$ Y.

Noisy-MAX is the general form of Noisy-OR in the non-Boolean type variables. A node has three states when non-Boolean, that is, higher, lower, and middle, with which the noisy-OR model cannot work. Suppose that non-Boolean variables X_1, X_2, …, X_n lead to the effect variable Y. The conditional probability table is established using the noisy-MAX model which is shown as follows.

$$P(Y = y|Xi = xi, Xj = 0[\forall j, j = i]) = q_{i,}^{xi} j = 1, \ldots, n \tag{5.3}$$

$$P(Y = y|X) = \begin{cases} P(Y \leq 0|X) & y = 0 \\ P(Y \leq y|X) - P(Y \leq y-1|X) & y > 0 \end{cases} \tag{5.4}$$

where X represents parent node configurations of Y; $X = x_1, x_2, \ldots, x_n$, and $P(Y = 0|X_1 = 0, \ldots, X_n = 0) = 1$.

All parametric model building is done offline. The conditional probability table is obtained by establishing a parametric model offline, which can be directly used for online fault diagnosis and inference. The result of the OOBN model is obtained in the form of probability. For different fault diagnosis systems, different probability thresholds should be developed correspondingly. Usually, the threshold is selected as 70–90% (Cai et al. 2017). In the proposed OOBN model, a conservative threshold (70%) is chosen to ensure that all possible situations are discussed in the next step. The fault obtained by the probability of more than 70% after reasoning can be considered as may have occurred and will be screened for concurrent fault analysis based on D-S evidence theory.

5.2.2 D-S Reason Modeling of Concurrent Fault Diagnosis

In Bayesian theory, parent nodes and sub-nodes can only be established as discrete or continuous variables after discrete processing. Sensor signals of complex electrical systems, especially pressure signals and flow signals, usually have complex continuous timing features. Disclosure of continuous signals can make these continuous timing features lost. At the same time, the discretization of continuous features can also make the probability reasoning of parent nodes and sub-nodes only be modeled into a linear relationship. The concurrent fault of the system is usually a non-linear superposition. These factors make OOBNs cannot be used directly to diagnose concurrent faults. In this part, D-S evidence theory is used for further evidence fusion to diagnose a concurrent fault when fully using the characteristics and non-linearity characteristics. After the information fusion of D-S evidence theory, the reliability and accuracy of the comprehensive diagnosis results are improved greatly. The D-S evidence theory is used for further evidence fusion to establish a concurrent fault diagnosis model which can fully utilizing the nonlinearity characteristics.

The concurrent fault reasoning model established based on D-S evidence theory is shown in Fig. 5.3. The concurrent fault reasoning method based on D-S theory evidence is as follows. Suppose that a set Θ includes all of the possible events. All elements of the set Θ are mutually exclusive and poor. Among them, 2^{Θ} is the power set of Θ, and A is an album of 2^{Θ}. Assume a function m: $2\Theta \rightarrow [0, 1]$ that meets the following conditions

$$m(\phi) = 0 \tag{5.5}$$

$$\sum_{A=\Theta} m(A) = 1 \tag{5.6}$$

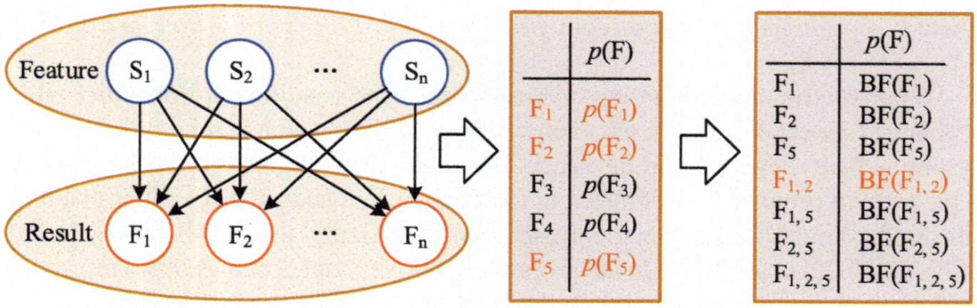

Fig. 5.3 Fault diagnosis modeling based on D-S evidence theory

where m can be called the belief function allocation of Θ, and A is a focal element of belief function allocation m. $m(\text{A})$ is the belief value of A, which is the degree of support for related evidence.

Assuming the faults $F1$, $F2$ and $F5$ are possible faults screened out of OOBNs. $p(F1)$, $p(F2)$, and $p(F5)$ are the probability of fault obtained by OOBN reasoning, respectively. The events contained in each fault label include two states: fault has occurred (F^f) and fault has not occurred (F^n), that is,

$$\begin{cases} F_1(F_1^f, F_1^n) \\ F_2(F_2^f, F_2^n) \\ F_5(F_5^f, F_5^n) \end{cases} \tag{5.7}$$

All possible fault combinations in concurrent fault reasoning are

$$F(F_1, F_2, F_3, F_1F_2, F_1F_5, F_2F_5, F_1F_2F_5) \tag{5.8}$$

Evaluate the belief degree of each fault combination, recorded as

$$\begin{cases} Bef(F_1) = BF_1 \\ Bef(F_2) = BF_2 \\ Bef(F_5) = BF_5 \\ Bef(F_1, F_2) = BF_{1,2} \\ Bef(F_1, F_5) = BF_{1,5} \\ Bef(F_2, F_5) = BF_{2,5}. \\ Bef(F_1, F_2, F_5) = BF_{1,2,5} \end{cases} \tag{5.9}$$

Because OOBNs are almost impossible to output if a fault has occurred under non-fault conditions, the empty set is not considered here to add possible events. That is

$$Bef(\phi) = 0 \tag{5.10}$$

The final diagnosis results are obtained by evaluating the function for various events obtained by a combination of arrangement. The distribution of the belief degree function of discrete variables can be obtained based on historical data and expert experience. The belief degree function of traditional evidence with continuous variables is usually a linear distribution. Affected by the fluid stagnation, this strategy does not apply to diagnosing the faults of hydraulic system. This work gives a new degree of belief evaluation function as follows.

$$\begin{cases} m(A) = \dfrac{p(Vo - Vr)}{1 + e^{-(Vo - Vr - q)}} \\ m(\dot{A}) = 1 - m(A) \end{cases} \tag{5.11}$$

where *Vo* is the observation values variable evidence, and *Vr* is the reference values variable evidence. *p*, *q* are optional parameters given by expert experience. The proposed belief degree evaluation function is shown in Fig. 5.4.

The evaluation function of the belief degree of concurrent fault is given as shown in Fig. 5.5. The belief degree of the evidence used for different fault diagnosis is evaluated. The belief degree function of a single fault couples the concurrent fault belief degree function. Based on this model, the evaluation results of the belief degree results of each evidence can be obtained.

Generally, the evidence of reasoning for fault is not a single piece of evidence, especially in diagnosing concurrent fault. Therefore, the belief degree functions are necessary to be fused by the selected evidence. The specific rules of evidence are shown below.

$$m(A) = \frac{1}{1-k} \sum_{A_i \cap B_j = A} m_1(A_i) m_2(B_j) \tag{5.12}$$

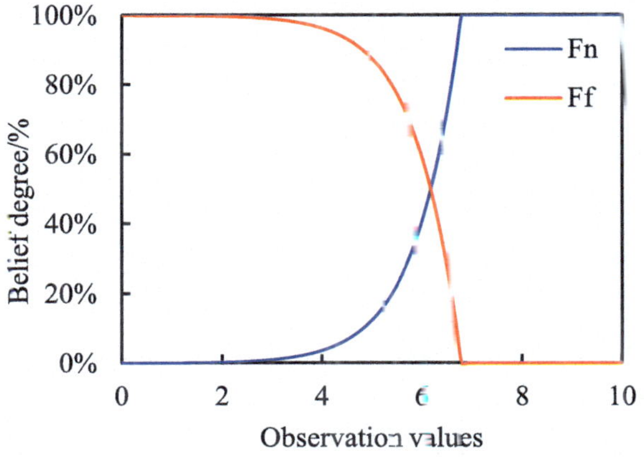

Fig. 5.4 The proposed belief degree evaluation function

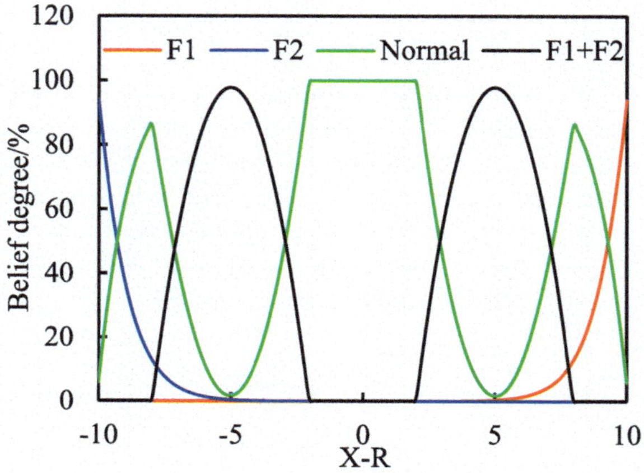

Fig. 5.5 The relationship between the observation value and the belief degree of the diagnosis result

where m_1 and m_2 are the original evidence that needs to be fused, m is the fused evidence, and A_i and B_j is the fault composition element.

5.2.3 *Fault Identification Rules Based on Belief Degree*

After the previous steps, we can get a quantitative single fault's probability and belief degree evaluation results. The strategy proposed in this section transforms these quantitative indicators into the result of fault reasoning for operators to decide. Generally, the probability fault diagnosis results are selected as the probability threshold as the criteria for fault identification. However, this strategy is limited when dealing with the complex situation of concurrent faults. Through several rounds of selection, testing, and optimization, the optimal threshold with higher diagnostic accuracy is selected as the fault identification rules. To cope with the concurrent fault diagnosis, fault identification rules, probability and diagnostic beliefs are established as follows.

(1) The fault determined has occurred when the probability that the state of nodes in the fault layer is a fault greater than 90% and the belief degree is greater than 80%.
(2) The fault determined has occurred when the probability that the state of nodes in the fault layer is fault greater than 80% and lesser than 90%, and the belief degree is greater than 90%.
(3) The fault determined may have occurred when the probability that the state of nodes in the fault layer is fault greater than 80% and lesser than 90%, and the belief degree is greater than 80% and less than 90%.
(4) The fault determined may have occurred when the probability that the status of nodes in the fault layer is fault greater than 70%, and the belief degree is greater than 30% & lesser than 80%.

For the 3rd case below, more indicators from evidence theory are used for further discussion. Support degree (SD), certainty degree (CD), decision degree (DD), and credibility degree (CRD) between evidence are used to further distinguish.

The definition of support degree SD is shown as follows

$$SD(i) = \sum_{j=1, j \neq i}^{n} sim(m_i, m_j) \tag{5.13}$$

where m_i, m_j are two belief degree evaluation functions of Θ, $sim(m_i, m_j)$ is the similarity degree and is shown as follows

$$sim(m_i, m_j) = 1 - difBetP_{m_i}^{m_j} \tag{5.14}$$

where $difBetP_{m1}^{m2}$ is the Pignistic probability distance and is shown as follows

$$difBetP_{mi}^{m_j}(A) = \frac{\sum_{A \subseteq \Theta} \left| BetP_{m_i}(A) - BetP_{m_j}(A) \right|}{2} \tag{5.15}$$

where $BetP_{m1}$ *and* $BetP_{m2}$ is the Pignistic probability function and is shown as follows

$$BetP_m(A) = \sum_{B \subseteq 2^{\Theta}, A \in B} \frac{1}{|B|} \frac{m(B)}{1 - m(\phi)} \tag{5.16}$$

The definition of certainty degree CD is shown as follows

$$CD(i) = \sum_{j=1}^{n} m_j(\theta_j) \tag{5.17}$$

The definition of decision degree DD is shown as follows

$$DD(i) = \left| \max(m_i(\theta_j)) - \min(m_i(\theta_j)) \right| \tag{5.18}$$

The definition of credibility degree CRD is shown as follows

$$CRD(i) = SD(i) + CD(i) \cdot DD(i) \tag{5.19}$$

This work defines a weight for evidence for the quantitative evaluation diagnosis and the correction results. The weight for evidence i can be defined as follows.

$$w(i) = \frac{CRD(i)}{\sum_{j=1}^{n} CRD(j)} \tag{5.20}$$

For the 4th case below, an extreme situation needs to be discussed. That is, there are two or more types of faults in the Θ to meet the proposed requirements. This is extremely rare and has not occurred in subsequent case studies. However, such a phenomenon may occur in theory. Therefore, the conflict of evidence evaluation is used to deal with this problem.

Suppose there are n pieces of evidence, the average weight of this evidence is recorded as $\hat{w}$

$$\hat{w} = \frac{1}{n} \tag{5.21}$$

When the weight of the evidence is less than $\hat{w}$, the evidence is considered conflict. A coefficient $K(i)$ is established to modify the pieces of evidence which have a conflict between them

$$K(i) = \frac{w(i)}{\hat{w}} \tag{5.22}$$

The method of using the correction coefficient for conflict correction is given as shown as follows

$$\begin{cases} m'(\theta_j) = K(i) \cdot m(\theta_j) \\ m'(\Theta) = 1 - K(i) \cdot (1 - m(\Theta)) \end{cases} \tag{5.23}$$

5.3 Case of Subsea BOP System

5.3.1 Analysis of the Cause of Concurrent Fault

An experimental platform built in this work is as shown in Fig. 4.12, which has been introduced in Sect. 4.4.1. The BOP control system is not in a working state most of the time. Because the hydraulic system has closed characteristics, the fault characteristics of the component cannot appear when they are not working. In order to ensure the high reliability operation of the equipment, regular testing is performed on the BOP system to prevent the system failure caused by faults. The fault of the BOP system is usually diagnosed when being tested. During the working of BOP system, the observation system for fault detection and diagnosis is composed of pressure sensors and flow sensors. Fault detection is done through comparing the difference between the measured values and with reference values. This work focuses on the fault diagnosis of its electro-hydraulic control system.

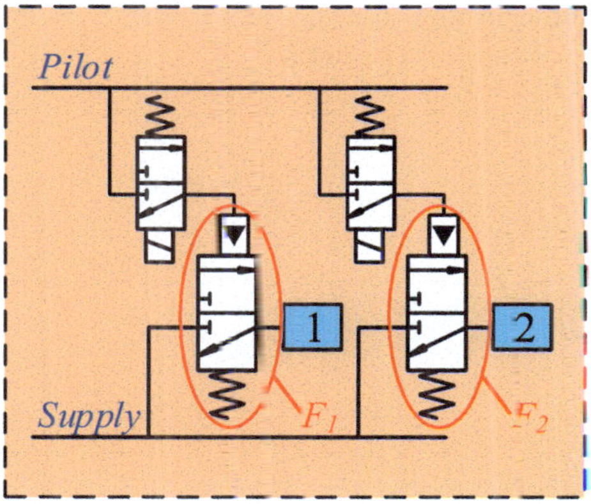

Fig. 5.6 Concurrent fault caused by common-cause failure (SPM_1 +SPM_2)

5.3.1.1 Concurrent Fault Caused by Common-Cause Failures

Common-cause failures occur when both redundant components fail from a single event. Common-cause failures often appear in systems with parallel structure and redundant structures. The BOP system is a typical system with multiple redundant and multi-channel characteristics. Because the BOP is the key safety grid to prevent blowout accidents and has a highly robust requirement, multiple modular redundant technology is widely used in its control system. Each link of the control process is designed to be redundant to ensure the robust requirements of the system. In addition, most of the control functions of the BOP system are completed by parallel components. Therefore, it is considered to be the most frequent cause of the BOP control system due to fault. A typical case of concurrent fault caused by common-cause failures is shown in Fig. 5.6. SPM_1 and SPM_2 are used to provide driving oil under work pressure for different BOP control functions respectively. During the system running, under the influence of factors such as external impact, pressure disturbance and abnormal load, SPM_1 and SPM_2 may have a synchronous degradation process. The phenomenon of SPM_1 and SPM_2 valves under this factor's influence is considered a concurrent fault caused by cascading failure.

5.3.1.2 Concurrent Fault Caused by Cascading Failure

In a multilevel system, a single fault may cause load changes of other components under complex conditions, and other related components are quickly destroyed. This phenomenon is called cascading failure. The cascading fault usually occurs in the following two situations. One situation is that the cascading effect occurred during the group's unsuccessful period. Another situation is that the cascading fault occurs in a short period of time. The subsea BOP control system is a typical multilevel system. During operation, control signals pass through the CPU, remote I/O, CCSVs, SPM

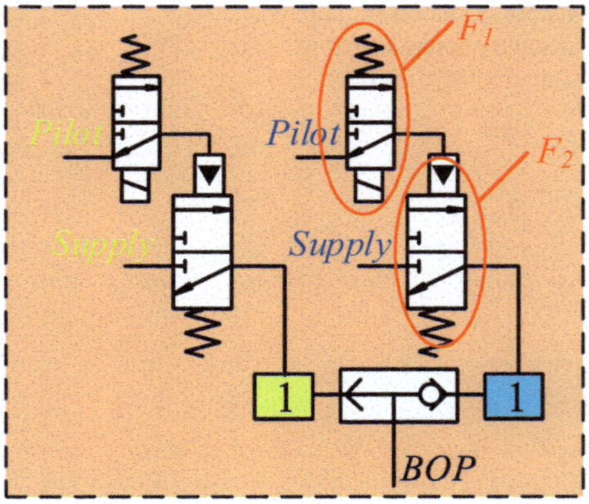

Fig. 5.7 Concurrent fault caused by cascading failure (CCSV+SPM)

valves, and shuttle valves. The signal form is passed through electrical signals, low-pressure oil signals, high-pressure oil signals, and finally converted into mechanical actions. A typical case of concurrent fault caused by cascading failure is shown by Fig. 5.7. CCSV and SPM valves are the most important components of the BOP control system. CCSV converts the electrical signal into a low-pressure pilot oil signal. The SPM valve converts the pilot oil pressure signal into a high-pressure working oil signal. When the coil of the CCSV receives a digital signal from the remote I/O of PLC, the magnetic force generates the valve core to move and open the CCSV. The pilot oil can flow into the SPM's driver. The valve core of the SPM valve moves driven by pilot pressure. The working accumulator device's pressure supply oil passes and performs the corresponding tasks. When CCSV has failed, its valve core cannot reach the specified position under the coil's drive. This causes a disturbance in the output of CCSV. The instability of the pilot signal causes the SPM valve to open and close repeatedly. The degeneration of the valve core of the SPM valve is accelerated. The phenomenon of CCSV and SPM valves under this factor's influence is considered a concurrent fault caused by cascading failure.

5.3.1.3 Concurrent Fault Caused by Independent Factors

Another factor that causes a concurrent fault is the independent factor of probability. The BOP control system is a complex multi-component system. The huge number of components makes the two components have to be considered simultaneously under different factors. The probability of this phenomenon in a concurrent fault is the lowest. A concurrent fault under the influence of an independent factor is shown in Fig. 5.8. PV_1 reduces the control of the control end of the liquid control pressure valve. DO_B belongs to one of the board cards of remote I/O. The cause of PV_1 and DO_B fault is not considered to have any correlation. However, there is a certain probability that PV_1 and DO_B will occur simultaneously. The phenomenon of

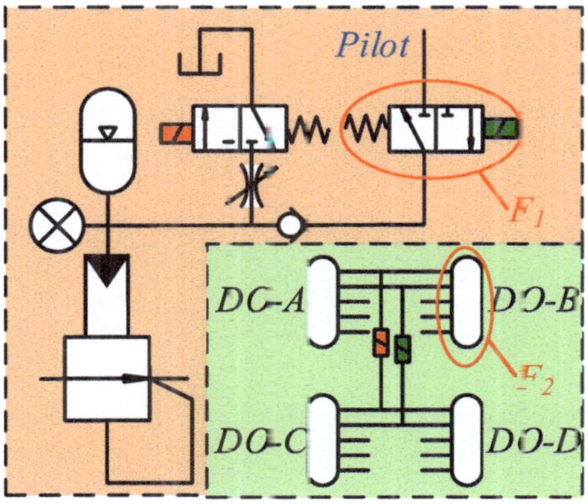

Fig. 5.8 Concurrent fault caused by independent factors (PV_1+DO_B)

PV_1 and DO_B valves under this factor's influence is considered a concurrent fault caused by independent factors.

5.3.2 *Results and Discussion*

5.3.2.1 Diagnosis Result of Single Fault

The diagnostic capability of a model for a single fault is discussed in this part. The three most common faults (CCSV, SPM valve and pilot accumulator) are selected to demonstrate the proposed method. Thirty samples were randomly selected from each defective species. The diagnosis result of the CCSV fault is shown in Figs. 5.9 and 5.10. According to the established fault identification, the fault of the 29 samples was diagnosed correctly. As shown in the figure, the fault probability of the sample 15 was 73.29%, and the belief degree of the fault has been occurred was 65.88%. It still needs to meet all the rules diagnosed with the sample. As a result, sample 15 is considered as a misdiagnosis. A more in-depth analysis of the fault signal reveals that the sample signal is affected by an external impact. CCSV operates under relatively low pilot pressure. The relative errors caused by noise and impact are relatively large, which is the main reason for the misdiagnosis of CCSV.

The result of the CCSV fault diagnosis is shown in Figs. 5.11 and 5.12. It can be seen from the figure that the probability of faults from all samples obtained by reasoning is greater than 85%. The confidence level of all diagnosis results is greater than 90%. According to the establishment of fault identification, the fault of all 30 samples is accurately diagnosed due to the high working pressure of the SPM valve. In a subsea BOP control system, the working pressure is generally 5–10 times higher than the pilot pressure. Therefore, the impact of the pressure signal used to diagnose the SPM valve is weak.

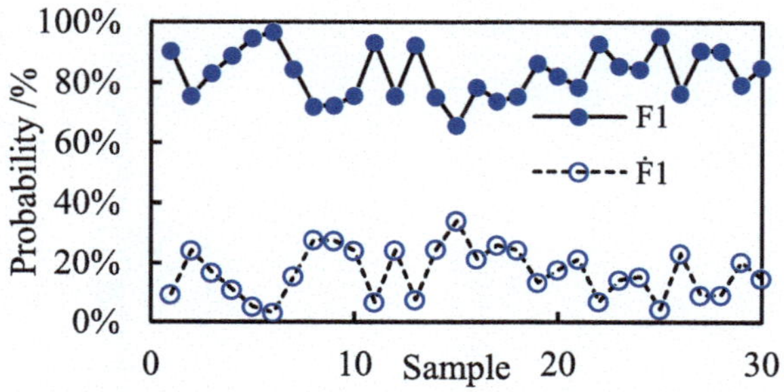

Fig. 5.9 CCSV fault probability

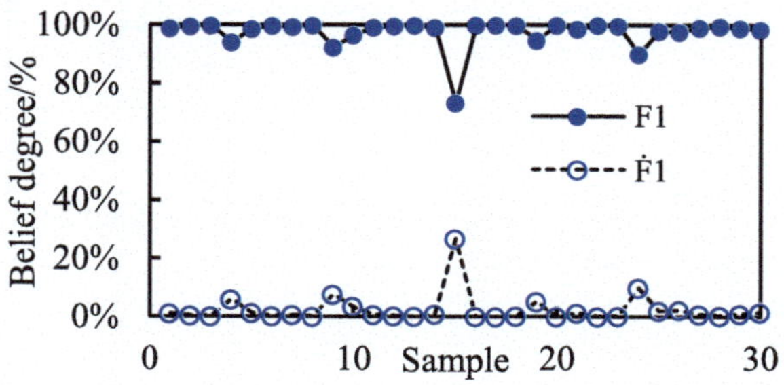

Fig. 5.10 Belief degree of CCSV fault

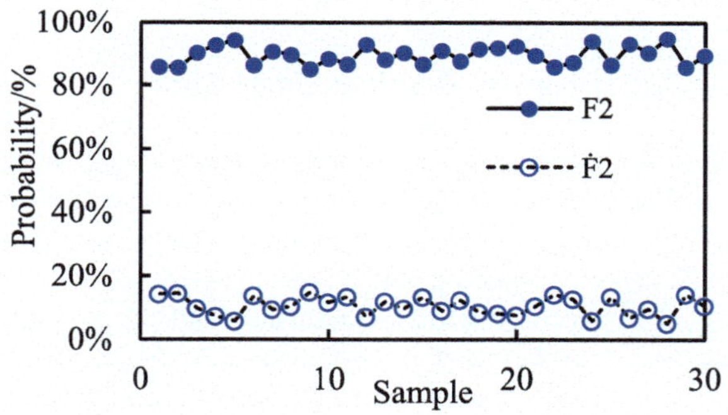

Fig. 5.11 SPM valve fault probability

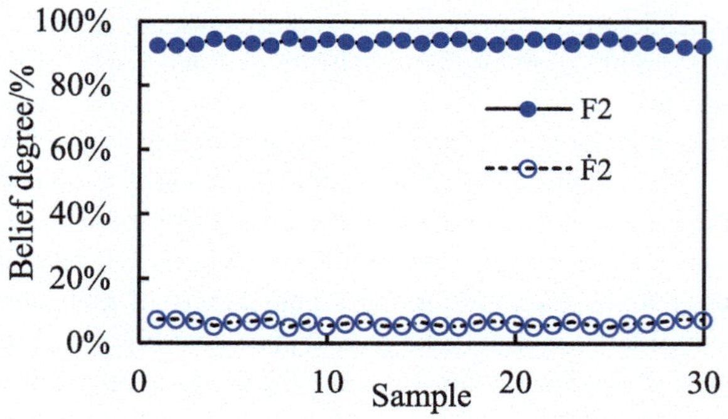

Fig. 5.12 Belief degree of SPM valve fault

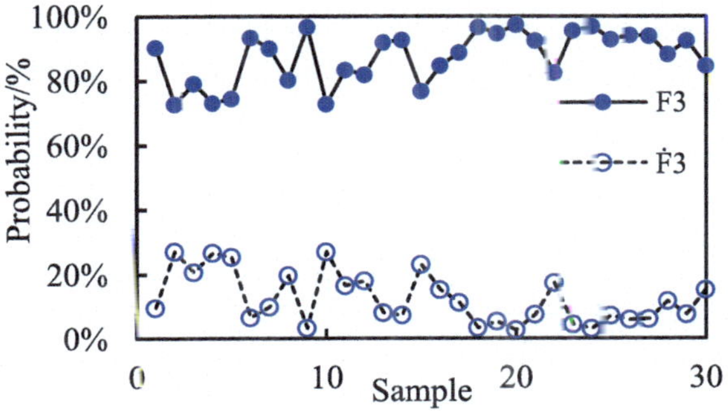

Fig. 5.13 Pilot accumulator fault probability

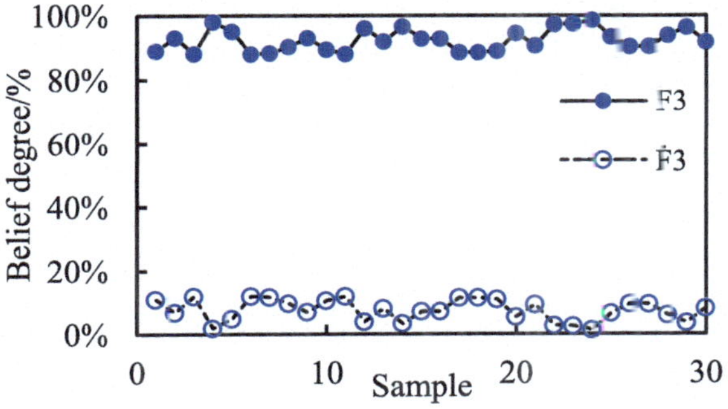

Fig. 5.14 Belief degree of pilot accumulator fault

The result of the pilot accumulator fault diagnosis is shown in Figs. 5.13 and 5.14. It can be seen from the figure that the probability of faults from all samples obtained by reasoning is greater than 85%. The belief degree of all diagnosis results is greater than 80%. According to the established fault identification, all the faults of the 30 samples is accurately diagnosed.

5.3.2.2 Diagnosis Result of Concurrent Fault

This part discusses the diagnostic ability of a model for a single fault. Three concurrent faults mentioned in Sect. 5.3.1 are selected to demonstrate the proposed method. Diagnosis results caused by common-cause failure (CCSV_1+ CCSV_2) are given by Figs. 5.15 and 5.16. It can be seen from the figure that the fault of only two samples has yet to be diagnosed correctly due to fault caused by common-cause failure, a branch path or a redundant component in the system. These components are independent monitoring and diagnosis. The coupling relationship between concurrent fault characteristics is weak. At the same time, a large probability of priority may be set up by a concurrent fault caused by the common-cause failure. Therefore, the proposed model performs excellently when diagnosing a concurrent fault caused by common-cause failure.

Figures 5.17 and 5.18 present the diagnosis results caused by a cascade failure (SPM valve+CCSV). It can be seen from the figure that the fault of only six samples is correctly diagnosed. The concurrent fault caused by cascading failure of class disorders typically occurs when completing the same task collaboratively. Therefore, the component components have a common monitoring system. Hence, the features of the concurrent fault caused by the cascading failure are coupling strongly. The use of evidence theory effectively explains this coupling. Fault characteristics can be compensated after a few cases by analyzing misdiagnosis samples. Given the constraints of underwater space and information transmission capacity, only a limited number of observers are used to monitor underwater systems. Increasing the new observer node can effectively reduce diagnostic errors. Therefore, the concurrent fault caused by cascading failure can be considered a factor during sensor optimization. Some related work on optimal sensor placement has been confirmed to help further improve the performance of fault diagnosis (Zhang et al. 2022; Kohtz et al. 2024).

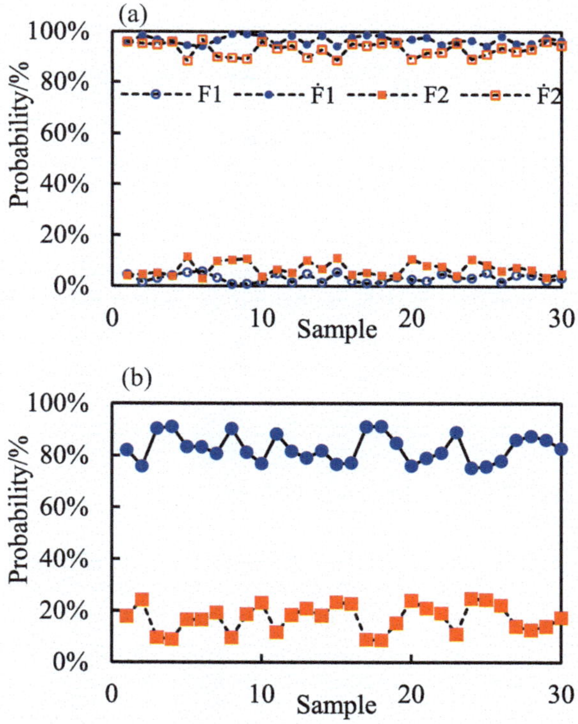

Fig. 5.15 Fault probability of fault CCSV_1+CCSV_2. **a** Single fault: F1- CCSV_1, F2-CCSV_2; **b** concurrent fault: CCSV_1+CCSV_2

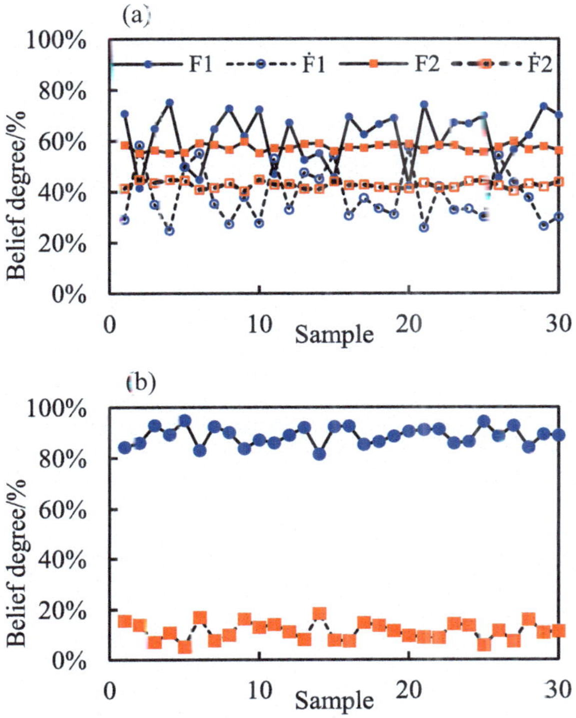

Fig. 5.16 Belief degree of fault CCSV_1+ CCSV_2. **a** Single fault: F1- CCSV_1, F2-CCSV_2; **b** concurrent fault: CCSV_1+CCSV_2

The diagnosis results caused by independent factors (PV_1+DO_B) are shown in Figs. 5.19 and 5.20. It can be seen from the figure that the fault of all samples is correctly diagnosed. In most cases, concurrent components caused by random factors are independent. The established observations were also completed independently. The characteristics of the faults are completely independent. Therefore, the concurrent fault caused by a random factor is equivalent to a single-point parallel fault diagnosis. This type of concurrent fault is considered the easiest to diagnose.

5.3.2.3 The Accuracy of the Proposed Method

In this work, there are 23 kinds of single-point faults, 12 kinds of concurrent faults caused by common-cause failure, 12 kinds of concurrent faults caused by cascading failure, and 12 kinds of concurrent faults caused by independent factors simulated in the experiment. With historical data as a reference, the number of samples of four types of faults is 200, 100, 60, and 20, respectively. The distribution of the number of samples is similar to the frequency of faults in engineering records. The

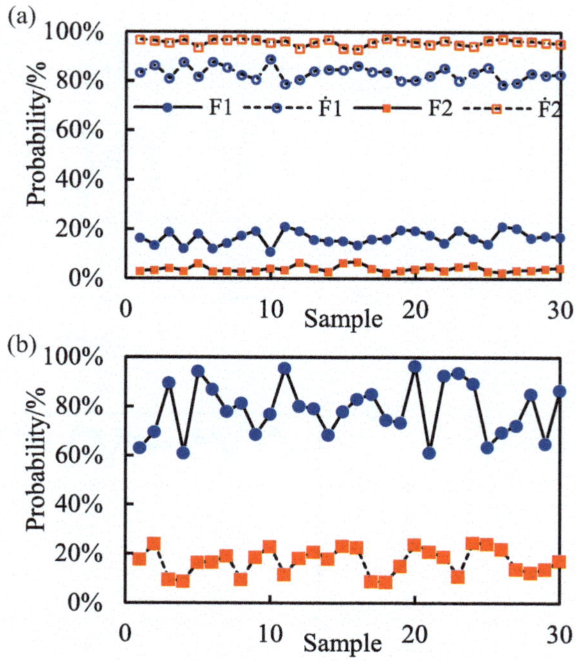

Fig. 5.17 Fault probability of fault SPM+CCSV. **a** Single fault: F1-SPM, F2-CCSV; **b** concurrent fault: SPM+CCSV

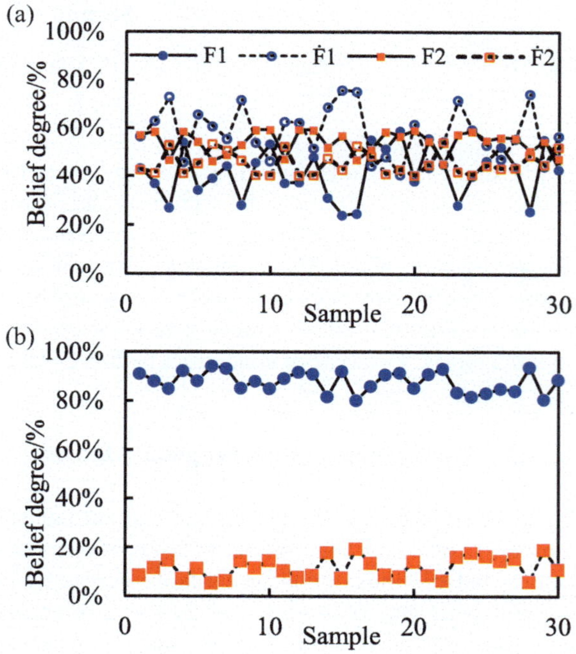

Fig. 5.18 Belief degree of fault SPM+CCSV. **a** Single fault: F1-SPM, F2-CCSV; **b** concurrent fault: SPM+CCSV

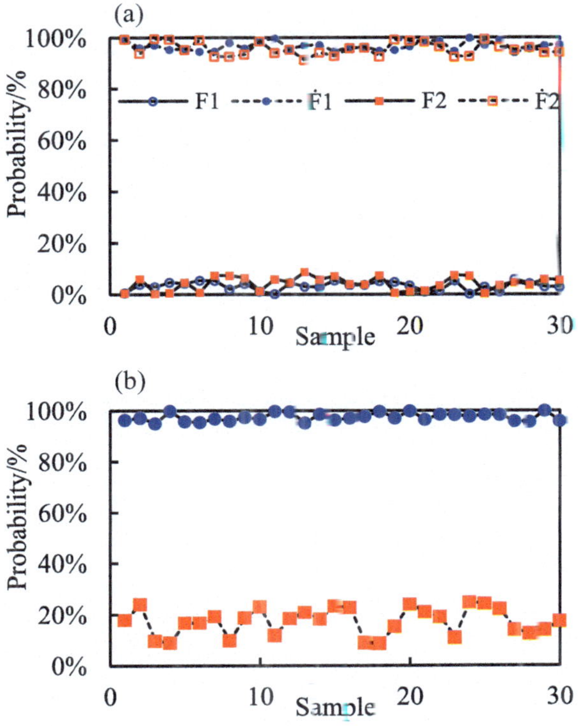

Fig. 5.19 Fault probability of fault PV_1+DO_B. **a** Single fault: F1- PV_1, F2- DO_B; **b** concurrent fault: PV_1 +DO_B

accuracy of the fault diagnosis of all samples is shown in Fig. 5.21. The accuracy of the diagnosis of single-point fault is 95.30%. The accuracy of the diagnosis of concurrent faults is 92.13%. Among them, the diagnostic accuracy of concurrent faults caused by common-cause failure, cascading failure and independent factors is 92.75%, 90.42%, and 94.17%, respectively. In general, the diagnosis accuracy of all faulty samples is 94.29%. The proposed method shows excellent accuracy performance during the diagnosis of faults, especially the diagnosis of concurrent faults.

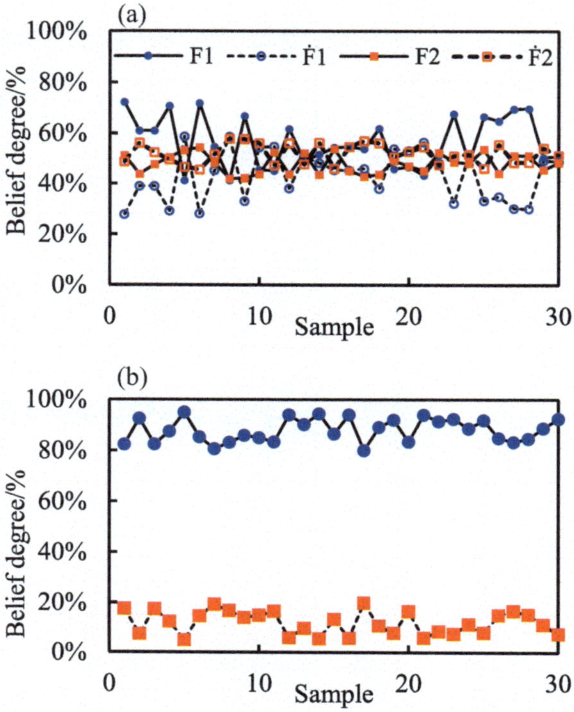

Fig. 5.20 Belief degree of fault PV_1+DO_B. **a** Single fault: F1- PV_1, F2- DO_B; **b** concurrent fault: PV_1+DO_B

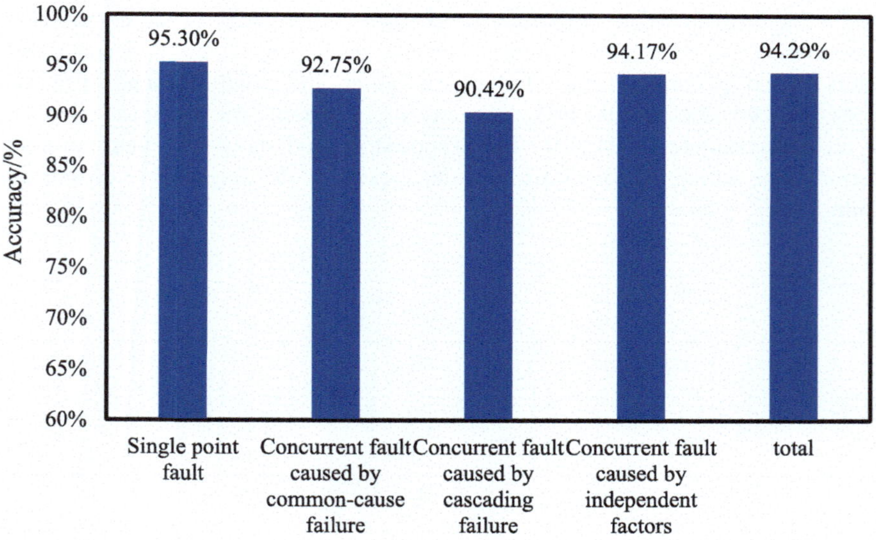

Fig. 5.21 The accuracy of the fault diagnosis of all samples

Chapter 6
Intelligent Fault Diagnosis for Subsea Production System

Abstract To get a fault diagnosis system with long-term stable performance, an intelligent full-stage stable fault diagnosis method for subsea production system is proposed. A data-driven fault diagnosis model and a model-based fault diagnosis model are simultaneously established. Among them, the model with better performance is used in the current stage of fault diagnosis. The digital twin model is used to provide virtual data to data-driven models, which shortens the training cycle of the model. The data from the South China Sea is used to study the performance of the model. The results showed that the model-based diagnostic model has better performance when there is insufficient training data, and better performance when there is sufficient data. The results also show the long-term stable diagnostic performance of the proposed method.

6.1 Introduction

As mentioned before, the data-driven method and the model-based method demonstrates show their pros and cons. For example, the data-driven method boasts adaptability due to its up-to-date training data, but its performance suffers when training data is lacking. On the other hand, the model-based method demonstrates strong diagnostic capabilities initially, but its effectiveness diminishes as the system degrades. During the early stages, the system is stable and reliable with minimal faults and insufficient fault data, resulting in limited training data available. Typically, a subsea Christmas tree is the core component of a subsea production system, which is designed to have a service life of 30 years underwater. Over its service period, the system performance undergoes changes, which can result in a gradual decline in the effectiveness of traditional fault diagnosis methods. Therefore, a full-stage stable diagnosis method is required by a subsea production system.

To maintain consistent fault diagnosis performance over an extended period, a combined approach using both data-driven and model-based methods is required. For long-term stable monitoring of subsea production systems, a fault diagnosis framework that integrates both digital and model-driven techniques is introduced in

B. Cai et al., *Intelligent Operation and Maintenance for Subsea Production Systems*,
https://doi.org/10.1007/978-981-96-2677-9_6

this chapter, alongside a dynamic digital twin model to expedite the training process. A real-world case study from the South China Sea is used to evaluate the effectiveness of the proposed method, and the results indicate that it maintains stability over a long duration.

6.2 Intelligent Full-Stage Stable Fault Diagnosis Method

The intelligent full-stage stable diagnosis method is shown in Fig. 6.1. In the picture, V is valve of subsea production system, P is pressure transmitter, H is hidden layer of the data-driven fault diagnosis model, AD is accuracy of data-driven fault diagnosis model and AM is accuracy of model-based fault diagnosis model. In the process of oil and gas production, the subsea production system cannot be observed. In addition, it takes improvements to the fault diagnosis model. On the one hand, it provides a display tool for results of fault diagnosis. On the other hand, more data is provided for model training.

In Fig. 6.1, the data usually include the operation data such as the data of sensors, the state of switch and the operation state. A digital twin model should be connected to the physical system to get the data timely. The feedback in Fig. 6.1 usually refers to the result of fault diagnosis. In this chapter, the service process of the subsea production systems is divided into two stages, the early stage and the late stage. In the early stage, the subsea production system is stable and there are few failures. In this case, the performance of data-driven model is poor due to a lack of data for training, especially fault data. In late stage, the training data is sufficient. This can lead to better performance of data-driven method. In fact, the division of these two stages is not strict because the performances the systems are different. Usually, the accuracies of the two fault diagnosis methods in the past year are calculated. When the accuracy of the data-driven method is 5% larger than that of model-based method. It can be considered that the second stage begins. It is not difficult to see that, the model-based methods are better in the early stages, while data-driven methods are better in the later stages. This fault diagnosis method attempts to get a long-term stable performance throughout the service process.

6.2.1 The Method of Establishing a Digital Twin Model

The fluid flows are controlled by the valves and guidance of the pipeline in the hydraulic system. A model for describing the fluid state of hydraulic systems is a necessary prerequisite for establishing a digital twin. Bernoulli Equation is usually used to calculate the fluid state and the performance of it is excellent (Li et al. 2023). To simulate a subsea production system dynamically, three parameters should be included in the digital twin model, that is pressure loss, control, and fault parameter. These parameters are updated in real-time to ensure that the digital twin model is

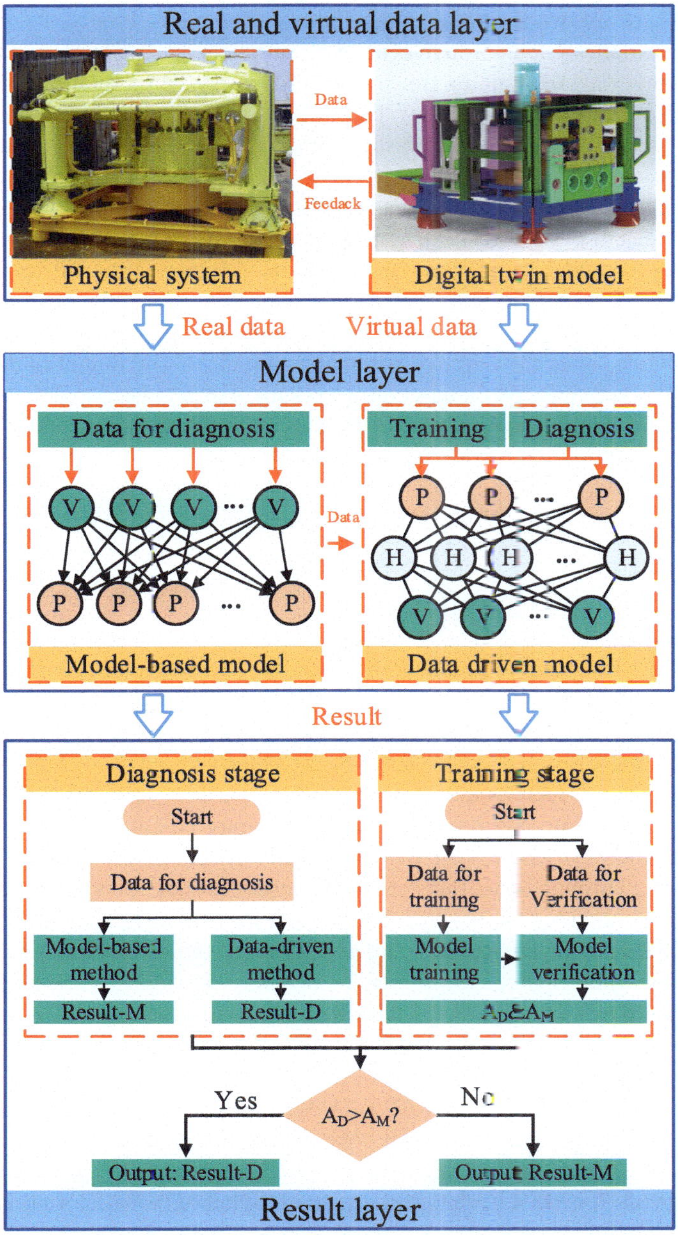

Fig. 6.1 Intelligent full-stage stable diagnosis method

accurate. There are two functions of the digital twin models. Firstly, it can simulate the operation of the system through generating sufficient data for training data-driven models. In this way, the training time can be reduced. When a sudden change leads to a decrease in the performance of data-driven methods, digital twin models can help it achieve good performance faster. Another function is to display diagnostic results and system data.

Two circuits related to fluid are included in a subsea production system. One is the production circuit. It used for transporting subsea oil and gas. They are complex and difficult to model using Bernoulli equation. Another is the subsea hydraulic control circuit. For environmental protection, seawater is hydraulic medium of most subsea production systems. Their function is to provide power for valve switches. When modeling them, the required parameters are clear. The digital twin model is established to simulate hydraulic control systems. For a subsea hydraulic system, the valve is the boundary point, and the state of its front and rear circuits is updated through the Bernoulli equation. For a certain valve, the digital twin model before and after the valve is shown in (6.1):

$$p_i + \rho g h_i + \frac{1}{2}\rho v_i^2 = \left(p_{i-1} + \rho g h_{i-1} + \frac{1}{2}\rho v_{i-1}^2\right)\zeta_{i-1}\tau_{i-1}\varphi_{i-1} \tag{6.1}$$

where p_i is the pressure of the i th valve, ρ is the density of fluid, g is acceleration of gravity, v_i is flow rate of fluid near the ith valve. ζ, τ, φ are loss, control, and fault parameters, respectively.

It can be seen from (6.1) that ζ, τ and φ are added to the Bernoulli Equation. In fact, local pressure loss and the pressure loss along the way are needed in a traditional Bernoulli Equation. However, it is impossible to get them from the field. Because they change over time and location, and they cannot be calculated or measured. During the process of production, the mentioned loss, control, and fault parameters are important. Through them, the state of the subsea production system can be well simulated. It is notable that two characteristics of the established digital twin model makes it different from traditional simulations. Firstly, these parameters are adjusted in real-time, which updates the digital twin model dynamically. Secondly, these changes are not pre-set, they change with changes in the actual system. Among these parameters, ζ is loss parameter and it can be calculated with data during normal operation. In fact, a fixed ζ is proposed for the construction of digital twin models in existing work (Attaianese et al. 2023). A dynamically changing ζ can reduce the error of the digital twin model and improve the performance of fault diagnosis. τ is the control parameter, and the calculation method is shown in (6.2). φ is the state of the local circuit. It is used to change the state of digital twin model. In this way, the digital twin model can run in an abnormal state to follow an abnormal state of the physical system.

$$\tau = \begin{cases} 0.9 + \varepsilon & \textit{the valve is opened} \\ 0.2 - \sigma & \textit{the valve is closed} \end{cases} \tag{6.2}$$

where ε and σ are adjusting parameters used to fine-tune the control parameters under different conditions. Their change will not exceed 0.05, which is very small compared with the control parameters. The adjustment parameter is mainly used to describe slight changes in the system. The conditions of subsea are much more complex due to high pressure and the compressibility of the involved fluids. Due to the multiplication of three parameters, small changes can cause significant changes in the model. To avoid over adjustment, only one parameter is fine tuned.

Among the three parameters, the control parameter has almost only two states, 0 or 1. Because the switch of the valve is Boolean controlled. The fault parameters are updated by the fault diagnosis model, and they are initially set to 1, defaulting to no faults occurring. In this case, the loss of parameters is easily obtained by calculating the average value. I fact, in actual working conditions, the data of the system is obtained in real time. Every time the actual data is obtained, the three parameters of the digital twin model are updated. That allows the digital twin model to dynamically simulate the actual system. The update frequency of digital twin model depends on the actual data collection frequency.

According to the current digital twin model, virtual model, physical model and their connections should be included (Dui et al. 2023). For subsea production systems, the physical model and connection are clearly defined. Virtual model should include virtual display models and dynamic simulation algorithms. In this work, the dynamic simulation algorithm is mainly used to interact with the fault diagnosis model, so it is introduced.

6.2.2 Model-Based Fault Diagnosis Method

Subsea production systems in different regions are almost completely different, which results in that the data from a subsea production system cannot be used for another one. In this case, the data used for training a model is insufficient when a subsea production system is first put into service. In addition, significant changes usually occur after major maintenance or relocation. This can also lead to a decrease in the performance of the data driven fault diagnosis model. Model based fault diagnosis methods can ensure effective diagnostic performance at this time.

For a subsea production system, the diagnosis model based on fault tree is commonly used. With the development of diagnosis methods, BN replaces it. Because the output results of it are expressed in the form of probability (Woo et al. 2023). The results characterize the likelihood of the failure of a component. In this chapter, it is a good indicator for evaluating φ. Usually, structure modeling, parameter modeling, and fault reasoning are main steps for establishing a BN-based fault diagnosis model (Cai et al. 2017).

Figure 6.2 shows a structure of a fault diagnosis model. In a subsea production system, the pressure transmitters (PT), and control valves (CV) are the main components. Through control system, the values of PT are got and the fault state of CV should be judged through them. In a fault diagnosis model based on BN, causality is the basis for structural modeling. Changes of state of CV can cause pressure changes. Thus, the CVs should be the parent node and the PT should be the child node. What's more, each node contains two states: normal (N) and abnormal (A).

Parameter modeling is the second step in establishing a fault diagnosis model based on BNs. By giving the prior probabilities and the condition probabilities, the connection between the parent nodes and child nodes is established. Obtaining by experience and training are two methods that are usually be used. The model-based method is used when the training data is insufficient. Thus, it can not be trained. Usually, the drawings are easily obtainable. Through them, the control process is clear. In this case, the conditional probability is easily obtained (Dui et al. 2024). In fact, prior probabilities have little impact on the model. They would be updated when the information is input into the fault diagnosis model. And then the posterior probability is calculated by combining the conditional probability. In this chapter, the prior probability of failure is 5%, and the normal probability is 95% because the probability of valve failure is low. According to the relationship of hydraulic circuit components, the probability of an associated abnormal PT is 95% when a valve fails.

After the diagnosis model is constructed, it can be used for fault diagnosis. For a node X with parent nodes U and children nodes Y, the conditional probability from U to X is defined as $\pi_X(U)$, and the conditional probability from Y to X is defined as $\lambda_Y(X)$. For some known evidence E, the joint marginal for the family of X with the parent nodes U_i and children nodes Y_i is given by.

$$P(XU, E) = \lambda_E(X)\Theta_{X|U} \prod_i \pi_X(U_i) \prod_j \lambda_{Y_j}(X) \tag{6.3}$$

where $\lambda_E(X)$ is evidence indicators and $\lambda_e(X) = 1$ if $x \epsilon X$ is consistent with evidence $e \,\epsilon\, E$ and zero otherwise.

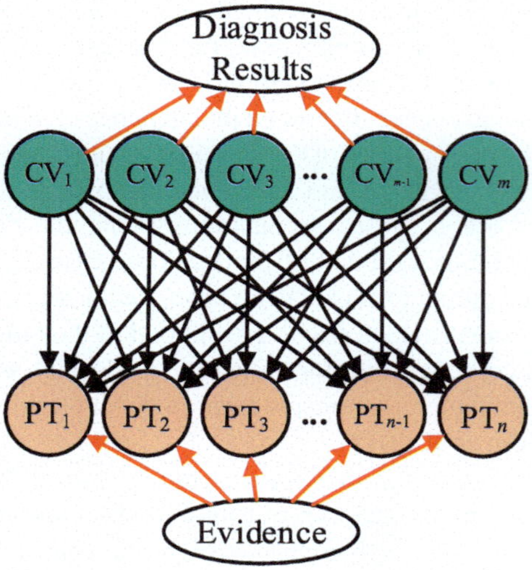

Fig. 6.2 Model-based fault diagnosis model

When a posterior probability is obtained, it is assigned to φ as:

$$\varphi = \begin{cases} 1 - P(CV) & P(CV) \geq 0.65 \\ 0.9 + \kappa & P(CV) < 0.65 \end{cases} \tag{6.4}$$

where κ is the adjusted parameter for fine-tuning the digital twin model, and $P(CV)$ is the probability of the corresponding valve failure.

6.2.3 Data Driven Fault Diagnosis Method

The characteristic of BRNN is that it can use a small amount of data to obtain a model. This is suitable for subsea production system where there is less data. In fact, when there is sufficient data, the diagnostic performance is also good. In this method, the training data accumulates over time. The less training data is required, the earlier the model can meet the usage standards, that is, the accuracy is higher than model-based methods (Dui et al. 2024). With the operation of the subsea production systems, data becomes sufficient and can be used for training fault diagnosis models. Compared to model-based method, the data-driven method can dynamically update the training database to ensure good dynamic diagnostic performance of the model (Liu et al. 2024). Bayesian Regularization Neural Network (BRNN) is used to establish a data driven fault diagnosis model and the structure is shown in Fig. 6.3. The value of PT can be got form the control system. Through them, that state of CV is got. Thus, in this fault diagnosis model, the PT is input node and the CV is output node.

In a logic regression, the maximum likelihood method is used to estimate the parameters to obtain the cost function. It is believed that the value of θ should

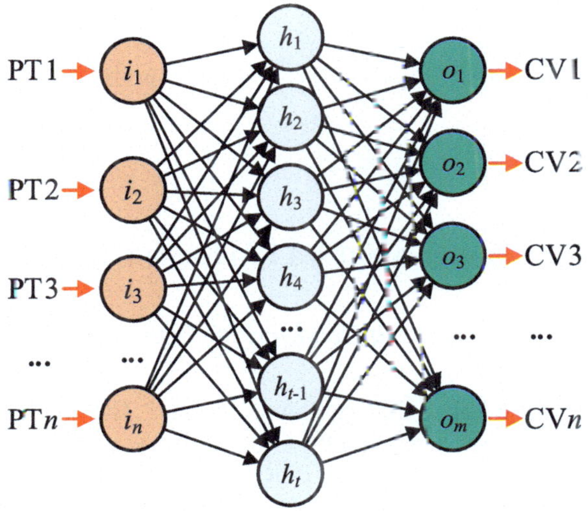

Fig. 6.3 Data driven fault diagnosis model

maximize the likelihood function and minimize the cost function, as shown below.

$$\theta_{ML} = \arg\max_{\theta} \prod_{i=1}^{m} p(y^{(i)}|x^{(i)}; \theta) \tag{6.5}$$

where m is the number of nodes, y is output nodes and x is input nodes. It can be found that in a maximum likelihood method, θ is a parameter and it is unknown. In fact, affected by many factors, the distribution of parameters is usually not random. In the Bayesian Regularization, θ follows a certain distribution $p(\theta)$. $p(\theta)$ is called prior probability. For a training set $S = (x^{(i)}, y^{(i)})_{i=1}^{m}$, the posterior probability is calculated by

$$\begin{aligned} p(\theta|S) &= \frac{p(S|\theta)p(\theta)}{p(S)} \\ &= \frac{(\prod_{i=1}^{m} p(y^{(i)}|x^{(i)}, \theta))p(\theta)}{\int_{\theta} (\prod_{i=1}^{m} p(y^{(i)}|x^{(i)}, \theta))p(\theta)d\theta} \end{aligned} \tag{6.6}$$

where $p(y^{(i)}|x^{(i)}, \theta)$ is learning model and the calculation method is

$$p(y^{(i)}|x^{(i)}, \theta) = h_{\theta}(x^{(i)})^{y^{(i)}}(1 - h_{\theta}(x^{(i)}))^{1-y^{(i)}} \tag{6.7}$$

The calculation method of $h_{\theta}(x(i))$ is shown below.

$$h_{\theta}(x^{(i)}) = 1/(1 + \exp(-\theta^{T} x^{(i)}))^{3} \tag{6.8}$$

The new x can be inferred through the distribution of θ as shown below.

$$p(y|x, S) = \int_{\theta} p(y|x, \theta)p(\theta|S)d\theta \tag{6.9}$$

6.3 Case of a Subsea Production System

6.3.1 A Subsea Production System in the South China Sea

Figure 6.4 shows the case in this chapter. The data is got from a subsea production system that working at the South China Sea. The offshore part and the subsea part are the main components of it. In the offshore part, the hydraulic power unit (HPU) is the main component. It is used for providing pressure for the subsea component.

The accumulators in subsea can be fully charged with it. For subsea components, the power is provided by the accumulators which enables the subsea components to response quickly. The electronic power unit (EPU) is another offshore device. The main function of it is to provide electricity for subsea equipment. Converting the signals is another function of it. The signals from the main control station (MCS) can be transmitted to subsea by it. With the control of MCS, the different signals are transferred to the subsea components. And then, a subsea distribution unit (SDU) is used to receive the signal and distribute it to eight subsea Christmas trees (XT). For a XT, the Subsea Control Module (SCM) is control core. The production valves are controls by it. These valves are installed on the body of the XT and are used to control the production process of oil and gas. At the pressure inlet of SCM, the possibility of pressure supply failure is reduced through a hydraulic selector valve that is used to select a usable hydraulic supplies circuit. Production Main Valve (PMV), Production Wave Valve (PWV), Annual Main Valve (AMV), Annual Wave Valve (AWV), and Translator Valve (XOV) are the main components of a XT. In the process of oil production, the PMV and PWV are opened. They are used to control the process of transferring the oil in the central manifold. After the oil or gas from the eight XTs is collected, they would be transferred to the offshore. When the annular space pressure is higher than normal, the AMV and XOV would be opened for pressure relief. AAV is used for workover and chemical input. During the process of oil production, the PMV, PWV and AMV, is usually opened, and the XOV, and AWV is closed. Due to the production valve being controlled by the control valve, and the pressure transmitter being installed at the outlet of the production valve, the value of pressure transmitter is affected by the control valve. The correspondence of them can be seen in Fig. 6.4. PT01A, PT1C, PT05, PT 06 and PT07 are usually in a high-pressure state, while PT08 and PT09 are usually in a low-pressure state.

In this case study, operational data in September 2021 and October 2021 are collected. To study the changes of the system over time, data in October and November 2022 are also collected. In a short period of time, the changes in a subsea production system can be almost ignored. Considering factors such as data quality and storage costs, the data is collected every two hours. In this case, 13 sets of data are collected every day. In order to study the dynamic changes of the Christmas tree, ten days is set as a stage. Through it, the data is divided into six stages. It should be point out that the data is collected when a maintenance is just completed. Thus, it can be ensured that there is no fault in these data. The data that include fault comes from records, which are input into data sequences. However, the fault data are little. Other fault data is needed. In this chapter, fault is defined as the degree of failure, which varies from 15 to 90%. In this way, some invalid data similar to recorded fault data is generated. The simulated faults are leaks of valves, resulting in a decrease in pressure. The number of normal and fault data for each stage is listed in Table 6.1. In this system, the data obtained is pressure in PT01A, PT01C, PT05, PT06, PT07, PT08 and PT09. The component being diagnosed is the valve QDV, CV5, CV6, CV7, CV8 and CV9.

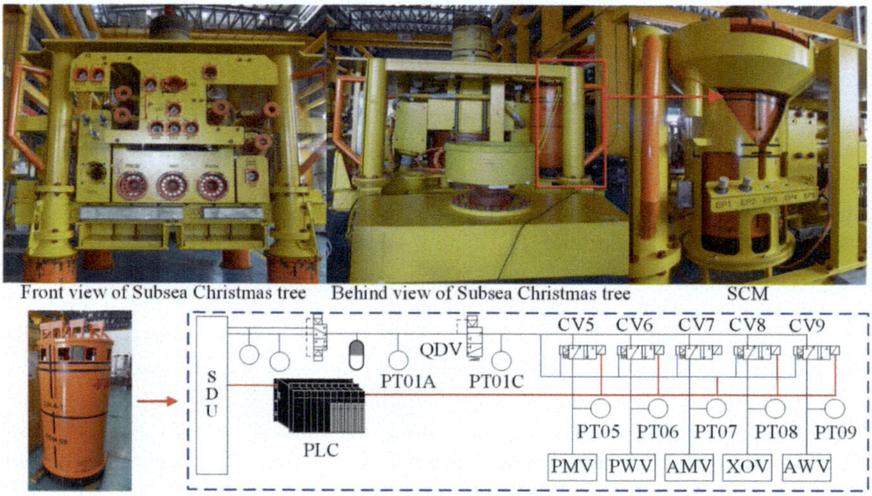

Fig. 6.4 A subsea production system in South Sea of China

Table 6.1 Data from subsea production system

Year	Stage	Date	Normal data	Data with fault
2021	1	09.14–09.23	130	46
	2	09.24–10.03	130	46
	3	10.04–10.12	117	46
2022	4	10.11–10.20	130	46
	5	10.21–10.30	130	46
	6	10.31–11.09	130	46

6.3.2 *Output of the Dynamic Digital Twin Model*

Digital twin is a virtual image of physical system. The main function of a digital twin is to obtain signals from physical systems and simulate it. Different from simulation, dynamic and real-time running is the characteristics of a digital twin model. Response time and error are commonly used to evaluate the performance of a digital twin model. In this research, the data from digital twin model is used for training the data-driven fault diagnosis model. The quality of data is an important indicator for evaluating the digital twin model. As described, the digital twin model includes fault parameters, loss parameters, and control parameters. The control parameter and fault parameter are updated according to system control command and the results of the fault diagnosis model. The loss parameters would be affected by the environment, component degradation, and status of oil and gas. A fixed parameter can lead to significant errors in long-term operation. To study the changes in long-term operation, the performance of digital twin models with fixed loss parameters and dynamic loss parameters is compared. The parameters of digital twin model used in case study is shown in Table 6.2. The valves in this system are all electromagnetic directional valves. It only relies on two states, that is open and close. When the valve is opened, all fluid passes through and pressure is almost completely transmitted, so the control parameter is close to 1. Similarly, when the valve is closed, the control parameters are close to 0. In the initial stage, all components default to no faults, so the fault parameter is 1. The loss parameters are calculated based on normal operating data and adjusted according to the actual situation. The average of 30 sets of data are calculated and used to calculate the loss parameter.

The field data of subsea production system is shown in Fig. 6.5. The data in seven sensors are obtained from an offshore in the South China Sea. The value is the average of all collected data in the day. Among them, the value of PT08 and PT09 is small, and they are displayed by the vertical axis on the right while the rest are measured by the left ordinate. It indicates that most pressure is between 220-240psi, while some are between 20-30psi. Besides, the pressure in 2021 is significantly different from that in 2022, with a difference of about 4psi. This difference indicates that the subsea production system changes over time with the influence of much factors. From a one-year time span, this change is acceptable for a subsea production system and would not cause system malfunctions. For each sensor, the pressure fluctuates within a range. It is can be seen that the trends are similar. Because the pressure measure by PT01A is the input of the system, and the pressure in other positions depends on the input pressure. In this way, all changes of the pressure are similar to the pressure changes of PT01A. It can also be seen that the trend of PT08 and PT09 is different. The reason is that the output pressure of the normally closed valves is depended on the load. The pressure at this position does not change with the input pressure.

The output of a digital twin model with a fixed loss parameter is shown in Fig. 6.6 and the parameters are shown in Tab. II. In this model, the input is the value of PT01A and all of the others are output. The loss parameter is obtained by calculating the average of all loss parameters in 2021. After the data of PT01A is input into digital twin model, the digital twin model output data of the other sensors. The error is measured by calculating the error between the digital twin model and the actual value. When the error of a day is required, the average value of the error is calculated. The data would be used to train data-driven methods. It shows that, the trends in 2021 and 2022 are different. In 2021, the trend is similar with the real data, but in 2022, there is a large error between the real data and the output of the digital twin model. The fixed loss parameter makes the trends of the other are similar with the trend of the

Table 6.2 The parameters used in case study

	QDV	CV5	CV6	CV7	CV8	CV9
ζ	0.9951	0.9903	0.9876	0.9918	0.9736	0.9857
τ	1	1	1	1	1	1
φ	1	1	1	1	0.1851	0.1255

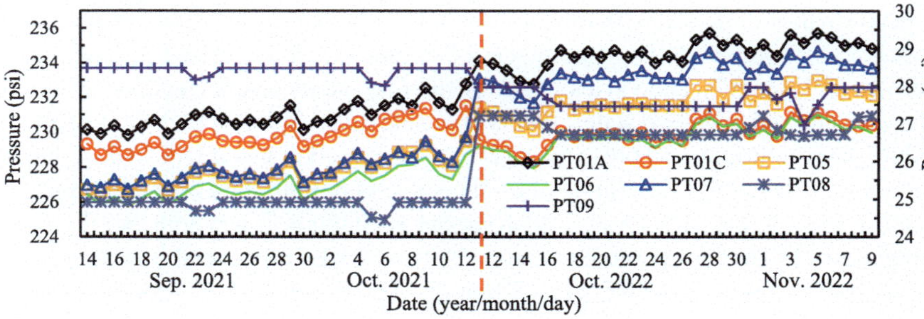

Fig. 6.5 Field data of subsea production l system

PT01A. However, this is different from the truth. This leads to a significant difference in output from actual values in 2022. It should be noted that, the valve of PT08 and PT09 is different. A fixed small loss parameter leads to an almost unchanged output. These data cannot be used for training of the fault diagnosis model. Because they are almost identical.

To get a more dynamic digital twin model, a real-time updated loss parameter is proposed. In the proposed digital twin model, the loss parameter is updated by calculating the average loss parameters in the previous day. The output is shown in Fig. 6.7. Obviously, the output of it is more similar to real data. In fact, the performance of digital twin model with fixed loss parameters is good in 2021, and that is similar to the digital twin model with dynamic loss parameters. However, the differences between them are quite obvious in some details. For example, in real data, it shows that there is a slight change in their values around September 22, 2021 and October 5, 2021. These are not simulated by a digital twin with a fixed loss parameter. But, it is well simulated by a digital twin with a dynamic loss parameter.

In 2022, the differences between the two models are clearly reflected. Obviously, the output of digital twin models with dynamic parameters is more similar to the actual output. In fact, the digital twin model with dynamic loss parameters is more delayed than the real data change. In the research, this delay is one day because, the

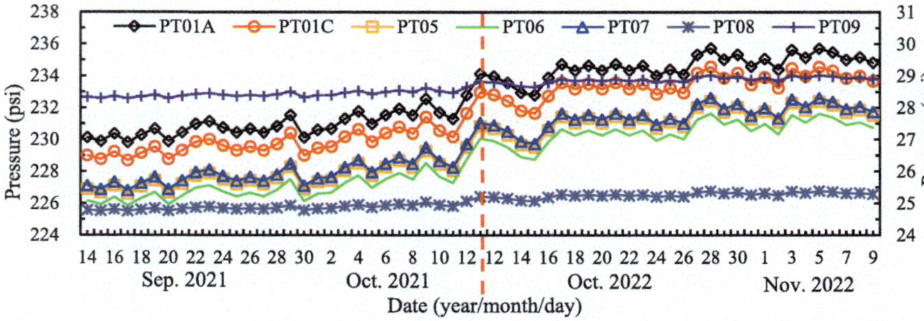

Fig. 6.6 Output of a digital twin model with a fixed loss parameter

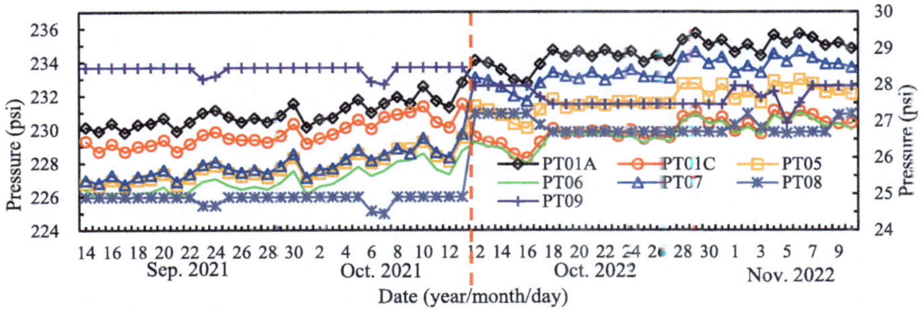

Fig. 6.7 Output of a digital twin model with a dynamic loss parameter

loss parameter is updated by calculating the average loss parameters in the previous day. This delay depends on the time interval between adjacent data points. As the frequency of data point collection increasing, the delay would be reduced, correspondingly, the computational workload would increase. The changes in a subsea production system are slow, a delay of one day can meet needs of field.

The errors of the two models are calculated as shown in Fig. 6.8. It is notable that the two fixed parameters and two dynamic parameters are the same. The bar chart represents the numerical magnitude of the error, while the line chart represents the percentage of error. They are obtained by calculating the average difference of all of data.

Overall, the error at each position is below 2 psi, less than 5%. It is can be seen that the error of the digital twin model with dynamic loss parameters is smaller than that of the digital twin model with fixed loss parameter rs. The error is reduced by about 1 psi, with a maximum reduction of 3%. In fact, the reduction of error is not significant in 2021, however, it is large in 2022. This is because the loss parameters are calculated based on data from 2021 and that results in a good performance of

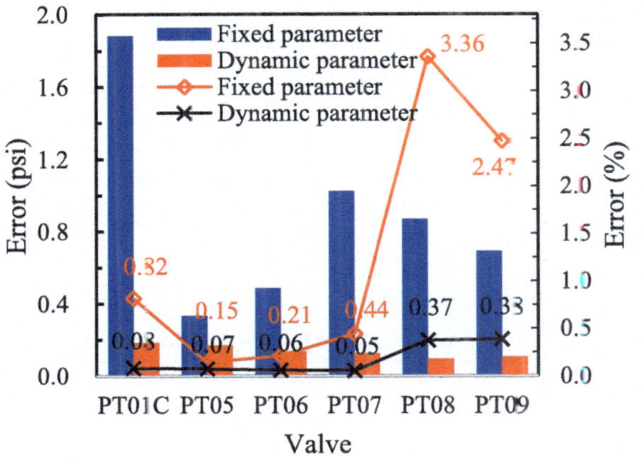

Fig. 6.8 Error of digital twin model

the digital twin model in 2021. However, it can be seen from Fig. 6.5 that there is a significant change in data after a year. This leads to a poor performance of the digital twin model with fixed loss parameters in 2022. According to the result, it shows that a dynamically updated key parameter greatly improves the digital twin model.

6.3.3 Performance of the Fault Diagnosis Model

The performance of the two fault diagnosis models is shown in Fig. 6.9. The data used for training is accumulated. At each stage, the data from all previous stages is used to train the data-driven model. When it is combined with digital twin model, the data from digital twin model are also used for training. It shows that the accuracy of the model-based method is large than 96% at the first three stage. In this stage, the data using for training is little. Thus, the accuracy of data-driven method is low. However, the decrease in accuracy is significant when it is used in 2022. The accuracy below 90%. In contrast, the accuracy of data-driven method increases gradually over time. More significantly, the accuracy of the fault diagnosis model with the digital twin model is higher than that of the fault diagnosis model without the digital twin model at each stage. In the sixth stage, the accuracy of them is similar. Through that, it shows that in this method, the digital twin model cannot improve the accuracy of the data-driven model.

By providing data for training, the digital twin model accelerates the training process. In this way, a fault diagnosis model with good performance can be obtained earlier. A noteworthy phenomenon is that there is a significant decrease of accuracy at the fourth stage. This is because the data of the third and fourth stages are separated by one year. That results in a visible decrease. In fact, the changes in a subsea production system are small in a short time. The decrease in accuracy is not significant when

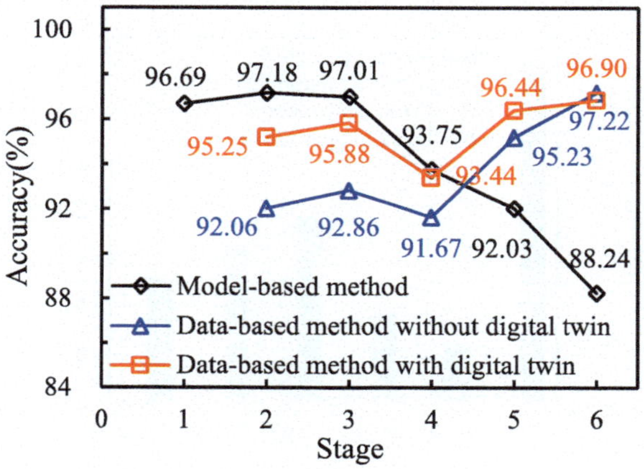

Fig. 6.9 Accuracy of difference fault diagnosis models

continuous data is used. Over time, the trends of the accuracy of several models are similar to this result. In the proposed fault diagnosis framework, the model with higher accuracy is selected. Thus, the accuracy of this method can maintain above 96% in all stage. The result show that the proposed method can combine the advantages of data-driven and model-based method. It can select the most effective method for diagnosing faults at each stage. The accuracies are all more than 93% at each stage.

To further research, the data are grouped by fault type and year. Figure 6.10 shows the results of model-based method. The horizontal axis is the actual faults, while the vertical axis is the identified type of faults. It shows that when the 2021 data is diagnosed, the erroneous results are concentrated on the incorrect identification of fault in QDV. This is because QDV is associated with all sensors, the relevant data is easily misidentified. When the data of 2022 is diagnosed, many normal data are identified as faults. In fact, through a large amount of experimental, it can be found that a high diagnostic accuracy is contradictory to a good compatibility. When a fault diagnosis model is established based on data of 2021, high accuracy is required. At this point, a small change in the 2021 data would be considered a fault. However, As the system runs, minor changes are common. This leads to that many normal data are identified as faults.

When the data of 2022 is diagnosed, the results are shown in Fig. 6.11. It should be pointed that the model is used for diagnosing the data of 2021 is trained just with the data of 2021.From the results, it shows that when the data of 2021 is diagnosed, the misdiagnosis results are scattered. This is because the insufficient training data resulted in an imperfect model. Thus, some data is not identified correctly. When the model is used to diagnose the data of 2022, the situation is improved. Although there are still some normal data that are identified incorrectly, the decrease in this

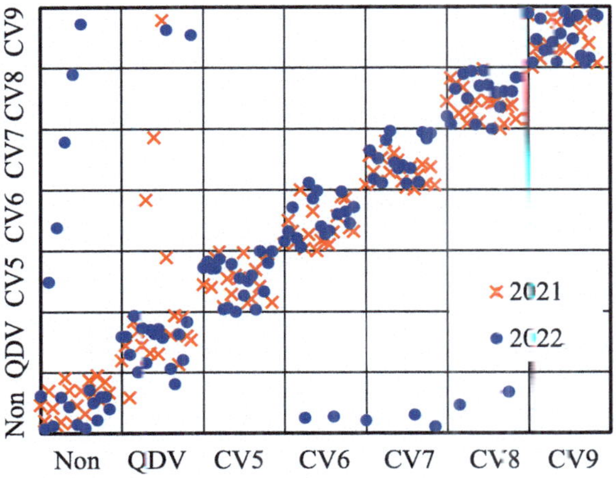

Fig. 6.10 Result of model-based fault diagnosis model

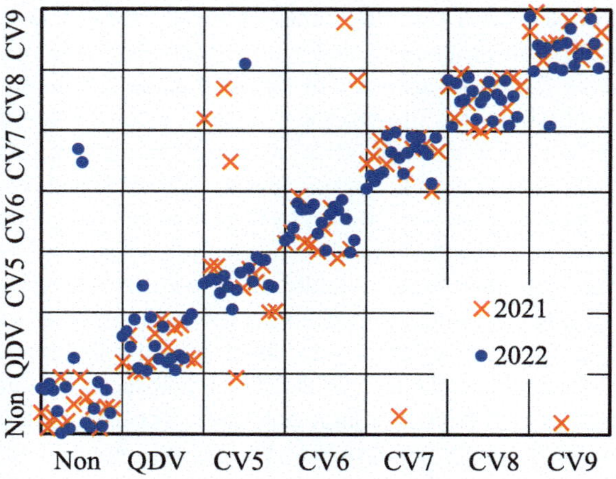

Fig. 6.11 Result of data-driven fault diagnosis model

quantity is significant. The results show that the combination of these two methods can effectively ensure the stable performance of the diagnostic model throughout the entire stage.

Part I
Concluding Remarks

Part I *Fault diagnosis* delves into advanced methodologies for fault diagnosis, exploring innovative models and techniques tailored to the unique challenges posed by subsea environments. A digital twin-driven approach is proposed to handle composite faults, while cross-validation techniques are utilized to enhance the accuracy of diagnosing minor faults. Additionally, integration of virtual and real fault data poses challenges, so a DT-assisted fault diagnosis framework is introduced for this purpose, ensuring that virtual data is closely integrated with real data. Optimal sensor placement is another key consideration, and a sensor placement method optimized by discrete particle swarm algorithm is proposed to improve the performance of fault diagnosis in the hydraulic control system. This part also addresses the challenges of diagnosing concurrent faults in electric-hydraulic systems, proposing models that can handle multiple fault occurrences simultaneously. Lastly, intelligent full-stage stable fault diagnosis method is covered, focusing on long-term stable diagnostic performance. The comprehensive methodologies presented in this part are validated by real-world cases and experimental data, demonstrating their practical applicability to subsea production systems. This holistic approach to fault diagnosis forms a foundation for the intelligent operation and maintenance of the subsea production systems.

Part II
Fault Prognosis

Fault prognosis plays a crucial role in ensuring the performance and efficiency of subsea production systems, where the harsh operational conditions and inaccessibility make accurate prediction of component states vital. Part II *Fault Prognosis* focuses on RUL prediction as a key tool for optimizing maintenance strategies and preventing unexpected failures.

Subsea production systems often suffer from data scarcity or imbalanced datasets, making RUL prediction particularly challenging. To address this issue, various advanced models are introduced, capable of dealing with both small sample data and multi-sample data, thereby bridging the gap between limited information and accurate predictions. In addition, multiple causes often contribute to the degradation of subsea equipment, requiring prediction models that integrate various causes to capture a holistic view of system states.

Another challenge in fault prognosis arises from the complex, multi-stage nature of degradation, where components experience different phases of degradation over their service life. Hybrid models that account for this multi-stage degradation provide a more dynamic and accurate approach to RUL estimation. Furthermore, subsea systems often involve dependent components, where failure of one component may accelerate the degradation of others. Models that consider these failure dependences and cascading effects are essential for realistic and reliable RUL predictions. Beyond component interactions, the dual effects of degradation and external shocks on system states are also explored, offering a detailed view of how these factors integrate to impact overall system states.

An important aspect of RUL prediction is the dynamic nature. As conditions change or new data becomes available, RUL models must be updated accordingly to ensure continued accuracy. This re-prediction capability allows for continuous monitoring and adjustment, reflecting the evolving operational states of subsea equipment. Finally, the integration of industrial incomplete information into fault prediction models shows better adaptability for uncertainty and dynamic range analysis.

By grounding these theoretical approaches in real-world applications, such as subsea valves, pipelines, Christmas trees, and hydraulic control systems, this part presents a comprehensive framework for effective fault prognosis, advancing both the academic understanding and practical application of RUL prediction in subsea production systems.

Chapter 7
RUL Prediction with Small Sample Data and Multi-sample Data

Abstract This chapter introduces two distinct approaches for RUL prediction, addressing the challenges of small sample data and multi-sample data in subsea equipment maintenance planning. For small sample data, a method is proposed that integrates data and models using intermittent monitoring data to establish a health index (HI) for subsea equipment. By expanding the HI dataset and applying life analysis on the Weibull based distribution, the method converts the HI into reliability metrics, overcoming data scarcity to predict the RUL of subsea valves. This approach provides a probabilistic framework for RUL prediction under limited data conditions, enhancing model fitting and prediction accuracy. For multi-sample data, a hybrid model-data-driven method is introduced, combining Kalman Filter (KF) with DBN. The DBN-KF fusion improves the state estimation of the system by reducing estimation and observation errors while accounting for the uncertainties in degradation and environmental parameters. This method enables more accurate RUL prediction by integrating these uncertainties into the model, which is then applied to subsea Christmas tree valves. Both methods are demonstrated through case studies, offering tailored solutions for different data availability scenarios, enhancing the reliability of RUL prediction and supporting maintenance decision-making.

7.1 Introduction

Prognostics and health management (PHM) leverages advanced technologies such as data analytics and machine learning to monitor and predict the health of engineering equipment. A critical component of PHM is the prediction of RUL (Ochella et al. 2021), which enables enterprises to strategically plan maintenance, extend equipment service life, and enhance overall operational efficiency (Ma et al. 2023). Effective fault prediction underpins essential processes like failure analysis, resilience modeling, risk management, and maintenance decision-making. Accurate RUL predictions facilitate better maintenance planning, minimize unexpected downtime, and ultimately improve production efficiency. However, significant challenges hinder

B. Cai et al., *Intelligent Operation and Maintenance for Subsea Production Systems*,
https://doi.org/10.1007/978-981-96-2677-9_7

accurate RUL prediction, particularly concerning the use of sample data, including small sample data and multi-sample data.

When relying on small sample data, model uncertainty can increase (Wu et al. 2023), adversely affecting generalization to new datasets and making predictions less reliable. Additionally, noise from irrelevant or erroneous information in small sample datasets can complicate the model training process, further diminishing prediction performance. Researchers have begun exploring hybrid methods that combine data-driven and model-driven approaches to address these challenges. Techniques such as feature engineering, appropriate model selection, and cross-validation are critical to improving the reliability of RUL predictions in scenarios characterized by limited data. This chapter proposes an innovative data-model-linked RUL prediction method specifically tailored for small sample monitoring data. By utilizing intermittent monitoring data (e.g., monthly or quarterly) to construct a health index (HI), the method addresses the limitations associated with long-interval small sample data. The HI serves as a bridge between the monitoring data and reliability analysis through the Weibull distribution, enhancing the model's ability to predict RUL. This research is driven by the need to improve the accuracy of predictions despite the challenges posed by insufficient data. It addresses key issues such as model uncertainty, data scarcity, and the impact of noise, providing a more robust framework for making reliable maintenance and operational decisions.

Regarding multi-sample data, significant challenges arise from the presence of noise and variability, which can degrade prediction accuracy for RUL. Even with larger datasets, the interference from irrelevant or erroneous information can obscure true patterns, leading to unreliable predictions. To address these challenges, this chapter also proposes a novel hybrid RUL prediction method that combines DBN with the Kalman Filter (KF). This innovative approach leverages the strengths of both models: DBN enhances the observation process, improving the accuracy of system state estimations and reducing prediction errors throughout the equipment lifecycle. By fully considering uncertainties in degradation and environmental parameters, the method effectively manages the randomness inherent in degradation processes. The integration of DBN within the KF framework enhances adaptability in addressing system uncertainties, providing a robust solution for engineering applications that require reliable RUL predictions in the face of multi-sample data challenges.

7.2 Data-Model-Linked RUL Prediction Model

A data-model-linked RUL prediction method for small sample monitoring data is proposed, as shown in Fig. 7.1. This study targets small sample data with time labels from monitoring devices. This type of data is very common in the industry due to storage and detection costs, resulting in long monitoring periods for equipment, which leads to a limited number of data samples available for assessing the equipment's health status. First, collect small sample monitoring information, which includes small sample data, historical faults, expert information, related database information and similar equipment degradation data. Then, a modified sigma function is estimated through the system characteristics. This function is used to complete the mapping of small sample monitoring data to the HI. Bayesian network (BN) is introduced to complete the solution of parameter uncertainty and random features to obtain more comprehensive evaluation results. Next, the HI is expanded using a Gaussian mixture model (GMM), and the expanded data is input into the estimation model. First, the GMM is trained using the collected small sample data. During training, the GMM attempted to be optimized to find the best combination of Gaussian distributions to fit the data. The Expectation–Maximization (EM) algorithm is commonly used for training. Then, the trained GMM model is used to generate synthetic data points. This is achieved by sampling from each learned Gaussian distribution. Finally, the generated synthetic data is evaluated to ensure that they accurately reflect the characteristics of the original data. The HI of the system is converted into a reliability index, and the characteristics of the Weibull life distribution are used to estimate the RUL of the equipment. So far, this method has completed using the difference between the monitoring data and the best state to predict the RUL of the equipment. The details of the algorithm are shown below.

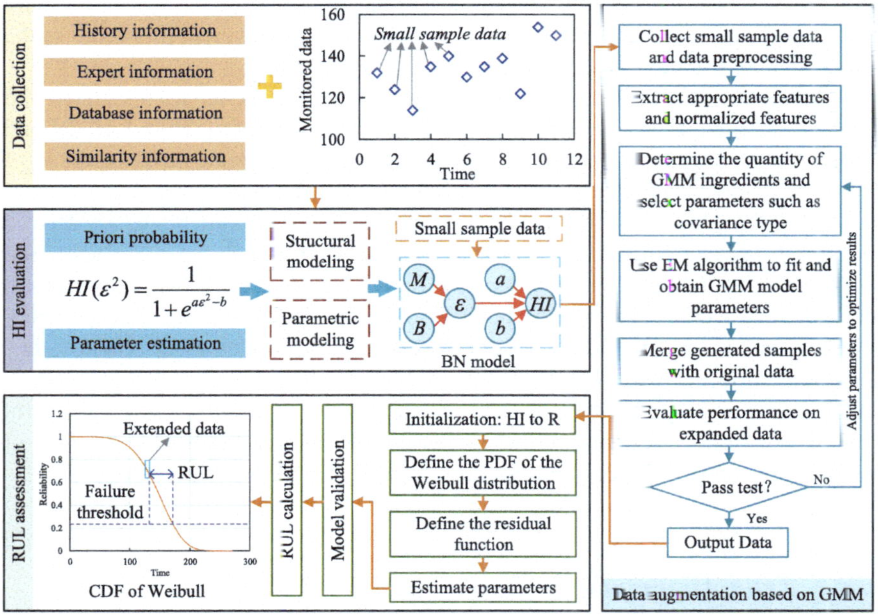

Fig. 7.1 Proposed RUL prediction framework

7.2.1 BN-Based HI Assessment

In general, the assessment of HI involves extracting fault features from the gathered data, analyzing the data in real-time or periodically through model training, and subsequently calculating the corresponding index. Usually, this index is between 0 and 1. However, there is a large amount of uncertainty in the degradation of equipment with complex working conditions, and it is difficult to evaluate HI using limited small sample data. In response to this difficulty, this work first uses sigmoid functions to convert monitoring data into a unified health index of 0–1 (Jiujian et al. 2021), and then uses BN to model parameter uncertainty. The modified sigmoid function is used to construct HI for small sample data.

$$HI(\varepsilon^2) = \frac{1}{1 + e^{a\varepsilon^2 - b}} \tag{7.1}$$

where ε is the difference between the monitoring value and the best state value. Usually, the difference for each monitoring value is obtained by subtracting the monitored data from the best state value. The a and b are parameters related to the equipment, which can be estimated based on the best state value and the failure threshold. The best state value and the failure threshold are inherent properties of the device and are usually related to design parameters.

Subsequently, the prior information and HI calculation model are integrated and mapped into BN, which mainly includes the construction of structural model and parameter model. The role of BN in HI calculation is to integrate the uncertainty of limited data and weaken the impact of randomness on HI estimation. The structure model of BN is obtained from Eq. (7.1). The nodes of the structure model are derived from the parameters in Eq. (7.1), and the connections between the nodes are based on the computational relationships of the parameters in Eq. (7.1). The structural model mainly includes monitoring node M, optimal state node B, difference node ε, estimated nodes a and b, and evaluated node HI. After the nodes are discretized, they are given a priori probability, and the conditional probability table (CPT) of BN is transformed according to Eq. (7.1). Through this BN, the monitoring data at different times are input to estimate the HI at the corresponding time, and the conversion of small sample monitoring data into small sample health index is completed.

7.2.2 Data Augmentation Based on GMM

The GMM is a model for modeling probability distributions, often used to model multidimensional data. GMM represents the distribution of data through a linear combination of multiple Gaussian distributions (Mebarki et al. 2022). The probability density function (PDF) of a Gaussian distribution describes the probability of a data point falling near a certain value. For a univariate Gaussian distribution, its PDF is:

$$p(x; \mu, \sigma) = \frac{1}{\sqrt{2\pi\sigma^2}} \exp\left(-\frac{(x-\mu)^2}{2\sigma^2}\right) \tag{7.2}$$

where μ is the mean and σ is the standard deviation.

For a multivariate Gaussian distribution, its PDF is:

$$p(x; \mu, \Sigma) = \frac{1}{2\pi^{D/2}|\Sigma|^{1/2}} \exp\left(-\frac{1}{2}(x-\mu)^T \Sigma^{-1}(x-\mu)\right) \tag{7.3}$$

where D represents the dimension of the data, x is a D-dimensional vector of data points, $\boldsymbol{\Sigma}$ is the covariance matrix.

GMM is a linear combination of multiple Gaussian distributions, and its PDF is:

$$p(\boldsymbol{x}; \boldsymbol{\theta}) = \sum_{i=1}^{K} w_i \cdot p(\boldsymbol{x}; \boldsymbol{\mu}_i, \boldsymbol{\Sigma}_i) \tag{7.4}$$

where $\boldsymbol{\theta}$ includes the parameters of all Gaussian distributions, w_i is the weight of each component, and K is the number of Gaussian distributions.

The parameters of GMM are usually estimated by the EM algorithm, which iteratively performs E-step (expectation step) and M-step (maximization step) to maximize the likelihood function. Use the fitted GMM model to generate new samples through sampling. For each new sample, first choose a Gaussian distribution i, according to the weight w_i to make a selection. Then, samples are generated from the selected Gaussian distribution:

$$i \sim Discrete(\boldsymbol{w}) \tag{7.5}$$

$$x_{new} \sim \mathbf{N}(\boldsymbol{\mu}_i, \boldsymbol{\Sigma}_i) \tag{7.6}$$

where $Discrete(\boldsymbol{w})$ represents discrete distribution sampling according to weight distribution, $N(\boldsymbol{\mu}_i, \boldsymbol{\Sigma}_i)$ represents multivariate Gaussian distribution sampling.

The generated new samples are added to the original training set to form an expanded training set.

$$\text{New training set} = \{x_1, x_2, \ldots, x_N, x_{new1}, x_{new2}, \ldots\} \tag{7.7}$$

7.2.3 RUL Assessment

Due to the similarity between the sigmoid function involved in calculating HI and the Weibull distribution involved in reliability calculation, and the commonality between the two in characterizing the health level of equipment, including value ranges, failure thresholds, etc., this method determines that HI and reliability can be equivalently converted. The Weibull distribution is a probability distribution that describes the time when a positive definite event occurs, and it is often used in reliability analysis. The Weibull distribution has two parameters: λ (scale parameter) and k (shape parameter). For two-parameter estimation of the Weibull distribution, the least squares method (LSM) can be used to estimate the two parameters.

The PDF of the Weibull distribution is:

$$f(t;\lambda,k)=\frac{k}{\lambda}\left(\frac{t}{\lambda}\right)^{k-1}e^{-(t/\lambda)^k} \tag{7.8}$$

where t is random variable, λ is scale parameter, and k is shape parameter.

In LSM, the objective is to minimize the sum of squared residuals between observed data and the fitted Weibull distribution. For the Weibull distribution, this involves optimizing the scale parameter λ and shape parameter k such that the following objective function is minimized:

$$\min_{\lambda,k}\sum_{i=1}^{n}(y_i-f(x_i;\lambda,k))^2 \tag{7.9}$$

The trust region reflective algorithm is a numerical calculation used to implement the minimization process. Its basic idea is to approximate the objective function through a local quadratic model in each iteration and find the minimum value within a "trust region". If the local quadratic model is no longer accurate near the boundaries of the trust region, the algorithm adjusts the size of the trust region and repeats the iteration (Yang et al. 2017). In each iteration, a quadratic model is used to approximate the objective function, usually expressed as:

$$m_k(\boldsymbol{p})=f_k+\nabla f_k^T\boldsymbol{p}+\frac{1}{2}\boldsymbol{p}^T\nabla^2 f_k\boldsymbol{p} \tag{7.10}$$

where $\boldsymbol{p}$ is the change amount of the parameter vector, f_k is the objective function value of the current iteration, ∇f_k is the gradient of the objective function, and $\nabla^2 f_k$ is the Hessian matrix of the objective function.

In each iteration, the algorithm dynamically adjusts the size of the trust region to ensure that the model's approximation of the objective function remains accurate within the trust region. The size of the trust region may be adjusted based on the fit quality of the current model. Then, an appropriate step size is chosen within the trust region to update the parameter vector. Finally, iteratively finds the minimum value of the objective function by repeatedly adjusting the size of the trust region, selecting the step size, and updating the parameter vector.

Subsequently, the confidence intervals of the Weibull distribution parameters are calculated through the Bootstrap method, and the process is shown in Fig. 7.2. The core idea of the Bootstrap method is to repeatedly extract samples from the original samples with replacement to form new Bootstrap samples, and then estimate the distribution of parameters or calculate confidence intervals based on these Bootstrap samples. The basic steps of the Bootstrap method include inputting the original sample data, initializing the number of resampling B, repeating the Bootstrap process B times, extracting new samples in each process, calculating statistics, saving statistics, and finally calculating the confidence interval of the statistics and outputting it.

Assuming that the reliability of the equipment at time τ is R_τ, then RUL can be expressed as the time period from time τ to the time of system failure, which can be expressed as:

$$RUL = \lambda[(-\ln R_f)^{\frac{1}{k}} - (-\ln R_\tau)^{\frac{1}{k}}] \tag{7.11}$$

where R_f is reliability failure threshold.

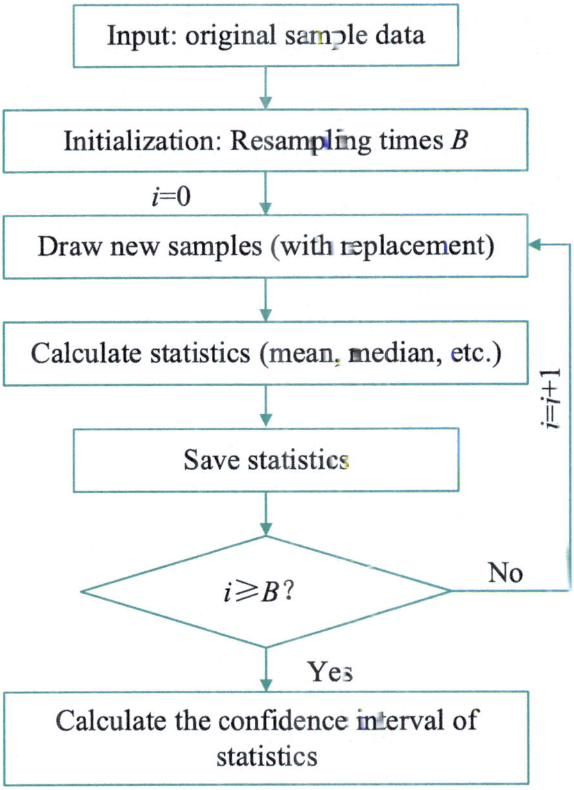

Fig. 7.2 Flow chart of Bootstrap method

7.3 Hybrid DBN-KF-Based RUL Prediction Model

7.3.1 Hybrid DBN-KF-Based Method

To reduce the influence of measurement error and variable operating parameters on the prediction results, a novel hybrid DBN-KF-based RUL prediction method is proposed. The framework of hybrid DBN-KF-based method is shown in Fig. 7.3. The advantage of the proposed improved KF algorithm is that it strengthens the ability of KF to predict linear degradation and provides an uncertain solution. The hybrid method mainly includes three stages: state prediction, reasoning, and state update.

State prediction is a traditional way, that is, using the state of the last time to estimate the current state. F_t. B_t, u_t, and Q_t are the four important parameters of state prediction, where F_t is the state transition matrix, B_t is the control matrix, u_t is the control vector, and Q_t is the noise of the prediction model. Based on the state value of the previous moment, the covariance of the previous moment, the parameters of the previous moment (F_t, B_t, u_t, Q_t), system state $\hat{x}_t$ and covariance matrix P_t are the first to be predicted. The state space equation of system degradation is used for the estimation of the prediction stage.

DBN is mainly used to calculate the observation of KF. In reasoning, the DBN integrating the uncertainty of degradation parameters and environmental parameters is constructed based on the state space equation of system degradation. In DBN, the

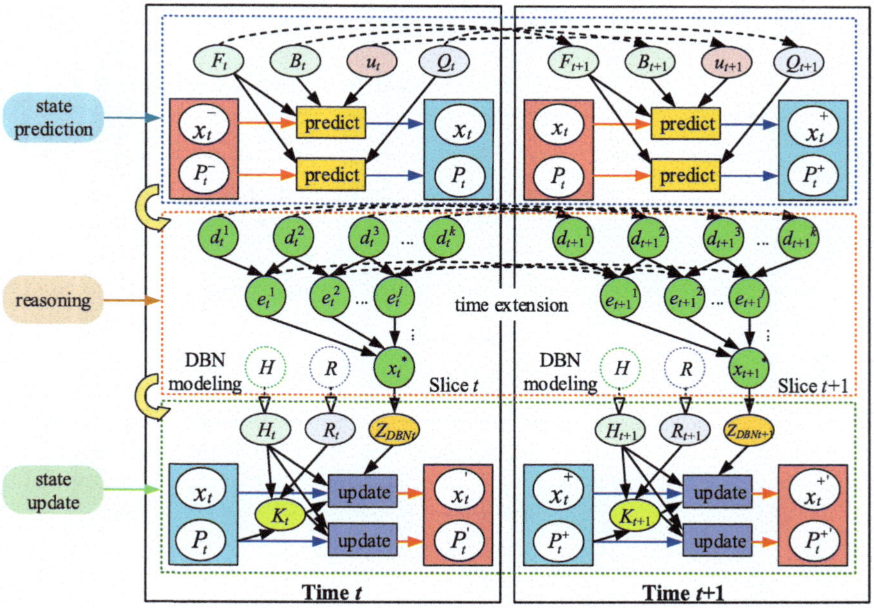

Fig. 7.3 Framework of hybrid DBN-KF-based method

first layer is the variable layer, that is, d layer, which represents the basic parameters that affect the system degradation. The uncertainty problem is quantified as a random distribution in the variable layer. The e layer is the mediating layer, which represents the intermediate state in the system state estimation. In addition, the system observation value (x_t^*) is calculated by DBN. Transformation matrix, H, represents the relationship between the observed state and the system state variables, and is obtained from the state space equation during the system observation. Generally, a specific observation system has a corresponding transformation matrix. DBN modeling is the modification of observation values, and part of its structural model and parameter model is the mapping of transformation matrix. When estimating the observation value, DBN considers the uncertainty of all parameters. Observation noise matrix, R, is obtained from the calculation deviation of DBN.

In the state update phase, Kalman gain K_t is introduced to upgrade the system state. Optimal estimation is performed by balancing the relationship between the estimated value, the monitored value, and the true value. Parameter changes can be updated by specifying the state transition model in the dynamic extension of DBN. The updated state is implemented in the prediction of the next time, iterative calculation is carried out, and the optimal estimation of system state is expanded.

(1) State prediction

The KF is used to obtain the optimal prediction between estimated value and measured value. Measured value is also called observed value. State prediction can be described as

$$\hat{x}_t^- = F_t\hat{x}_{t-1} + B_t u_t, \tag{7.12}$$

where $\hat{x}_t^-$ is the predicted value of the state at t, that is, a prior state estimate, and $\hat{x}_{t-1}$ is the optimal estimation of the state at $t-1$, also known as a posteriori state estimation; F_t is the basis for state inference from two adjacent times and shows how to infer the current state from the previous state; B_t indicates how the control quantity u_t acts on the current state. Accordingly, the covariance matrix can be expressed as

$$P_t^- = FP_{t-1}F^T + Q, \tag{7.13}$$

where P is the covariance matrix and P is the uncertainty of each time in the system; the prediction model itself is not absolutely accurate, and Q represents its noise.

(2) Reasoning

The main objective of DBN in the reasoning is to build an observation estimation model that integrates uncertainty information. First, the BN based on the degradation model is established. The BN model mainly includes three types of nodes: hypothesis, information, and mediating nodes. Hypothesis nodes are the variables that are either impossible or too costly to observe directly. Information nodes are the variables that can be observed. Mediating nodes are introduced for a special purpose that may be to properly reflect the independence properties in the domain, facilitate the acquisition of conditional probabilities, reduce the number of distributions to acquire for the network.

The conditional probabilities tables (CPT) of DBN are obtained by converting from the physical model (empirical formula) according to the following steps. First, the sub-nodes are discretized. Then, using the given physical model, the relationship between all variable nodes is determined. Finally, Monte Carlo simulation is used to sample the parent node. Finally, CPT is obtained by calculating the various states and corresponding probabilities.

The static network is extended to DBN through parameter change model and physical degradation formula. If two fundamental suspicions are fulfilled, (1) the alter of conditional likelihood with time is steady; (2) the energetic alter of likelihood has property of Markov, at that point DBN demonstration can be built.

The initial condition of DBN can be

$$P(X_0) = \prod_{i=1}^{c} P(X_0^i|Pa(X_0^i)) \tag{7.14}$$

The conditional probability of transition of adjacent time slices can be

$$P(X_t|X_{t-1}) = \prod_{i=1}^{c} P(X_t^i|Pa(X_t^i)) \tag{7.15}$$

The posterior probability of DBN node, that is, the joint probability between joint time slices and kernel time slices, can be

$$P(X_{0:t}) = P_{B_0}(X_0) \prod_{i=1}^{t} P_{B\rightarrow}(X_t|X_{t-1}), \tag{7.16}$$

where $X_{0:t} = \{X_0, X_1, \ldots, X_t\}$ is the ith node of $X_t{}^i$ at time t, $P_{B\rightarrow}$ is the conditional probability transition between two adjacent nodes of DBN, and $Pa(X_t{}^i)$ is the parent node of $X_t{}^i$.

Assuming the probability distribution of the degradation parameter (d^1–d^k) is specified in the estimation of the observed value, the DBN in Fig. 7.4 can be constructed according to conditional probability theory. Mediating nodes may run through many layers. e^1–e^j and g^1–g^3 in Fig. 7.4 are mediating variables used for observation calculation. The results of the observations are updated step by step over time. Generally, the time-varying characteristics of degenerate parameters drive time expansion.

(3) State update

According to the analysis of the uncertainty results of DBN, the standard deviation of the calculation results is obtained, which is transformed into observation noise, and

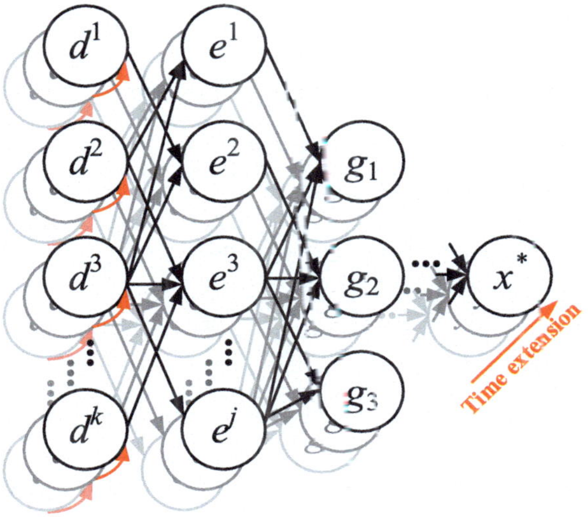

Fig. 7.4 Time extension of DBN

the mean value of the calculation results is transformed into the observation value. Then, the observation equation in the updating of KF can be expressed as

$$z_{DBN\,t} = H_t x_t + \xi, \tag{7.17}$$

where z_{DBN} is the supplementary information obtained by DBN; H_t is the transformation matrix; ξ is the noise during observation, and its matrix form is usually expressed as R.

If observation vector z can be divided into the following vector forms,

$$\mathbf{Z}_l = \begin{bmatrix} z_{DBN1} \\ z_{DBN2} \\ z_{DBN3} \\ \vdots \\ z_{DBNl} \end{bmatrix} \tag{7.18}$$

Then, the corresponding noise sub vectors are uncorrelated, that is,

$$\underset{\xi}{E}\left\langle \xi_i \xi_j^T \right\rangle = \begin{cases} 0, & i \neq j \\ R_i & i = j \end{cases} \tag{7.19}$$

Thus, matrices H and R can be similarly divided into

$$\boldsymbol{H}_l = \begin{bmatrix} H_1 \\ H_2 \\ H_3 \\ \vdots \\ H_l \end{bmatrix}, \boldsymbol{R}_l = \begin{bmatrix} R_1 & 0 & 0 \cdots & 0 \\ 0 & R_2 & 0 & \cdots & 0 \\ 0 & 0 & R_3 & \cdots & 0 \\ \vdots & \vdots & \vdots & \ddots & \vdots \\ 0 & 0 & 0 \cdots & R_l \end{bmatrix} \tag{7.20}$$

Moreover, the corresponding information matrix and covariance matrix are

$$Y_t = \sum_{i=1}^{t} H_i^T R_i^{-1} H_i = Y_{t-1} + H_t^T R_t^{-1} H_t \tag{7.21}$$

$$P_t = Y_t^{-1} = \{P_{t-1}^{-1} + H_t^T R_t^{-1} H_t\}^{-1} \tag{7.22}$$

Using the inverse matrix correction formula, the measurement update equation of covariance matrix can be expressed as

$$P_t = P_t^- - \underbrace{P_t^- H_t^T [R_t + H_t P_t^- H_t^T]^{-1}}_{K_t} H_t P_t^- \tag{7.23}$$

Kalman gain, K_t, is also a matrix, which is the weighting matrix of the residual and the intermediate result of filtering calculation. K_t first weighs the size of the predicted state covariance matrix and the observed value matrix, and considers whether it is more inclined to believe in the prediction model or the specific observation model. K_t can be expressed as

$$K_t = P_t^- H_t^T [R_t + H_t P_t^- H_t^T]^{-1} \tag{7.24}$$

The recursive update equation of the corresponding estimate is

$$\begin{aligned} \hat{x}_t &= P_t \left\{ \sum_{i=1}^{t} H_i^T R_i^{-1} z_{DBNi} \right\} \\ &= \{P_t^- - K_t H_t P_t^-\} \left\{ \sum_{i=1}^{t-1} H_i^T R_i^{-1} z_{DBNi} + H_t^T R_t^{-1} z_{DBNt} \right\} \\ &= \{I - K_t H_t\} \left\{ P_t^- \sum_{i=1}^{t-1} H_i^T R_i^{-1} z_{DBNi} + P_t^- H_t^T R_t^{-1} z_{DBNt} \right\} \\ &= \{I - K_t H_t\} \{\hat{x}_t^- + P_t^- H_t^T R_t^{-1} z_{DBNt}\} \\ &= \hat{x}_t^- - K_t H_t \hat{x}_t^- + \{I - K_t H_t\} P_t^- H_t^T R_t^{-1} z_{DBNt} \end{aligned} \tag{7.25}$$

The observation matrix in the state updating of KF is replaced by DBN result and expressed as

$$\hat{x}_t = \hat{x}_t^- + K_t(z_{DBNt} - H_t\hat{x}_t^-) \tag{7.26}$$

7.3.2 RUL Calculation

RUL is an assessment of future working time, usually based on the final performance (Hu et al. 2023). RUL is calculated by degradation state, x, time, t, and fault threshold, x_{th}, which is determined by the failure modes of specific objects. If cumulative damage is used to represent the performance index, RUL can be expressed as

$$RUL = \inf\{t : x(t) \geq x_{th} | x(0) < x_{th}\} \tag{7.27}$$

The prediction results of RUL also contain a large amount of uncertainty information, which is often expressed in the form of probability density function (PDF), as shown in Fig. 7.5. The mean value of RUL probability distribution is often compared with the actual RUL to analyze the accuracy of the prediction method.

7.4 Cases of Subsea Valves

Subsea valves play a crucial role in the oil and gas industry by controlling the flow of fluids in oil wells or natural gas wells to maintain the stability of the production system. Typically installed on the seafloor, these valves operate in complex environmental and high-pressure conditions. Their functions include opening, closing, or adjusting fluid flow to ensure the normal operation of the production system. Due to the extremely high installation and maintenance costs associated with subsea valves, predicting their RUL becomes of paramount importance. Utilizing advanced

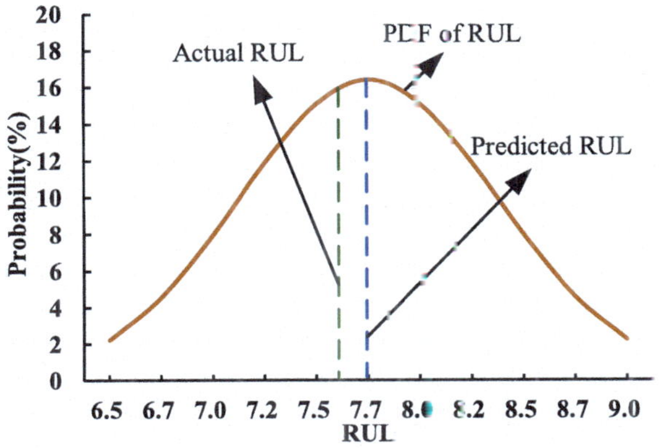

Fig. 7.5 Distribution and value of RUL

sensors and monitoring technologies allows real-time tracking of the valve's status and performance. Based on data analysis and predictive models, the prediction of the valve's RUL becomes possible. This not only aids in early detection of potential failures, reducing the risk of production interruptions and accidents but also optimizes maintenance schedules, leading to an overall reduction in operational costs (Tobon-Mejia et al. 2012). However, in the oil and gas industry, the status recording intervals of subsea valves are relatively large, which results in highly discrete monitoring data. Therefore, the monitoring data of subsea valves can be well applied to the proposed method for obtaining health status and RUL.

The subsea Christmas tree can be defined as a combination of valves and fittings with production and injection functions to control the flow of products, chemicals, water, or gas into the oil well, with valves, pipes, and joints. For example, the production main valve and the production wing valve are used to control the oil transportation, and the two chemical injection valves accurately control the injection flow of glycol and inhibitor. During normal production, other valves can be used to balance the pressure in the upper and lower space of the pipe hanger. Therefore, the main components of the subsea Christmas tree are the subsea valves, and degradation of the valves is one of the main failure forms of the subsea Christmas tree. The accurate prediction of its RUL is an important guarantee to ensure the normal operation of the subsea Christmas tree. In this work, a subsea valves case is selected to demonstrate the proposed method.

7.4.1 Case with Small Sample Data

(1) Subsea valve monitoring data

Figure 7.6 shows the evaluation process for this case. The status data of the valve is collected according to the sensor and transmitted to the monitoring platform. Based on the monitoring data of small samples, the proposed method is used to estimate the HI of the valve and calculate the RUL. The calculation results are verified based on small sample reliability tests. Use the high-pressure chamber to simulate the real high-pressure environment of the valve, conduct timed tail-cutting tests on the valve, count the failure of the valve, and finally obtain the reliability and life data of the valve. These data are used for accuracy comparison with this method and other state-of-the-art methods.

It is common to use pressure data to assess valve health, which helps monitor valve performance and detect potential problems in a timely manner. The pressure on both sides of the valve, as well as the velocity and volume of fluid passing through the valve, are usually measured by installed sensors. Before a valve is put into use, a baseline is usually established to record the pressure data and performance parameters of the valve under normal operating conditions, and the optimal pressure value and failure pressure value are obtained through design standards and expert experience. Valve testing is usually periodic testing, and this case uses periodic testing results

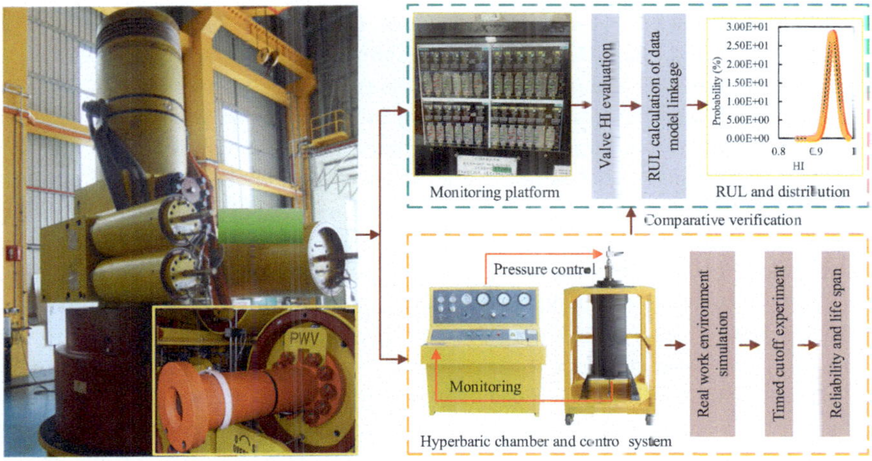

Fig. 7.6 Valve RUL estimation and verification framework

for analysis, which is shown in Fig. 7.7. The blue markers represent the pressure data recorded by periodic detection. The gray dotted line represents the historical monitoring records, while the brown dotted line represents the pressure data to be predicted. The red dashed line represents the optimal state value of the valve, indicating 100% performance when the valve pressure is in this state, while the green dashed line represents the failure pressure threshold of the valve, indicating that when the pressure drops to this value, the valve is no longer considered to have pressure holding capacity, that is, it fails.

(2) HI assessment of valve

The HI of the valve is obtained based on the pressure data detected. Firstly, when the difference between the valve pressure and the optimal state is 0, it indicates

Fig. 7.7 Pressure monitoring data of valve

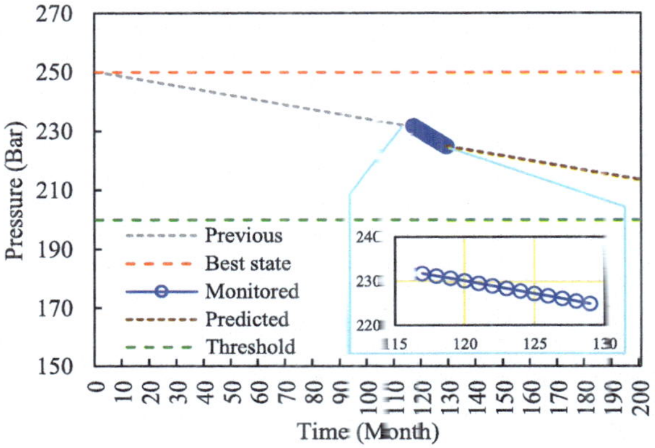

that the HI has reached its maximum value of 1; When the pressure threshold of the valve is reached, the difference from the optimal state is around 50, and the HI approaches 0. Use this feature to complete the parameter estimation of a and b in (1). The BN of valve HI evaluation is shown in Fig. 7.8. The BN mainly consists of four parts. First, the parameter b is estimated using the characteristic that the smaller the difference, the closer HI is to 1. It involves node $HI_{\to 1}$ with HI close to 1, node $\varepsilon_{\to \min}$ with difference 0, and parameter node b. The blue arrow indicates the estimation process of the parameter b. After determining the parameter b, parameter a is estimated using the characteristic that the greater the difference, the closer the HI is to 0. It involves parameter b node, $HI_{\to 0}$ node with HI close to 0, pressure threshold node F, best state node B, node $\varepsilon_{\to \max}$ with the difference close to the maximum, and parameter node a, where node F and node B are used to determine the maximum value of the difference and its distribution. The orange arrow depicts the node relationship during the evaluation process of parameter a. Subsequently, the evaluation of HI at the current moment is completed using parameter a, parameter b and monitoring data. This calculation process is represented by a green arrow and involves the best state node B, monitoring node M, difference node ε, parameter a node and parameter b node. Finally, the monitoring data at different moments are evaluated as HI respectively, where n represents the number of monitoring data.

The parameters related to the HI calculation are listed in Table 7.1. The probability information of parameter a and parameter b is obtained through the BN calculation. Due to differences in valve design and working environment, there are slight differences in the optimal state and pressure threshold of different valves. Therefore, this work gives a normal distribution for this type of valve to represent the uncertainty of parameters. In addition, the monitoring data is obtained by sensors

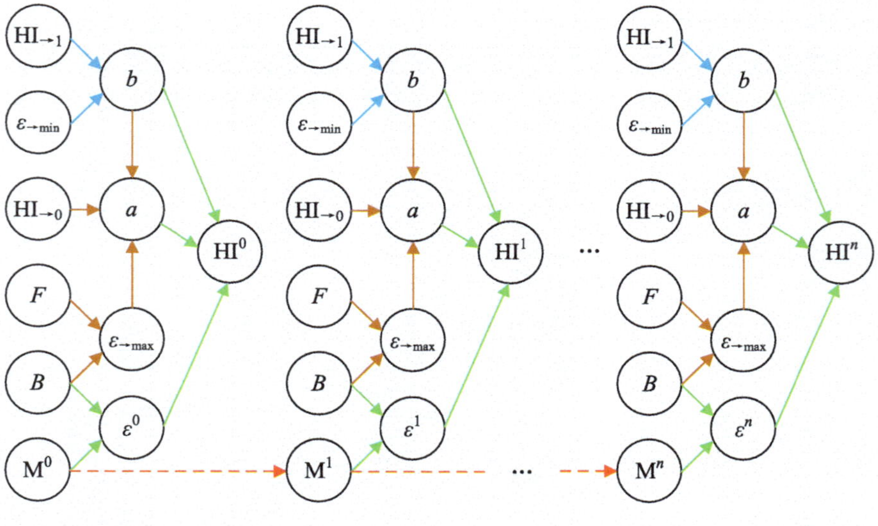

Fig. 7.8 BN model for HI evaluation

Table 7.1 The parameters related to the HI calculation

Parameters	Distribution	Mean	Standard deviation
Best state	Normal	250	5
Pressure threshold	Normal	200	5
Monitoring data	Normal	–	5
a	Normal	0.0032	0.001
b	Normal	4	0.5

and usually contains a certain amount of white noise, so this chapter also gives a normally distributed monitoring value.

The relationship between the difference between the pressure detection value and the optimal state and HI is shown in Fig. 7.9. The small sample monitoring data is converted into HI by calculating the difference with the best state value. As the difference gradually increases, the rate of HI reduction accelerates. A more microscopic HI variation within the range of 16–28 differences is given, which is the range of differences for detecting pressure data. The overall distribution of HI corresponding to the detected pressure ranges from 0.86 to 0.96. When considering parameter uncertainty, the characteristics of BN are utilized to estimate the HI distribution, as shown in Fig. 7.10. The figure shows the probability distribution of HI from the initial detection time to the 12th month. As time goes on, the HI of the valve gradually decreases, and its probability distribution explains this well. In addition, the overall distribution of HI is approximately a normal distribution, gradually diverging from concentration over time, indicating that the uncertainty of the prediction results is gradually increasing.

(3) RUL calculation of valve

The mean of HI estimated using BN is used for data augmentation. Figure 7.11 shows the generated HI data. In order to ensure that the generated data meets expectations, Kolmogorov–Smirnov (KS) test is usually performed. The KS test is a non-parametric

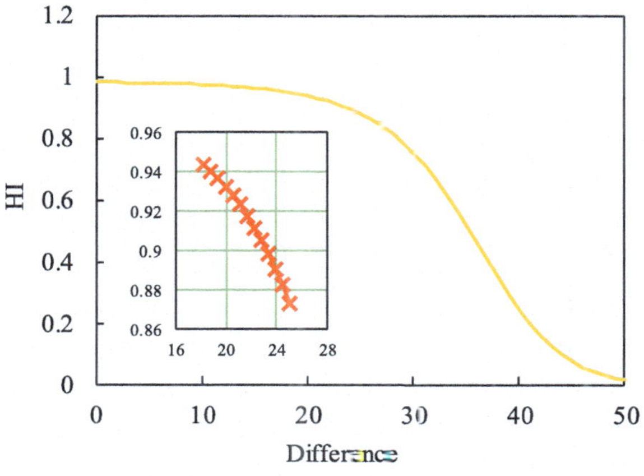

Fig. 7.9 The relationship between pressure difference and HI

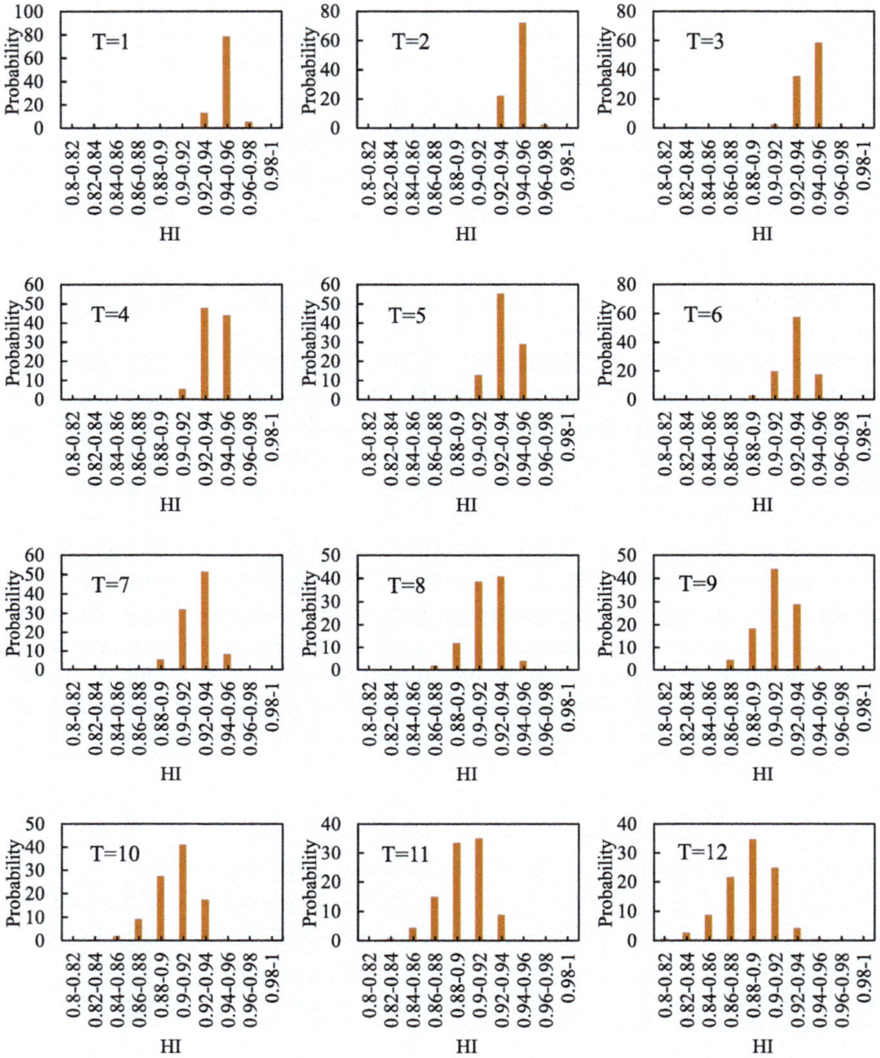

Fig. 7.10 The probability distribution of HI

statistical method used to test whether two samples come from the same continuous distribution. The relevant parameters involved in the KS test are shown in Table 7.2. The p value represents the significance level. If it is much greater than 0.05, it means that the test result is H0, and the generated data and the original data are from the same distribution.

The expanded HI data are then converted into reliability data and used for two-parameter estimation of the Weibull distribution. Furthermore, the 95% confidence intervals of the two parameters are estimated by the Bootstrap method. The calculated

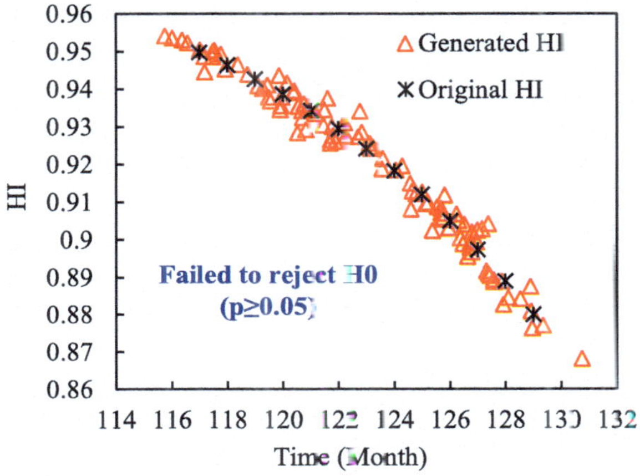

Fig. 7.11 Data augmentation based on GMM

Table 7.2 Relevant parameters of KS test

Parameters	Description	Value
H0 hypothesis	Original data and generated data come from the same distribution	–
H1 hypothesis	Original data and generated data come from different distributions	–
p	The significance level for rejecting the null hypothesis	0.999848
ks	The maximum difference between two cumulative distribution	0.095385

Weibull parameters are listed in Table 7.3. Figure 7.12 shows the estimated cumulative distribution function of reliability. After 120 months, the reliability begins to decline rapidly and reaches 0.5 around 150–160 months. The 95% confidence intervals of λ and k obtained before data expansion are [158.98, 162.68] and [9.06, 9.88], while the 95% confidence intervals of λ and k obtained using the expanded data are [159.43, 162.18] and [9.13, 9.75]. The difference is very obvious in the figure. Data augmentation provides more confident intervals for Weibull parameter estimates, and accuracy is improved. Figure 7.13 shows the PDF of reliability. 150–160 months has the highest probability. It can also be seen that the PDF estimated using the augmented data has a smaller range.

Figure 7.14 shows the performance comparison of different RUL prediction methods on valve. It can be observed that the proposed method performs well in

Table 7.3 Weibull parameters and confidence intervals

Parameters	λ	k
Estimation	160.77	9.43
95% confidence interval (before expansion)	[158.98, 162.68]	[9.06, 9.88]
95% confidence interval (after expansion)	[159.43, 162.18]	[9.13, 9.75]

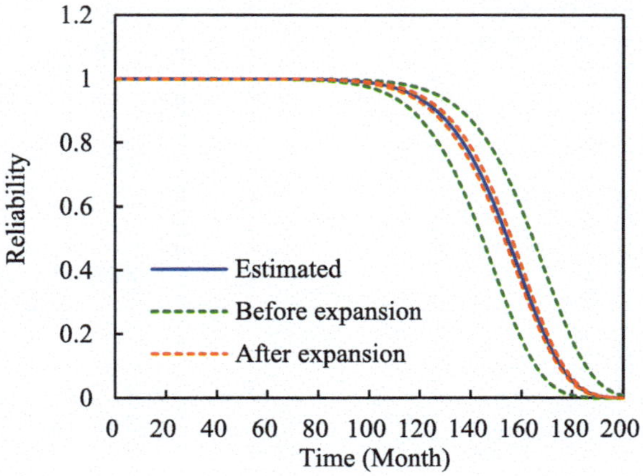

Fig. 7.12 The cumulative distribution function of reliability

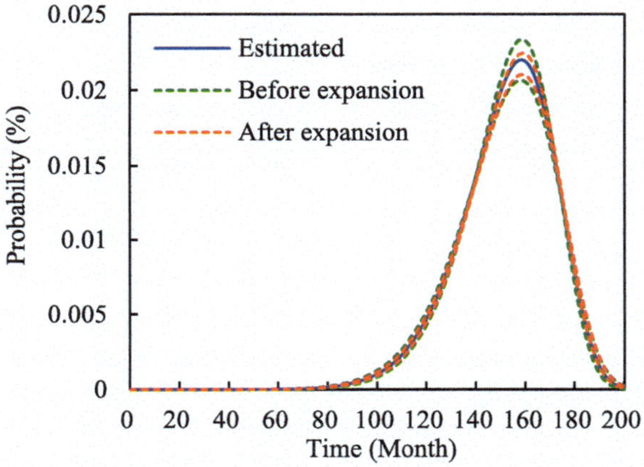

Fig. 7.13 The PDF of reliability

RUL prediction and is very close to the life span of the experimental data (163–206 months), with a predicted value of 169 months. This demonstrates that proposed method is highly reliable in accuracy and can effectively estimate the RUL of the valve. The performance of other methods is also striking compared to the proposed method. For example, the predicted RUL of the long short-term memory (LSTM) method is 151 months, which is relatively low; the predicted value of the convolutional neural network (CNN) method is 146 months, the autoregressive integrated moving average (ARIMA) method is 216 months, the polynomial fitting is 125 months, and the exponential fitting is 135 months. There is a large gap between the predicted values of these control groups and the experimental data. The proposed method may strike a balance among these aspects, making it superior in RUL prediction. In addition, among the control methods, the higher prediction value of the ARIMA method may be related to its sensitivity to time series trends, while the

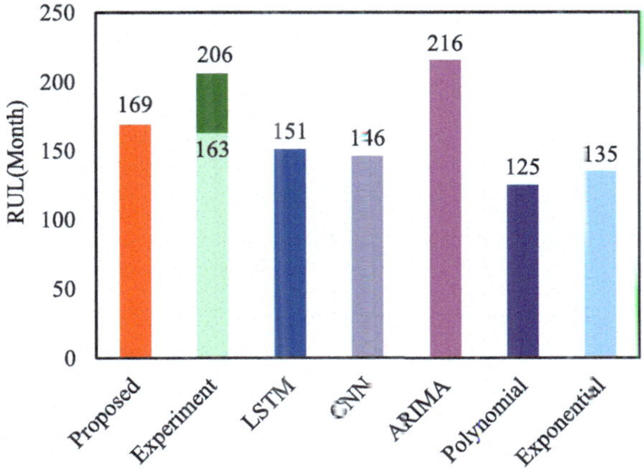

Fig. 7.14 RUL predicted by different methods

results of polynomial fitting and exponential fitting may be too simplified and fail to capture the complex dynamic changes of the system.

(4) Modeling parameter analysis

Random errors permeate the entire process of HI evaluation, such as noise in state monitoring data and probabilistic parameters in degradation assessment, introducing uncertainty into the evaluation of HI (Shi et al. 2019). Therefore, conducting sensitivity analysis of parameters *a* and *b* of HI evaluation is necessary. The HI is constructed based on the difference between monitoring data and the optimal state, *a* and *b*, therefore the analysis of the impact of parameters on HI is carried out. When $b = 4$, the relationship between the difference and HI corresponding to different *a* is shown in Fig. 7.15. When the difference is less than 20, the impact of parameter *a* on HI evaluation is very small. When the difference is greater than 20, the larger the value of *a* under the same difference, the smaller the calculated HI. When *a* remains unchanged and *b* is 2, 3, 4, 5, and 6, the relationship between the difference and HI obtained is shown in Fig. 7.16. It can be seen that when the value of *b* is different, there is a significant deviation in the corresponding HI when the difference is very small, even when the difference is 0. Under the same difference, the larger the value of *b*, the larger the calculated HI.

Similarly, the impact of different *a* and *b* on the reliability evaluation results has been studied, as shown in Figs. 7.17 and 7.18. Using the same set of monitoring data, estimate the Weibull distribution parameters of the corresponding parameter combinations based on the HI corresponding to Figs. 7.15 and 7.16, and then obtain the corresponding reliability under different parameters. There is almost no difference in the reliability evaluated within the first 100 months under different values of *a* set in this case, and 100 months later, the larger the value of *a*, the faster the reliability decreases. Due to the uncertainty of the optimal state value, there is a fluctuation range in the calculation of parameter *b*. When the difference is 0, the HI is determined solely by parameter *b*. As a result, varying *b* values will lead to slightly different HI when

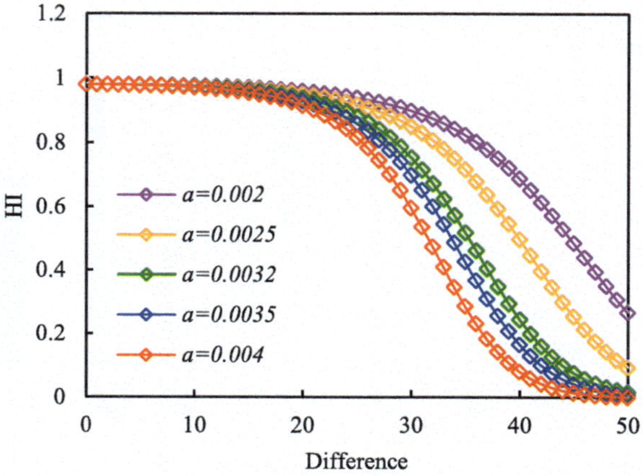

Fig. 7.15 The impact of parameter *a* on HI evaluation

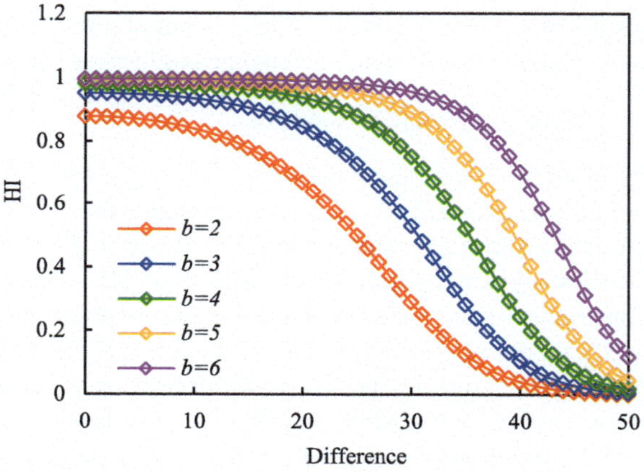

Fig. 7.16 The impact of parameter *b* on HI evaluation

the difference is 0. For the different *b* values set in this case, the impact of *b* on the reliability estimation results in the first 60 months is very small. Differences begin to appear after 60 months, and the larger the value of *b*, the higher the reliability.

7.4.2 Case with Multi-sample Data

(1) Wiener Process

Wiener process is an important independent incremental procedure, sometimes called Brownian motion process (Li et al. 2023), which has been used to describe the degradation of subsea Christmas trees for many years. The current value of the process is all the information needed to make its future prediction. Wiener process has independent increment. The probability distribution of the process in any time

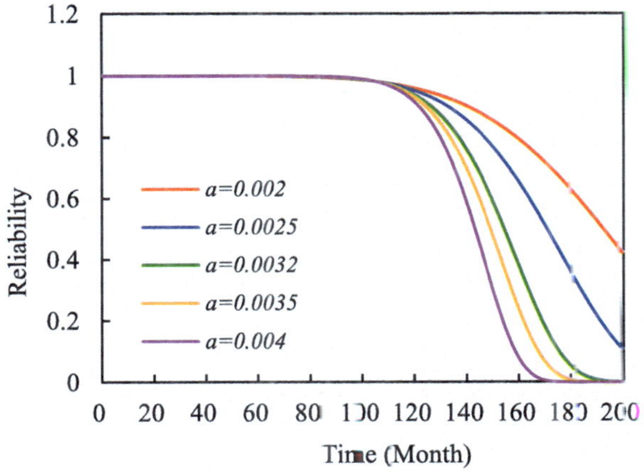

Fig. 7.17 The impact of parameter *a* on reliability estimation

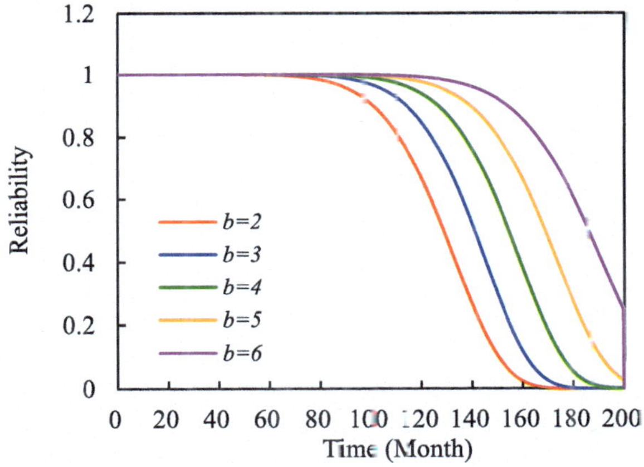

Fig. 7.18 The impact of parameter *b* on reliability estimation

interval is independent of its probability in any other time interval. It has the increment of normal distribution at any finite time, and the variance is positively correlated with the time interval. The main parameters of Wiener process include drift coefficient, λ, and diffusion coefficient, σ_B. Wiener process can be described as

$$W(t) = W(0) + \lambda t + \sigma_B BM(t) \tag{7.28}$$

where $W(0)$ is the initial state, and $BM(t)$ is the Brownian motion with time t.

Considering the natural obstructions within the information collection preparation, the measured value, $Z(t)$, can be expressed as

$$\begin{cases} Z(t) = W(t) + \zeta \\ \zeta \sim N(0, \sigma_\xi^2) \end{cases} \tag{7.29}$$

where ζ is the error in the measurement.

(2) Parameter Estimation

Assuming a device has been monitored n times, and each monitoring includes i time points, $X_{n,ti}$ is the state value at time point t_i in the nth monitoring. $\boldsymbol{X}_n = \{\Delta X_{n,t1}, \Delta X_{n,t2}, \ldots, \Delta X_{n,ti}\}'$, where $\Delta X_{n,ti}, X_{n,ti} - X_{n,ti-1}$. $\boldsymbol{T}_n = \{\Delta T_{n,t1}, \Delta T_{n,t2}, \ldots, \Delta T_{n,ti}\}'$, where $\Delta T_{n,ti}$ is the interval between two time points in the nth monitoring. The estimated parameters mainly include drift coefficient and diffusion coefficient, that is, $\Theta = (\lambda_1, \lambda_2, \ldots, \lambda_n, \sigma_B^2)$. Wiener process has a normal distribution increment in any finite time, and its variance has a linear relationship with the time interval. Therefore, $\boldsymbol{X}_n$ follows a multivariate normal distribution with the mean of $\lambda_n \boldsymbol{T}_n$ and covariance of $\sigma_B^2 \,\mathrm{diag}(\Delta T_{n,t1}, \Delta T_{n,t2}, \ldots, \Delta T_{n,ti})$ (Xu et al. 2019), where diag $(\Delta T_{n,t1}, \Delta T_{n,t2}, \ldots, \Delta T_{n,ti})$ represents a diagonal matrix whose diagonal elements are $\Delta T_{n,t1}, \Delta T_{n,t2}, \ldots, \Delta T_{n,ti}$.

In this work, the maximum likelihood algorithm is used for parameter estimation. Maximum likelihood algorithm is one of the commonly used parameter estimation methods, which has good application to degraded data in this case. The likelihood function based on Wiener process (Xu et al. 2019) can be expressed as

$$\begin{aligned}\ln L(\Theta|X) = &-\frac{1}{2}\ln 2\pi ni - \frac{1}{2}\ln \sigma_B^2 ni - \frac{1}{2}\sum_{n=1}^{i} \ln|\Phi_n| \\ &- \frac{1}{2\sigma_B^2}\sum_{n=1}^{i} (X_n - \lambda_n X_n)'|\Phi_n|^{-1}(X_n - \lambda_n X_n)\end{aligned} \tag{7.30}$$

where ni is the total monitoring points of the equipment, and $\Phi_n = \mathrm{diag}(\Delta T_{n,t1}, \Delta T_{n,t2}, \ldots, \Delta T_{n,ti})$.

The extreme value can be computed by calculus as

$$\frac{\partial}{\partial \lambda_n} \ln L(\Theta|X) = -\frac{-T_n'\Phi_n^{-1}X_n + \lambda_n T_n'\Phi_n^{-1}T_n}{\sigma_B^2} = 0 \tag{7.31}$$

$$\frac{\partial}{\partial \sigma_B^2} \ln L(\Theta|X) = -\frac{ni}{2\sigma_B^2} + \frac{1}{2(\sigma_B^2)^2}\sum_{n=1}^{i}(X_n - \lambda_n T_n)'\Phi_n^{-1}(X_n - \lambda_n T_n) = 0 \tag{7.32}$$

The parameter set, $\Theta = (\lambda_1, \lambda_2, \ldots, \lambda_n, \sigma_B^2)$, can be expressed as

$$\hat{\lambda}_n = -\frac{T_n'\Phi_n^{-1}X_n}{T_n'\Phi_n^{-1}T_n} \tag{7.33}$$

$$\hat{\sigma}_B^2 = \frac{\sum_{n=1}^{i}(X_n - \lambda_n T_n)'\Phi_n^{-1}(X_n - \lambda_n T_n)}{ni} \tag{7.34}$$

The mean and variance of the drift coefficient are

$$\hat{\mu}_{\lambda} = \frac{1}{N}\sum_{n=1}^{n}\hat{\lambda}_{n} \tag{7.35}$$

$$\hat{\sigma}_{\lambda}^{2} = \frac{1}{N}\sum_{n=1}^{n}\left(\hat{\lambda}_{n} - \hat{\mu}_{\lambda}\right)^{2} \tag{7.36}$$

(3) Hybrid DBN-KF-based Method with Wiener Process

The premise of KF calculation is to obtain the state space model of the degradation model. The primary task of the hybrid DBN-KF-based method with Wiener process is to transform the Wiener process into state space model as

$$\begin{cases} x_t = x_{t-1} + \lambda_{t-1}t + \sigma_B BM(0, \Delta t) \\ y_t = x_t + \xi \end{cases} \tag{7.37}$$

Without considering the uncertainty, $\lambda_t = \lambda_{t-1}$. Therefore, the state space model can be transformed into

$$\begin{cases} \begin{bmatrix} x_t \\ \lambda_t \end{bmatrix} = \begin{bmatrix} 1 & \Delta t \\ 0 & 1 \end{bmatrix}\begin{bmatrix} x_{t-1} \\ \lambda_{t-1} \end{bmatrix} + \begin{bmatrix} \sigma_B^2 \Delta t \\ 0 \end{bmatrix} \\ y_t = \begin{bmatrix} 1 & 0 \end{bmatrix}\begin{bmatrix} x_{t-1} \\ \lambda_{t-1} \end{bmatrix} + \xi \end{cases} \tag{7.38}$$

which can be simplified as

$$\begin{cases} X_t = F_t X_{t-1} + B_t \\ y_t = HX_{t-1} + \xi \end{cases}, \tag{7.39}$$

where $X_t = [x_t, \lambda_t]'$, $F_t = [1, \Delta t; 0, 1]$, $B_t = [\sigma_B^2, 0]'$, and $H = [1, 0]$.

The predicted noise is

$$Q_t = \begin{bmatrix} \sigma_B^2 \Delta t & 0 \\ 0 & 0 \end{bmatrix} \tag{7.40}$$

The initial condition is

$$X_0 = \begin{bmatrix} 0 \\ \mu_{\lambda 0} \end{bmatrix} \tag{7.41}$$

$$P_0 = \begin{bmatrix} 0 & 0 \\ 0 & \sigma_{\lambda 0}^2 \end{bmatrix} \tag{7.42}$$

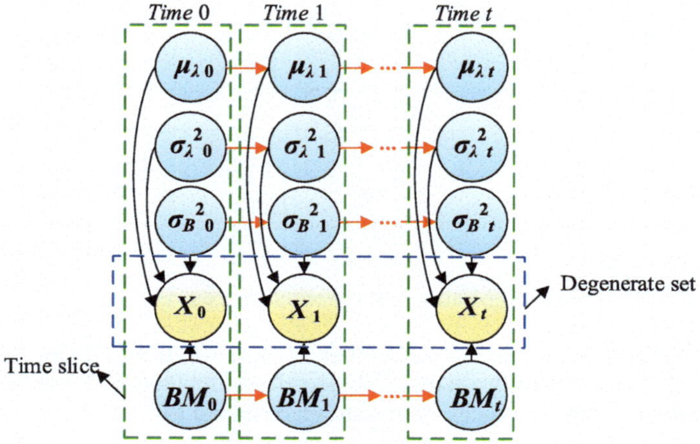

Fig. 7.19 Degradation of Wiener process based on DBN

The degradation of Wiener process based on DBN is shown in Fig. 7.19. The main nodes of DBN are μ_λ, ${\sigma_\lambda}^2$, ${\sigma_B}^2$, BM, and X. These nodes represent the main parameters of the Wiener process. Through time expansion, the degradation of each time is calculated, and the degradation set of the Wiener process is obtained to determine the variance and the observation noise. The degenerate set of X is obtained through time extension calculation. The construction of hybrid DBN-KF-based method with Wiener process is shown in Algorithm 1.

Algorithm 1: Hybrid DBN-KF-based method with Wiener process

Inputs: Data set, n group (online observing, online database, chronicled information, etc.)
Outputs: The updated degradation based on Wiener process, $\boldsymbol{X}$

```
for i = 0 to n do
    Calculate Θ by (7.22) and (7.23)
end for
    Calculate mean and variance of λ by (7.24) and (7.25)
    Substitute the calculated Θ into (7.26);
for j = 0 to t do
    State prediction by (7.28)
    Calculate observation, z_DBN, by DBN
    State update by (7.12)–(7.15)
    Obtain optimized estimation, x_t
end for
    Obtain the updated degradation, X (x_1, x_2 … x_t)
```

Using the hybrid DBN-KF-based method, the state of the current time, x_t, can be calculated. The PDF of RUL based on Wiener process can be described as

$$f(t) = \sqrt{\frac{(x_{th} - x_t)^2}{2\pi \sigma_B^2 t^3}} \exp\left(- \frac{(x_{th} - x_t - \overline{\lambda} t)^2}{2\sigma_B^2 t} \right) \tag{7.43}$$

The degradation of valves refers to the statistical data on the failure rate of subsea Christmas tree in the OREDA handbook (DNV 2015). After evaluation, the life range of subsea Christmas tree valves is concentrated around 200 months. The failure rate of the valve is estimated as 0.008 by random sampling, which is equivalent to the degradation parameter in the Wiener process. The variation range is completed by Monte Carlo simulation. The monthly degradation of ten subsea valves is shown in Fig. 7.20. The ten valves have the same material, size, and stress environment, so their corresponding parameters are the same, as shown in Table 7.4. In the degradation modeling of DBN, the parameter, ${\sigma_B}^{2*}$ is a random parameter that replaces ${\sigma_B}^2$ in the uncertainty analysis and represents the floating of the environment parameters and state parameters.

(4) Results and Discussions

1. Cumulative degradation analysis

The parameters of the initial KF are considered constant, that is, $\lambda_{t-1} = \lambda_t = 4.1 \times 10^{-3}$, and ${\sigma_B}^2{}_{t-1} = {\sigma_B}^2{}_t = 9.9014 \times 10^{-4}$. The hybrid DBN-KF-based method incorporates more uncertainties, and a variable diffusion coefficient, ${\sigma_B}^{2*}$, is used to weigh this change. The parameters, λ and ${\sigma_B}^2$, fluctuate around the mean with

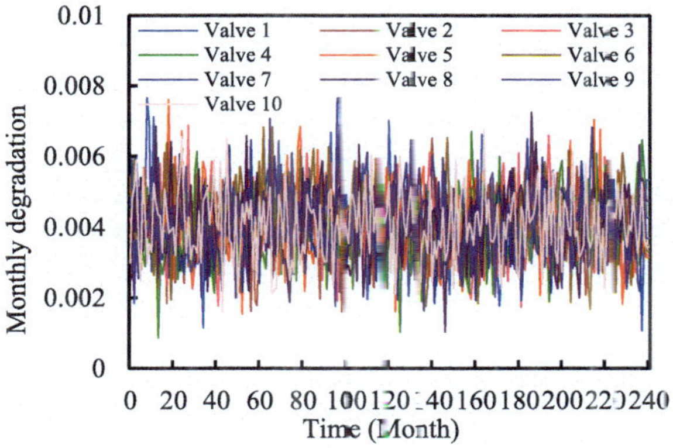

Fig. 7.20 Monthly degradation of 10 subsea valves

Table 7.4 Parameters of degradation calculation

Parameter	Distribution	Value
λ	Normal	[0.0041, 0.0041, 0.0041,0.0040, 0.0042, 0.0042, 0.0040, 0.0041, 0.0042, 0.0041]
μ_λ	–	0.0041
${\sigma_\lambda}^2$	–	5.7311e−09
${\sigma_B}^2$	–	9.9014e−04
${\sigma_B}^{2*}$	Normal	(9.9014−04, 1.3e−05)
R	–	1×10^{-6}

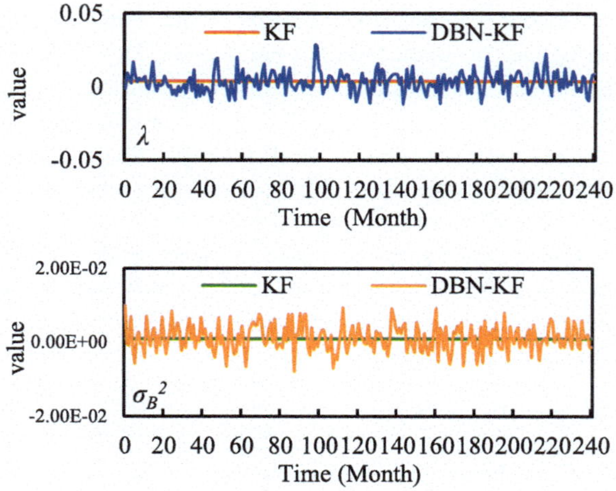

Fig. 7.21 Monitoring status of parameters λ and ${\sigma_B}^2$ over time

the variances of 5.7311×10^{-9} and 1.3×10^{-5}, respectively, which represents an uncertain analysis in degradation. The floating range of parameters λ and ${\sigma_B}^2$ over time is shown in Fig. 7.21. In the proposed method, the parameters change with time. Compared with other traditional stochastic process modeling, the randomness analysis using DBN is more intuitive and can provide the probability distribution of model parameters at any prediction stage.

The cumulative degradation of KF and hybrid DBN-KF-based method is shown in Fig. 7.22. The optimization is an improved selection based on estimation and observation, which is realized by KF filtering. The general drift of cumulative degradation is straightly expanding, and the distinction from the starting to the conclusion is not self-evident, showing the expectation blunder and perception mistake is moderately minor. The model rapidly comes to the ideal state of expectation.

The DBN-KF method has the same observation noise as the KF method, but with smaller fluctuations due to the use of DBN for noise handling. While the overall trend of cumulative degradation calculated by the DBN-KF method is similar, the degradation speed differs slightly, with the DBN analysis result being a mean value that comprehensively considers probability. KF, on the other hand, estimates each degradation amount randomly and without considering the mean value under randomness, leading to deviations in the degradation path between the two. Although DBN considers global uncertainty, it can sometimes be overly conservative in local estimation. for example, in the 70th to 90th months, the DBN calculation results deviate significantly, which the algorithm acknowledges and responds by considering the estimated value more reliable. Compared to DBN, our proposed method has greater advantages in global stability and error control.

Figure 7.23 compares actual degradation, KF, DBN, and hybrid DBN-KF-based methods. Actual degradation is generally faster, with the proposed method showing better accuracy and closer proximity to actual degradation. The KF calculation is consistent with expectation at early stages but deviates significantly after 90 months. The proposed method improves prediction accuracy by integrating uncertain parameters with DBN and reducing prediction error through KF. The proposed method is

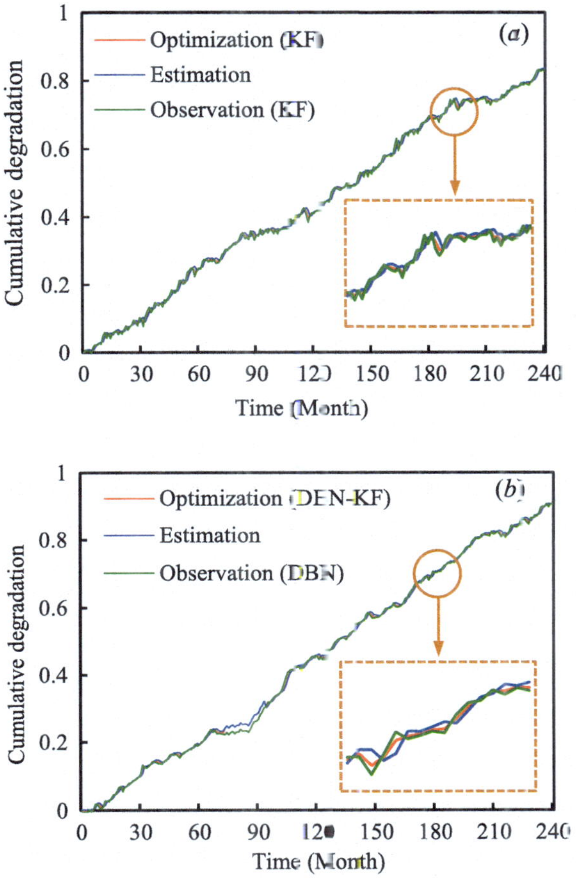

Fig. 7.22 Degradation of **a** KF and **b** hybrid DBN-KF-based method

significantly improved compared with the KF method. The overall degradation of the DBN-KF method is similar to that of DBN, but local differences indicate the proposed method is closer to the true degradation.

Figure 7.24 shows the deviation between actual degradation and predictions from optimization (DBN-KF), optimization (KF), observation (DBN), and estimation. The deviation of observation (KF) is initially small but gradually increases, reaching a maximum of 0.17 after approximately 100 months. Optimization (DBN-KF), estimation, and observation (DBN) have similar trends with deviations fluctuating within 0.1 and showing stable overall trends. The proposed method (DBN-KF) has the smallest maximum deviation (< 0.1) and closest proximity to actual degradation. Although the degradation trend predicted by DBN alone is similar to the proposed method at the final stage, deviation analysis throughout the cycle demonstrates small prediction deviation for the proposed method in the medium-term of prediction (60th-140th months), whereas DBN alone has large fluctuating prediction accuracy. Overall, the proposed method exhibits better stability in predicting the entire lifecycle.

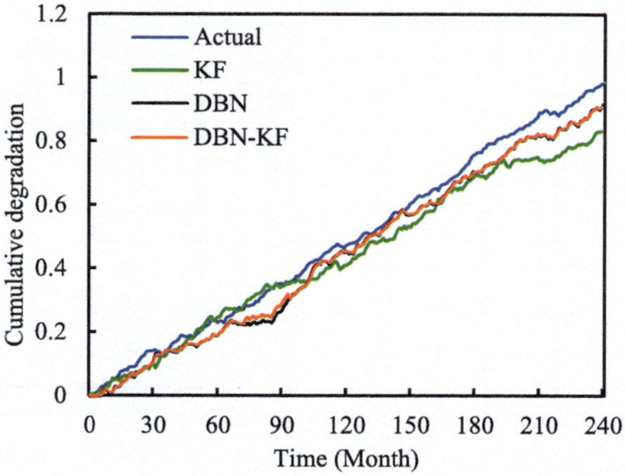

Fig. 7.23 Comparison of different simulation results and actual degradation

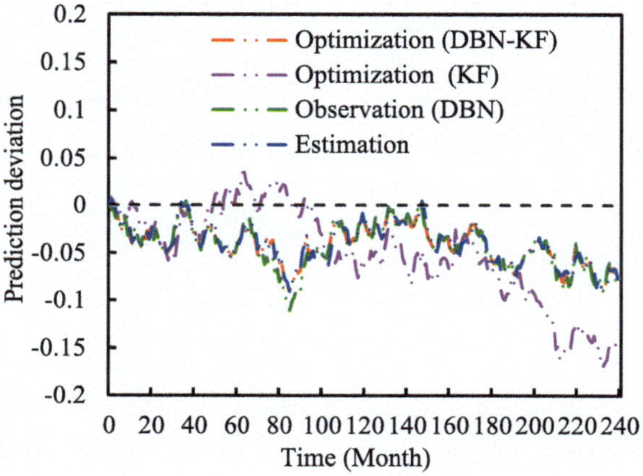

Fig. 7.24 Prediction deviation between simulation value and actual degradation

To quantify the advantages of the proposed method, error analysis of different methods and actual degradation are implemented, and the results are shown in Table 7.5. Sum of squares error (SSE), mean absolute error (MAE), mean square error (MSE), and root mean square error (RMSE) are used to measure the error size of the method. The proposed method has the minimum error under each indicator, indicating it has a higher accuracy over the entire life cycle. Correspondingly, compared with the KF method, the accuracy improvement of the proposed method is significant, with each index reduced by 63.9, 41.6, 64.2, and 40%. Compared with the DBN method, the proposed method is also more excellent in global error control, with indicators reduced by 28.1, 11.9, 28.6, and 15.2%.

2. Degradation parameters analysis

To explore the influence of observation noise on the prediction results, a set of observation noises with different magnitudes is input into the DBN-KF algorithm. The

Table 7.5 Error analysis of different methods

Methods	SSE	MAE	MSE	RMSE
KF	2.6927	0.0891	0.0112	0.1057
DBN	1.3534	0.059	0.0056	0.0749
DBN-KF	**0.9727**	**0.052**	**0.004**	**0.0635**

Bold indicates the error of the proposed method for comparison with other methods

percentage difference of estimation and observation represents the degree of deviation between optimization and the two, as shown in Fig. 7.25. When the observed noise is 1×10^{-5}, the difference of the initial stage fluctuates greatly. The algorithm evaluates that the noise is greater than the prediction noise of the system, so the estimated value is more reliable and has a smaller deviation in the whole process. On the contrary, when the observation noise is 1×10^{-8}, because the observation noise is smaller, and the algorithm considers the observation more reliable, the difference of the observation is smaller, and the estimation has larger fluctuations. When the observation noise is 1×10^{-6} (default) or 1×10^{-7}, only few symmetrical fluctuations exist at the initial stage, and the rest of the time is stabler overall, which shows the DBN-KF algorithm quickly weighs the observation and the estimation under this noise, reaching the state of stable prediction.

Considering the results of the current input, sensitivity analysis measures the extent to which the findings of any parameter can affect the belief in the predicted results. The quantitative results of the sensitivity analysis of parameters are more meaningful for degradation control and life design. Variance reduction and variance of belief are two important indicators of sensitivity. Parameter sensitivity analysis in the 180th month (the near-failure stage is used as a typical sensitivity analysis) is conducted, as shown in Fig. 7.26. At the stage approaching aging, the calculation result of the RUL is the most sensitive to the drift coefficient, λ and the least sensitive to the diffusion coefficient, σ_B, because the drift coefficient mainly affects the degradation speed and is the most important factor of RUL. The diffusion coefficient controls the floating of the degradation, but it has a small effect on the overall degradation trend. In the sensitivity analysis of thresholds, the failure threshold is a normally distributed variable. The sensitivity of the threshold is obtained by detecting the impact of small changes in the threshold on the prediction results. The threshold, x_{th}, also has a small effect on the RUL results, but the sensitivity is relatively weak.

3. RUL analysis

The prediction distribution of RUL is implemented at the initial stage of prediction (the first month), as shown in Fig. 7.27. The main parameters of the PDF include estimated drift coefficient and diffusion coefficient as well as the failure threshold (0.8) determined according to the valve characteristics. In this case, the failure threshold is obtained based on the experience of engineers and experts at the offshore oil development site, and is determined as a fixed value to indicate the valve is deemed to have failed when the failure rate reaches this value for the first time.

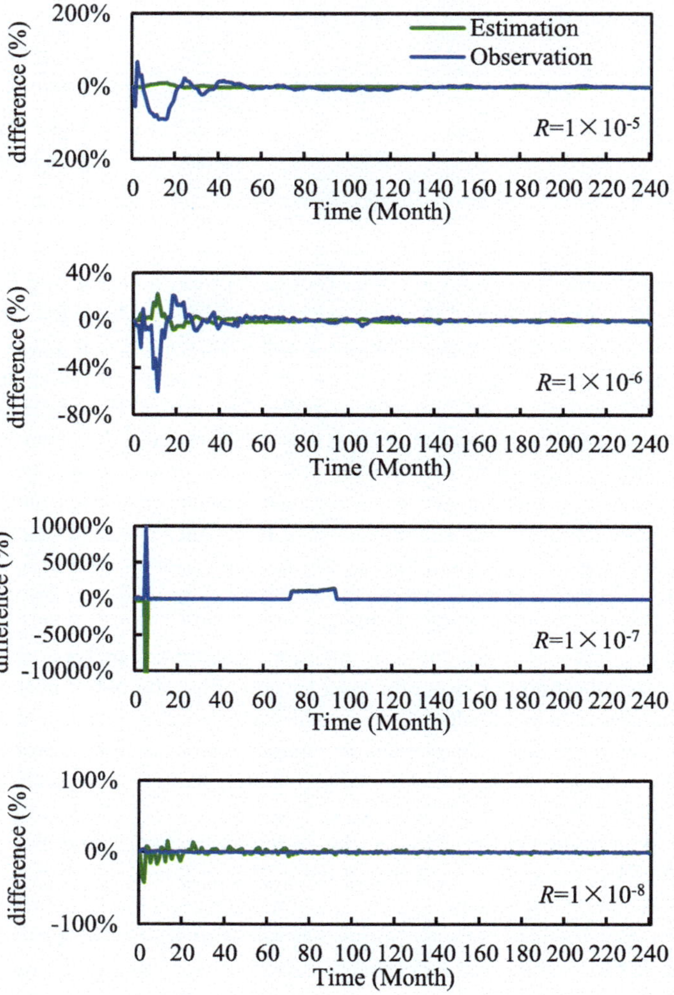

Fig. 7.25 Percentage difference of estimation and observation

When the valve degrades at this speed and the cumulative degradation reaches the failure threshold, the valve is deemed to have failed. The prediction results estimated according to KF, DBN, and DBN-KF are concentrated in about 190–200 months. Despite a slight difference between the four, the results calculated by the DBN-KF method are closest to the actual RUL distribution, including the mean, variance, or corresponding probability of the distribution.

The RUL calculated based on the hybrid DBN-KF-based method has been given more attention because it is closer to the actual value. The RUL distribution based on the proposed method is calculated on the 0, 50th, 100th, 150th, and 200th month, as shown in Fig. 7.28. At the beginning of the prediction stage, the prediction results

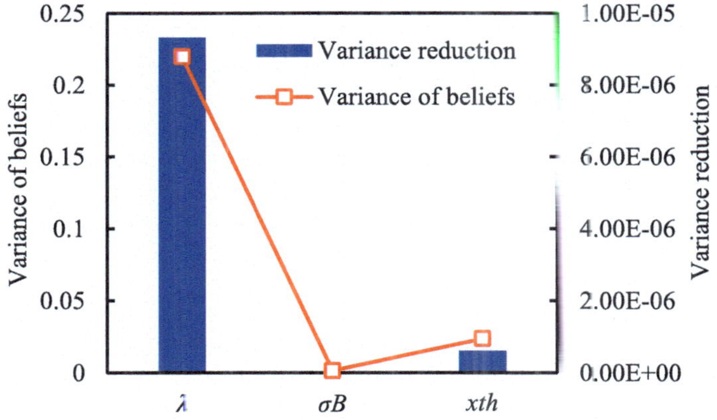

Fig. 7.26 Parameter sensitivity analysis results

Fig. 7.27 RUL probability density distribution for the first month

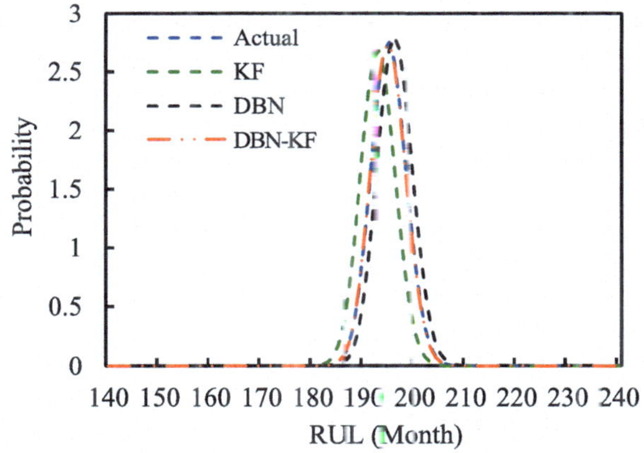

of the model are uncertain due to the lack of available degradation information, so the distribution range of RUL is large. As more degradation information is obtained, the confidence of the predicted RUL distribution increases. The whole degradation trajectory is linear.

Referring to the method of Fig. 7.28, the trend of RUL over time can be calculated. Generally, predicted RUL decreases with time. A large difference is noted in the overall degradation before and after adding the DBN data. Optimization (KF), optimization (DBN-KF), and actual degradation are used to calculate the trend of the RUL over time, as shown in Fig. 7.29. The predicted value of RUL degrades approximately linearly with time. About the one hundredth month is a critical point. Within one hundred months, the RUL calculated according to the optimization (DBN-KF) is higher than the result of the optimization (KF), and the opposite is true after one hundred months. Overall, the optimization (DBN-KF) is closer to the actual RUL, especially in the second half of the prediction. In addition, the predicted RUL of

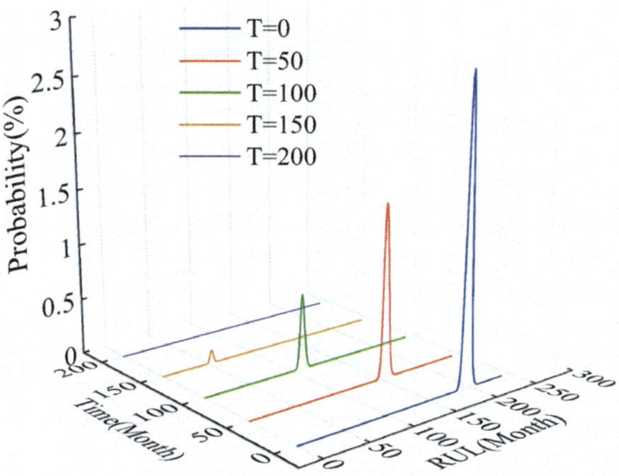

Fig. 7.28 RUL probability density distributions of hybrid DBN-KF-based method in the 0, 50th, 100th, 150th, and 200th month

optimization (DBN-KF) dropped to 0 in about 200 months, whereas it was around 230 months with optimization (KF). The actual degraded RUL decreased to 0 in 190–200 months, and the comparison found the optimized value obtained from DBN-KF method is more reliable.

When the predicted life is 0, it is considered to be the RUL value of the equipment. The RUL value of actual degradation, KF, DBN, and hybrid DBN-KF-based method is calculated according to the RUL change in Fig. 7.29. The RUL value of actual degradation, KF, DBN, and hybrid DBN-KF-based method is shown in Fig. 7.30. The error between the life prediction only through KF and the actual RUL is very large, more than 40 months. The RUL calculated by the improved KF is 200, which is the best prediction value among the three methods. The proposed method's ability to improve prediction accuracy is considered meaningful, even if the improvement is small. In offshore oil exploration, a two-month improvement in accuracy can

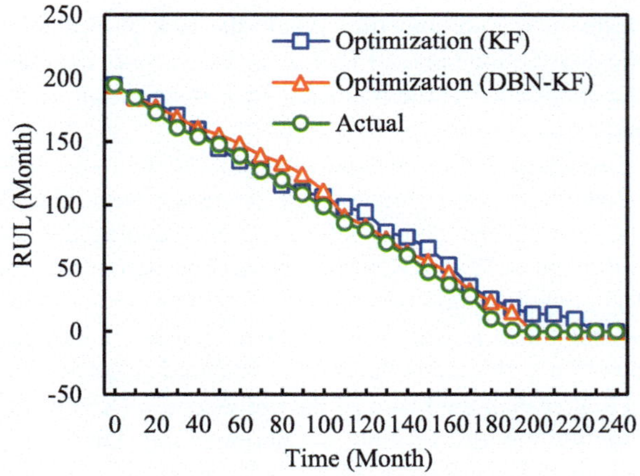

Fig. 7.29 Predicted RUL trend over time

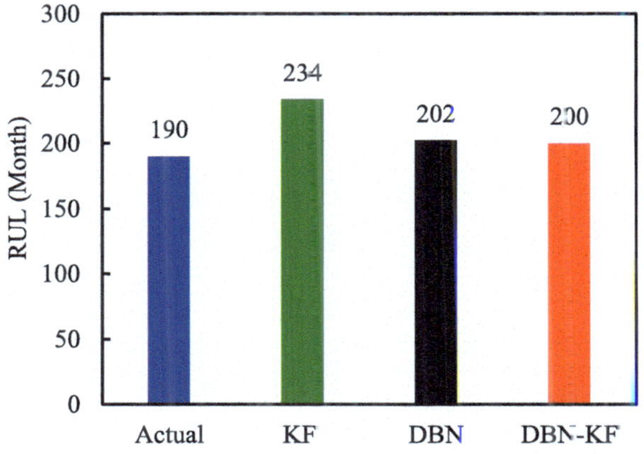

Fig. 7.30 RUL value of actual degradation, KF, DBN, and hybrid DBN-KF-based method

significantly enhance enterprise economic benefits. Moreover, compared to the DBN method, the proposed method provides a more reliable basis for maintenance decisions in the medium-term forecast period due to smaller errors over the entire life cycle, solving the problem of poor local accuracy in DBN prediction.

Chapter 8
RUL Prediction with Multiple Causes

Abstract In complex and dynamic environments, the degradation of structural systems typically results from multiple factors rather than a single one. This chapter proposes a hybrid physics-model-based and data-driven methodology for RUL prediction in structural systems, considering the influence of multiple degradation causes in dynamic complex environments. Using DBNs, the method models the degradation processes caused by various factors, such as fatigue, corrosion, sand erosion, and internal waves, which contribute to system uncertainty. By integrating theoretical or empirical physical models with DBNs, the approach addresses data insufficiency and builds a comprehensive RUL estimation framework. The RUL is calculated based on the time difference between the detection point and the predicted failure point, determined by a performance failure threshold. The model can be updated with sensor data and expert knowledge to refine RUL estimates in real-time. The methodology is demonstrated through its application to subsea pipelines in offshore oil and gas production systems, highlighting its effectiveness in accounting for multiple degradation factors in complex environments.

8.1 Introduction

Several significant challenges persist in the accurate estimation of RUL, particularly for structural systems. These systems often face degradation due to multiple factors, including fatigue, corrosion, and environmental stressors, making the modeling of their failure processes complex. Traditional physics-based methods struggle to capture the intricate degradation mechanisms due to the difficulty of developing precise models for multifaceted systems (Cui et al. 2016). Conversely, data-driven approaches, while capable of handling large datasets and identifying patterns, frequently suffer from the lack of sufficient high-quality data, leading to potential inaccuracies in RUL predictions. Thus, the current landscape reveals a pressing need for more effective methods that can integrate both physics-based insights and real-time data analysis.

B. Cai et al., *Intelligent Operation and Maintenance for Subsea Production Systems*,
https://doi.org/10.1007/978-981-96-2677-9_8

This research is motivated by the desire to overcome these limitations through the innovative application of a DBN framework. By combining physics-based modeling with data-driven techniques, this hybrid approach aims to enhance the robustness and accuracy of RUL predictions. The integration of DBNs allows for the representation of uncertainties and the influence of various degradation factors, facilitating a more comprehensive understanding of structural health. This dual approach not only addresses the shortcomings of existing methods but also provides a more versatile tool for practitioners in the field of maintenance management.

This chapter presents a novel hybrid estimation method for RUL prediction that integrates DBNs with both physics-based and data-driven approaches. The proposed framework is designed to effectively address the complexities associated with the degradation of structural systems, particularly in challenging environments such as subsea pipelines. Through a detailed case study, the method's applicability and effectiveness are demonstrated, showcasing its potential to significantly enhance CBM strategies. The insights gained from this research contribute to advancing the field and offer practical solutions for improving operational reliability and reducing maintenance costs.

8.2 RUL Estimation Model Based on Multiple-Cause Integration

8.2.1 Modeling Methodology

DBNs have a strong capability to represent the temporal and causal relationships among variables. A novel DBN-based hybrid physics-model-based and data-driven method is proposed to estimate the RUL of structure systems. The proposed methodology uses DBNs to model the degradation process of structure systems with the influence of multiple causes. The modeling methodology consists of two main phases, namely, single-cause modeling phase and multiple-cause integration phase, as shown in Fig. 8.1.

In the single-cause modeling phase, the causal relationships among performance variables within single-cause networks are modeled with BNs and then extended to DBNs considering the dynamic change of structural performance. The structural performance could include the reliability, availability, structure safety, or even crack depth, determined by the estimated objects. These causes may affect one another. The interrelated performance variables within different single-cause networks are linked using arcs to represent the influence relationships, denoted by blue lines and arrows in Fig. 8.1.

In the multiple-cause integration phase, the estimated structural dynamic performance under the influence of all causes is integrated to calculate the total performance. The performance variables in the single-cause networks may also affect or be affected by the performance variables in the integrated multiple-cause network.

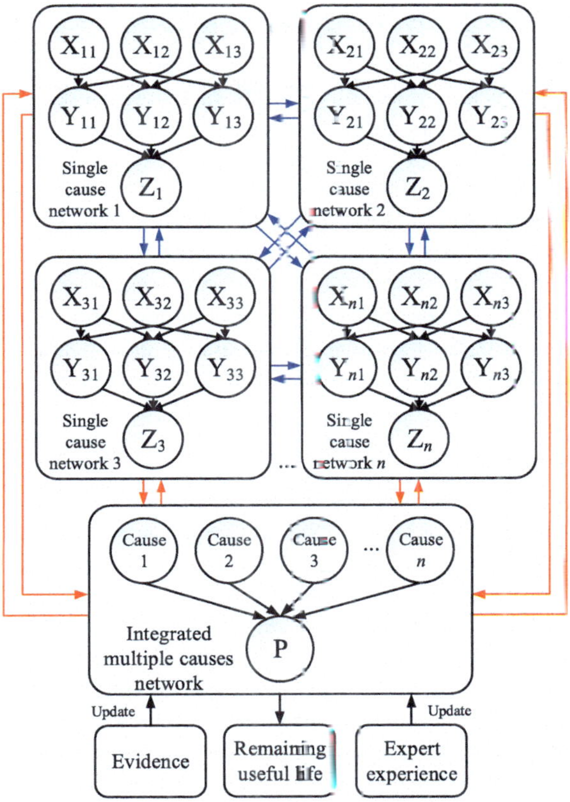

Fig. 8.1 DBN-based RUL estimation methodology

The interrelated variables are also linked using arcs, denoted by red lines and arrows in Fig. 8.1.

The two-phase RUL estimation methodology includes three significant procedures: structure modeling of DBNs, parameter modeling of DBNs, and estimation of RUL. First, the transforming method from the physical models of structure systems is used for DBN structure construction. Second, the prior and conditional probability tables (CPTs) are obtained using the distributions of performance variables and transforming methods from the physical models, respectively. Finally, the RUL value of the structure is calculated on the basis of the estimated dynamic performance. Sensor data and expert experience can be input into the evidence nodes of DBNs to update the RUL value.

8.2.2 Structure Modeling of DBNs

BN structure modeling methods, including cause-and-effect relationships, mapping algorithms, and structuring learning, have been summarized in our previous work (Cai et al. 2019). For structure systems, the detailed cause-and-effect relationship and sufficient training data are generally difficult to achieve, and no one-to-one correspondence between a certain original RUL estimation method and DBN model exists. In such situations, the structure of DBNs can be achieved by transforming from the physical models of structure systems (Straub 2009; Groden and Collette 2017). The nodes of DBNs represent the variables of the physical model, and the arcs of DBNs represent the causal relationship among variables.

A general physical model is used as an example:

$$p^t = f(x_1, x_2, \ldots, x_n, e_1, e_2, \ldots, e_n, p^0, e_p, t) \tag{8.1}$$

where p^t is the physical performance of the structure, such as crack depth, reliability, availability, or safety; p^0 is the initial physical performance; x_1, x_2, ..., x_n are the performance variables; e_1, e_2, ..., e_n are the corresponding evidence variables; and e_p is the evidence variable of physical performance.

At a fixed time, the performance variables $x_1, x_2, \ldots, x_n$ directly affect the physical performance p; thus, the performance variable nodes are connected to the performance node within a time slice by using inner slice arcs. Performance variables x_1, x_2, ..., x_n and physical performance p are detectable, as denoted by evidence variables e_1, e_2, ..., e_n and e_p; thus, the performance variable and physical performance nodes are connected to the corresponding evidence nodes by using inner slice arcs. If the performance variables x_1, x_2, ..., x_n are independent, then they are not linked together, as shown in Fig. 8.2. On the contrary, they can be linked on the basis of the practical physical relationships.

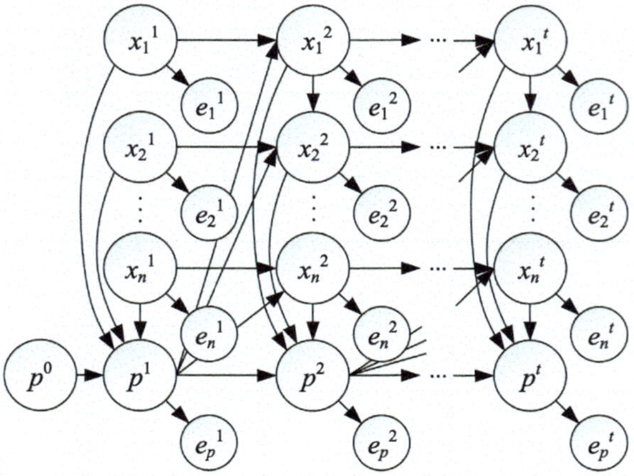

Fig. 8.2 DBNs for the general physical model

The time slices are connected by temporal links, i.e., inter-slice arcs, to constitute a full dynamic model. Performance variables $x_1, x_2, \ldots, x_n$ and physical performance p in the current time affect their own states in the next time; thus, the performance variable and physical performance nodes between adjacent time slices are connected using inter-slice arcs. In addition, performance p in the current time affects the performance variables in the next time; thus, the performance node in the current time and performance variable nodes in the next time are connected using inter-slice arcs.

8.2.3 Parameter Modeling of DBNs

Parameter modeling of DBNs includes determinations of prior probabilities of parent nodes in the initial moment and the conditional probability between parent and child nodes. For discrete variables, the conditional probabilities between parent and child nodes within a time slice and between adjacent time slices can be represented by CPTs.

The prior probabilities of DBNs for RUL estimation are obtained from the distributions of performance variables that correspond to parent nodes in the initial moment. The distribution can be normal, exponential, lognormal, and so on, on the basis of practical physical situations. These corresponding nodes are subsequently discretized using discretization approaches to form discrete prior probabilities.

The CPTs of DBNs are obtained by transforming from physical models through the following steps. The child nodes are first discretized in addition to the previous discretized performance variable of parent nodes. The relationships among all performance and performance variable nodes are determined based on the physical models provided. The parent nodes are subsequently sampled with suitable approaches, such as Monte Carlo simulation, particle filtering, and stochastic filtering. The CPTs are finally obtained using the physical relationships with sufficient sampled values and corresponding probabilities.

8.2.4 Estimation of RUL

The RUL of a structure is calculated based on the estimated dynamic performance. If the performance represents structural reliability or availability, then the RUL generally decreases over time. If the performance represents structural crack depth, then the RUL generally increases over time. The RUL value is determined by a failure threshold, that is, a critical value, which is determined by the practical physical failure mechanism of the estimated structure.

As shown in Fig. 8.3, in a certain detection point t_d, the performance variables of the structure can be detected using various sensors, and the dynamic performance can be updated through the proposed DBN model. For this structure, the performance

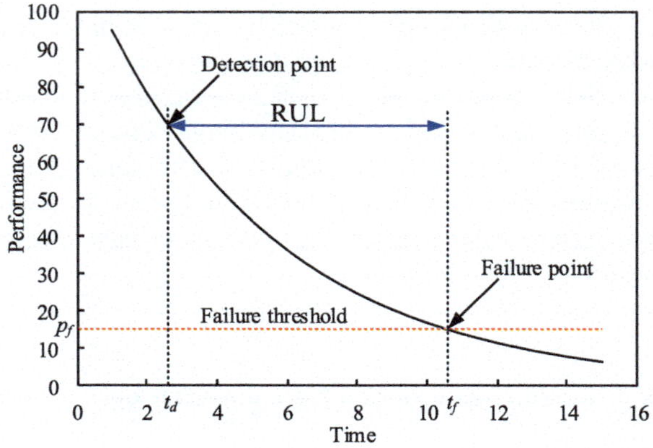

Fig. 8.3 RUL calculation model

decreases over time. Failure point t_f is the moment that the performance reaches critical value p_f. The RUL value in detection point t_d is exactly the period between failure point t_f and detection point t_d.

8.3 Case of Subsea Pipelines

8.3.1 Influence of Multiple Causes on Subsea Pipelines

Subsea pipelines are safety critical structures in the offshore oil and gas industry. They are used for many purposes in developing subsea oil and gas resources. For example, they can be used to transport oil and gas from wellheads to platforms, inject water or chemicals from platforms to wellheads, or transfer products from production systems to process and distribution facilities. Subsea pipelines are vulnerable to environmental-assisted cracks under harsh working conditions and are therefore subjected to continuous degradation (Cheng and Chen 2017). Crack can cause the failure of subsea pipelines and primarily determines the RUL. Many causes can lead to the occurrence and propagation of cracks, such as fatigue, corrosion, sand erosion, and internal waves.

Subsea pipelines are influenced chronically by cyclic stresses, and mechanical deterioration caused by fatigue loads is a significant factor of crack. The fatigue characteristic is generally related to many factors, such as material properties, stress and strain history, and initial geometry of crack. Corrosion refers to the deterioration of a metal through its electrochemical interactions with its environment. Cracks can be induced in a corrosive environment. Subsea pipelines are more susceptible to internal corrosion than to external corrosion because they generally transform the corrosive oil, gas, and fluids with high temperature and pressure. Internal corrosion in pipelines is influenced by many factors, including temperature, carbon dioxide content, water chemistry, flow velocity, oil or water wetting, and the composition and surface condition of steel. Erosion of pipelines may be caused by the direct and random impingement of particles and fluid. Sand erosion is the most common source of erosion problems in subsea oil and gas pipelines because small amounts of sand entrained in the produced fluid can result in significant erosion and erosion–corrosion damage. Sand erosion can lead to severe cracks in the internal pipe wall, especially at fittings that redirect the flow, such as elbows and tees. The reason is that the particles in the fluid can possess sufficient momentum to traverse the fluid streamlines and impinge the pipe wall (Bai and Bai 2012). Internal waves can be consistently generated in continuous and stably stratified fluids. Internal waves are typically the most energetic high-frequency events in the coastal ocean, which can generate strong currents and turbulence. Therefore, they can seriously destroy subsea structures, such as oil and gas pipelines and platforms.

This work adopts subsea pipelines (tees except where noted) as the estimated object to demonstrate the proposed DBN-based RUL estimation methodology. The combined influences of multiple causes, including fatigue, corrosion, erosion, and internal waves (internal waves except where noted) are considered. Crack depth of subsea pipelines is selected to be the physical performance of the structure to calculate the RUL value.

8.3.2 RUL Modeling for Subsea Pipelines

(1) *Fatigue Modeling*: The physical performance model of fatigue crack growth is represented by the famous Paris law as

$$\frac{dD}{dn} = C(\Delta K)^M \tag{8.2}$$

where D is the crack depth, n is the number of applied load cycle, dD/dn represents the crack growth rate, C and M are the empirically determined constant parameters related to material properties, and ΔK is the stress intensity factor, which can be empirically represented as

$$\Delta K = \Delta S\sqrt{\pi D} \tag{8.3}$$

$$\Delta S = \lambda\Gamma\left(1 + \frac{M}{k}\right)^{\frac{1}{M}} \tag{8.4}$$

where Γ is the Gamma function and λ and k are the scale and shape parameters of Weibull distribution, respectively.

The crack depth in the nth stress cycle can be achieved from the basic Paris law when the initial crack depth is assumed to be D^0 as follows (Luque and Straub 2016):

$$D(n) = \left[(D^0)^{1-M/2} + \left(1 - \frac{M}{2}\right) C \Delta S^M \pi^{M/2} n \right]^{1/(1-M/2)} \quad (8.5)$$

On the basis of the transformation method from physical performance models to DBNs described previously, the fatigue crack growth model is represented by the DBNs in Fig. 8.4. C, M, λ, and k are the variable nodes, and D is the performance node. The parameter distributions and corresponding values are obtained from historical data and expert knowledge. For different structure systems, the distributions are generally the same, but the values are totally different. The values in this work are determined using historical data and expert knowledge. Variable D^0 is the exponential distribution with a rate parameter of 1. Variable M is the normal distribution with a mean of 2 and standard deviation of 0.2. Variable λ is the lognormal distribution with a mean of 1 and standard deviation of 0.2. Variable k is the uniform distribution with a value of 0.8. Lastly, variable C is correlated with and therefore determined by M.

(2) *Corrosion Modeling*: In the oil and gas industry, many corrosion prediction models have been developed by companies. The Shell model for carbon dioxide corrosion proposed by Waard is the most widely used (de Waard et al. 1991). Starting from a "worst-case" corrosion rate prediction, the model applies correction factors to quantify the influence of environmental parameters and corrosion product scales formed under various conditions. Considerable new knowledge and information have been added to the model to develop new ones. The resistance model considers the influences of mass transport and fluid velocity on corrosion rate, and it is selected to be the physical performance model of carbon dioxide corrosion for subsea pipelines.

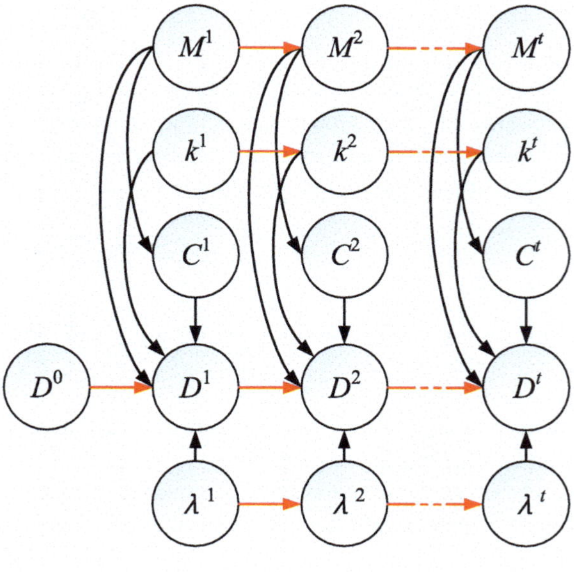

Fig. 8.4 DBNs for degradation caused by fatigue

$$V_c = \frac{1 - R_c}{\frac{1}{V_r} + \frac{1}{V_t}} \tag{8.6}$$

where V_c is the corrosion rate; V_r is the reaction rate, which is determined by temperature, carbon dioxide pressure, and pH; V_t is the mass transfer rate, which is determined by velocity and liquid film thickness; and R_c is the corrosion protection factor because of the internal corrosion inhibitors and internal coatings in subsea pipelines.

For the subsea pipeline steel containing 0.18% C and 0.08% Cr, the reaction rate V_r and mass transfer rate V_t could be determined by

$$\log(V_r) = 4.93 - \frac{1119}{T_p + 273} + 0.58\log(n_C p_o) \tag{8.7}$$

$$V_t = 2.45\frac{V_l^{0.8}}{d_p^{0.2}}n_C p_o \tag{8.8}$$

where T_p is the subsea pipeline fluid temperature, n_C is the fraction of carbon dioxide in the gas phase, p_o is the operating pressure, V_l are the liquid flow velocity, and d_p is the pipe diameter.

The corrosion model as a DBN is transformed from the resistance model, as shown in Fig. 8.5. T_p, n_C, p_o, V_l, and d_p are the variable nodes, V_r and V_t are the intermediate nodes, and V_c is the performance node.

(3) *Sand Erosion Modeling*: Sand erosion of subsea pipelines with oil and gas is related to the kinetic energy of sand fragments. Many parameters affect the sand erosion rate, such as fluid velocity, fluid density, and size of the sand fragments. The physical process is only partially understood, and the prediction of critical conditions needs to rely on empirical or semi-empirical models. For the subsea pipelines, the physical performance model of sand erosion is represented by the Salama model as follows (Salama 2000):

$$E_p = \frac{(1 - R_p)V_m^2 d_s}{S_p d_p^2 \rho_m} \tag{8.9}$$

where E_p is the sand erosion rate; d_s is the particle diameter; d_p is the pipe diameter; V_m is the mixture velocity determined by $V_m = V_l + V_g$; ρ_m is the mixture density determined by $\rho_m = (\rho_l V_l + \rho_g V_g)/V_m$; V_l and V_g are the liquid and gas flow velocities, respectively; ρ_l and ρ_g are the liquid and gas flow densities, respectively; S_p is the geometrical constant, which is 25,000 for tees, 12,000 for ells, and 2000 for elbows with oil and gas (Bai and Bai 2012); and R_p is the erosion protection factor because of the internal coatings in subsea pipelines.

Fig. 8.5 DBNs for degradation caused by corrosion

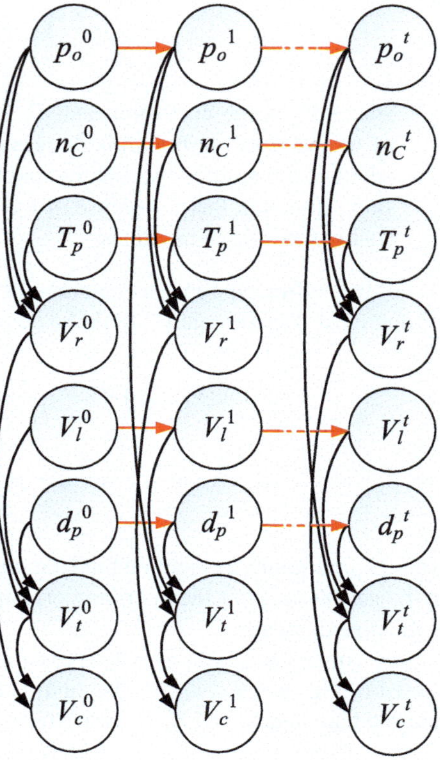

The sand erosion model as a DBN is transformed from the Salama model, as shown in Fig. 8.6. d_s, d_p, V_m, S_p, and ρ_m are the variable nodes, and E_p is the performance node.

(4) *Internal Wave Modeling:* The physical performance model of crack growth caused by internal waves is represented by the modified Paris law. The model involves the concept of equivalent stress intensity factor range (Huang, Moan et al. 2008), which is considered to be the driving force of crack growth.

$$\frac{da}{dN} = c[(M_R M_P F \Delta\sigma\sqrt{\pi a})^m - (\Delta K_{th0})^m] \tag{8.10}$$

$$M_R = \begin{cases} (1-R)^{-\beta_1} & (-5 \le R < 0) \\ (1-R)^{-\beta} & (0 \le R < 0.5) \\ (1.05 - 1.4R + 0.6R^2)^{-\beta} & (0.5 \le R < 1) \end{cases} \tag{8.11}$$

where a is the crack caused by internal waves, da/Dn represents the crack growth rate, c and m are the empirically determined constant parameters, M_R is the correction factor for the loading ratio, and M_P is the correction factor for the loading sequence

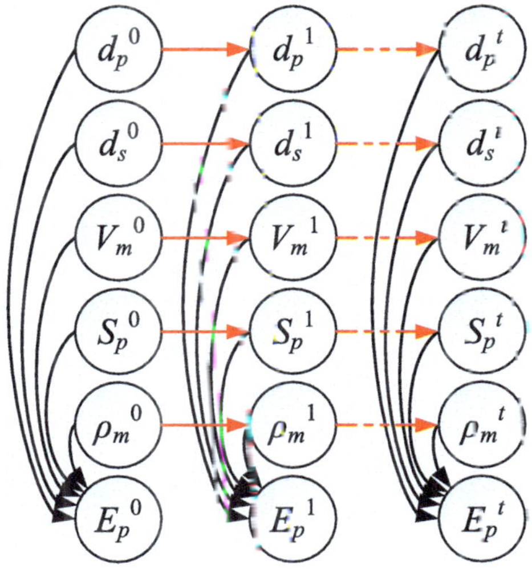

Fig. 8.6 DBNs for degradation caused by erosion

interaction. F is the shaping factor function, which depends on the specific point at the crack front, and its value is set to a constant in this work. $\Delta\sigma$ is the stress range of internal waves, and Table 8.1 presents the values and corresponding probabilities of occurrence in three sea areas. ΔK_{th0} is the threshold stress intensity factor range at $R = 0$, β and β_1 are the shaping exponents, and R is the stress ratio.

On the basis of the transformation method from physical performance models to the DBNs described previously, the modified Paris model is represented by the DBNs in Fig. 8.7. c, m, F, M_P, $\Delta\sigma$, ΔK_{th0}, R, β, and β_1 are the variable nodes, M_R is the intermediate node, and a is the performance node.

(5) *Entire DBNs for RUL Estimation:* The four developed DBNs for the degradation of subsea pipelines with four causes are integrated to form the entire DBNs, as shown in Fig. 8.8. Corrosion rate V_c and sand erosion rate E_p are transformed into the crack depth under the influences of carbon dioxide corrosion and sand erosion. Performance nodes D, V_c, E_p, and a in the four DBNs are linked to the overall performance node A, and the total crack depth A of tees is used to

Table 8.1 Stress range of internal waves in three sea areas

Internal wave I (Mpa)	Probability (%)	Internal wave II (Mpa)	Probability (%)	Internal wave III (Mpa)	Probability (%)
5.5	8	3.5	8	0.55	8
5.7	16	3.7	16	0.57	16
5.9	30	3.9	30	0.59	30
6.1	26	4.1	26	0.61	26
6.3	14	4.3	14	0.63	14
6.5	6	4.5	6	0.65	6

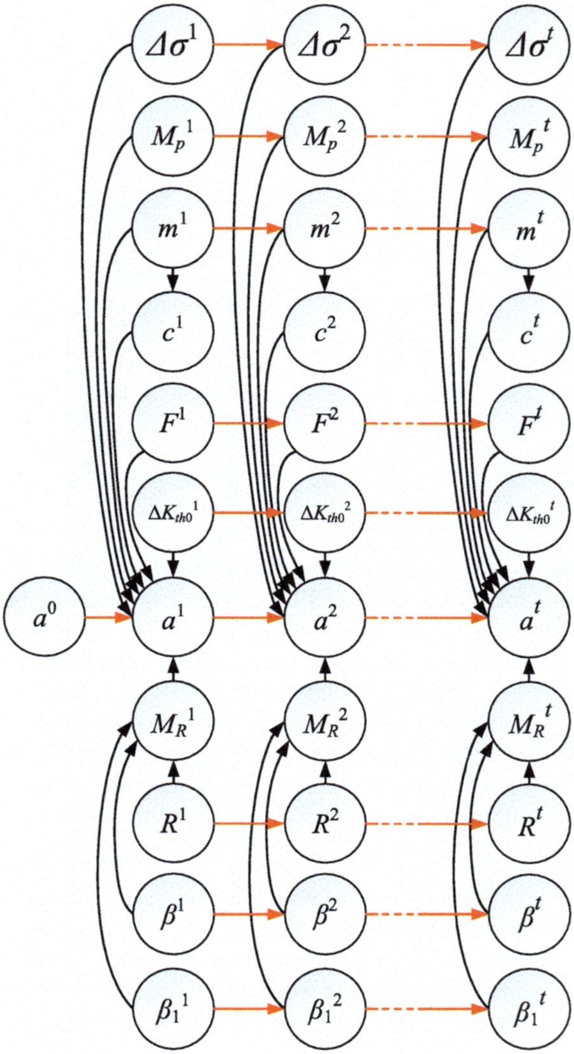

Fig. 8.7 DBNs for degradation caused by internal waves

calculate the RUL of the subsea pipelines. In RUL estimation, the performance indexes of each subnetwork must be unified, that is, using the same time scale to describe the change in crack depth. For example, the crack depth for fatigue and internal waves is obtained by the product of crack growth rate and the number of stress cycles. The depth for corrosion and sand erosion is obtained by the product of velocity and time. Tables 8.2 and 8.3 list the required parameters of the variable. Some of the parameters are referred to the *Subsea Engineering Handbook* (Bai and Bai 2012), and other parameters are obtained on the basis of historical data and expert experience. For the estimated tees, these parameters

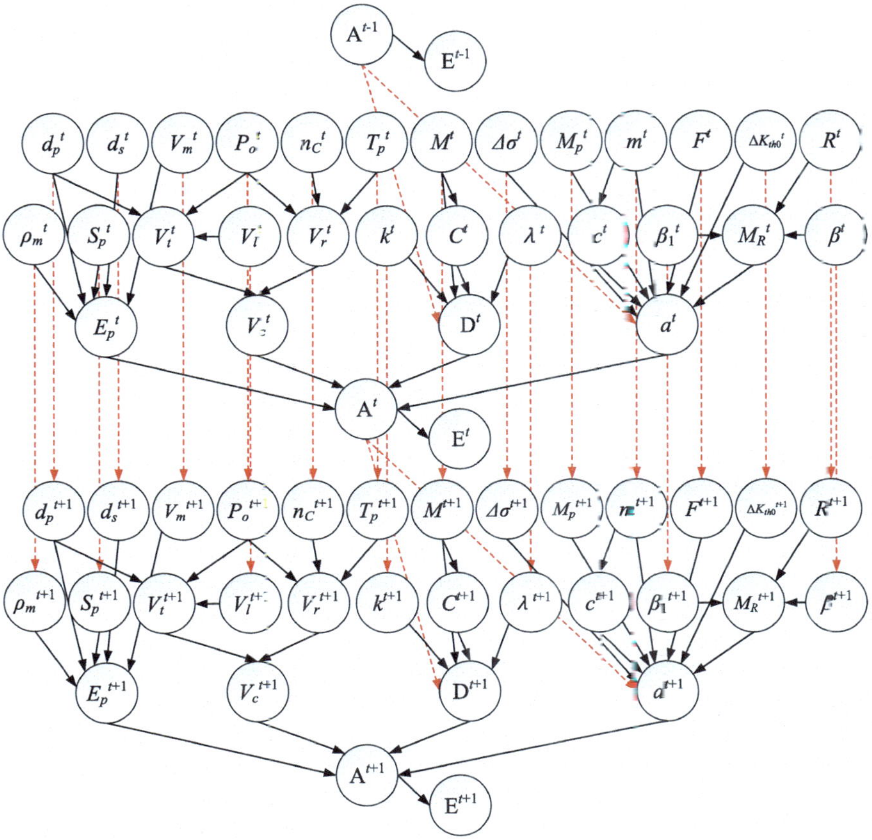

Fig. 8.8 Entire DBNs for degradation caused by multiple causes

are obtained from engineering practices. The overall performance node A is connected to the evidence node E. When the crack depth of subsea pipelines is detected using various non-destructive testing technologies, the evidence node E at this moment is updated, the entire DBNs can also be updated, and the real-time RUL value is obtained.

8.3.3 Results and Discussions

(1) *RUL Calculation*: For the developed DBNs in Fig. 8.8, the exact inference algorithm is used to update the probability distributions of posterior crack depth, as shown in Fig. 8.9. The peak of crack depth probability moves to the right over time, that is, the tees have a high probability for the existence of deep cracks over time. Therefore, the RUL of tees declines over time.

Table 8.2 Parameters of the DBN-based RUL model (1)

Variable	Distribution	Mean	Standard deviation
D^0 (mm)	Exponential	1	1
M	Normal	2.5	0.2
λ	Lognormal	1	0.2
k	Deterministic	0.8	–
n_C (mol%)	Deterministic	0.5	–
p_o	Deterministic	52	–
c	Deterministic	1.58×10^{-11}	–
m	Deterministic	3	–
F	Deterministic	20	–
M_P	Deterministic	1	–
M_R	Deterministic	1	–
ΔK_{th0} (Mpa $\sqrt{m}$)	Deterministic	10 lePara>	–
R	Deterministic	0	–
β	Deterministic	0.3	–
β_I	Deterministic	0.5	–

Table 8.3 Parameters of the DBN-based RUL model (2)

Parameter	Value	Probability (%)	Value	Probability (%)	Value	Probability (%)
T_p (°C)	20	30	22.5	40	25	30
V_I (m/s)	1	20	2	50	3	30
d_p (mm)	40	23.3	50.8	56.2	65	20.5
d_s (μm)	300	30	350	50	400	20
V_m (m/s)	1	20	2	50	3	30
ρ_m (kg/m^3)	800	32.2	910	49.2	1000	18.6

The crack depth must be quantified to calculate the RUL value. The sum of products of the crack depths and corresponding probabilities is used to represent the crack depth value. Figure 8.10 plots the depth values in the first 21 years. The crack depths of tees under the influences of multiple and single causes increase over time. The crack depth initially increases linearly and slowly and then rapidly and almost exponentially under the influence of multiple causes. The internal waves and fatigue have greater effects on the crack depth than corrosion and sand erosion, especially in the later stage. The reason is that the applied loads of internal waves and fatigue on tees can make the existing microcracks grow rapidly. The internal waves and fatigue curves intersect around the 14th year, that is, fatigue has greater effects than internal waves in the early stage and smaller effects later. The internal waves contribute considerably to crack growth in the later stage of subsea pipelines.

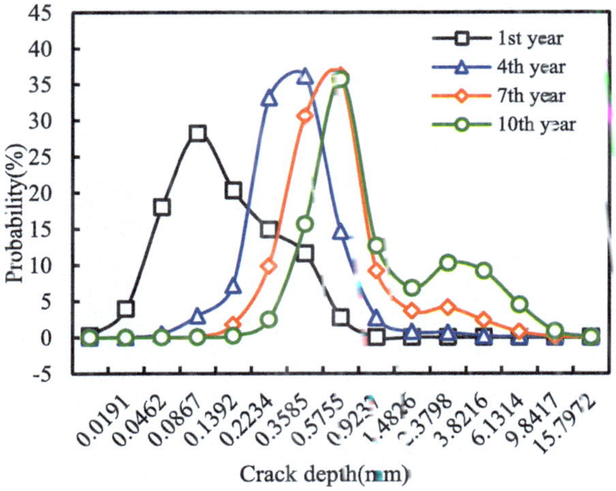

Fig. 8.9 Crack depth probability distribution over time

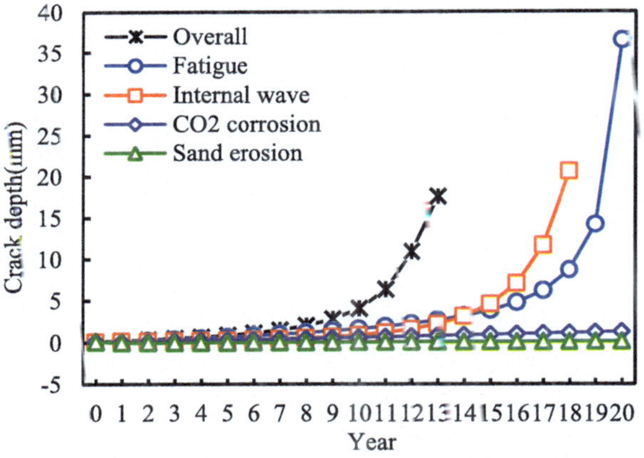

Fig. 8.10 Crack depth growth over time

A failure threshold of 16.5 mm is selected, and the detection point is set to 0 in this work. Crack depth represents the performance of subsea pipelines. Based on the RUL calculation model in Fig. 8.3, the RUL values of tees with different causes are calculated, as shown in Fig. 8.11. The RUL values of tees under the influences of single fatigue and internal waves are 19.1 and 17.5 years, respectively. The RUL values of tees with single corrosion or sand erosion is considerably higher than 20 years, whereas that of tees with multiple causes is 12.8 years. In addition, the sensitivity analysis is conducted by changing the threshold within a specified range, as given in Fig. 8.12. The RUL value increases from 12.59 years to 12.94 years when the threshold increases from 15.5 mm to 17.5 mm. Therefore, the RUL value is not so sensitive to the threshold.

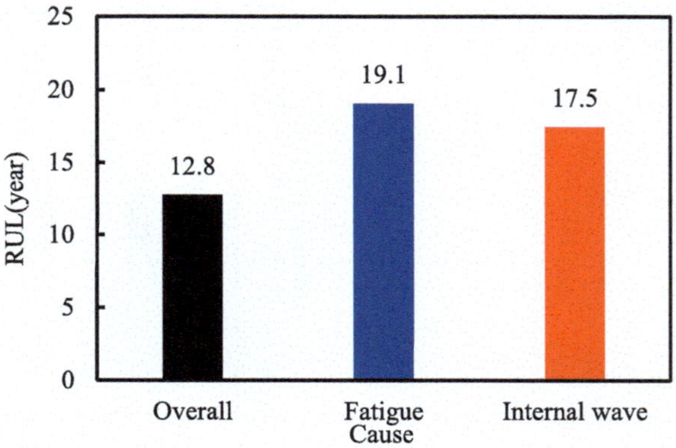

Fig. 8.11 RUL value under the influences of different causes

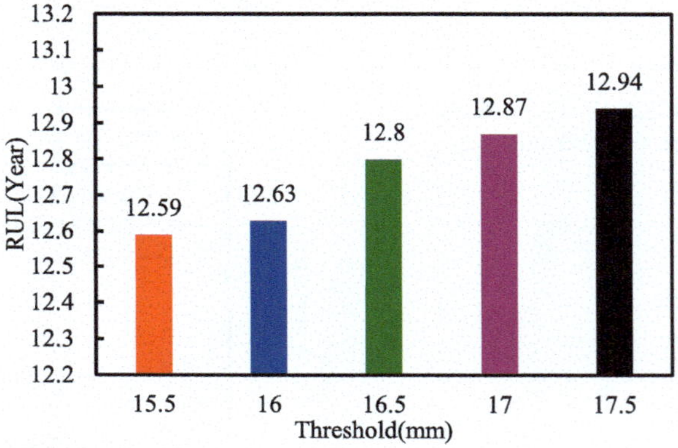

Fig. 8.12 Sensitivity analysis results of the threshold

(2) *RUL Updating with New Evidence*: The information observed by the expert and that detected by the sensor can be input into the DBNs as evidence. The variable of the evidence node is therefore instantiated, that is, given a fixed value. In this work, the practical crack depth of tees can be detected by non-destructive testing technologies during the overhaul period (no expert experience in the current case), and it can be input into the evidence node E of the DBN-based RUL model. The new crack depth and RUL value of subsea pipelines can be obtained. Three new crack depths (0.139, 0.575, and 2.38) are detected for three tees in the 5th year. The three evidence values, that is, Evidence 1 with 0.139, Evidence 2 with 0.575, and Evidence with 2.38, are input into the DBNs. Figure 8.13 shows that the crack depths for the tees with Evidences 1 and 2 increase slowly because the new evidence suggest a smaller crack depth than the expected one of 0.904. Similarly, the crack depth for the tees with Evidence 3 increases rapidly because this evidence suggests a greater crack depth that the expected one in the 5th year.

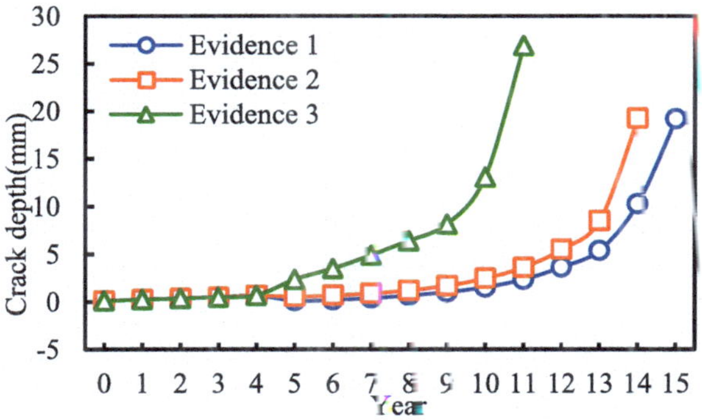

Fig. 8.13 Crack depth growth updating with new evidence

The RUL values of tees with Evidences 1 and 2 increase from the expected value of 12.8 to 14.7 and 13.8, and the value for Evidence 3 decreases to 10.3.

(3) *Parameter Sensitivity Analysis*: The sensitivity analysis of BNs measures the degree to which findings at any node can influence the beliefs at another node, considering the findings currently entered. The measures can be in the form of variance of beliefs and variance reduction. The parameter sensitivity analysis of variable and performance nodes in the 3rd, 5th, and 10th years are conducted, as shown in Fig. 8.14.

In the early stage, i.e., the 3rd and 5th years, the crack depth is sensitive to the mixture velocity, which is determined by the oil and gas flow velocities. In the later stage, i.e., 10th year, the crack depth is sensitive to the stress range of internal waves and the mixture velocity of oil and gas. The reasons are that the initial crack depth of tees is small, and the loads of fatigue and internal waves cannot greatly aggravate the degradation, whereas the mixture velocity of oil and gas considerably affects the sand erosion, which contribute considerably to crack depth growth. Similarly, when the crack depth is large, internal waves can rapidly aggravate the degradation and therefore dominate the crack depth growth in the later stage. The other variables, namely, pipe diameter d_p, particle diameter d_s, operating pressure p_o, fluid temperature T_p, and fatigue parameter M, have small sensitivities.

(4) *Influencing Factor Analysis*: Many factors influence the crack depth growth rate. Figure 8.15 shows that with the increase in initial depth, the crack depth grows rapidly. The crack depth increases to 32.6 mm in the 12th year for the initial depth of 0.4418 mm compared with the crack depth of 5.96 mm for the initial depth 0.0100 mm. Thus, the RUL of subsea pipelines with large initial depth decreases greatly after 10 years. Therefore, the initial crack depth should be detected and controlled strictly when installing the subsea pipelines. Figure 8.16 shows that the higher the strength of internal waves, the larger the crack depth, especially in the later stage of tees. Therefore, the overhaul cycle should be shorted when strong internal waves occur frequently for the subsea pipelines serving more than 10 years.

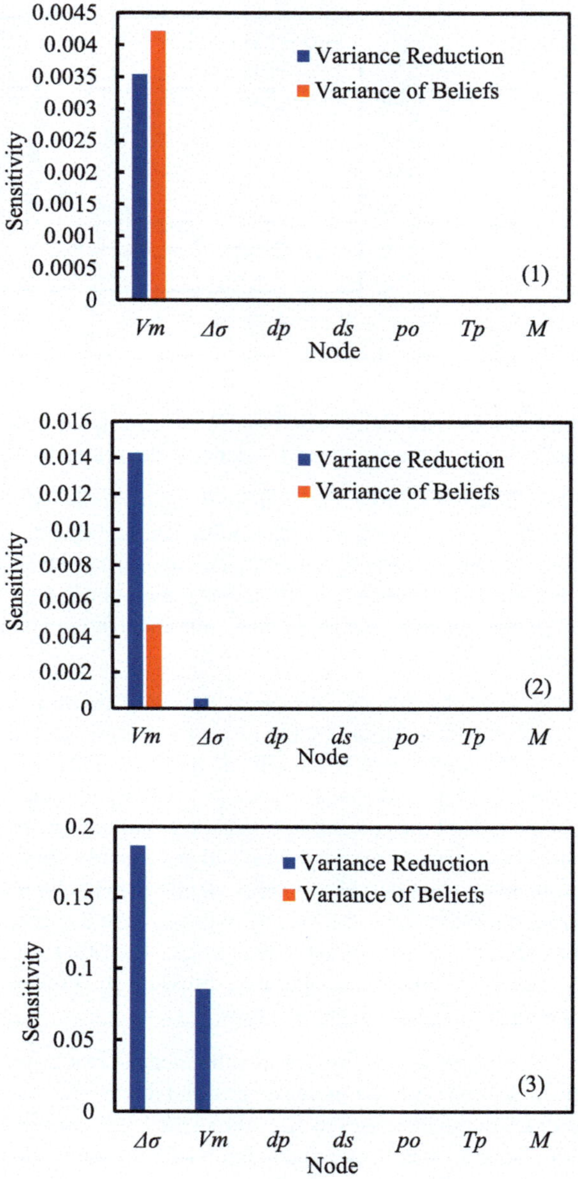

Fig. 8.14 RUL value updating with new evidence. Parameter sensitivity analysis results: (1) 3rd year, (2) 5th year, and (3) 10th year

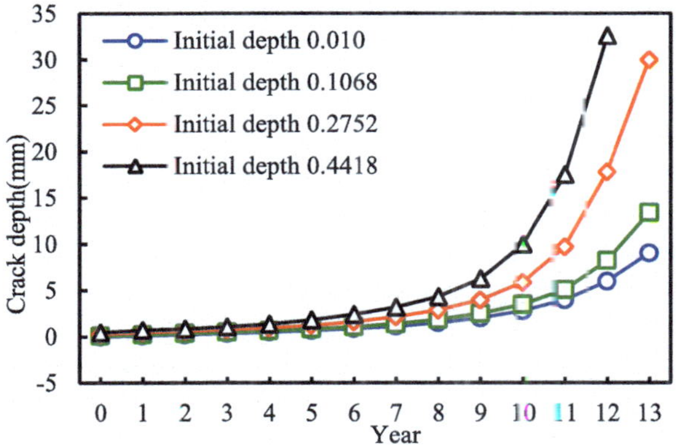

Fig. 8.15 Influences of initial depth on crack depth growth

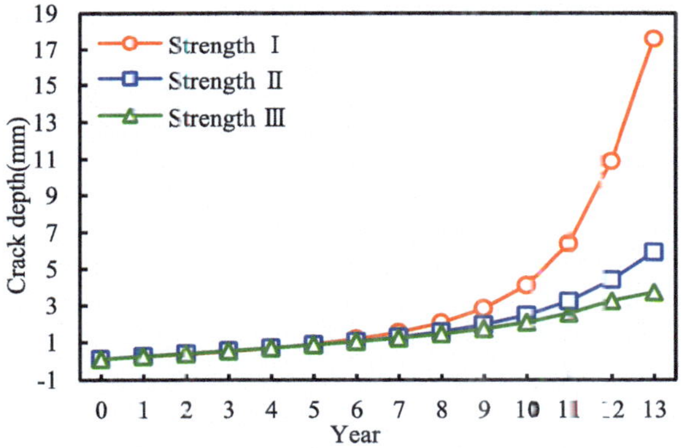

Fig. 8.16 Influences of the strength of internal waves on crack depth growth

The three significant parts of subsea pipelines are tees, ells, and elbows. Crack depth growths of ells and elbows over time are researched and compared with that of tees. Figure 8.17 shows that the tees have the smallest crack depth, the elbows have the largest depth over time, and the ells are in between, indicating that the tees have the biggest RUL value. This situation occurs because the tees and ells are generally composed of corrosion- and erosion-resistant materials and have high RULs.

(5) *Model validation*: The validation of DBN models are crucial to RUL evaluation because they provide reasonable confidence to the assessment results. Validation includes determining whether a model accurately reflects reality, and it can be accomplished using real data, simulated data, and contrastive modeling (Cai et al. 2019). Contrastive modeling is adopted in the current work. The crack depth curve with fatigue (no literature related to multiple-cause influence is reported) was researched using Bayesian inference (Karandikar et al. 2012). By

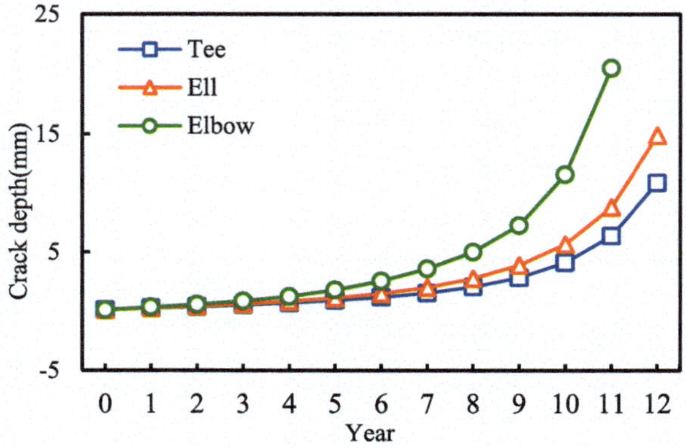

Fig. 8.17 Crack depth growth of three components of subsea pipelines

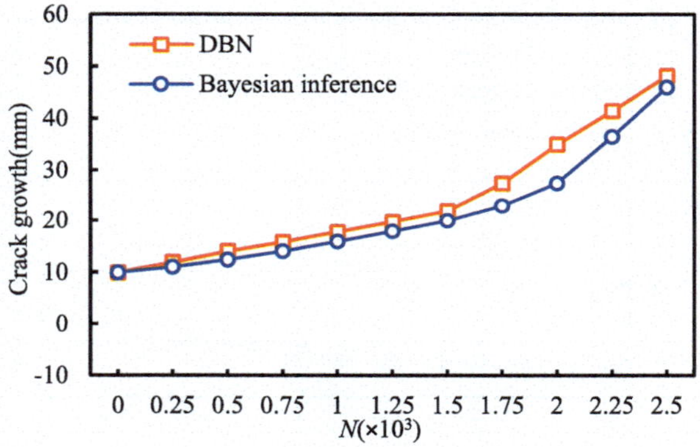

Fig. 8.18 Comparative results of crack depth with DBNs and Bayesian inference

using the same data, the DBN-based RUL estimation model is established on the basis of the proposed hybrid methodology, and the results are compared in Fig. 8.18. The trend of the crack depth is similar to the increase in the number of stress cycles, *N*. This finding validates that the proposed method is correct.

(6) *RUL software development*: A RUL calculation software is developed using the proposed RUL estimation methodology, as shown in Fig. 8.19. The distributions and parameters can be manually input into the software, and data detected by the sensors can be automatically input as evidence to update the real-time RUL curves.

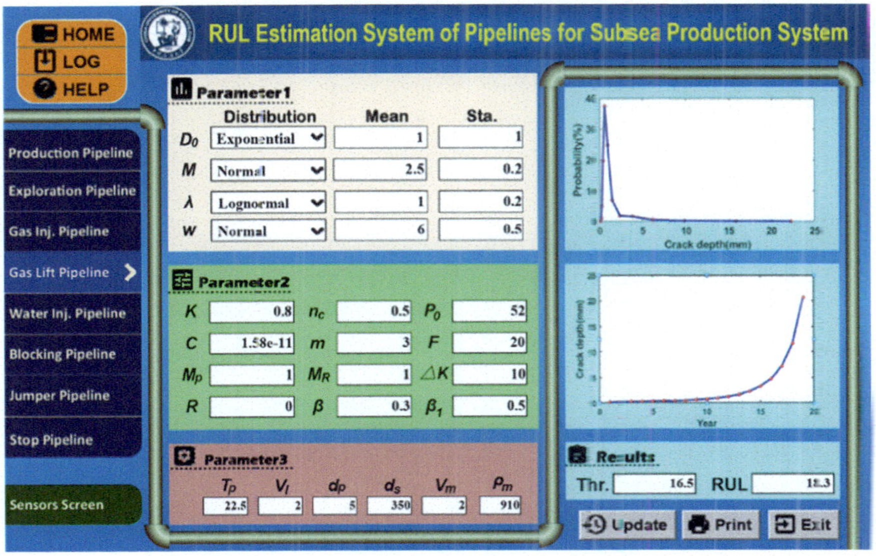

Fig. 8.19 One of the developed human–machine interfaces for RUL estimation

Chapter 9
RUL Prediction with Hybrid Multi-stage

Abstract Addressing challenges posed by operational uncertainty, noise in signal acquisition, and limited real-time data, this chapter presents a hybrid multi-stage methodology for RUL prediction in control systems. The approach combines a variant of the unscented Kalman filter (UKF) with DBNs to improve uncertainty analysis in predicting nonlinear degradation processes. In the early stages, the dynamic UKF models estimate the distribution of random faults and process noise, match the system's degradation stage, and acquire operational data. The degradation process is then optimized through cyclic iteration, calculating the system's covariance and optimal estimates. This process compensates for inaccuracies in measuring real system degradation, enhancing the precision and robustness of RUL predictions. The methodology is validated through application to a subsea Christmas tree with electro-hydraulic compound control, demonstrating its effectiveness in improving RUL prediction under complex operational conditions.

9.1 Introduction

Current RUL prediction methodologies are broadly categorized into model-driven (Lei et al. 2018; Teng et al. 2020), data-driven (Motahari-Nezhad and Jafari 2021; Pillai and Vadakkepat 2021), and hybrid approaches (Zang et al. 2021). Each category presents distinct limitations: model-driven methods often rely on precise mathematical models that may necessitate oversimplified assumptions, while data-driven methods require substantial amounts of data for effective training, which can lead to increased time costs. Hybrid methods seek to leverage the strengths of both approaches by incorporating system degradation mechanisms and health state evolution derived from data. However, challenges remain, particularly in scenarios with limited sample data or high-dimensional input spaces. Furthermore, existing models often overlook the uncertainties stemming from operational processes and measurement noise, which can compromise the reliability of RUL predictions.

The motivation behind this research stems from the need to address these gaps, particularly the instability caused by model uncertainty and the inadequacy of

B. Cai et al., *Intelligent Operation and Maintenance for Subsea Production Systems*,
https://doi.org/10.1007/978-981-96-2677-9_9

existing methods to quantify uncertainty effectively. A novel hybrid multi-stage RUL prediction methodology is proposed, integrating dynamic Bayesian analysis into the Unscented Kalman Filter (UKF). This approach aims to analyze the RUL of uncertain and nonlinear degenerate systems while effectively handling the nonlinear transfer of mean and covariance. The introduction of DBNs enhances the robustness of predictions by allowing for the incorporation of prior knowledge and a more nuanced understanding of system dynamics.

In summary, this chapter presents a comprehensive methodology to improve RUL predictions in control systems where real-time measurement data is often inaccessible under complex operational conditions. The proposed ensemble technique addresses the uncertainty in degradation data by modeling random faults and process noise, thereby enhancing measurement stability. Additionally, it bridges the gap in model uncertainty quantification when employing data-driven models within the UKF framework. This methodology not only enhances the accuracy and robustness of RUL predictions but also provides valuable insights for decision-making in complex industrial environments.

9.2 RUL Prediction Model Based on DUKF

A hybrid multi-stage RUL prediction framework based on statistical law considering random faults is proposed. The schematic diagram of the proposed methodology is shown in Fig. 9.1.

- Firstly, according to the statistical law of the existing degradation data of complex system, the effective health information is extracted to estimate the degradation rate of the system, and process noise is calculated according to system running status. Practice has proved that the failure rate of most equipment is a function of time. Therefore, system stages can be divided according to failure rates.
- In the process of system operation, the inevitable occurrence of random phase, with random strength of the fault. Establish a DBNs model according to the system fault type, and the distribution of random faults at each stage of system operation is solved by the DBNs to deal with the uncertainty of the system degradation process.
- Then, the dynamic UKF algorithm is performed for each stage and the random fault is considered to simulate the process of system degradation in order to predict system performance. Bayesian analysis results are incorporated into the UKF algorithm as parameters to solve the probability density distribution of the nonlinear degradation process function approximating the system. The nonlinear transmission problem in the process of system state transition is solved by UT, and the noise is filtered by cyclic iteration so that the predicted value of the entire system state approaches its actual measured value. The optimal prediction process of system degradation is obtained.

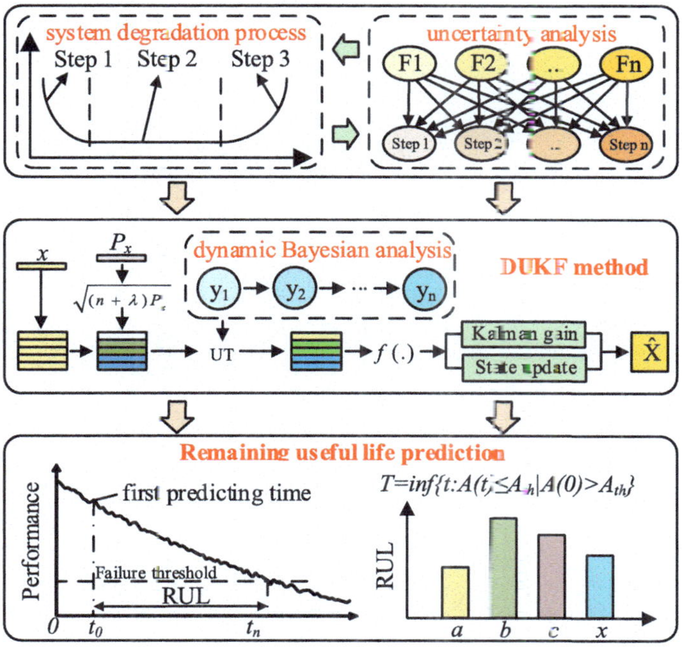

Fig. 9.1 Schematic diagram of DUKF prediction methodology

- Finally, the predicted value is compared with the actual monitoring value and the predicted value of DBNs to obtain the corresponding relationship between the predicted value and the actual value, so as to make a reasonable estimate of the future degradation of the system and deduce the remaining lifetime under the definition of the first arrival time.

9.2.1 *Modeling of the Degradation*

General mechanical structure control systems are composed of series, parallel and redundant structures (Cai et al. 2018), and most are redundant systems consisting of multiple components (Dui et al. 2022). A series parallel redundant system is considered here. Subsystems a, b, and c are connected in parallel. They are then connected in series with d, d and e are redundant systems, and x is the system degradation data. The most common mechanical systems consist of these parts, and establish the DBNs model according to the structure of the system. The block diagram of the system can be given in Fig. 9.2.

According to the statistical law of the existing degradation data, the effective health information is extracted to estimate the degradation rate of the system, and the process noise is calculated according to system running status. Considering the

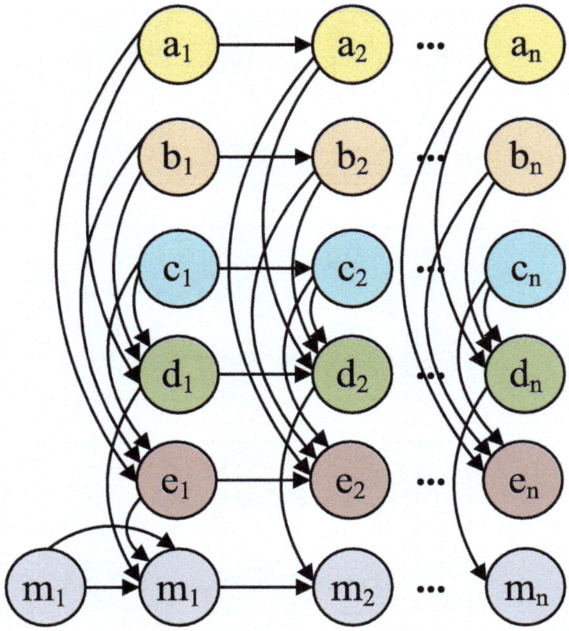

Fig. 9.2 The block diagram of the system

uncertainty of subsystem failure, the DBNs are used in the degradation performance simulation, and the performance of the system in continuous time is estimated. In the DBN model, Bayesian inference is used to infer the performance of the system in future time slices. After the system operates for a while, the state changes, as shown in Eq. (9.1), where h is the time frame, which can be defined as the year, month or day depending on needs. The variation can also be the crack depth, reliability, wear and corrosion, failure rate, etc.

$$x = (x_1, x_2, \ldots, x_h) \tag{9.1}$$

According to the monitoring value of the working state of the system the degradation process of the whole system is divided into stages. The modeling process is illustrated by

$$\begin{cases} X_t^j = f_j(x_t^j, \theta, W_t^j, V_t^j) + M_t^j \\ Z_t^j = h_j(y_t^j, \theta, W_t^j, V_t^j) + N_t^j \end{cases} \tag{9.2}$$

where

Xj t is the state j of the system at time t, representing the degradation data change process of the system.

Zj t is the predicted value of DBNs at time t in stage j, representing the predicted result of DBNs.

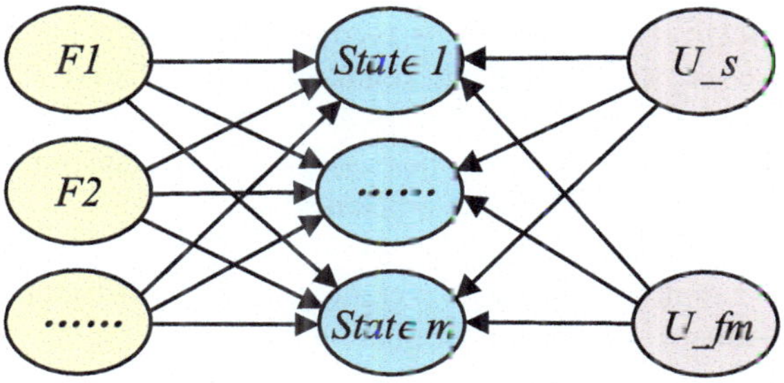

Fig. 9.3 BNs for random fault distribution

θ	is the model parameter vector.
W_t and V_t	represent uncertain faults at time t in the process of system operation, which may occur randomly at any stage of system operation.
$M_t \sim M\ (0, Q_t)$	is the random process noise, and $N_t \sim N\ (0, R_t)$ is the random observation noise.

The uncertainty factors in the multi-stage degradation system are analyzed, and the system state transition process is calculated based on the influencing factors of each stage. The solution process is shown in Fig. 9.3. The uncertain faults in the system operation are generated randomly, and the timing and intensity of the uncertain faults W_t and *Vt* are uncertain. DBNs are used to calculate the distribution law, estimate the stage of their possible distribution, and calculate the probability of occurrence. Assuming that the system has *m* operation stages, the unknown reasons affecting the functioning of system are F_1, F_2, ..., U_s, which size is *W*, and the probability of occurrence in stage *i* is *ai*. *U_fm* is an unknown failure mode, which size is *V*, and the probability of occurrence in stage *i* is *bi*, then the probability of occurrence of uncertain fault in stage *i* of the system is *Si*.

9.2.2 Proposed DUKF Methodology

According to the variables obtained above, the system state transition equation established for each stage is shown as follows.

$$X(k) = f(x(k), \theta, S_i) + M(k) \tag{9.3}$$

$$Z(k) = h(x(k), \theta, S_i) + N(k) \tag{9.4}$$

The state model describes the law of equipment degradation by constructing the relationship between degradation state and the degradation data. According to the known degradation data, the degradation law of the equipment in service in the future can be inferred, the parameters of the state model can be estimated. The RUL can be effectively and accurately predicted by introducing the dynamic Bayesian analysis process. The DUKF algorithm also uses UT to deal with the problem of nonlinear transfer of mean and covariance. If both the equation of state and the observation equation are nonlinear, there is no so-called state transition matrix. The methodology has high computational efficiency.

Bring Eqs. (9.3) and (9.4) into the DUKF algorithm, get the sigma point, which is sampling points with weight in the prior distribution, corresponding to the value estimated by UT transformation, and approach the posterior probability density of the state by a series of sigma points, so ensure the accuracy of the estimated value. The uncertainty information contained in the DBNs are incorporated into the DUKF algorithm. Eq. (9.5) is the UT change equation, determining the sigma point and its weight by UT transformation.

$$X^i = \begin{cases} \hat{x}(i = 0) \\ \hat{x} + \sqrt{(n+\lambda)P}_i (i = 1, 2, \ldots, n) \\ \hat{x} - \sqrt{(n+\lambda)P}_i (i = n+1, \ldots, 2n) \end{cases} \tag{9.5}$$

where n is the dimension of the input quantity, and $n + 1$ is the number of sigma point sets. X^i is the time state vector of the i point obtained by sampling, and P is the state variance matrix.

The weight corresponding to the sampling point is shown as follows.

$$w_m^i = \begin{cases} \lambda/(n+\lambda)(i = 0) \\ \lambda/2(n+\lambda)(i = 1, 2, \ldots, 2n) \end{cases} \tag{9.6}$$

$$w_c^i = \begin{cases} \lambda/(n+\lambda) + 1 - \alpha^2 + \beta(i = 0) \\ \lambda/2(n+\lambda)(i = 1, 2, \ldots, 2n) \end{cases} \tag{9.7}$$

where *wi m* represents the weight of the mean, *wi c* represents the weight of the variance; $\lambda = \alpha^2(n + \rho) - n$ is the scaling parameter, α is the scale factor, β is used to integrate the prior information of the random variable x.

Bring Eq. (9.5) for the UT transformation to obtain the sigma point matrix of the X column vector at time k.

$$X_k^i = \begin{cases} \hat{X}_k (i = 0) \\ \hat{X}_k + \sqrt{(n+\lambda)P}_k (i = 1, 2, \ldots, n) \\ \hat{X}_k - \sqrt{(n+\lambda)P}_k (i = n+1, \ldots, 2n) \end{cases} \tag{9.8}$$

Substitute the matrix Eq. (9.8) into Eq. (9.3) to obtain the state of the system.

$$X_{k+1|k} = f(X_k^i) \tag{9.9}$$

Then, the state matrix at time $X_{(k+1|k)}$ is obtained. The weighted mean value of $X_{(k+1|k)}$ and the updated covariance matrix are obtained from Eqs. (9.6) and (9.7).

$$\hat{X}_{k+1} = \sum_{i=0}^{2n} w_m^i X_{k+1|k}^i \tag{9.10}$$

$$P_{k+1} = \sum_{i=0}^{2n} w_c^i (X_{k+1|k}^i - \hat{X}_{k+1})(X_{k+1|k}^i - \hat{X}_{k-1})^T + Q \tag{9.11}$$

According to the updated state matrix Eqs. (9.10) and (9.11), a new set of sigma points is generated using the UT transformation again, and the point set is substituted into Eqs. (9.3) and (9.4). The predicted value of dynamic Bayesian analysis is systematically representative by double UT transformation, so as to obtain Eqs. (9.12) and (9.13).

$$X_{k+1|k}^i = \begin{cases} \hat{X}_{k+1|k} \; (i = 0) \\ \hat{X}_{k+1|k} + \sqrt{(n+\lambda)P}_k (i = 1, 2, \ldots, n) \\ \hat{X}_{k+1|k} - \sqrt{(n+\lambda)P}_k (i = n+1, \ldots, 2n) \end{cases} \tag{9.12}$$

$$Z_{k+1}^i = h(X_{k+1|k}^i) \tag{9.13}$$

The process of data interaction between DBNs and UKF is shown in Fig. 9.4. The predicted value of dynamic Bayesian analysis for system degradation data is input into DUKF as a parameter. The weight and variance are solved in DUKF algorithm, and the prediction process of DBNs analysis is optimized through iterative cycle. Therefore, the uncertainty associated with the RUL estimation is incorporated into the decision-making process. Subsystem performance varies over time, and the interaction of information can achieve data fusion. More accurate results can be obtained by combining the advantages of both by cyclic iteration.

Likewise, the measured values of the new sigma points generated by Eq. (9.12) are weighted to obtain the predicted mean value and covariance, as shown below.

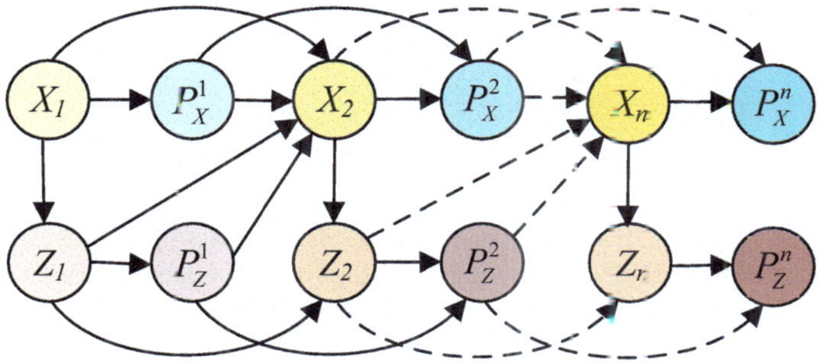

Fig. 9.4 Data interaction between DBNs and UKF

$$\hat{Z}_{k+1|k} = \sum_{i=0}^{2n} w_m^i Z_{k+1|k}^i \tag{9.14}$$

$$P_{zz} = \sum_{i=0}^{2L} w_c^i (Z_{k+1|k}^i - \hat{Z}_{k+1})(Z_{k+1|k}^i - \hat{Z}_{k+1})^T + R \tag{9.15}$$

$$P_{\mathrm{xz}} = \sum_{i=0}^{2n} w_c^i (X_{k+1|k}^i - \hat{X}_{k+1})(Z_{k+1|k}^i - \hat{Z}_{k+1})^T \tag{9.16}$$

According to Equations as follows, the Kalman gain coefficient K can be obtained. The state and covariance matrix of the system can be updated by the Kalman gain coefficient K.

$$K_{k+1} = P_{xz} P_{zz}^{-1} \tag{9.17}$$

$$\hat{X}_{k+1|k+1} = \hat{X}_{k+1|k} + K_{k+1}(Z_{k+1} - \hat{Z}_{k+1}) \tag{9.18}$$

$$P_{k+1|k+1} = P_{k+1|k} - K_{k+1} P_{zz} K_{k+1} \tag{9.19}$$

According to the results of the previous iteration of the algorithm, $\hat{X}_{(k+1|k)}$ is the prior estimate of the optimal estimate of the system at $k + 1$, and $P_{(k+1|k)}$ is the prior estimate of the covariance at $k + 1$. Z_{k+1} is the prior prediction at $k + 1$, and $\hat{X}_{(k+1|k+1)}$ is the posteriori estimate of the optimal estimate at time $k + 1$. $P_{(k+1|k+1)}$ is the posteriori estimate of the covariance at time $k + 1$. This will update the output $\hat{X}_{(k+1|k+1)}$ as the optimal prediction state of the system, and $\hat{X}_{(k+1|k+1)}$, $P_{(k+1|k+1)}$ continues to iterate like the initial value.

The DUKF algorithm can estimate the next time only through the current state of the system, and the algorithm does not need a large computing unit and storage unit, which has high efficiency.

9.2.3 RUL Prediction

The flow chart of algorithm for predicting the RUL of a single subsystem of the system, as shown in Fig. 9.5. The DUKF methodology is used to obtain the priori estimates and the covariance of state variables and observation variables. The cyclic iteration is carried out to make the predicted value of system state closer to its real measured value. The optimal prediction of the system degradation process is obtained by updating the model parameters to complete the evaluation of the degradation state of the system. Judge the state, and if the system enters the stage of accelerated degradation, use the optimal prediction value to predict the remaining life of the system until it reaches the system fault threshold. The process of system degradation is obtained from the structural relationship of each subsystem.

The RUL of the system is calculated based on the estimated dynamic performance. While performance represents structural reliability or availability, it usually decreases over time. RUL is determined by the fault threshold A_{th}, that is, the critical value, which is determined by the actual physical fault mechanism of the system. Assuming the actual performance of the system is A, that is, when $A \le A_{th}$, the system fails, and the system service life can be defined as follows, and inf is the largest lower bound of all those lower bounds. It is the estimated normal operating time of the system from the current time to the fault threshold. When the performance degradation product of the system reaches the A_{th}, the system is judged to have failed.

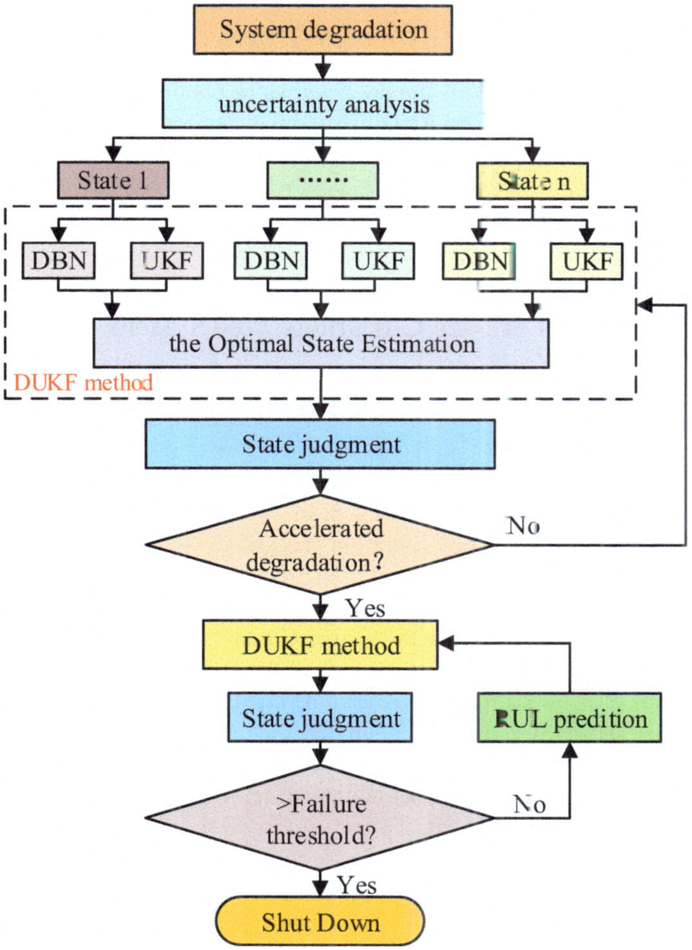

Fig. 9.5 Flowchart of the algorithm

$$T = \inf\{t : A(t) \leq A_{th} | A(0) > A_{th}\} \tag{9.20}$$

In order to quantitatively compare the prediction performance of the proposed model, the commonly used performance indicators: the relative accuracy (*RA*) and the RUL prediction error (*PE*) are used to judge the accuracy of the system. Eq. (9.21) is the methodology for calculating *RA* and Eq. (9.22) is the methodology for calculating *PE*. They quantify the accuracy of the prediction methodology according to specific time indicators.

$$RA_t = 1 - (X_t - \widehat{X}_t)/X_t \tag{9.21}$$

$$PE = \left| RUL_t - RUL_p \right| \tag{9.22}$$

RA_t is the relative accuracy at time t, PE is the prediction error of RUL, X_t is the value observed at time t, $\hat{X}_t$ is the predicted value obtained by the methodology. RUL_t is the real RUL value and RUL_p is the predicted value of RUL. For the *RA* indicator, the *RA* score range is [0, 1], where the full score is 1. The closer to 1, the more accurate the RUL prediction results can be; For the *PE* index, the closer they are to 0, the higher the accuracy of the estimate.

9.3 Case Study of Subsea Christmas Tree System

9.3.1 Introduction of Subsea Christmas Tree Control System

As an important offshore oil exploitation facility, the safe and efficient production of the subsea Christmas tree system is the basis of offshore oil exploitation. If there are any faults and leakages in the operation process, it will seriously affect the safe production activities of the entire oil field. The subsea Christmas tree control system takes on the important task of ensuring the safe operation of the subsea oil and gas fields. It is the bridge between the subsea Christmas tree system and the upper facilities. The subsea Christmas tree control system mainly controls the valves and throttles installed on the Christmas tree and the manifold, realizes the reception and transmission of data between water and underwater, monitors temperature, pressure and other data. It is urgent to predict the RUL of the subsea Christmas tree control system before it fails and to carry out maintenance, in order to avoid accidents and reduce loss of life and property.

At present, the electro-hydraulic composite control system of the subsea Christmas tree is widely used in subsea oil production. Therefore, take the system as an object to study and analyze the subsea Christmas tree, and the specific structure of the control system is shown in Fig. 9.6. The composition and operation process of the subsea Christmas tree control system is complex, the service time is long, and the time cost of obtaining the field observation data is high and difficult. Therefore, it is suitable for the proposed methodology. For a typical subsea Christmas tree with electro-hydraulic composite control, the electronic control equipment mainly consists of: the master control station (MCS), the electrical power unit (EPU), the chemical injection unit (CIU), the subsea distribution module (SDM) and related sensors. The hydraulic equipment mainly includes: the hydraulic power unit (HPU), the static umbilical termination assembly (SU), the dynamic umbilical termination assembly (DU), the subsea control module (SCM), and valves with hydraulic actuators, etc. The subsea control module includes combined pressure and temperature sensors, flow sensors, fluid level sensors, hydrocarbon leak sensors, pressure sensors, and so on. These sensors and control modules monitor Christmas tree performance data and make maintenance schedule if fault is detected, and this information is included in OREDA.

9.3.2 RUL Prediction of Subsea Christmas Tree Control System Based on DUKF

9.3.2.1 Degradation Modeling for Subsea Christmas Tree Control System

The OREDA database covers reliability data from a wide range of equipment used in oil and gas exploration and production. According to the statistics in the OREDA database, the overall degradation performance of each subsystem of the subsea Christmas tree control system in the service stage roughly conforms to the exponential degradation model, and its failure rate is constant for each degradation stage. Use the fault data from OREDA to calculate the failure rate, the parameters of the DUKF algorithm, and the distribution of random fault and process noise.

To estimate the failure rate λ, *Number of failures m* and the cumulative service time to be observed τ. The maximum likelihood prediction λ is,

$$\lambda = Number\ of\ failures/Aggregated\ time\ in\ service = m/\tau \tag{9.23}$$

Its reliability can be expressed as,

$$R(t) = \exp\left(-\int_0^t \lambda(t)dt\right) = \exp(-\lambda t) \tag{9.24}$$

According to historical degradation data provided by OREDA (DNV 2015), the degenerative process can be divided into three stages to prepare for the subsequent prediction work (Cai et al. 2021). Suppose *h* stands for time slice, and each time slice corresponds to six months. That is, h = 1–5 is the infant mortality phase which is the

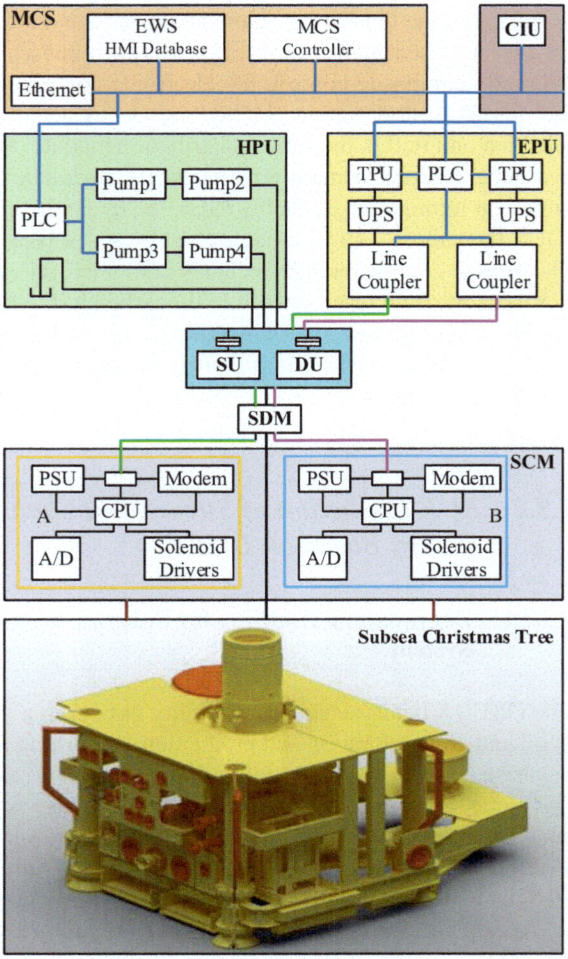

Fig. 9.6 Electro-hydraulic subsea Christmas tree control system

initial operation stage of the system, and the degradation rate is rapid; h = 6–18 is the random failures phase, the degradation rate tends to be gentle, which is the stable operation stage of the system and belongs to the area of accidental failure; after h = 19, it is the wear out failures phase of the accelerated degradation stage, which is faster than the initial stage. This is the stage of accelerated degradation of the system and belongs to the period of depletion failure. The main types of faults of the three stages are different.

Take the HPU as an example: the main faults of infant mortality phase are: Low output/performance, High output, No immediate effect, Internal/External leakage-utility medium, Internal/External leakage-process medium and Other failure mode. The main faults of random failures phase are: Internal/External leakage-utility medium, Internal/External leakage-process medium, Plugged/choked, No immediate effect Insufficient power, Other failure mode and Unknown failure mode. The main faults of wear out failures phase are: Insufficient power, No immediate effect, Unknown failure node and Other failure mode. Among them, Unknown failure node and Other failure mode occur in the three stages of operation of the system, but the probability of occurrence is difficult to analyze accurately. Therefore, it is necessary to analyze the uncertain faults occurring in the three phases of each subsystem.

First, DBNs are used to model the distribution of unknown severity and failure mode of each subsystem of the system, solve the probability distribution and conduct uncertainty analysis. Distribution of random failure rates of subsystems in different phases is shown in Fig. 9.7. According to the main failure modes and uncertainty analysis results of each stage of the subsea tree control system, the degradation rate of the key subsystem of the subsea tree control system is calculated, as shown in Fig. 9.8. On this basis, the degradation process of the system is calculated.

The state transition equation of the system is constructed according to the degradation model of the subsea Christmas tree control system. X_t represents the state of the system at time t and v_k represents the random noise in the degradation process by observing the degradation process of each subsystem, as shown below,

$$
\begin{aligned}
X_{t1} &= \exp(\lambda_1 \cdot t) + v_{k1}, v_{k1} \sim N(0, \sigma_{v1}) \\
X_{t2} &= \exp(\lambda_2 \cdot t) + v_{k2}, v_{k2} \sim N(0, \sigma_{v2}) \\
X_{t3} &= \exp(\lambda_3 \cdot t) + v_{k3}, v_{k3} \sim N(0, \sigma_{v3})
\end{aligned} \tag{9.25}
$$

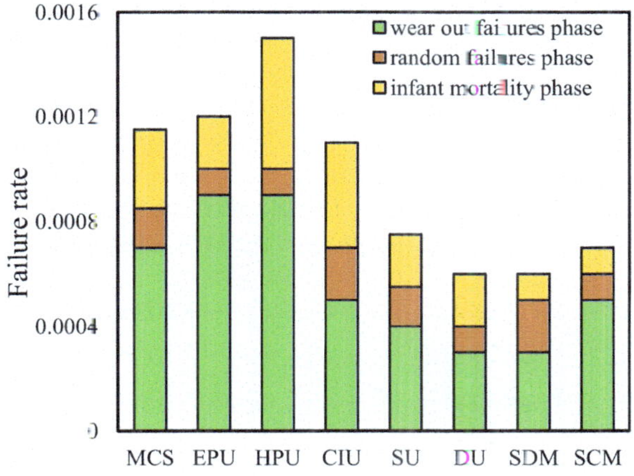

Fig. 9.7 Distribution of random failure rates of subsystems in different phases

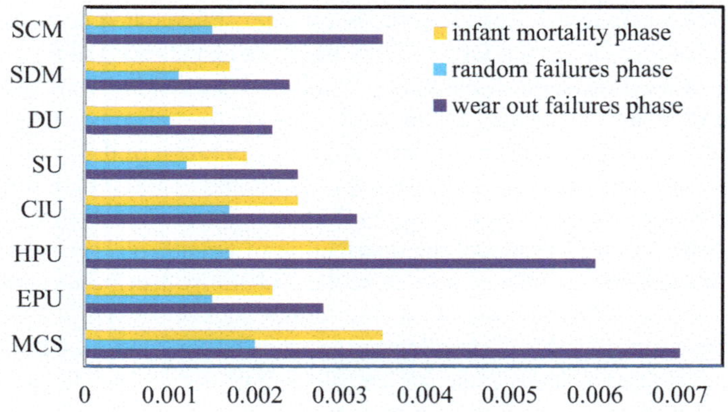

Fig. 9.8 Degradation rate of key subsystems of subsea Christmas tree control system

9.3.2.2 DUKF Methodology for Subsea Christmas Tree Control System

According to the structure of the system, the DBN model of the subsea Christmas tree control system is constructed with a period of six months as the time frame to predict the state of the three stages of the system. As shown in Fig. 9.9 *I-m* is the infant mortality phase, which is the initial operating stage of the system and belongs to the early failure period; *R-f* is the phase of random failures, which is the phase of stable operation of the system and belongs to the period of accidental failure; *W-o* is the wear out failures phase, which is the accelerated degradation stage of the system and belongs to the loss failure period. It is calculated according to the failure rate of each operating stage of the system calculated by the dynamic Bayesian analysis results.

Dynamic Bayesian analysis prediction is carried out for the system to obtain the prediction data $\{y_1, y_2, y_3\}$, Y_t represents the state of the system predicted by the dynamic Bayesian analysis results at time *t*, as shown in Eq. (9.26). The prediction equations are established according to the data respectively, and w_k represents the process noise in the DBNs prediction process for each subsystem.

$$\begin{aligned} Y_{t1} &= \exp(\lambda_1 \cdot t) + w_{k1}, w_{k1} \sim N(0, \sigma_{w1}) \\ Y_{t2} &= \exp(\lambda_2 \cdot t) + w_{k2}, w_{k2} \sim N(0, \sigma_{w2}) \\ Y_{t3} &= \exp(\lambda_3 \cdot t) + w_{k3}, w_{k3} \sim N(0, \sigma_{w3}) \end{aligned} \tag{9.26}$$

Prediction data is fed into the DUKF algorithm, and the number of time slots, the variance of noise in the prediction process, the starting point of the prediction and the initial value of the model parameters are set. The initial mean value of the state quantity is $\tilde{x}_0$. The initial variance matrix of the state quantity is P_0.

UT is performed on all the data using symmetric sampling of $n + 1$ sigma points, and constructs a set of sigma points $\left\{\chi_k^0, \chi_k^1, \ldots, \chi_k^{2n}\right\}$ around $\tilde{x}$. The corresponding

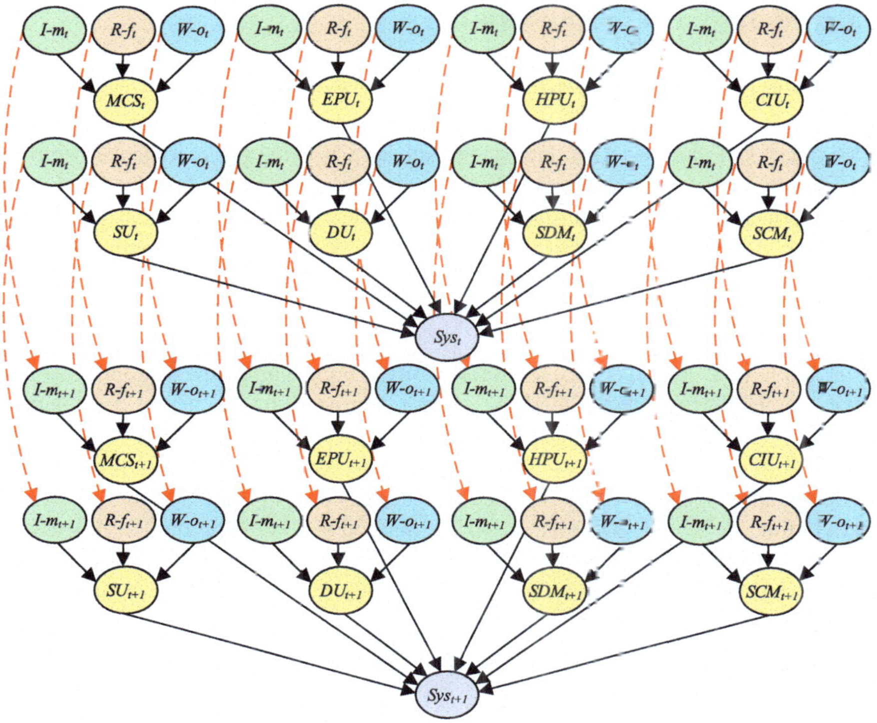

Fig. 9.9 DBNs model of subsea Christmas tree control system

mean weights $\{w_m^0, w_m^1, \ldots, w_m^{2n}\}$ and the variance weights $\{w_c^0, w_c^1, \ldots, w_c^{2n}\}$ are solved, and the sigma point set is modified proportionally. Then, the data is updated based on the weight, the sigma point set is retransmitted, and the priori estimates $\tilde{x}_{k+1}, \widetilde{P}_{k+1}$ of the state variables and the covariance are calculated. According to the priori estimate, a new sigma points set $\{\chi_k^0, \chi_k^1, \ldots, \chi_k^{2n}\}$ is obtained by again using the UT transform. The predicted value is more representative through two UT transformations. The new set of sigma points is returned and weighted to obtain the priori estimate and the variance of the observation variables.

Finally, the measured values are updated and the Kalman gain is resolved. The optimal estimate of the degradation process of the subsea Christmas tree control system is obtained.

9.3.3 Results and Discussions

9.3.3.1 System Reliability Prediction

The actual degradation (AD) data of the subsea Christmas tree control system is directly obtained from the monitored historical data in the OREDA database. The degradation process of various subsystems of the system is shown in Fig. 9.10, indicating the AD data of each unit of the subsea Christmas tree control system, the changes of system reliability predicted by the DBNs and DUKF. The subsea Christmas tree control system have long service time and high reliability requirements, so the degradation amount of a single subsystem is small, the degradation curve is relatively flat, and the overall reliability is high. The prediction optimization process of DUKF is better, which basically coincides with the actual degradation data curve.

It can be seen that the first to fifth time slices are the initial degradation stage, and the degradation speed is fast in this stage. With the stable operation of one stage, the degradation rate of each unit changes. The enlarged part is the prediction process of the two methodologies when the observational data enters the state of accelerated degradation, the 16–20th time slice. It can be seen from the figure that there is an obvious inflection point in the observation data around the 18th time slice, that is, the degradation rate of each unit increases after 9 years of service, the system enters the stage of accelerated degradation.

Each subsystem constitutes a control system according to the actual structure of the system. The process of system degradation is shown in Fig. 9.11, which shows the prediction process of the change of the system reliability. According to the analysis, the control system enters the stage of accelerated degradation around the 18th time slice. It is evident that the prediction curve of the DUKF methodology almost coincides with the actual degradation data.

9.3.3.2 RUL Prediction

As the subsea Christmas tree control system shoulders the important task of ensuring the safe operation of underwater oil and gas fields, its reliability requirements are high. The fault threshold is determined by relevant documents and expert experience. Failure threshold may be one or more values, or a distribution (Cai et al. 2022). Fig. 9.12 shows the comparison of the actual degradation data, the prediction of the DUKF methodology and the DBNs methodology. According to the results of the prediction of the DUKF methodology, the RUL of DU is the longest from the stage of accelerated degradation, and the total service life can reach 18.19 years.

The MCS in the electric control system is operated by computers and connected to the microprocessor in the underwater electronic module, and the HPU contains a high and normal pressure pump to transport the high pressure and normal pressure liquid into the hydraulic circuit to the underwater through umbilical cable. Both

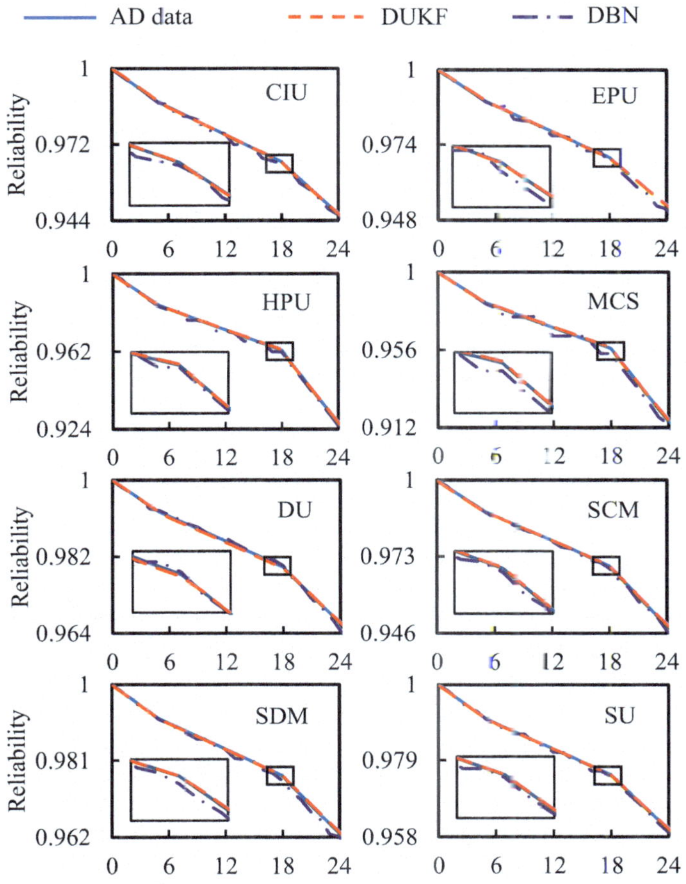

Fig. 9.10 Degradation process of each subsystem

have the characteristics of a high failure rate, a rapid decline in reliability and a short RUL. After entering the accelerated degradation stage, the remaining service life of MCS is only 1.24 years and that of HPU is 1.937 years. It is necessary to strengthen monitoring to prevent accidents. Overall, the remaining life of some underwater subsystems is slightly longer than that on water.

9.3.3.3 Prediction Effect Comparison

The relative accuracy of the two methodologies with respect to the actual degradation data is shown in Fig. 9.13. It shows that the accuracy of the DBNs estimation is highly volatile and irregular. The relative accuracy of the reliability of the DUKF prediction result is higher than that of the DBNs prediction, with low fluctuation, high

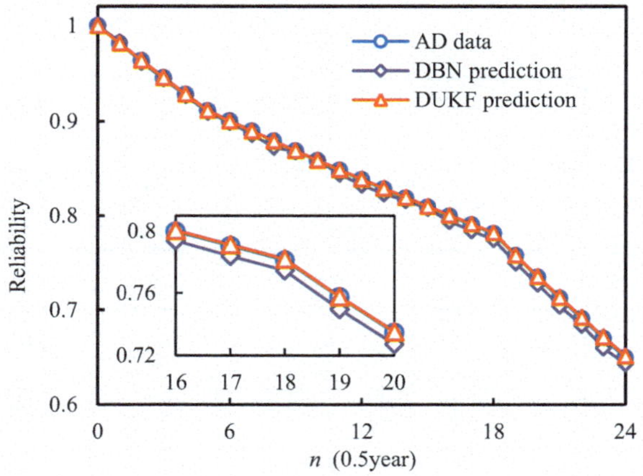

Fig. 9.11 Comparison of system degradation process

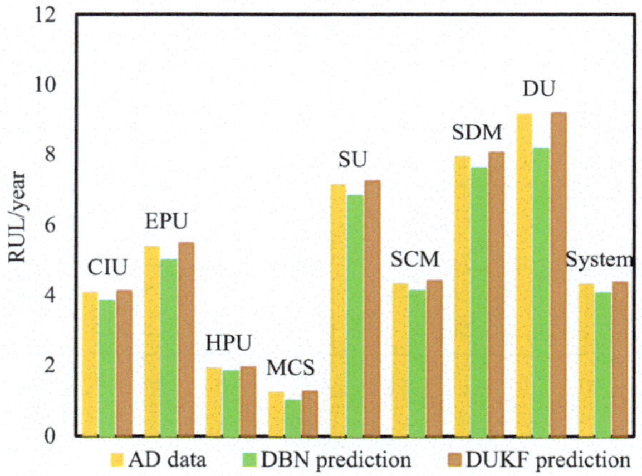

Fig. 9.12 Comparison of RUL prediction results

prediction convergence, strong ability to suppress random noise interference, and maintaining consistency between estimated data and observation data. The proposed hybrid method combines the historical degradation data with the DUKF methodology and can obtain results that are closer to the actual degradation data.

The comparison of the relative accuracy of the two methodologies for system prediction is shown in Fig. 9.14. From a comprehensive perspective in Figs. 9.13 and 9.14, shows that the estimated relative accuracy of the proposed DUKF methodology is higher, and the data analysis shows that its relative accuracy is increased by an order of magnitude. For the whole system, the average prediction error of DUKF is only 0.165 times that of the DBNs prediction. The prediction error of RUL prediction and actual degradation data of subsea Christmas tree control system subsystems using

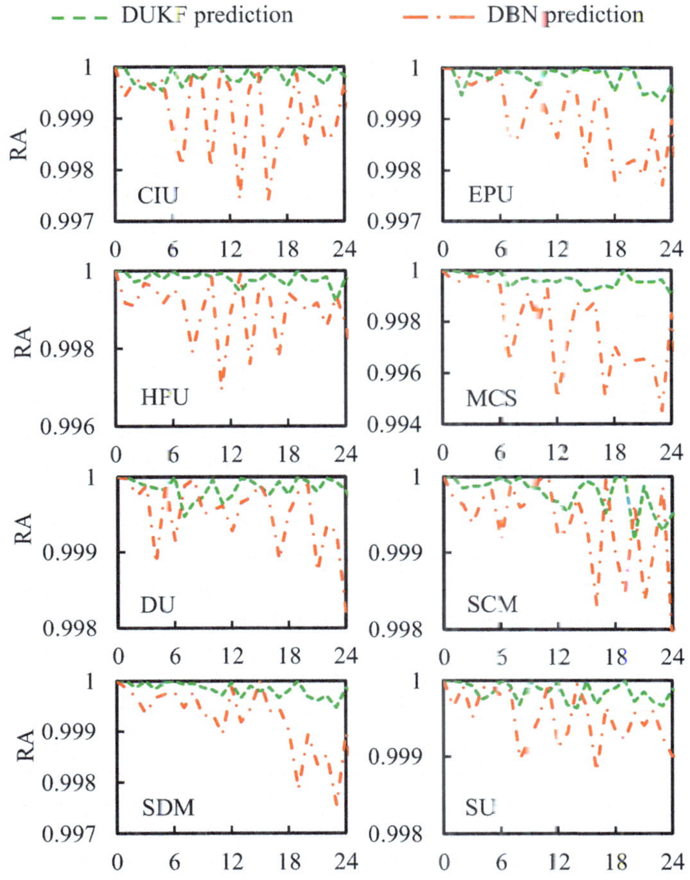

Fig. 9.13 Comparison of relative accuracy of RUL prediction

DBN and DUKF respectively is shown in Fig. 9.15. It can be seen from the above that the results of the prediction of the DUKF methodology is closer to the real degradation of the system, and that the prediction of RUL is more accurate. Fig. 9.15 also verifies this point.

The estimate of the subsystems of the subsea Christmas tree control system, the prediction error of DBN is 2.29 to 148.71 times that of the DUKF methodology, and the prediction error of the whole control system is as high as 4.53 times, which shows that the cumulative error predicted only by DBN is very large, while the cumulative error problem can be solved well by the DUKF methodology, and its prediction error is very low. It has stronger random noise inhibition.

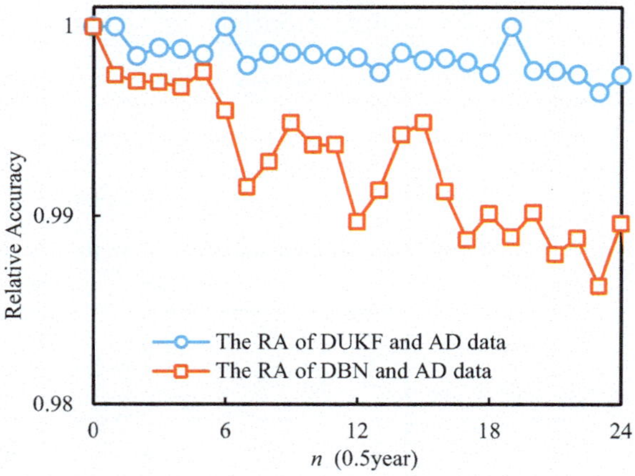

Fig. 9.14 Comparison of system relative accuracy

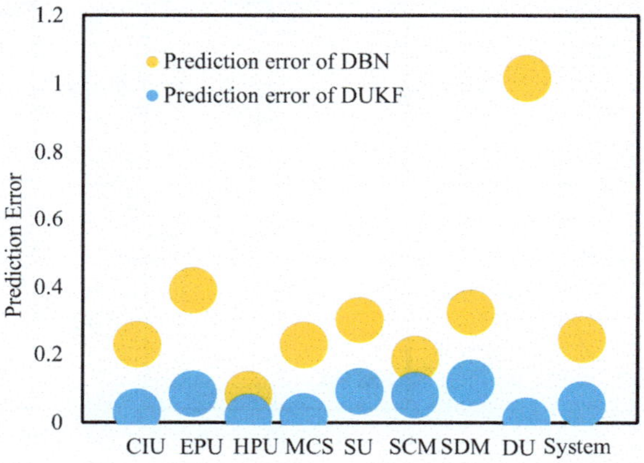

Fig. 9.15 Comparison of RUL prediction errors

9.3.3.4 Model Updating with Evidence

Due to the complex working environment of subsea Christmas trees, the environmental impacts are numerous, such as typhoons, earthquakes and tsunamis. Changes in the external natural environment will impact on the subsystems, which will lead to degradation or even failure of the subsystems. When an impact occurs, it means the model needs to be updated, and DBNs can supplement and update information in the process of dynamic prediction. The information added to the DBNs is called "evidence", which is performed through the evidence node. The yellow box of SCM is impacted in the fifth year, and the rate of degradation will be improved. It enters the stage of accelerated degradation in advance. External shocks lead to random failures occurring at a certain stage, so overall the failure rate of the system in each stage remains constant. The evidence node was introduced to calculate the distribution of the degradation rate, as shown in Fig. 9.16.

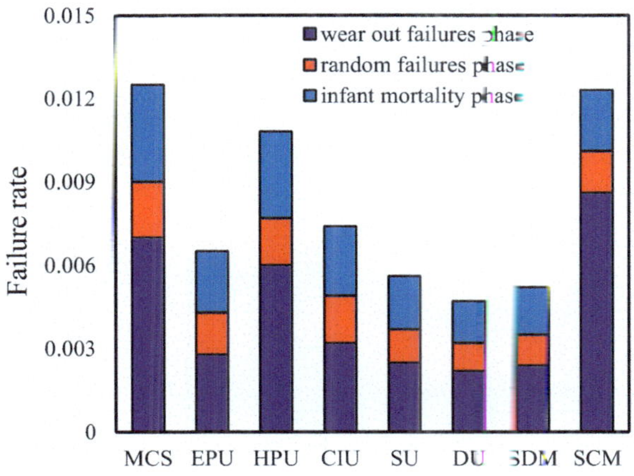

Fig. 9.16 Distribution of failure rates in different phases with evidence

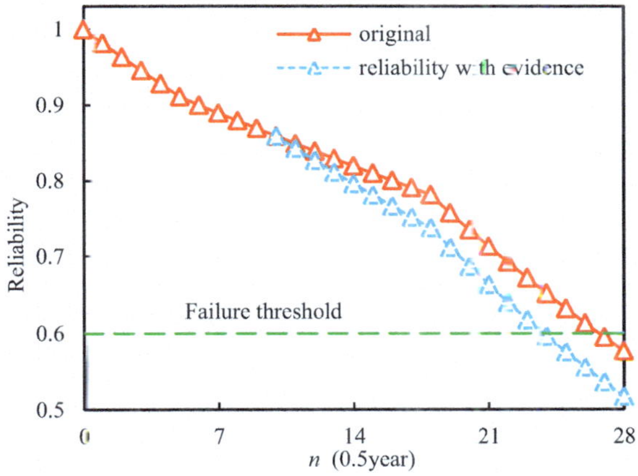

Fig. 9.17 Reliability change under the influence of shock

The system is predicted by the rate of degradation of each subsystem. The SCM yellow box is impacted in the fifth year, and the degradation rate is reinforced. The impact on system reliability is shown in Fig. 9.17, resulting in a reduction in the life of SCM by 5.91 years and the life of the control system by 1.45 years, as shown in Fig. 9.18. Combined with the evidence, this obvious change is easy to find, which helps in the timely maintenance of the system.

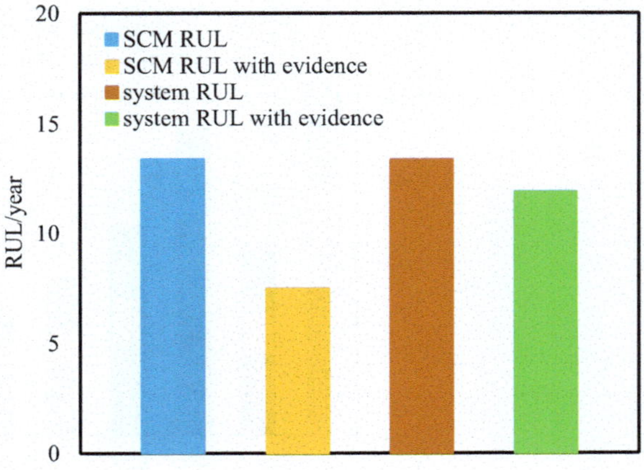

Fig. 9.18 RUL change under the influence of shock

Chapter 10
RUL Prediction with Failure Dependence

Abstract Failure dependence including degradation interactions and cascading failures pose challenges for RUL estimation in complex systems. For degradation interactions, a multi-stage model-based RUL estimation approach is proposed. This method accounts for the interactions between components caused by uneven working pressures, where even identical components can experience different degradation processes. Interaction coefficients and multi-stage degradation models are key to estimating RUL, calculated as the time difference between healthy operation and failure. The approach is validated on an electro-hydraulic subsea Christmas tree, demonstrating its effectiveness in evaluating system performance and RUL under interaction effects. For cascading failure, a novel RUL estimation methodology is introduced, focusing on multilevel systems. Using DBNs, the approach models cascading failures based on the position and function importance of system nodes. By monitoring node states and iteratively calculating performance impacts across system levels, the number of failure nodes and overall system performance are assessed. This method is validated through a subsea transportation system with a three-level network. Both approaches offer tailored solutions for different failure scenarios, enhancing RUL prediction accuracy and supporting more effective system maintenance strategies.

10.1 Introduction

Numerous methods have been proposed to improve RUL predictions, including machine learning approaches, deep neural networks, and probabilistic models such as particle filtering and BNs. While these methods have shown promise in various applications, such as bearings and electronic components, most research has focused on individual systems and neglected the failure dependence effects in interconnected and multilevel systems. The failure dependence effects can be characterized by the degradation interactions between components and the cascading failures, which significantly influence degradation rates and failure rates.

B. Cai et al., *Intelligent Operation and Maintenance for Subsea Production Systems*,
https://doi.org/10.1007/978-981-96-2677-9_10

Recent studies (Sayed et al. 2019) have begun to explore the interactions among components, particularly in complex systems. However, existing modeling techniques that account for these interactions often lack generalizability and adaptability, as they are tailored to specific situations or objects. The ambiguity surrounding the mechanisms of these interactions complicates the modeling process, leading to a gap in comprehensive frameworks that can effectively address them. Furthermore, practical applications often overlook the influence of uncertainty parameters on modeling outcomes, highlighting a blind spot in the research. One of the purposes of this chapter is to advance the understanding of interaction modeling within CBM by employing BNs and DBNs as probabilistic knowledge representation tools. BNs and DBNs have demonstrated effectiveness in fault diagnosis and reliability evaluation, as well as in managing uncertainty in the degradation process. This research will address the challenges of interaction modeling by providing a flexible framework that incorporates uncertainty and facilitates model updating. By doing so, it aims to enhance the accuracy of RUL estimation and improve maintenance decision-making, ultimately contributing to the development of more reliable and efficient CBM systems.

Cascading failures, which occur when the failure of one component leads to a chain of failures across subsystems, are common in critical infrastructure such as power grids, oil transportation networks, and wireless sensor networks (Fu et al. 2019). These cascading events introduce additional complexities, including the load redistribution and interdependencies, which complicate the accurate prediction of system RUL. The motivation behind this work arises from the need to address these complexities and improve RUL estimation for multilevel systems subjected to cascading failures. Although BNs and DBNs have been widely applied in PHM for RUL prediction, their application in modeling cascading failures remains underexplored. Existing models do not sufficiently account for the interactions between subsystems or the propagation of failures through complex networks. To bridge this gap, this chapter also introduces an innovative approach to RUL estimation by incorporating cascading failure mechanisms into the DBN framework. The proposed method offers a comprehensive solution for modeling the degradation process and predicting the RUL of complex systems under the influence of cascading failures. By resolving the limitations of existing methods, this work contributes to the improvement of predictive maintenance strategies, ensuring more accurate and reliable system health management.

10.2 RUL Estimation Model Considering Degradation Interactions

This chapter contributes a multi-stage model-based RUL estimation approach using DBNs, and the interaction among components is considered. DBNs mainly complete three tasks: (1) carry out the degradation modeling of components and systems; (2) identify the degradation stages of components, and then realize the modeling of

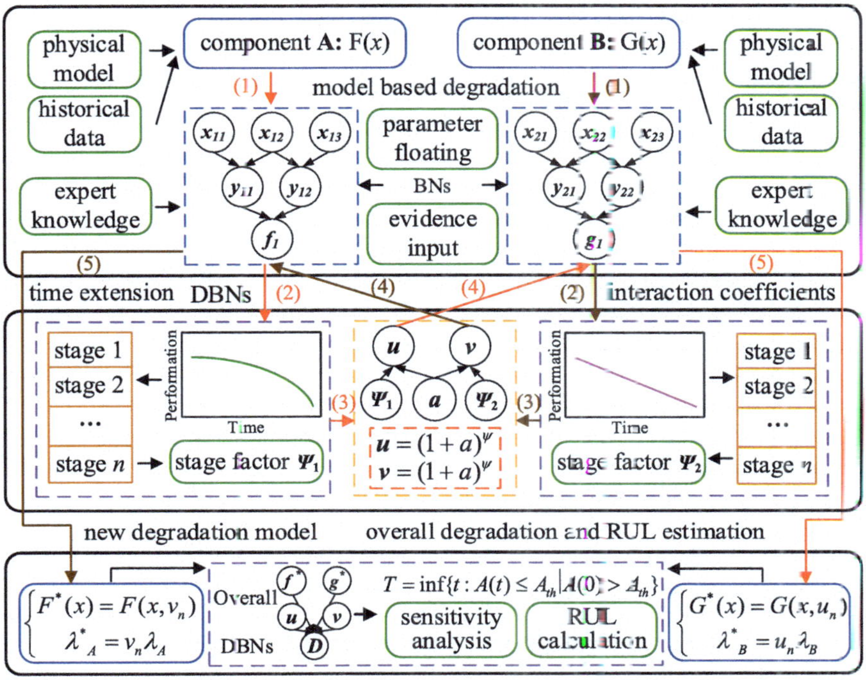

Fig. 10.1 Multi-stage model-based RUL estimation framework

interactions; (3) complete the analysis of parameter uncertainty in the degradation process. A framework of proposed approach is shown in Fig. 10.1.

The multi-stage model-based RUL estimation approach mainly includes 5 major steps: (1) Components degradation process modeling. Information available to build a degradation model is integrated into three categories: physical model, historical data and expert knowledge. (2) Degradation stages dividing. Combining component performance and expert experience, the degradation process is divided into multiple stages, and the influence of each degradation stage on the interactions is quantified. (3) Interaction coefficients calculating. A new concept is defined, called "interaction coefficients", which is used to solve the problem of interaction modeling. The interaction coefficients are calculated from the component characteristics and the degradation stage. (4) New degradation process modeling. A degradation model that considers interactions influence is constructed by interaction coefficients. (5) RUL estimation. An overall degradation model is built by DBNs, and then RUL calculation and sensitivity analysis are performed. More details are shown as follows.

10.2.1 Degradation Models of Components

Macroscopic degradation processes of components are obtained by empirical formulas or large amounts of historical data, for example, degradation of electronic components is usually considered an exponential degradation, and the degradation of hydraulic components can be subject to the Weibull degradation model (Roos et al. 2020). BNs and DBNs are adaptive to probabilistic information greatly, and they are advantageous for obtaining accurate models of degradation. The BNs are built on the basis of the performance degradation models $F(t)$ and $G(t)$, where the parent node x and the intermediate node y are parameters influencing the performance, and the child node f and g are the performance nodes of the components. Parameters may not be fixed values in practical conditions, but float around a certain value. Uncertain information in modeling can be solved through BNs. Performance is obtained by BNs that incorporates uncertain information. In addition, when a sudden event occurs, external information needs to be input into the degradation model, that is, input evidence information to update the model, which also needs to be done based on BNs.

10.2.2 Degradation Models Considering Interactions Influence

The primary goal of obtaining degradation models considering interactions influence is to deal with the interactions between components. This difficulty is solved through multi-stage division and assignment of interaction coefficients. A degradation effect coefficient is defined in (Hafsa et al. 2015), which is used to build a degradation model of multi-component interaction. In this work, an improved algorithm based on the above work is proposed, which considers the influence of multi-stage interaction and uncertainty. The improved algorithm can better describe the degradation acceleration process and deal with some parameter floating problems. First, the entire degradation process can be divided into n stages through the training samples with the health stage labels, which represent the severity of the system degradation failure. Degradation stages include lifetime with respect to the entire process of components (Compare et al. 2020). Next, assign a value to the "stage factor, $\boldsymbol{\Psi}$" of each degradation stage, indicating the influence of the component on another same component at different stages. The stage factor, $\boldsymbol{\Psi}$ can be expressed as

$$\Psi = (j_1, j_2, \ldots, j_n) \tag{10.1}$$

where j_n represents the influence of the component on the degradation of another same component when the component performance is in the nth degradation stage.

The relationship between interaction coefficients, $\boldsymbol{u}$ and $\boldsymbol{v}$, and stage factor, $\boldsymbol{\Psi}$ is defined as

$$u = (1 + a_h)^{\Psi} \tag{10.2}$$

$$v = (1 + a_g)^{\Psi} \tag{10.3}$$

where a_h and a_g are the parameter of component degradation characteristics.

Then in the nth degradation stage, the interaction coefficients $\boldsymbol{u}$ $(u_1, u_2, \ldots, u_n)$ and $\boldsymbol{v}$ $(v_1, v_2, \ldots, v_n)$ can be expressed as

$$(u_1, u_2, \ldots, u_n) = \{(1 + a_h)^{j_1}, (1 + a_h)^{j_2}, \ldots, (1 + a_h)^{j_n}\} \tag{10.4}$$

$$(v_1, v_2, \ldots, v_n) = \{(1 + a_g)^{j_1}, (1 + a_g)^{j_2}, \ldots, (1 + a_g)^{j_n}\} \tag{10.5}$$

Interaction coefficients, u_n and v_n directly affect the degradation rate. First, the degradation stages of $F(t)$ and $G(t)$ at time t is identified by BNs. Assuming that $F(t)$ is in the ith degradation stage at time t, then the stage factor of component A at that moment is u_i; $G(t)$ is in the kth degradation stage at time t, then the stage factor of component B at this moment is v_k, then at time $t + 1$, the stage factor u_i of component A at the previous moment is integrated into the degradation of component B at this moment. Similarly, the degradation factor v_k of component B at the previous moment is integrated into the degradation of component A at this moment. Through the expansion of DBNs, the dynamic identification of the degradation stage of the degradation model is realized, the stage factors are updated, and the degradation process of the entire life cycle is calculated. Degradation models considering interactions influence can be expressed as

$$\begin{cases} F^*(t) = F(t, v_n) \\ \lambda_A^* = v_n \lambda_A \end{cases} \tag{10.6}$$

$$\begin{cases} G^*(t) = G(t, u_n) \\ \lambda_B^* = u_n \lambda_B \end{cases} \tag{10.7}$$

where λ_A and λ_B are parameters that affect the degradation rate of the $F(t)$ and $G(t)$, λ_A^* and λ_B^* are parameters considering interactions, which are used to calculate a new degradation model.

10.2.3 Structure Modeling and Parameter Modeling of DBNs

Structure modeling methods mainly include cause-and-effect relationship, mapping algorithms, and structural learning. For complex system, diagram based on function is more accessible than other methods. In this approach, the structure of DBNs can actually be achieved by transforming from the function diagram of system. The performance model of each component is integrated into the performance degradation model of the system, which is on the basis of functional position of the component, and structure of systems for DBNs is determined. The nodes of DBNs represent the variables of degradation process of components, and the arcs of DBNs represent the causal relationship among variables.

For example, the degradation model of two components considering interactions is shown as follows

$$d^t = f(x_1, y_1, f, x_2, y_2, g, \boldsymbol{u}, \boldsymbol{v}, t) \tag{10.8}$$

where d^t is the performance of system at time t; x_1, y_1, x_2, y_2 are the variables of performance; f and g are the performance of components; $\boldsymbol{u}$ and $\boldsymbol{v}$ are the interaction coefficients.

The structural model considering interactions of two components is shown in Fig. 10.2. Nodes in DBNs are obtained from parameters in the equation. The black arrows indicate the performance calculation process; the red line between the 2 time slices indicates the change of the parameter over time, which shows the dynamic process of the parameter; the blue line between the 2 time slices indicates process of interactions influence by building a multi-stage model. The interaction coefficients act on another component, and the performance calculation is performed alternately in this way.

Parameter model includes prior probabilities of parent nodes in the initial moment, the conditional probabilities between the parent nodes and child nodes, and conditional probabilities between time slices. Parameter model is mainly obtained from

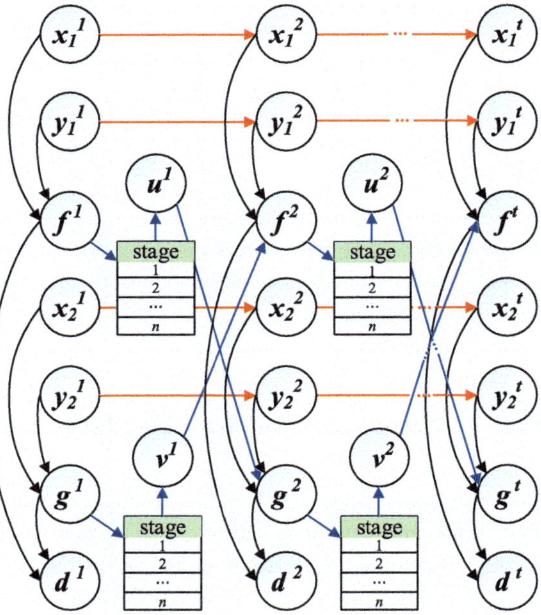

Fig. 10.2 DBNs of degradation model considering interactions

historical data, expert experience, and reasonable formulas. First, the nodes of variables are discretized. Then the distributions of variables are determined (normal distribution, lognormal distribution, exponential distribution, etc.). The key point to build parameter model is to achieve the conditional probability tables (CPTs). For some fuzzy causal relationships, parameter model is usually obtained through expert experience or previous sensor data. In this chapter, The CPTs of DBNs are obtained by transforming from physical models or equations. First, the relationships among all the performance nodes and performance variable nodes are determined based on the given degradation models; then the nodes are sampled with sample methods, such as Monte Carlo simulation. The CPTs are finally obtained from sufficient sampled values and corresponding probabilities.

10.2.4 RUL Estimation Based on Performance

RUL estimation is based on dynamic performance. For example, if performance represents reliability, availability, and serviceability of systems, performance usually decreases over time; if performance represents failure probability, performance generally increases over time. The RUL is obtained by the failure threshold A_{th}, that is, the critical value, which is determined by the actual failure mechanism of the systems. Assuming that the actual performance of the system is A, and the system fails when $A \leq A_{th}$, the RUL of the system can be defined as

$$T = \inf\{t : A(t) \leq A_{th} | A(0) > A_{th}\} \tag{10.9}$$

The RUL estimation method is shown in Fig. 10.3. The failure time is the moment that the performance first reaches the threshold. RUL can be divided into two categories: actual RUL and estimated RUL. The actual RUL refers to the RUL calculated by the entire performance monitoring or the real historical degradation data. The estimated RUL refers to the RUL calculated based on the predicted performance by establishing a degradation model and predicting the degradation trend. RUL considering interactions and RUL without interactions are two estimated RUL calculated through modeling, and the difference lies in whether interaction is considered. Obviously, the accuracy of RUL estimation is the primary goal.

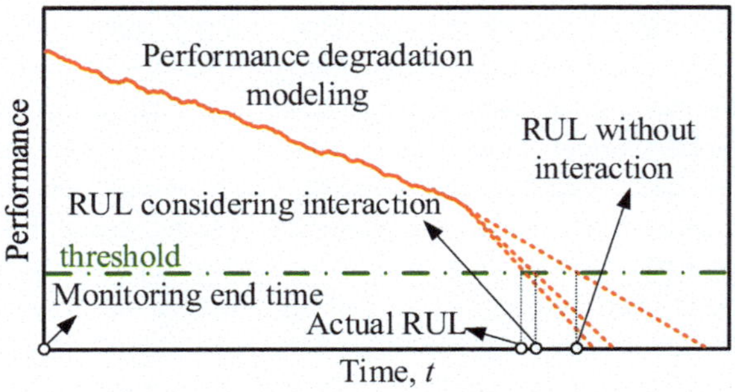

Fig. 10.3 RUL calculation method

10.3 RUL Estimation Model Considering Cascading Failure

10.3.1 Modeling Methodology

A methodology of RUL estimation for cascading failure of multilevel systems based on DBNs is proposed in this chapter. This methodology relies on the degradation model of multilevel systems. Using the position and function characteristics of multilevel systems, the position and function importance are proposed to build the degradation model of cascading failure, and then the RUL of multilevel systems is calculated. Figure 10.4 is a time slice model of cascading failure propagation, showing the bottom-up transmission process of a typical multilevel system. The degradation relationship of each node in each level of system is obtained from the parameter estimation of monitoring data and the filling of expert experience. It can be seen that the influence of normal nodes on the nodes of the next layer is connected by black lines, which indicates the normal effect of the multilevel systems; when the underlying node Z_1 fails, the cascading failure affects the nodes of the next layer, that is, the red line, representing the effect of cascading failures. When the nodes turn red, it means that the node fails under the influence of cascading failure and begins to affect the nodes of next layer. In the multilevel systems, each level system includes *n* nodes which complete the similar task, which belongs to the parallel relationship; the sub-systems at all levels transfer upward, which is a typical series structure. The transfer function is completed by the nodes in the subsystem, and any node in subsystems can be connected to all nodes in the adjacent subsystems. The failure of nodes or links will lead to performance degradation, which will lead to the failure of the original connection structure (Fu et al. 2019). When the cascading failure occurs, performance degradation of the next level will be affected based on location of the node and the importance of the function. The impact of cascading failure is transmitted slowly with time, and each node also degrading. Failure nodes in each level will affect all nodes in the next level until the network crashes. Through the transmission between levels, all nodes of the final level will be failed, which also means that the system has been unable to complete the normal work.

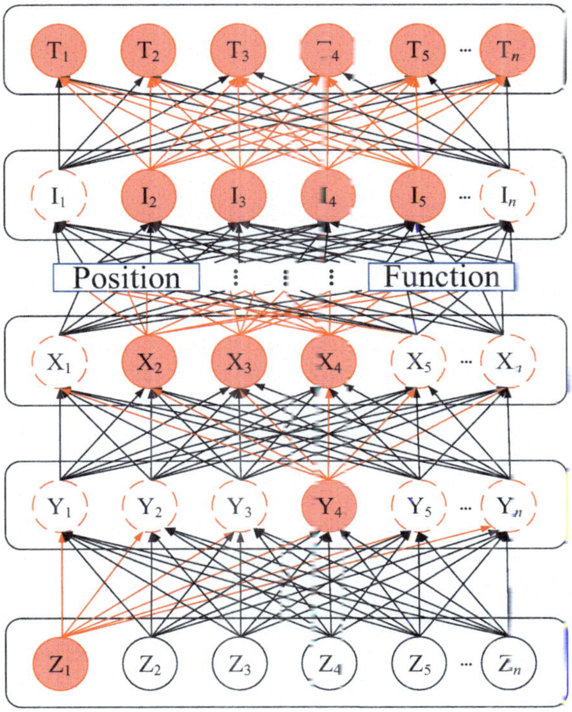

Fig. 10.4 Cascading failure model of multilevel systems

Generally, cascading failure usually does not occur immediately, but begins to spread after some nodes are transferred, which takes a period. From the acceleration of performance degradation of nodes to the gradual increase of failure nodes of multilevel network, it is also necessary to consider the performance degradation of network itself during this period, which makes the modeling of cascading failures difficult. A failure mode based on the joint effect of position importance and function importance on cascading failure is proposed in this section. Weight coefficients represent the joint influence of position importance and function importance on degradation of nodes, and indicate the proportion of the influence of position and function. It is determined by the characteristics of the system and the surrounding environment, that is, the contribution to the work and adaptability to the environment, covering safety, economy, and environment influence (weather, temperature, humidity, etc.). Cascading failure only occurs in the adjacent subnetwork, that is, the node failure of the first level subnetwork will only affect the relevant nodes in the next level subnetwork, and the cascading failure influence relationship is given as follows:

$$D_{y_{n+1}} = h(\delta, 1 + C_{ni}), \delta \in y_{n+1} \tag{10.10}$$

$$C_{ni} = \begin{cases} 0 & i\ normal \\ \varphi k_i + \tau z_i & i\ failed \end{cases} \tag{10.11}$$

where D_{yn+1} represents the cascading failure effect of nodes of layer n on layer $n+1$; y_{n+1} represents $n+1$-level subsystem; δ is the related node from $n+1$ layer; C_{ni} represents the cascading effect of node i in layer n; k_i and z_i represent the position importance and function importance of node i, respectively; φ and τ represent the weight coefficient of position importance and function importance, respectively.

The modeling process of influence relationship for cascading failure of multilevel systems is given as follows:

***Step* 1**: detect the working state of n-layer node at time t, and determine whether the node has failed according to the failure criteria;
***Step* 2**: determine the position importance and function importance of the node, and determine the weight coefficient of the position importance and function importance of the systems;
***Step* 3**: calculate the performance of related nodes of layer $n+1$ at $t+1$, and refer to ***step* 1** to determine whether the related nodes are invalid;
***Step* 4**: refer to ***step* 2** to determine the parameters related to cascading failure;
***Step* 5**: iteratively calculate until the last level, and determine the number of final failure nodes;
***Step* 6**: estimate the overall degradation and performance of the multilevel systems.

10.3.2 Mathematical Basis of DBNs

DBNs can extend BNs to the random models including time dimension. When building DBNs model, two basic assumptions need to be satisfied: (1) the change of conditional probability with time is stable; (2) the dynamic change of probability has property of Markov. The two assumptions can be expressed as an initial network and a transition network (Cai et al. 2016). The DBNs model can be built at any time according to these assumptions.

The probability distribution of the initial network is

$$P(X_0) = \prod_{i=1}^{n} P(X_0^i | Pa(X_0^i)) \tag{10.12}$$

The transition probability distribution of the transition network between adjacent time slices is

$$P(X_t|X_{t-1}) = \prod_{i=1}^{n} P(X_t^i | Pa(X_t^i)) \tag{10.13}$$

The joint distribution probability of each DBNs node is

$$P(X_{0:t}) = P_{B_0}(X_0) \prod_{i=1}^{t} P_{B\rightarrow}(X_t|X_{t-1}) \tag{10.14}$$

where X_0, X_t and X_{t-1} is the node at time 0, t and $t-1$, respectively; $X_{0:t} = \{X_0, X_1, \ldots, X_t\}$ is the ith node of $X_t^{\,i}$ at time t; $Pa(X_t^{\,i})$ is the parent node of $X_t^{\,i}$.

10.3.3 Structure Modeling and Parameter Modeling of DBNs

BNs structure modeling methods, including causality, mapping algorithms and structure learning. For multilevel systems, detailed causality and sufficient training data are generally difficult to obtain. In this case, the structure of the DBNs is obtained by transforming the physical model of the multilevel systems. The nodes of the DBNs represent the nodes of each level in the multilevel systems, and the directed arcs of the DBNs represent the functional relationships between the nodes.

In the RUL prognosis model of multilevel systems under cascading failure, the prior probabilities of all nodes in the multilevel systems network are usually given when the model is established, which are obtained through expert experience and historical data. The degradation of each node can be detected by the sensors at any time to update the prior probability. The conditional probability table (CPT) is an important basis of BNs reasoning. For some fuzzy causality, it is generally obtained by expert experience or previous data of sensors. The CPT can be determined by data training, but the process requires a large number of practical data. The more data, the more accurate the CPT will be. For the nodes with definite causal relationship, the CPT is usually expressed by standard values. In addition, if a certain formula between variables exists, it can also be transformed directly into CPT.

The performance degradation DBNs model under cascading failure is shown in Fig. 10.5. This is a dynamic model of performance degradation considering cascading failure of a general multilevel system. Y_{ni} represents the ith node of the nth level, and the performance p of the entire system is obtained by integrating the p_n (performance of each multilevel subsystem). The black connection lines represent the component relationships within the same DBNs time slice, the red connection lines represent the degradation relationship between the time slices, and the blue connection lines represent the cascading effect.

10.3.4 Estimation of RUL

The RUL prognosis methodology proposed in this chapter is based on the performance degradation model, which is shown in Fig. 10.6. Failure occurs when performance drops to the failure threshold for the first time. The failure threshold is determined by actual working conditions and expert experience. Failure threshold may

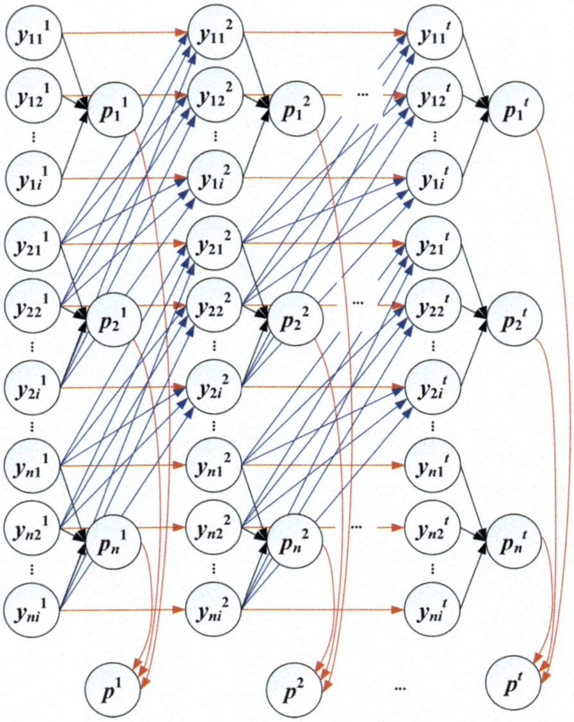

Fig. 10.5 Cascading failure DBNs model of multilevel systems

be one or more values, or a distribution. Therefore, the deviation between actual performance and estimated performance results in two RUL values: actual RUL and estimated RUL, which are calculated by detection time and failure time. The actual RUL is calculated based on the monitored historical data, while the estimated RUL is obtained from the degradation fitted by the inference method. The difference between the two will directly determine the accuracy of the prognosis result and the adaptability of the method. The sensors can obtain degradation data and ensure real-time correction of estimated performance, which makes estimated RUL dynamically approach actual RUL.

10.4 Case of Subsea Christmas Tree

10.4.1 Degradation Modeling of Subsea Christmas Tree

This section introduces the subsea Christmas tree as a case study. The description of the subsea Christmas tree refers to Sect. 9.3.1.

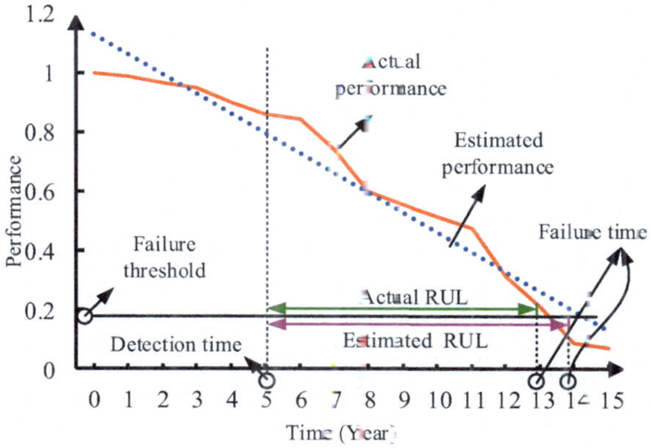

Fig. 10.6 RUL calculation model

10.4.1.1 Interaction Analysis of Electronic Control Systems

Electronic control systems mainly include EPU and subsea control boxes (A-box and B-box). The subsea control boxes are installed in watertight modules, responding to commands of SCM to control the subsea pipelines and valves. The parameters of the subsea temperature and pressure sensors are collected and fed back to MCS by the A-box and B-box. SCM is affected by EPU due to the serial structure; two boxes jointly control the electromagnetic reversing valves with parallel structure, and when a single control box fails, the working pressure and degradation process of another are increasing. The degradation of the electronic control system is consistent with the exponential degradation model (Cai et al. 2012), and its failure probability can be expressed as

$$Fe = 1 - e^{-dt} \tag{10.15}$$

where d represents the failure rate of degradation.

Assume that EPU, A-box and B-box are degraded by the exponential model of $Fe_e = 1 - e^{-det}$, $Fe_a = 1 - e^{-dat}$ and $Fe_b = 1 - e^{-dbt}$ respectively, and d_e, d_a, d_b represent the failure rate of the three. According to previous discussion, the multi-stage and interactions expression of exponential degradation are established, and the interactions between the control boxes can be expressed as

$$F^{*}e_{b\to a}(t) = 1 - e^{-davt} = 1 - e^{-da(1+a_2)^{j_b}t} \tag{10.16}$$

$$F^{*}e_{a\to b}(t) = 1 - e^{-dbut} = 1 - e^{-db(1+a_1)^{j_a}t} \tag{10.17}$$

where $F^{*}e_{b\to a}$ and $F^{*}e_{a\to b}$ indicate the interactions between the A-box and B-box; a_1 and a_2 are parameters of component degradation rate; j_a and j_b are stage factors which represent the influence of degradation stages of A-box and B-box, respectively.

Table 10.1 Stage factors of electronic control system

Failure probability	j_a	j_b
0–0.3	1	1
0.3–0.5	1.5	1.5
0.5–0.7	2	2
0.7–0.9	2.5	2.5
0.9–1	3	3

Table 10.2 Parameters of electronic control system

Parameter	Probability distribution	Mean	Standard deviation
d_e	Deterministic	0.1	–
d_a	Normal	0.2	0.05
d_b	Normal	0.3	0.05
a_1	Deterministic	0.05	–
a_2	Deterministic	0.05	–

In this chapter, the degradation parameters refer to OREDA handbook, and the parameters of relevant interaction coefficients are obtained by parameter training and parameter estimation through historical data in OREDA database (DNV 2015). The stage factors associated with the failure probability are shown in Table 10.1, and the parameters for degradation of electronic control system are shown in Table 10.2. The modeling of the electronic control system is described as shown in Fig. 10.7. The structure of DBNs is obtained from the methodology in Sect. 10.2.3.

10.4.1.2 Interaction Analysis of Hydraulic Systems

The HPU is a device designed to supply biodegradable water-based hydraulic oil or mineral hydraulic oil. The hydraulic control circuit mainly includes the tank, high and low pressure pump circuit, pressure regulating circuit, oil return circuit, etc. In this section, two low-pressure pumps with parallel structure are taken as an example to discuss the interactions of hydraulic system. The Weibull degradation is usually used to describe degradation of hydraulic components, and the probability density function of failure probability can be described as

$$W(t, k, \lambda, \theta) = \frac{k}{\lambda}\left(\left(\frac{t-\theta}{\lambda}\right)^{k-1} e^{-\left(\frac{t-\theta}{\lambda}\right)^k}\right) \tag{10.18}$$

where $t \geq 0$, $\lambda \geq 0$, $k \geq 0$, $\theta \geq 0$. λ represents the scale parameter, k is the shape parameter of the distribution and θ defines position of the distribution on the time axis. Assume that the predictions for all system components are completed at the

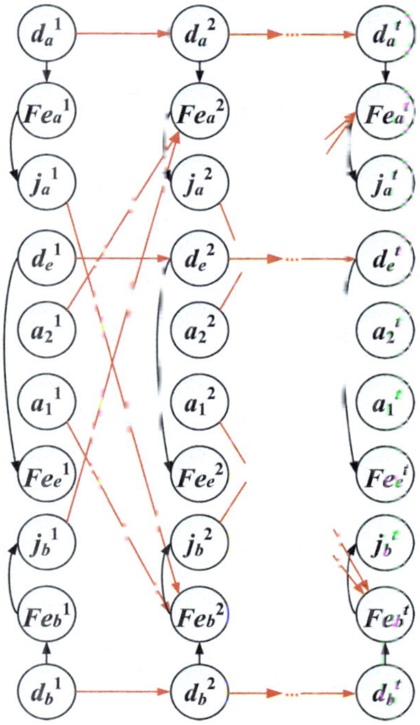

Fig. 10.7 The DBNs model of the electronic control system

same time, that is, $\theta = 0$. These parameters determine the probability distribution and signs of aging of the models. The corresponding failure probability cumulative distribution function can be expressed as

$$F_h(t, k, \lambda) = 1 - e^{-\left(\frac{t}{\lambda}\right)^k} \tag{10.19}$$

Similarly, in this work, the interaction coefficients act on the shape parameter. The degradation of A-pump and B-pump are Weibull degradation of λ_1, k_1 and λ_2, k_2 respectively, and their interactions can be expressed as

$$F^* h_{b\to a}(t, k, \lambda) = 1 - e^{-\left(\frac{t}{\lambda_1}\right)^{k_1 v}} = 1 - e^{-\left(\frac{t}{\lambda_1}\right)^{k_1 (1+a_4)^{j_d}}} \tag{10.20}$$

$$F^* h_{a\to b}(t, k, \lambda) = 1 - e^{-\left(\frac{t}{\lambda_2}\right)^{k_2 u}} = 1 - e^{-\left(\frac{t}{\lambda_2}\right)^{k_2 (1+a_3)^{j_c}}} \tag{10.21}$$

where $F^* h_{b\to a}$ and $F^* h_{a\to b}$ indicate the interactions between the A-pump and B-pump; a_3 and a_4 are parameters of component degradation rate; j_c and j_d are stage factors which represent the influence of degradation stages of A-pump and B-pump, respectively.

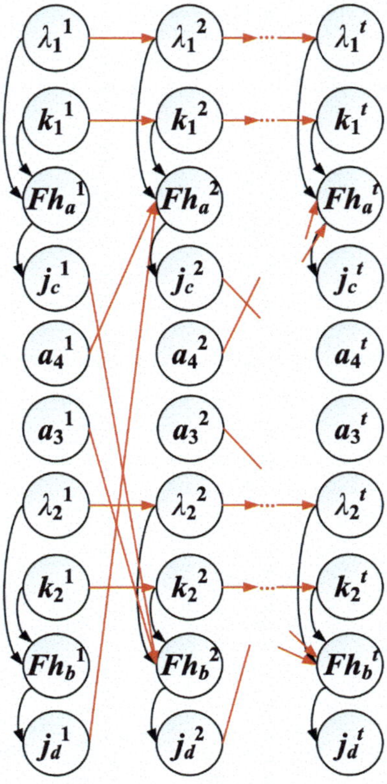

Fig. 10.8 The DBNs model of the hydraulic system

Table 10.3 Stage factors of hydraulic system

Failure probability	j_c	j_d
0–0.3	0	0
0.3–0.5	1.9	1.9
0.5–0.7	4.2	4.2
0.7–0.9	6.2	6.2
0.9–1	9.6	9.6

The modeling of the hydraulic control system is described as shown in Fig. 10.8, and Tables 10.3 and 10.4 provide the relevant parameters of degradation.

10.4.1.3 Interactions Modeling of Valves

Reversing valves are affected by electronic control system and the hydraulic control system. The degradation of valves can be assumed to Gamma process, which is usually recorded as $R_v = \Gamma(t, \alpha, \beta)$, where R_v is the reliability of valves, α is the shape

Table 10.4 Parameters of hydraulic system

Parameter	Probability distribution	Mean	Standard deviation
a_3	Deterministic	0.1	–
a_4	Deterministic	0.1	–
λ_1	Normal	2	0.1
k_1	Normal	3	0.2
λ_2	Normal	1.9	0.1
k_2	Normal	2.1	0.2

parameter, and β is the scale parameter (Ling et al. 2019; Zhang et al. 2021). Assume that $\{d(t), t \geq 0\}$ is a random Gamma process with the following characteristics.

(a) $d(0) = 0$;
(b) For any $\tau > t \geq 0$, the increment $d(\tau) - d(t)$ obeys the Gamma distribution, ie:

$$d(\tau) - d(t) \sim Ga[\alpha(\tau) - \alpha(t), \beta] \tag{10.22}$$

(c) $d(t)$ is a process with independent incremental, and $d(t_2) - d(t_1), \ldots, d(t_n) - d(t_{n-1})$ are independent of each other for any $t_1 < t_2 < \ldots < t_n (n > 2)$.

The performance of the valves depends on the hydraulic system and electric control system. Even if there is no interaction between valves, its performance is still affected by the interactions between hydraulic system and electronic control system. When the interactions are taken into account, the degradation rate of the valves may be faster. Therefore, the degradation analysis of valves can also reflect the importance of interactions to a certain extent. Assuming that $\beta = f(max(j_a, j_b), max(j_c, j_d))$ is obtained from historical data and expert experience, which represents the joint effect of the degradation level of the electronic control system and the hydraulic control system on the degradation of the electromagnetic reversing valves. The joint influence is achieved by changing the scale parameters in the Gamma process. Assuming that the reliability degradation of valves is Gamma process with $\alpha = 1$, and the value of β depends on the matrix below

$$\begin{pmatrix} 0.01 & 0.03 & 0.03 & 0.05 & 0.05 \\ 0.01 & 0.03 & 0.03 & 0.05 & 0.07 \\ 0.01 & 0.03 & 0.05 & 0.05 & 0.07 \\ 0.01 & 0.05 & 0.05 & 0.07 & 0.1 \\ 0.03 & 0.05 & 0.07 & 0.1 & 0.1 \end{pmatrix}$$

where the row of the matrix represents the effect of the degradation level of the electronic control system on the valves, and the column represents the effect of the reliability degradation level of the hydraulic system on the valves.

The modeling of the valves is described as shown in Fig. 10.9. The nodes j_a, j_b, j_c, and j_d are obtained from Figs. 10.7 and 10.8, in another word, the modeling of

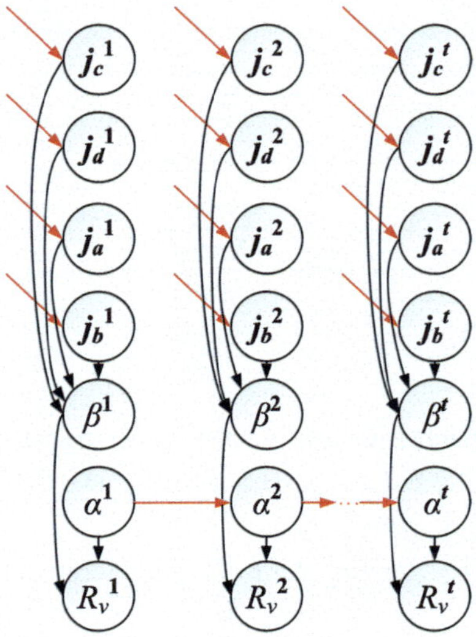

Fig. 10.9 The DBNs model of valves

valves is constructed by electronic control system and hydraulic system. The CPTs are obtained from the matrix mentioned above.

10.4.1.4 Reliability of System

The overall performance degradation model of the system is established according to the system structure after the performance degradation modeling of all components are finished. In addition, performance indicators must be unified to calculate the overall degradation of the system. The RUL is estimated with reliability degradation in this chapter. The reliability of the electronic control system and hydraulic system can be obtained from R_i (reliability) $= 1 - F_i$ (failure probability). The common series structure has components that are indispensable in the process of completing the work, and parallel structure has components that performs the same function. In subsea tree, the electronic control system and the hydraulic system have two components to complete the same function, namely, A-box, B-box, A-pump and B-pump; EPU, SCM, HPU and hydraulic valves control the pipeline of subsea tree jointly, so the four parts are integrated in series. The overall reliability of the system can be expressed as

$$R = Re \cdot Rh \cdot Rv \tag{10.23}$$

where Re, Rh, and Rv are the reliability of the electronic control system, hydraulic system and valves, respectively.

The modeling of DBNs for the reliability of system is described as shown in Fig. 10.10. The nodes Fe_e, Fe_a, Fe_b, Fh_a, Fh_b, R_v, R_e, R_h and R are the main child nodes, which are used to calculate the reliability of the system, while the parent nodes, that is, the parameter nodes, come from the degraded parameters of each subsystem.

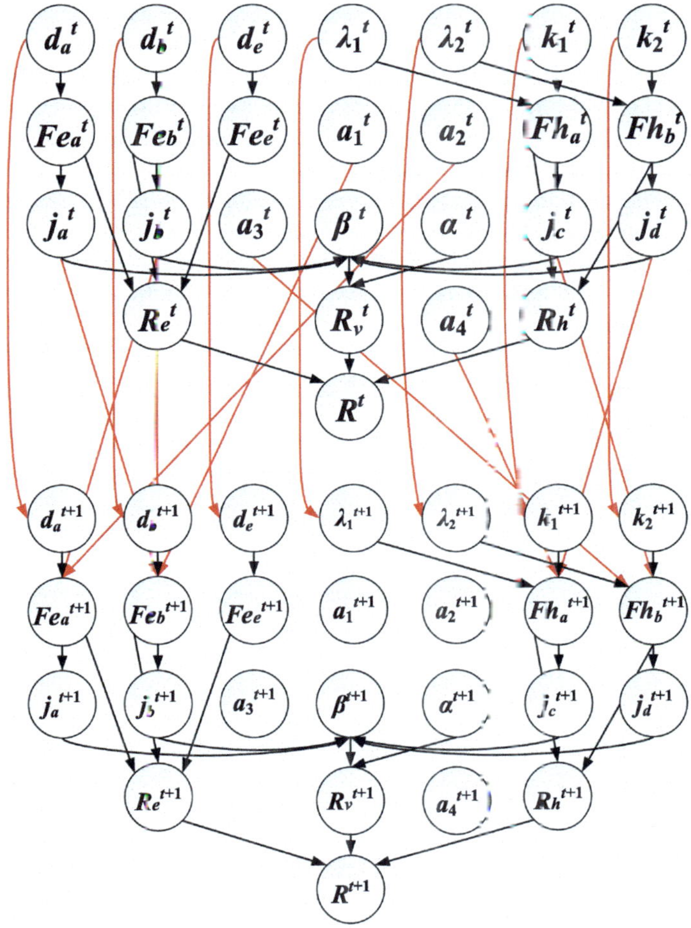

Fig. 10.10 The DBNs model of system reliability

10.4.2 Results and Discussions

10.4.2.1 System Reliability Prediction

The degradation rate of components depends on the degradation parameters of the model. Degradation parameters are not kept constant due to changing external environment and production load. During degradation modeling, degradation parameters are monitored in real time, as shown in Fig. 10.11. Degradation parameters, d_a, d_b, λ_1, k_1, λ_2 and k_2 fluctuated around 0.2, 0.3, 2, 3, 1.9 and 2.1, respectively.

The failure probability of the electronic control system is described in Fig. 10.12. Under the influence of interactions, the failure probability of EPU, A-box and B-box increased exponentially, and the B-box will fail first. In addition, the acceleration of failure process of the A-box in the middle and late stages is greater, because the degradation level of the B-box is higher, with a more significant influence on the A-box.

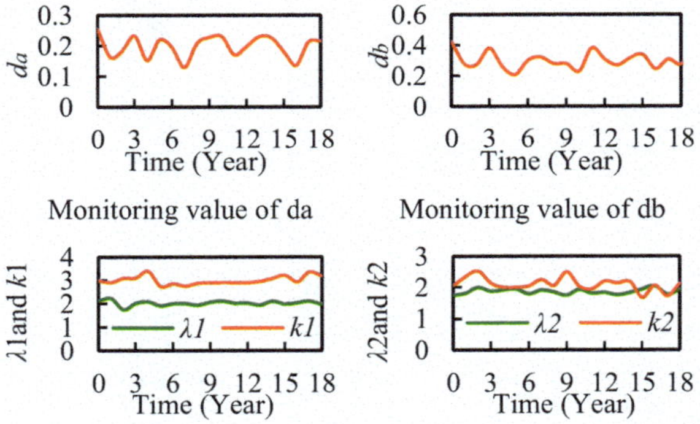

Fig. 10.11 Modeling parameter monitoring

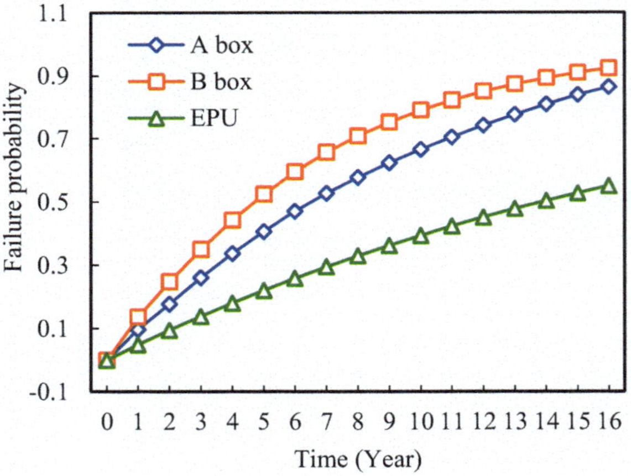

Fig. 10.12 The failure probability of the electronic control system

The failure probability of the two hydraulic pumps is shown in Fig. 10.13. Obviously, the degradation rate of B-pump is higher than that of A-pump. In addition, because the interactions modeling of hydraulic pump is based on the shape parameters of Weibull degradation, there is a great difference in the middle of the degradation of the two pumps. It can be found that in the 13th year, the failure probability of the two pumps starts to be the same until they fail completely.

The reliability degradation of the valves is shown in Fig. 10.14. Because the degradation of the valves is affected by the combination of the electronic control system and the hydraulic system, the degradation of the valves will also be indirectly affected by the interactions. The reliability without interactions is slower, and reliability obviously reduce faster considering the interactions. The mean value of each increment in the Gamma process is influenced by the degradation stages of electronic control system and hydraulic system, resulting in faster and faster reliability degradation.

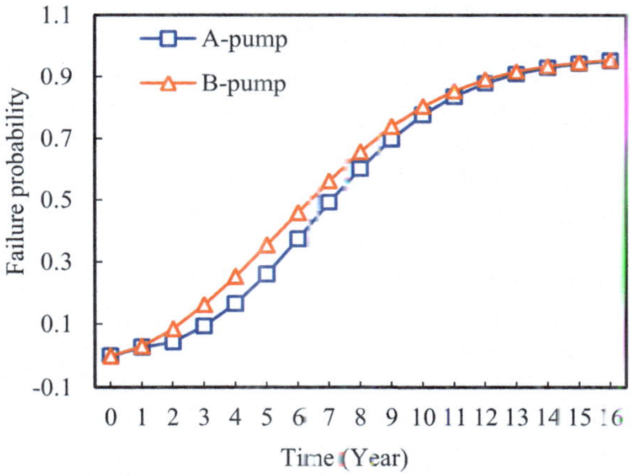

Fig. 10.13 The failure probability of the hydraulic system

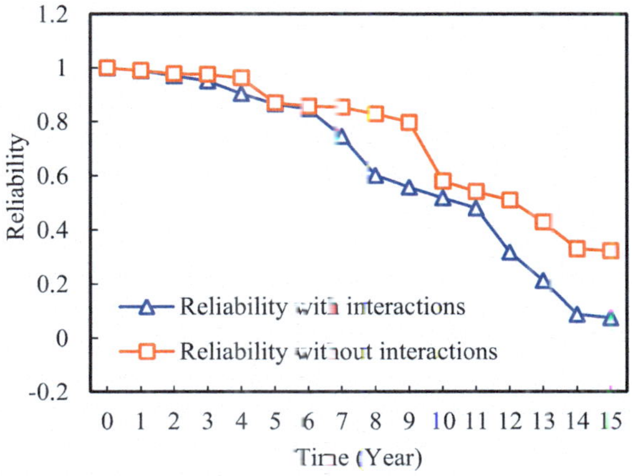

Fig. 10.14 The reliability of the valves

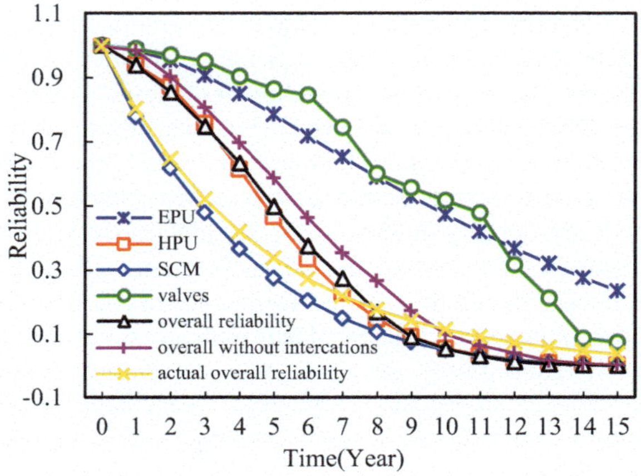

Fig. 10.15 Reliability of subsea Christmas tree system

The reliability of each component and overall reliability of system are shown in Fig. 10.15. There are three main indicators: overall reliability, overall reliability without interactions and actual overall reliability. The overall reliability is calculated by the established DBNs; the overall reliability without interactions can be obtained from overall reliability analysis without interactions modeling; the actual overall reliability can be calculated through the actual production data given in OREDA manual and the overall failure rate (24.88 per 10^6 hour) of Christmas tree system. The overall reliability decreases slowly at the beginning, and the fastest decline in the process from the 3rd year to the 8th year, where falls to 50% in the 5th year. The overall reliability without interactions is higher than the overall reliability considering interactions in the whole life cycle. It is worth noting that the overall reliability considering the interactions is closer to the actual overall reliability, indicating that the interactions solution is helpful to an accurate estimation of RUL.

The time-varying deviation between the reliability with interactions or without interactions and the actual reliability is shown in Fig. 10.16. In the early stage of degradation, the actual overall reliability degradation is faster, which is higher than the reliability of the two overall predictions. The overall reliability considering interaction is closer to the actual overall reliability in the early stage of degradation and has smaller deviation, which is conducive to RUL prediction. The deviation of overall reliability considering interactions or not is basically the same from 9th year to 10th year. After that, the overall without interactions has a smaller deviation, but this difference does not affect RUL prediction in a low reliability environment.

10.4.2.2 RUL Estimation

RUL of system can be obtained by degradation of reliability and failure threshold. According to the characteristics of subsea Christmas tree, the failure threshold is determined to be 0.2. The RUL estimated by the predicted system reliability is shown

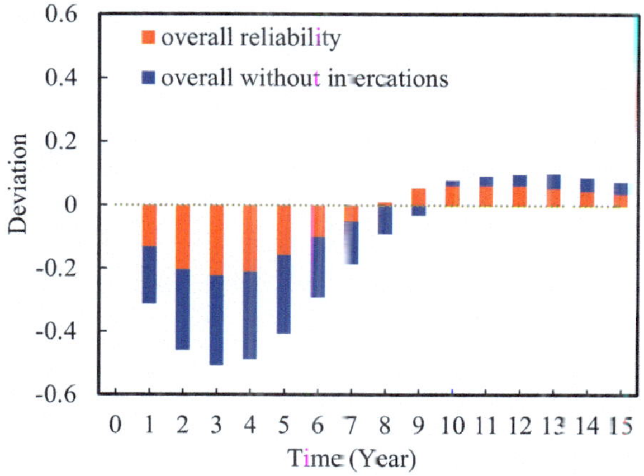

Fig. 10.16 Reliability deviation from actual degradation

in Fig. 10.17. The RUL estimated by actual overall reliability, overall reliability considering interactions and overall reliability without interactions are 7.3 years, 7.8 years, and 8.8 years, respectively. The RUL calculated by the method proposed in this chapter (overall reliability considering interactions) is closer to the actual RUL, which solves the problem of inaccurate prediction by traditional prediction methods. The premise of formulating the maintenance strategy is to obtain the RUL of the equipment at the current time. More accurate RUL provides a basis for the subsequent maintenance strategy formulation of subsea Christmas tree, including maintenance interval, maintenance method and maintenance cost.

Sensitivity analysis for threshold selection is achieved by changing it within a specified range. The threshold is actually determined according to the system characteristics and expert experience. The purpose of sensitivity analysis is to explore the impact of threshold changes on RUL estimation results. The thresholds are set to 0.1, 0.15, 0.2 (default), 0.25 and 0.3 and the corresponding RUL is calculated

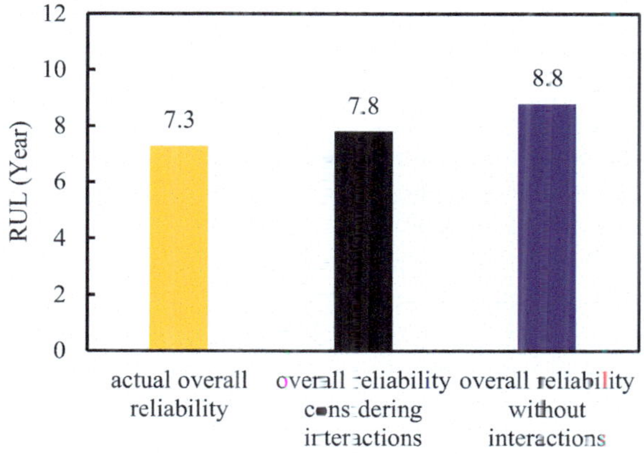

Fig. 10.17 RUL estimation results

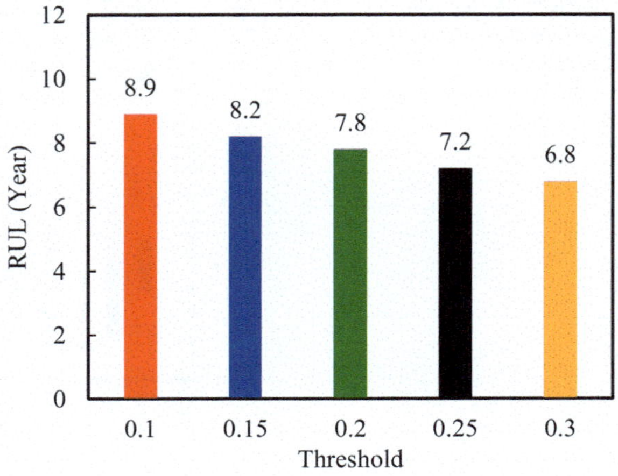

Fig. 10.18 Sensitivity analysis of threshold

as shown in Fig. 10.18, and RUL are 8.9 years, 8.2 years, 7.8 years, 7.2 years, and 6.8 years, respectively. It can be seen that the deviation of the threshold does not cause a wide range of fluctuations in the RUL. It is worth noting that when the threshold is set between 0.15 and 0.25, its RUL basically remains about 7.8 years, and the maximum deviation does not exceed 0.6 years, which is an acceptable range of change. The results show that when the threshold changes slightly, the change of RUL is not particularly obvious, indicating that the sensitivity of RUL calculation to the threshold is not high.

10.4.2.3 Uncertainty Analysis

In the monitoring of degradation parameters, it can be found that degradation parameters are random and constrained by conditions. Reliability computed using Bayesian networks is actually a probabilistic result, and the estimated reliability is actually a mean value, which may also lead to deviation in the calculation results when a small amount of sampling analysis is carried out. Considering that the degradation parameters obey random distribution, the reliability of calculation is not a fixed value, which leads to the uncertainty of degradation model, so uncertainty analysis is necessary. The probability distribution of reliability in the 2nd year, 4th year, 6th year and 8th year is shown in Fig. 10.19. Through the analysis of the mean value, the reliability gradually decreased in these four years. In the 2nd year, the system has very high reliability, and the reliability of all probabilities is greater than 0.65, and its reliability is around 0.85. The reliability decreases in the 4th year, and the probability distribution of reliability is relatively uniform in the 6th year. It can be seen that the mean of the reliability distribution is gradually moving to the left, that is, approaching in a smaller direction. By the 8th year, its reliability is very small, and the mean value is less than the threshold.

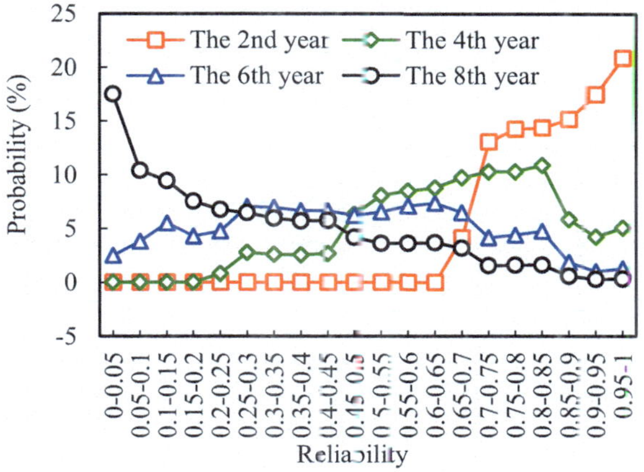

Fig. 10.19 Probability distribution of overall reliability

10.4.2.4 Model Updating with Evidence

Due to the complex working environment of subsea tree, there are a large number of environmental shocks, such as typhoons, earthquakes and tsunamis. Shock changes the degradation rate of the degradation model by affecting the degradation parameters in the performance evaluation, such as stress, temperature and humidity, and finally accelerates the degradation. It will cause the component to degrade or even fail when a shock is applied to one or more components. The focus of this part is to study the dynamic updating ability of this model, so it is assumed that a shock acts on the system. The disaster situation is simplified as a shock, that is, any external disaster is transformed into a shock to accelerate degradation. When a shock occurs, it means that the model needs to be updated. Traditional modeling methods are difficult to supplement this information in the process of dynamic prediction, but DBNs can realize this operation. The information added to DBNs is called "evidence", which is realized through the evidence nodes. The variable node in the DBNs can be connected to an evidence node. When a shock occurs, the parameters of the variable nodes can be changed by adjusting the parameters of the evidence nodes, thereby updating the entire DBNs. The estimation of the overall performance will be updated in the calculation of the next time slice of the DBN. Assuming that B-box and B-pump suffered a shock in the 4th year, causing the B-box to degrade in the form exponential model with -0.35 and the B-pump to degrade in the form of Weibull model with $\lambda_2 = 1.6$ and $k_2 = 2.1$, which is given by referring to the high degradation rate case in the OREDA database, and their parameter changes are shown in Fig. 10.20, and the dotted line represents the parameter monitoring value after the shock is updated. After inputting evidence, the reliability change of the system is shown in Fig. 10.21. It can be seen that the system reliability drops sharply from the 4th year, resulting in the reduction of service life of about 1 year. Combined with evidence information, this obvious change is easy to be found, which is conducive to the timely maintenance.

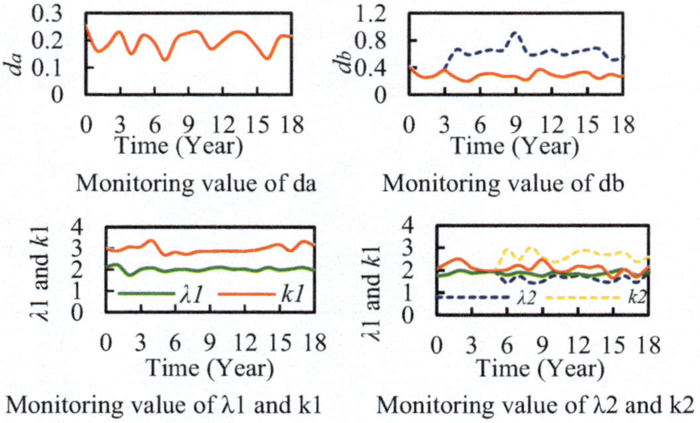

Fig. 10.20 Parameter changes under the influence of shock

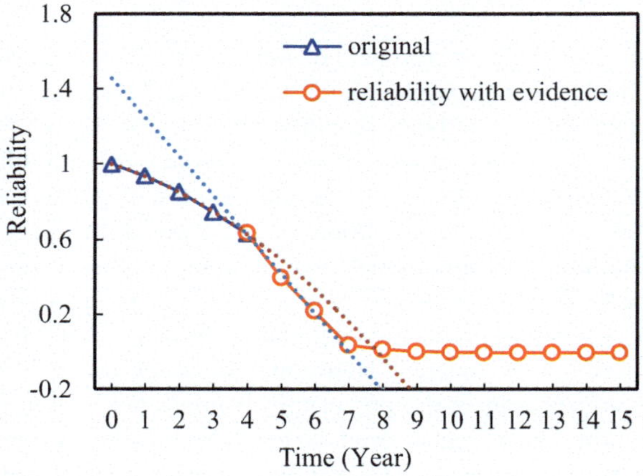

Fig. 10.21 Reliability change under the influence of shock

10.4.2.5 Parameter Sensitivity Analysis

Parameter sensitivity analysis refers to the evaluation of whether the combined results have changed significantly due to the influence of a single study. Parameter sensitivity analysis is very important to find the weak links of system components. The two important indicators of sensitivity analysis are variance reduction and variance of beliefs. The higher the two indicators of the parameter, the higher the sensitivity of the parameter to the overall results. Especially in the stage of low reliability, the importance ranking of each component is determined through sensitivity analysis, so as to make more efficient maintenance strategy. The sensitivity analysis results at the later stage of degradation (in the 6th year) are shown in Fig. 10.22. It can be

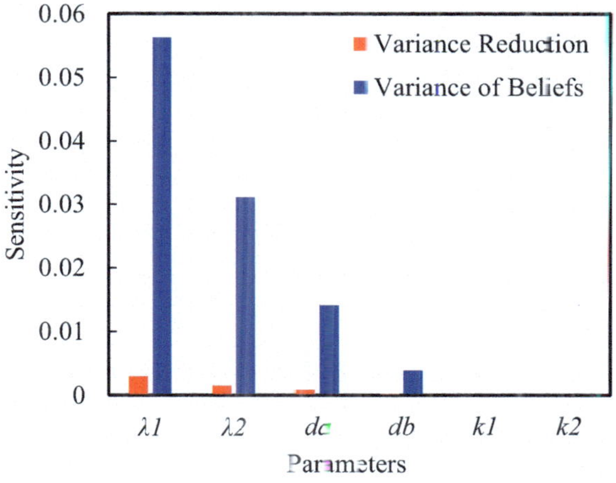

Fig. 10.22 Parameter sensitivity analysis in the later stage of degradation

seen that the scale parameters of the hydraulic system have the most obvious impact on the overall results, followed by the parameters of the electronic control system, indicating that the impact of the hydraulic system on the overall reliability in the later stage is higher than that of the electronic control system. Interestingly, A-box has a higher impact on the overall results than B-box. Similarly, A-pump has a higher impact on the overall results than B-pump. The reliability of A-box and A-pump is higher, but due to the interactions, the high degradation probability of B-box and B-pump leads to the accelerated degradation rate of A-box and A-pump, even if they are at a high reliability stage.

10.4.2.6 Model Validation

It is very important to make a reasonable validation of the DBNs model, which provides reasonable confidence for the evaluation of RUL. Validation involves determining whether the model accurately reflects the reality, which can be finished using real data, simulated data, and comparative modeling. In the current work, comparative modeling (mathematical methods and DBN based on multi-stage method) is used to study the engines and jacks with correlation effects. The DBN-based RUL estimation model is established by using the same data in (Hafsa et al. 2015), and the results are compared in the Fig. 10.23. It can be seen that the probability distribution of degradation obtained from DBNs-based RUL estimation approach is similar to mathematical derivation. Obviously, the two have a consistent trend, and the model is verified reasonably, proving that the proposed model is feasible.

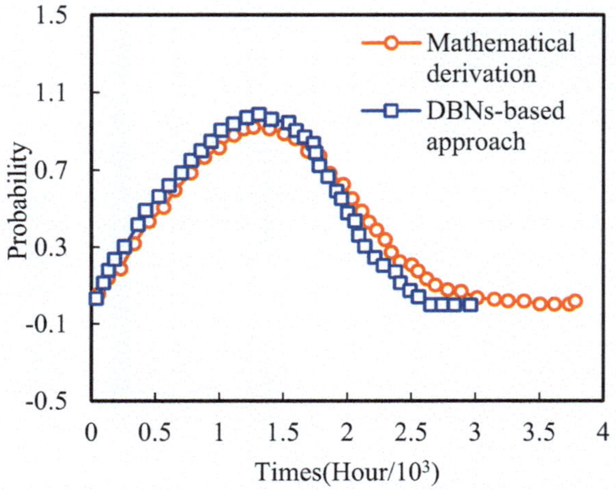

Fig. 10.23 Comparative results of mathematical derivation and DBNs-based approach

10.5 Case of Subsea Transportation Systems

10.5.1 Influence of Cascading Failure of Subsea Transportation Systems

In subsea oil extraction, the process of transporting crude oil from the production site to the surface for processing is called transportation of crude oil. Subsea oil and gas transportation system is shown in Fig. 10.24. The subsea transportation systems consist of oil pipelines, oil transfer stations and auxiliary production facilities. Long-distance pipelines and various chemical equipment, which have close relationship to the oil and gas production and transportation (Zhang and Hu 2013). Pipelines and transfer stations are the main body of the subsea transportation system. According to the user's situation and pipeline distance, the oil pipelines are equipped with gas stations, sub-transmission stations, metering stations and pigging stations. Pipelines are divided into main and branch pipelines according to functions. The branch pipelines are the pipelines around the main line, which are led out from the transfer station or distribution station of the main line and are connected to the control center of the main line network.

Figure 10.25 is a model of a typical subsea transportation system, which shows the transportation process of oil from subsea to the three areas: A, B, and C. The distance between the transfer stations at all levels determines the main and branch pipelines of the oil pipeline. In this case, area A is selected (three-level network, four nodes per level) for cascading failure analysis.

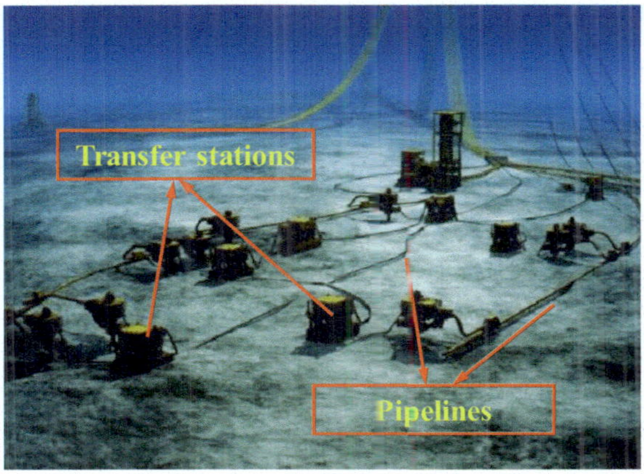

Fig. 10.24 Subsea oil and gas transportation system

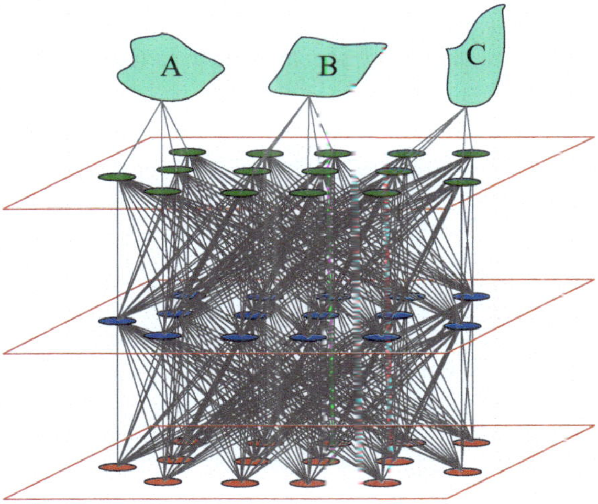

Fig. 10.25 Typical subsea oil and gas transportation systems

10.5.2 DBNs Modeling of Subsea Oil and Gas Transportation Systems

In actual production, it is easy to solve the failure of the pipelines because of good interchangeability of pipelines. In the transportation systems, the failure of the transfer station is mainly discussed. In order to simplify the modeling process in this case, the degradation of the transfer station and related equipment is integrated into a whole degradation, that is, the overall degradation conforms to the exponential degradation model, and its failure rate and reliability can be expressed as (Cai et al. 2012):

$$F_{ni} = 1 - e^{-\lambda_i t} \tag{10.24}$$

$$R_{ni} = 1 - F_{ni} = e^{-\lambda_i t} \tag{10.25}$$

where F_{ni} and R_{ni} represent the failure rate and reliability of the ith transfer station of the nth level; λ_i represents the degradation rate of the transfer station. When a transfer station fails, influence considering cascading failures on the next level can be expressed as:

$$R^{C}_{(n+1)i} = e^{-\lambda*(1+C)t} \tag{10.26}$$

where $R^{C}_{(n+1)i}$ represents the reliability of the ith transfer station of the nth level considering the cascading failure; C is the cascading effect coefficient, which is determined by Eq. (10.25).

The subsea transportation systems can be considered as a combination of a series system and a parallel system. The series system consists of n units, indicating that the system works normally when the n units are all working properly, and the parallel system means that the system fails when all n units fail. The models of the series system and parallel system are shown as follows:

$$R(t) = \prod_{i=1}^{n} R_i(t) \tag{10.27}$$

$$R(t) = 1 - \prod_{i=1}^{n} [1 - R_i(t)] \tag{10.28}$$

where $R(t)$ represents system reliability, n is the number of nodes, and $R_i(t)$ represents the reliability of the ith node. The DBNs model of the three-level network is established as shown in Fig. 10.26, showing DBNs expansion process from t to $t + 1$, y_{ni} represents the ith node of the nth level, and R represents the reliability of the system as a whole.

For the subsea transportation system, the transfer stations are the hubs of oil and gas transportation and are indispensable components for oil and gas production; the transfer stations are installed in deep sea environment, and there are earthquakes, tsunamis, and internal waves. The complex surrounding environment makes the failure of the transfer stations more likely to occur, and it also has an important impact on oil and gas production. In this case, the weight coefficients φ and τ of position importance and function importance are taken as 0.5, respectively, indicating that the impact of position and function in this degradation is the same. It is assumed that when the reliability drops to 0.3, a cascading failure occurs, but it is not considered a failure, and the failure of the systems is that the overall reliability of the systems drops to the failure threshold. The values and distributions of the degradation rate λ of the transfer stations at all levels are shown in Table 10.5. The position importance and function importance are determined by the position function, operating data and expert knowledge of the actual system, as shown in Tables 10.6 and

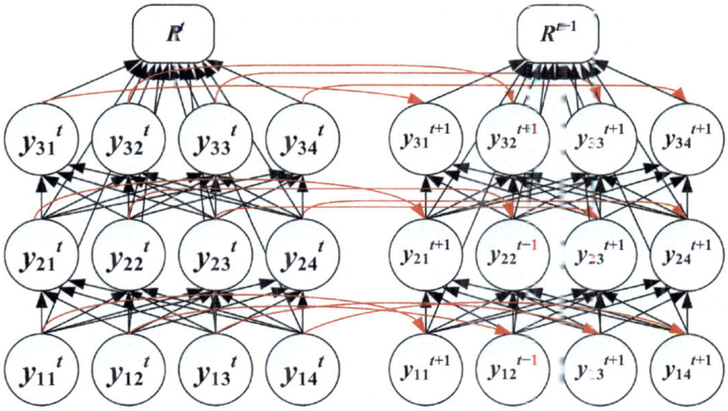

Fig. 10.26 DBNs model of cascading failure of subsea transportation systems in area A

10.7. Position importance and function importance are important factors affecting the cascading effect, and the two are expressed through position and function of nodes. Reliability degradation and RUL estimation for edge node (single-edge failures) and intermediate node (single-medium failures) in the same-level subsystem in single-node failures, and for same-level two-node failures and two-level two-node failures in double-nodes failures are discussed.

Table 10.5 Degradation rate of transfer stations at all levels

i	1	2	3	4
λ_1	Normal (0.1, 0.09)			
λ_2	Normal (0.08, 0.05)			
λ_3	Normal (0.05, 0.04)			

Table 10.6 The position importance of subsea transportation systems

i	$y_{(n+1)1}$	$y_{(n+1)2}$	$y_{(n+1)3}$	$y_{(n+1)4}$
y_{n1}	0.4	0.3	0.2	0.1
y_{n2}	0.25	0.4	0.25	0.1
y_{n3}	0.1	0.25	0.4	0.25
y_{n4}	0.1	0.2	0.3	0.4

Table 10.7 The function importance of subsea transportation systems

i	1	2	3	4
y_1	0.2	0.3	0.3	0.2
y_2	0.2	0.3	0.3	0.2
y_3	0.15	0.24	0.31	0.3

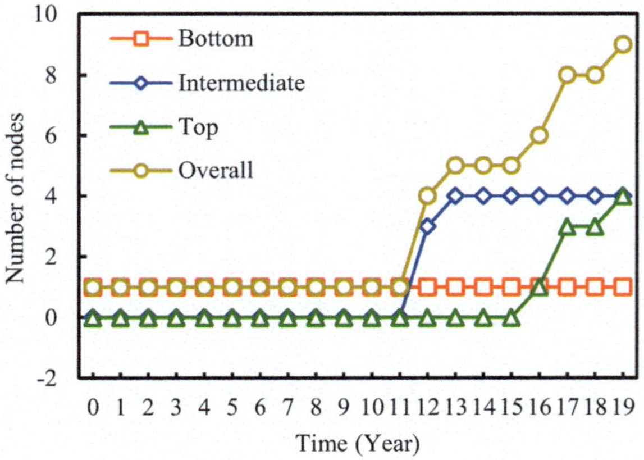

Fig. 10.27 Number of failed nodes of all levels with cascading failure of single edge node failure

10.5.3 Results and Discussions

10.5.3.1 Cascading Failure of Edge Node Failure

Figures 10.27 and 10.28 show the number of failed nodes and reliability degradation caused by the failure of the edge node at the bottom level. It can be seen that in the first 19 years, only one node at the bottom level fails, that is, the edge node at the bottom level that fails at the initial time; and from the 12th year, there are failed nodes at the middle layer, and since the 13th year, all of middle nodes suffer cascading failure. One node at the top level suffers cascading failure in the 16th year, and all of the nodes of top gradually fail in the next 3 years. For 19 years, 9 nodes fail in the multilevel systems, and the failure is passed level by level. It can be seen from Fig. 10.28 that the reliability of the systems at all levels has basically remained stable for the first 3–4 years, maintaining a high level of reliability, and gradually decreases with time from the 5th year. In a three-level system, the top-level always maintains a high level of reliability, while the bottom-level always has the lowest reliability. The overall reliability of the system reduces to 50% in the 11th year and to 20% in the 16th year.

10.5.3.2 Cascading Failure of Intermediate Node Failure

The number of failure nodes caused by the failure of the intermediate node of bottom is shown in Fig. 10.29. It can be seen that in the first 19 years, only one node of the bottom level fails, and the degradation of the bottom level is independent of each other without the impact of correlation. Compared with the mode of edge node failure, the middle-level nodes fail one year in advance, and in the 13th year, all the middle-level nodes suffer cascading failures. In the 16th year, the top-level nodes begin to have cascading failures, and in the 18th year, all the top-level nodes have

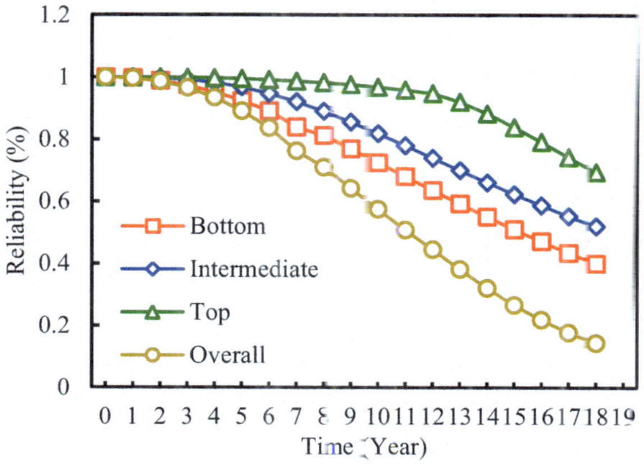

Fig. 10.28 System reliability of all levels with cascading failure of single edge node failure

cascading failures. There are 9 failed nodes in the system for 18 years. The reliability of each level of system under intermediate node failure is shown in Fig. 10.30. The reliability of the system at all levels maintains a high level in the first 3 years, and gradually decreases with time since the 4th year. The reliability of the middle-level and the bottom-level approximately linearly degraded, while the degradation rate of the top-level is slower in the first 12 years and accelerated significantly in the 13th year. The overall reliability of system reduced to 50% in the 11th year, and decreased to 20% in the 16th year, which is similar to the overall reliability of the mode of edge node failure.

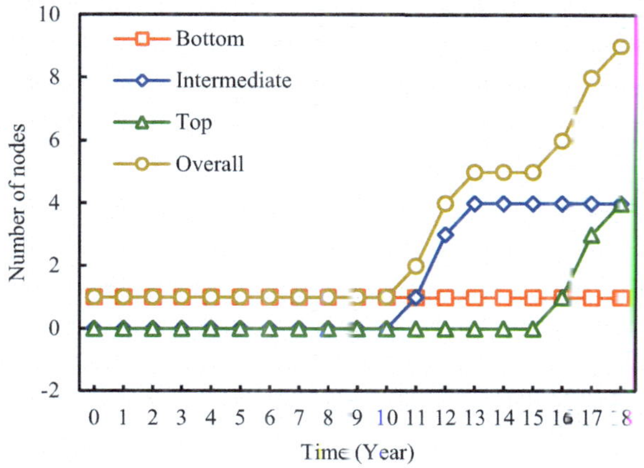

Fig. 10.29 Number of failed nodes of all levels with cascading failure of single intermediate failure

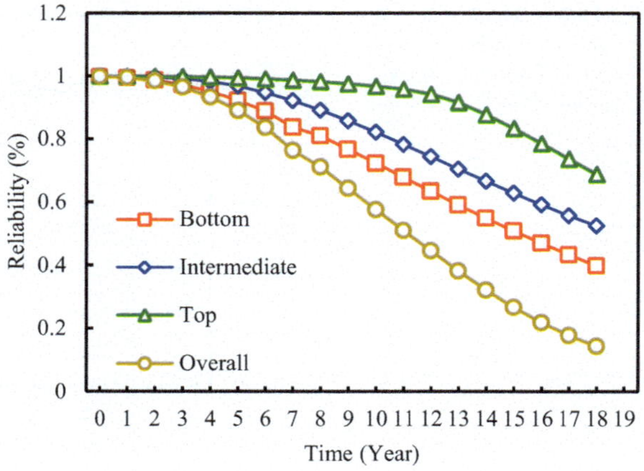

Fig. 10.30 System reliability of all levels with cascading failure of single intermediate failure

10.5.3.3 Cascading Failure of Same-Level Two-Node Failure

The number of failure nodes of each level system with the same-level two-node failure is shown in Fig. 10.31. Since it is assumed in this case that an edge node and an intermediate node of the bottom-level have failed, the number of failed nodes at the bottom layer is 2. In the 9th year, the first node with cascading failure appeared in the middle layer, all the nodes quickly failed with cascading failure within 2 years. The top-level began to cascade in the 15th year, and all nodes fail in the 18th year. The total number of failed nodes in the 18th year was 10, that is, 4 at the top, 4 at the middle, and 2 at the bottom. The reliability degradation of all levels of the same-level two-node failure is shown in Fig. 10.32. The reliability degradation trend of all levels of the system shows a significant difference from the 2nd year. The top-level system maintains a high level until the 10th year, the reliability of the middle-level system slowly decreases, and the reliability of the bottom-level system decreases rapidly. The overall reliability of system drops to 50% in the 9th year, and drops to 20% in the 14th year.

10.5.3.4 Cascading Failure of Two-Level Two-Node Failure

The two-level two-node cascading failure is based on the failure of the bottom-level edge node and the middle-level intermediate node. Figure 10.33 shows the number of failed nodes in a two-level two-node failure system with cascading effect. It can be seen that the number of failed nodes at the bottom and middle level is one in the initial stage, because the assumption in this case is that one edge node at the bottom level and one intermediate node at the middle level have failed. In the 12th year, the middle-level begins to suffer from cascading failures, and in the 14th year, all middle-level nodes failed. The top level has cascading failure in the 15th year, and all nodes have cascading failure in the 18th year. By the 18th year, the total number

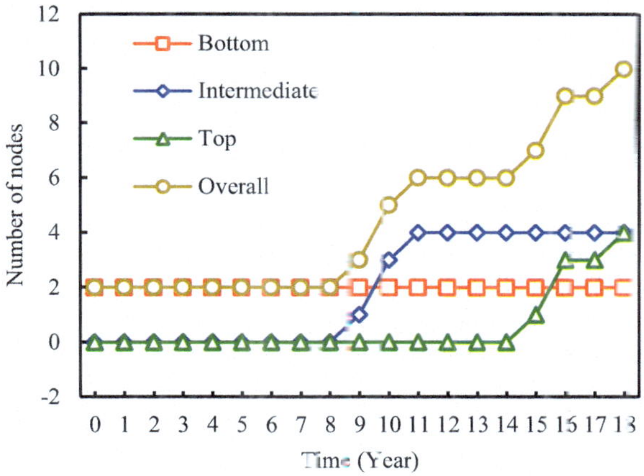

Fig. 10.31 Number of failed nodes of all levels with cascading failure of same-level two-node failure

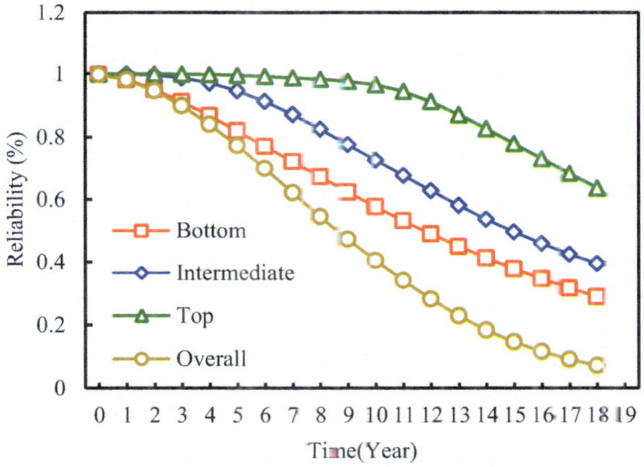

Fig. 10.32 System reliability of all levels with cascading failure of same-level two-node failure

of failed nodes in the system is 9. Figure 10.34 shows the reliability of a two-level two-node failure system with cascading effects at all levels. The reliability of all levels maintain high level in the first 3 years, and gradually decrease over time from the 4th year. The reliability of the middle and bottom levels tend to be consistent, while the reliability of the top-level system has always maintained a high level. The overall reliability of system reduced to 50% in the 10th year and to 20% in the 15th year.

10.5.3.5 Influence Analysis of Cascading Failure

The comparison of the number of failed nodes in different cascading failure modes is shown in Figure 10.35. It can be seen that the sequence of the four types of failure modes of cascading failures: cascading failure of same-level two-node failure,

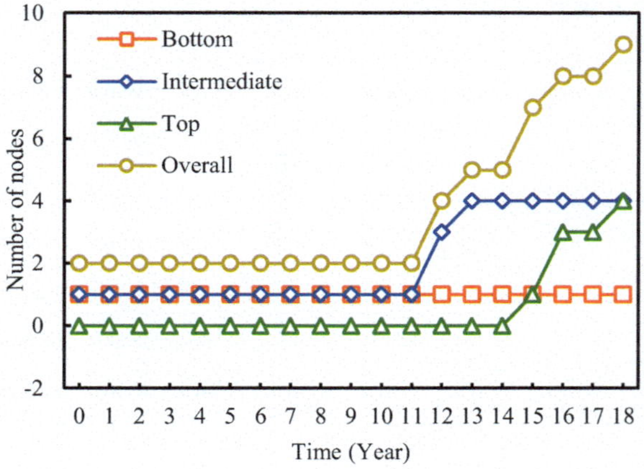

Fig. 10.33 Number of failed nodes of all levels with cascading failure of two-level two-node failure

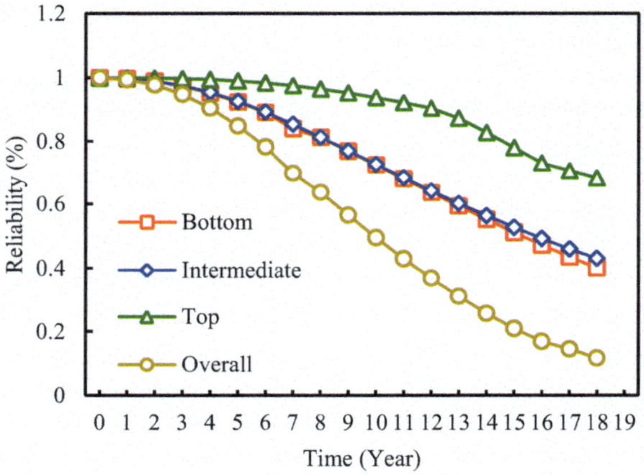

Fig. 10.34 System reliability of all levels with cascading failure of two-level two-node failure

cascading failures of intermediate node failure, cascading failures of edge node failure, and cascading failures of two-level two-node failure. The total number of failed nodes under the cascading failure of same-level two-node failure is 10, while the total number of failed nodes under the three others are 9. Figure 10.36 shows the comparison of reliability under different cascading failures and no cascading failure. Obviously, the overall reliability of the system when cascading failure is not considered is significantly higher than the reliability considering cascading failure. When cascading failure is not considered, the reliability decreases to 50% around the 14th year, and decreases to 20% around the 20th year. Reliability of system under the cascading failure of same-level two-node failure decreases most rapidly among several failure modes, with the lowest reliability. The reliability under the cascading failures of edge node failure and the cascading failure of intermediate node failure are very similar. This also shows that when a single node fails, the node location has little influence on the overall reliability of the system. Expert experience has an important impact on the reliability degradation of each failure mode, because the most important factors affecting cascading failures, function importance and position importance, are evaluated by experts. For single-node failures, the reliability degradation is slower, because the initial number of failures is less; the joint effect of the position importance and function importance of same-level two-node failures is higher than two-level two-node failures, because the node pressure distribution of same-level two-node failures causes a higher load on the other two normal nodes, resulting in faster degradation of the bottom-level nodes, and the impact of cascade failure more obvious.

In addition, research on the impact of function importance on cascading failures is implemented using the intermediate node failure mode as a case, by exchanging the functional importance of each layer of edge nodes and intermediate nodes, as shown in Table 10.8, that is, the functional importance of intermediate nodes is weakened.

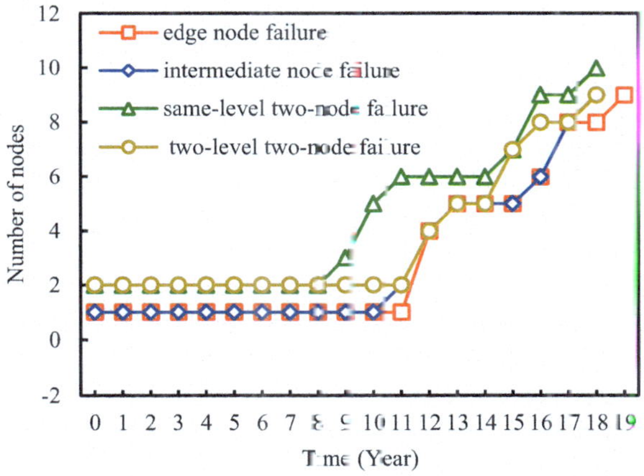

Fig. 10.35 Number of failed nodes under different cascading failures

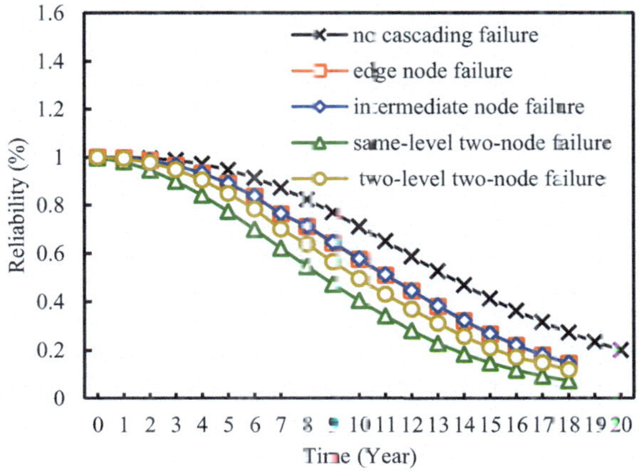

Fig. 10.36 System reliability under different cascading failure modes

Table 10.8 The function importance of subsea transportation systems considering the importance of functionality

i	1	2	3	4
y_{1*}	0.3	0.2	0.2	0.3
y_{2*}	0.3	0.2	0.2	0.3
y_{3*}	0.24	0.15	0.3	0.31

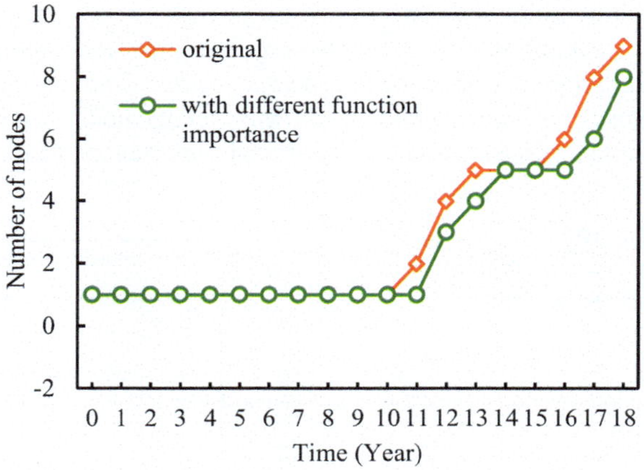

Fig. 10.37 Number of failed nodes with different function importance

Number of failed nodes and system reliability with different function importance are shown in Figs. 10.37 and 10.38. It can be seen that under the different function importance, the cascading failure of the system occurred one year later, and the total number of failed nodes is also smaller than the original model (intermediate node failure); it is obvious that the overall reliability of the system is also higher than the original model, especially in the middle part. The function importance of nodes around the failed node directly affects the strength of the cascading effect. When the functional importance of the closest node to the failed node is higher, the cascading failure of the system expands faster. Therefore, in actual engineering, timely maintenance of nodes around the failed node will effectively reduce the risk of failure of other nodes.

10.5.3.6 RUL Calculation

RUL of multilevel systems can be obtained by degradation of reliability and failure threshold. Failure threshold of the multilevel systems is a normal distribution with the mean of 0.2 and the standard deviation of 0.02, which is determined by historical data and expert experience. The values of RUL is shown in Fig. 10.39. It can be seen that the RUL without cascading failure and with cascading failure of edge node failure, cascading failure of intermediate node failure, cascading

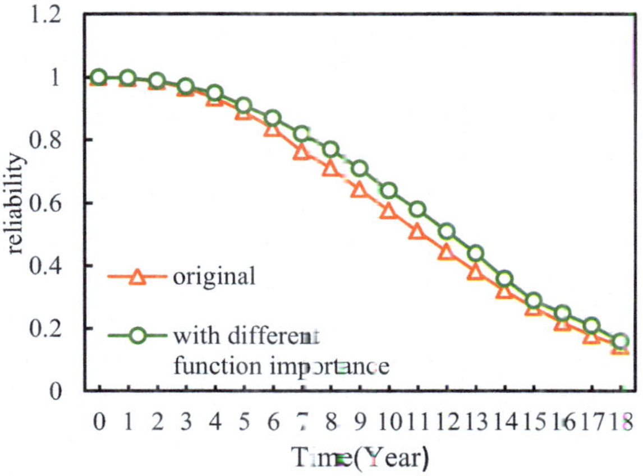

Fig. 10.38 System reliability with different function importance

failure of same-level two-node failure and cascading failure of two-level two-node failure are 20.1 years,16.8 years, 16.8 years, 13.6 years and 15.3 years, respectively. After considering the cascading failure, the RUL of the system reduced obviously. Compared with the degradation of the non-cascading failure mode, the deviation is up to 5 years, which shows the importance of considering cascading failure in actual industrial production. As the reliability under the single failure mode tends to be consistent, their RUL are also similar, being approximately 16.8 years. The RUL under the cascading failure of same-level two-node failure is the lowest among several failure modes. Such failures should also be prioritized in troubleshooting. In addition, the sensitivity analysis of the threshold should be taken into account after the threshold is determined, because it is related to the accuracy of the RUL prognosis. Fine-tune the threshold, and estimate the RUL of the system with the thresholds of 0.15, 0.175, 0.2, 0.225, and 0.5, respectively. Figure 10.40 shows that obtained RUL values are 15.05, 14.2, 13.6, 13.05, and 12.55, respectively. It can be seen that the threshold change is 0.35, and the change of the RUL is 2.5 years, indicating that RUL is not so sensitive to the threshold.

10.5.3.7 RUL for Evidence Input

In the process degradation of all nodes (including cascading failure), there is often some deviation between its estimated RUL value and actual RUL value. In actual production, real-time information of the nodes is available (Xu et al. 2009). The information that detected by the sensor can be input into the DBNs as evidence, which can bring RUL closer to actual value. The greater the number of detections, the more accurate the RUL obtained. Cascading failure of same-level two-node failure is used to verify the validity of the evidence with 3 times detections. Figure 10.41 shows the reliability of the multilevel systems in three detections, which were performed in the 2nd year, 4th year and 8th year, respectively. It can be seen that when three

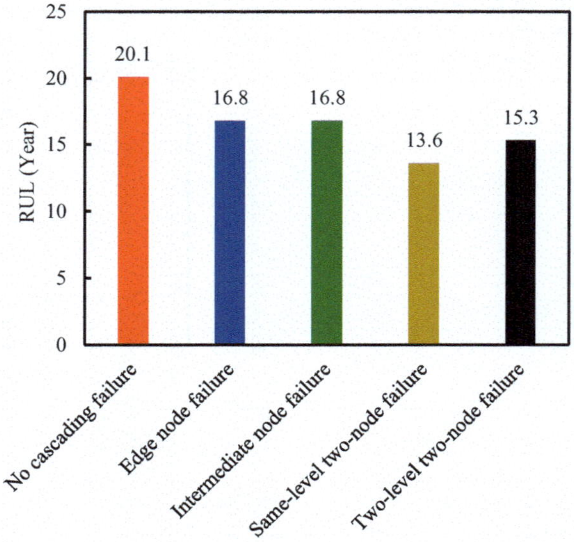

Fig. 10.39 RUL under different cascading failures

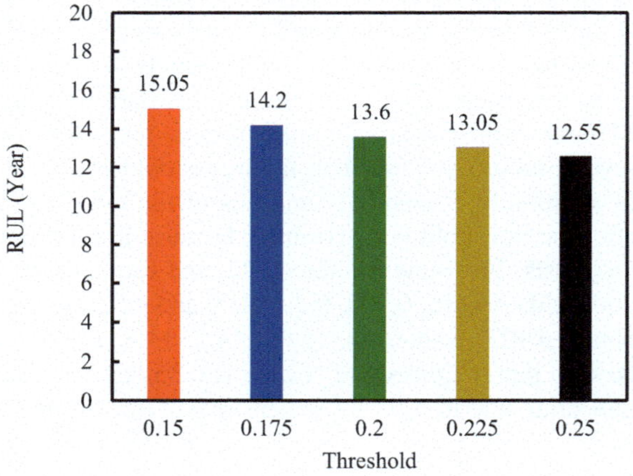

Fig. 10.40 Threshold sensitivity analysis

detections were performed, the reliability trend of system changed three times, that is, the degradation accelerated. The RUL of system after detections is shown in Fig. 10.42. Compared with the RUL (13.6) without detections, the RUL value after detections is significantly smaller. The actual RUL value and the estimated RUL value are 11 years and 11.8 years, respectively. The actual RUL is calculated based on the actual reliability, and the actual reliability is obtained through the calculation and fitting of the monitored historical data and empirical data. According to the performance evaluation results related to the transportation system in the subsea production system, the time scale and the performance scale (reliability, failure rate, etc.) are unified, and the actual reliability is obtained by combining calculation and

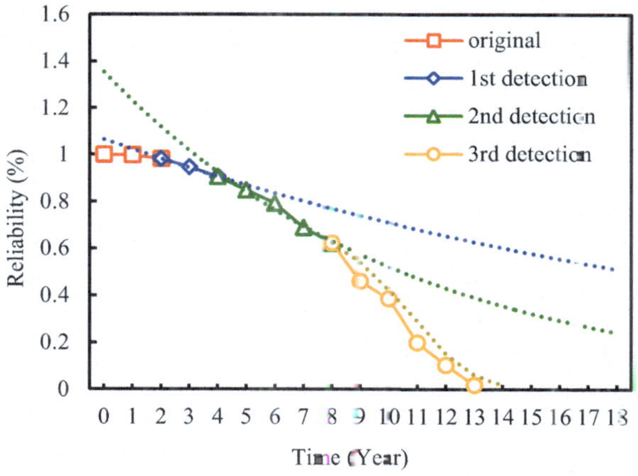

Fig. 10.41 Reliability of the multilevel systems with three detections

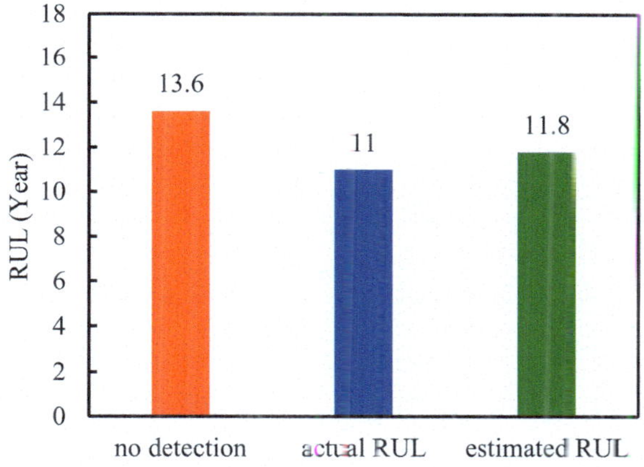

Fig. 10.42 RUL of multilevel systems with three detections

fitting (Chen et al. 2018; Chang et al. 2019). The actual RUL is different from the estimated RUL, and it provides a comparison of the estimated results. The estimated RUL is closer to the actual RUL, Therefore, proper evidence input can make RUL prognosis more accurate.

Chapter 11
RUL Prediction with Degradation-Shock Dependency

Abstract This chapter proposes a novel hybrid method for RUL prediction that addresses the challenges of modeling the effects of both external shocks and internal degradation, focusing on the dependency between these factors. By combining degradation models with field data, the method assesses RUL under the influence of system degradation and external shocks. A dependency factor is introduced to account for the interaction between degradation stages and shock intensity, improving system failure prediction. Comprehensive uncertainty analysis is performed, considering random degradation, unpredictable shocks, and the dynamic evolution of system degradation. The method is validated through numerical simulations on a hydraulic control system, demonstrating its robustness and accuracy in handling complex dependency modeling and uncertainty. The results show that the proposed approach effectively predicts the RUL of the hydraulic control system, while also identifying the vulnerability of its subsystems.

11.1 Introduction

Despite the advancements in RUL prediction methods, several challenges remain, particularly when dealing with systems subjected to both internal degradation and external shocks. Traditional data-driven and model-based approaches are often limited by their ability to handle complex interactions between degradation processes and shock events. Shocks can accelerate the degradation process or cause sudden system failure, complicating the prediction task. Additionally, the uncertainties surrounding the timing and magnitude of shocks, combined with incomplete data on system degradation, make it difficult to construct reliable models. Existing models often struggle to incorporate these dynamic interactions, leading to less accurate RUL predictions and suboptimal maintenance decisions.

The motivation for this study arises from the need to improve the precision of RUL prediction by addressing the dependency between degradation and shocks. This research proposes a novel hybrid approach that introduces a dependency factor, which quantifies the correlation between shock events and system degradation. This new

B. Cai et al., *Intelligent Operation and Maintenance for Subsea Production Systems*,
https://doi.org/10.1007/978-981-96-2677-9_11

indicator allows for a more accurate representation of how external shocks impact system failure. Furthermore, the model incorporates a comprehensive uncertainty analysis, dynamically updating the probability distributions of degradation stages and shock intensities. By integrating both random degradation and random shocks, this method provides a more holistic view of system health, improving prediction accuracy and robustness.

In summary, this chapter presents an innovative degradation–shock dependency model for RUL prediction, which enhances the understanding of the interplay between degradation and shock events. The proposed hybrid approach addresses the uncertainties present in traditional methods and offers a practical solution to the challenge of limited data in real-world applications. By providing a more accurate and comprehensive prediction model, this work contributes to more effective maintenance strategies and better system reliability, particularly in complex, high-risk environments.

11.2 RUL Estimation Model with Degradation–shock Dependence

A hybrid RUL prediction framework that considers degradation–shock dependency on the basis of multi–stage degradation and random shocks is proposed, as shown in Fig. 11.1. A new concept, dependency factor is introduced in this work; it is obtained by parameters, such as the stage of performance degradation, shock strength, and shock probability. The shocks effect (L_1, L_2,..., L_c) is the average effect of the frequency and intensity of shock occurrence on units over a period combined with expert experience to supplement unmonitored data and discretization schemes. The main objective of stage modeling is to divide the entire degradation process into n stages on the basis of the characteristics of system degradation. The effect of each stage on system degradation, namely, (s_1, s_2, ..., s_n), is evaluated and quantified through an analysis of historical data, offline databases, and expert experience. The dependency factor is calculated based on the shock effect at the current moment and degradation stage s at the previous moment. Dependency factors act on the degradation model and affect the rate of system degradation. RUL prediction is implemented based on DBNs and used for stage identification, quantification of shocks, dynamic evolution of degradation stages, and uncertainty analysis of the results. The shock effect ($L_1, L_2, \ldots, L_c$) is integrated as an l node, which is a probabilistic node, and used to evaluate performance degradation in DBNs. Different subsystems have different degradation modes (d, g, etc. represent degradation parameters in different degradation models), and the performance of each subsystem is ultimately integrated into overall performance p on the basis of the system characteristics. Then, on the basis of the evaluated system performance and failure threshold, the time when the system first reaches failure is determined, that is, the RUL of the system is calculated.

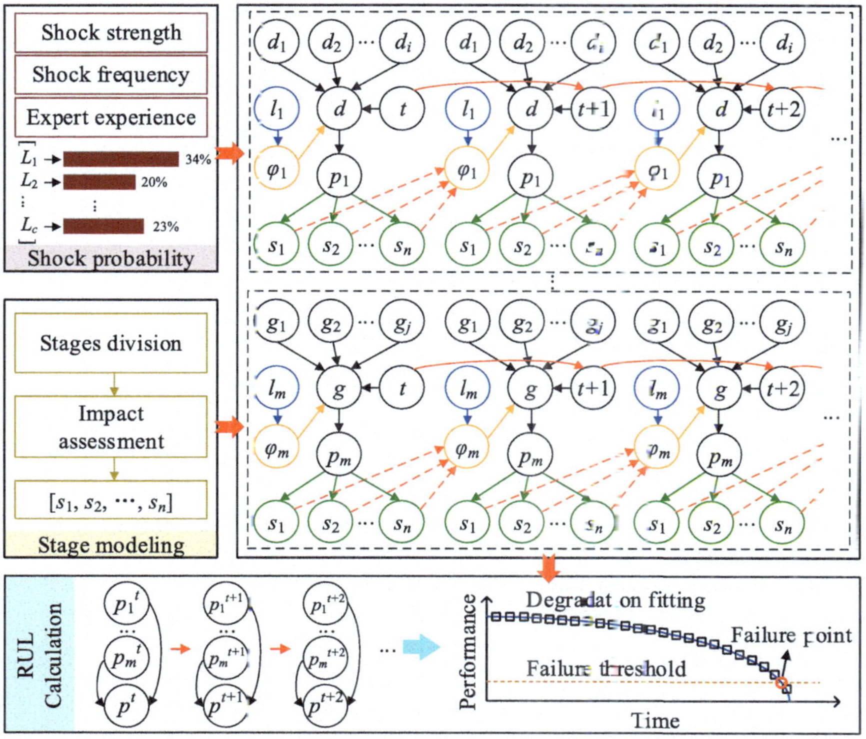

Fig. 11.1 RUL prediction framework that considers degradation–shock dependency

11.2.1 Dependency Modeling

Large equipment is subject to various shocks throughout the year. These shocks are classified into c different levels, namely, $L_1,\ldots, L_c$, with strengths of I_1, $I_2 \ldots I_c$. Then the probability of each strength shock occurrence is calculated based on the frequency of each level under long-term monitoring.

$$P(L_c) = \frac{num(L_c)}{\sum_{i=1}^{c} num(L_i)} \tag{11.1}$$

Afterward, $[L_1(I_1, P(L_1)), L_2(I_2, P(L_2)),\ldots, L_c(I_c, P(L_c))]$ is used to represent the strength corresponding to different shock levels (the effect on the overall degradation) and the probability of shock occurring at that level. Similarly, the performance changes throughout the entire life cycle are divided into n different degradation stages, and the effects of different degradation stages on the overall performance are evaluated through a correlation analysis of degradation and expert experience.

Degradation stages are discrete. The stage model can be presented as $[1(s_1), 2(s_2),\ldots, n(s_n)]$.

The dependency model is constructed based on multiple degradation stages and random shock effects. The dependency matrix factor, Φ, is used to represent the degradation–shock dependency relationship. It is expressed as follows:

$$\Phi = \begin{bmatrix} \mathbf{L}_1 \\ \vdots \\ \mathbf{L}_c \end{bmatrix} \times \begin{bmatrix} s_1 \\ \vdots \\ s_n \end{bmatrix}^T = \begin{bmatrix} \varphi_{11} & \varphi_{12} & \cdots & \varphi_{1n} \\ \varphi_{21} & \varphi_{21} & \cdots & \varphi_{2n} \\ \vdots & \vdots & \vdots & \vdots \\ \varphi_{c1} & \varphi_{c2} & \cdots & \varphi_{cn} \end{bmatrix} \tag{11.2}$$

where $\boldsymbol{L}_1,\ldots,\boldsymbol{L}_c$ are the effects of c intensities, including the probability of occurrence and the corresponding degradation acceleration effects; $s_1,\ldots, s_n$ are the stage factors of n degradation stages; and φ_{cn} represents the dependency factor at the shock of strength c and at the n-th degradation stage.

11.2.2 Performance Evaluation Based on DBNs

Many uncertain parameters are involved in the performance evaluation of large-scale systems, and they make the calculation increasingly complex. Bayesian networks (BNs) are considered effective tools for addressing uncertainty problems and widely used in fault diagnosis and prediction. The proposed dependency modeling method also involves modeling uncertainties, such as probability of random shocks, probability of degradation at different stages, dependency factor probability, and even different probability densities in the final performance evaluation. Usually, DBNs require the following steps for dynamic expansion.

The initial condition of DBNs can be established as

$$P(x_0) = \prod_{i=1}^{n} P(x_0^i | Pa(x_0^i)) \tag{11.3}$$

The conditional probability transition of adjacent time slices can be expressed as

$$P(x_t | x_{t-1}) = \prod_{i=1}^{n} P(x_t^i | Pa(x_t^i)) \tag{11.4}$$

The posterior probability of DBNs node, that is, the joint probability between joint and kernel time slices, can be established as

$$P(x_{0:t}) = P_{B_0}(x_0) \prod_{i=1}^{t} P_{B\to}(x_t | x_{t-1}) \tag{11.5}$$

where $x_{0:t} = \{x_0, x_1, \ldots x_t\}$ is the i-th node of $x_t^{\,i}$ at time t, $P_{B\rightarrow}$ is the conditional probability transition between two adjacent nodes of DBNs, and $Pa(x_t^{\,i})$is the parent node of $x_t^{\,i}$.

The performance degradation of a system is related not only to time but also to the performance at the previous moment and the shocks at the current moment, which are integrated into dependency factors. Therefore, DBNs for system performance can be constructed based on the following relationship:

$$p_t = f(t, d, \varphi(\boldsymbol{L}_1, \boldsymbol{L}_2, \ldots, \boldsymbol{L}_c, s_1, s_2, \ldots s_n)) \tag{11.6}$$

where d is the degradation-related parameter of the system.

From a system functionality perspective, the subsystems of a complex system often operate in series, with each subsystem contributing to the overall performance of the system. The relationship between the m subsystems and the overall performance can be described as follows:

$$p^t = \sum_{i=1}^{m} p_i^t \tag{11.7}$$

where p^t is the performance of the system at time t and p_i^t is the performance of the i-th subsystem at time t.

11.2.2.1 Parameter Estimation and Information Update

Historical degradation data typically includes pressure, flow, temperature, and other data monitored by sensors; historical fault data; and historical detection data. Historical performance degradation data are obtained through sensor data with degradation characteristics, and the degradation parameters of the system are derived. The performance value is determined by monitoring the deviation of the data from the optimal state. Usually, the optimal state is an inherent property of the device and represents the monitored parameters in the initial state when no degradation occurs. In the presence of n monitored feature sensor data in time period t_1–t_2, the performance of the system can be evaluated as

$$p_{t1-t2} = \frac{\sum_{i=1}^{n} fea_i}{nF_b} \tag{11.8}$$

where $p_{t1\text{-}t2}$ is the performance in time period t_1–t_2, fea is a critical feature that reflects performance degradation, and F_b is the optimal state during normal system operation.

The maximum likelihood estimation (MLE) algorithm is used in this method to update variable parameters. Let $C = (J, \mu)$ be a structure with J and parameters μ. Let Z be the variable in C. In addition, let $\boldsymbol{D}$ be a group of cases, where each case is the configuration of all variables in Z. For each $\boldsymbol{d}$ belonging to $\boldsymbol{D}$, the probability $P(\boldsymbol{d}|C)$ is called the likelihood of C for a given d. Assume that the cases in $\boldsymbol{D}$ are independent under a given model; then, the likelihood of C given $\boldsymbol{D}$ is

$$L(C|\mathbf{D}) = \prod_{\mathbf{d}\in\mathbf{D}} P(\mathbf{d}|C) \tag{11.9}$$

If the estimation conditional probability is calculated, then the likelihood model C_μ agrees with the structure but differs with respect to parameters μ. The MLE of parameter can be written as

$$\hat{\mu} = \arg\max_{\mu} L(C_\mu|\mathbf{D}) \tag{11.10}$$

Different classes of inference algorithms, such as junction tree or variable elimination, can be utilized to compute the posterior probability distribution of a specific variable by incorporating observed evidence e. After parameter estimation on offline data, online data can be used for model parameter updates and parameter correction. Usually, online data are inputted into the model in the form of evidence, and the network is updated through evidence nodes. Let the BNs in U be $\{x_1,\ldots,x_n\}$. These algorithms operate based on Bayes' theorem.

$$P(U|e) = \frac{P(e|U)P(U)}{P(e)} = \frac{P(e|U)}{\sum_U P(e|U)} \tag{11.11}$$

11.2.3 RUL Estimation Based on Performance

RUL estimation relies on the dynamic performance of a system. Performance can be indicative of the reliability, availability, and serviceability of systems. Typically, performance tends to decline over time in these aspects. However, when performance is measured by failure probability, it generally increases with time. RUL is determined by a critical value known as the failure threshold (*TH*), which is specific to the system's failure mechanism. In consideration of the actual performance of the system (p), RUL can be defined as the duration between the current state and the point at which the system fails.

$$T = \inf\{t : p(t) \le TH | p(0) > TH\} \tag{11.12}$$

The proposed method is implemented based on DBNs, and the flowchart is shown in Fig. 11.2. Shock and stage modeling are conducted synchronously, and both are converted into probabilistic nodes in DBNs, with the aim of constructing dependency factor nodes in DBNs. The dependency factor node is connected to the degradation parameter node of the performance degradation model to evaluate subsequent degradation trends. With the dynamic expansion of DBNs, the performance at each moment is re-identified as a new degradation stage for dynamic performance calculations. Then, the RUL of the system is obtained based on the predicted performance and failure threshold.

11.3 Case of Subsea Hydraulic Control System

The hydraulic control system in a subsea production system plays a crucial role in the operation and functioning of the system, as shown in Fig. 11.3. The hydraulic control system includes the main control station (MCS), electrical power unit (EPU), hydraulic power unit (HPU), and subsea control module (SCM). The production of subsea Christmas trees is controlled through the hydraulic control system. The main components of hydraulic control systems (i.e., electronic and hydraulic) are often subjected to external shocks and accompanied by their own performance degradation; they can be continuously monitored for a long time. Given that hydraulic and electronic components experience different degradation and shocks, the subsea hydraulic control system is divided into two subsystems, namely, the control system and the hydraulic system, for analysis. The two subsystems are connected in series on the basis of the working principle. The data sources for all modeling and analysis in this case are long-term monitoring data of the control system and hydraulic system sensors of the LH4-1 oilfield in the South China Sea. The status data of the platform is collected onsite, and information on the shocks and degradation stages is supplemented by LH4-1 onsite experts.

11.3.1 Control System Modeling

The degradation of the control system is consistent with that in the exponential degradation model (Cai et al. 2012), and the performance in consideration of shock–degradation dependency can be expressed as:

$$p_e^t = e^{-\varphi_e \lambda t} \tag{11.13}$$

where $p_e{}^t$ is the performance of the control system at time t, φ_e is the dependency factors of the control system, and λ represents the failure rate of the control system.

The MLE method is used to estimate the degradation rate parameters. The degradation rate parameter value that maximizes the probability of observed data appearing is determined by maximizing the likelihood function. The mean and standard deviation of the degradation parameters are obtained simultaneously to effectively represent the uncertainty in performance degradation. Similarly, the shock and stage effects are

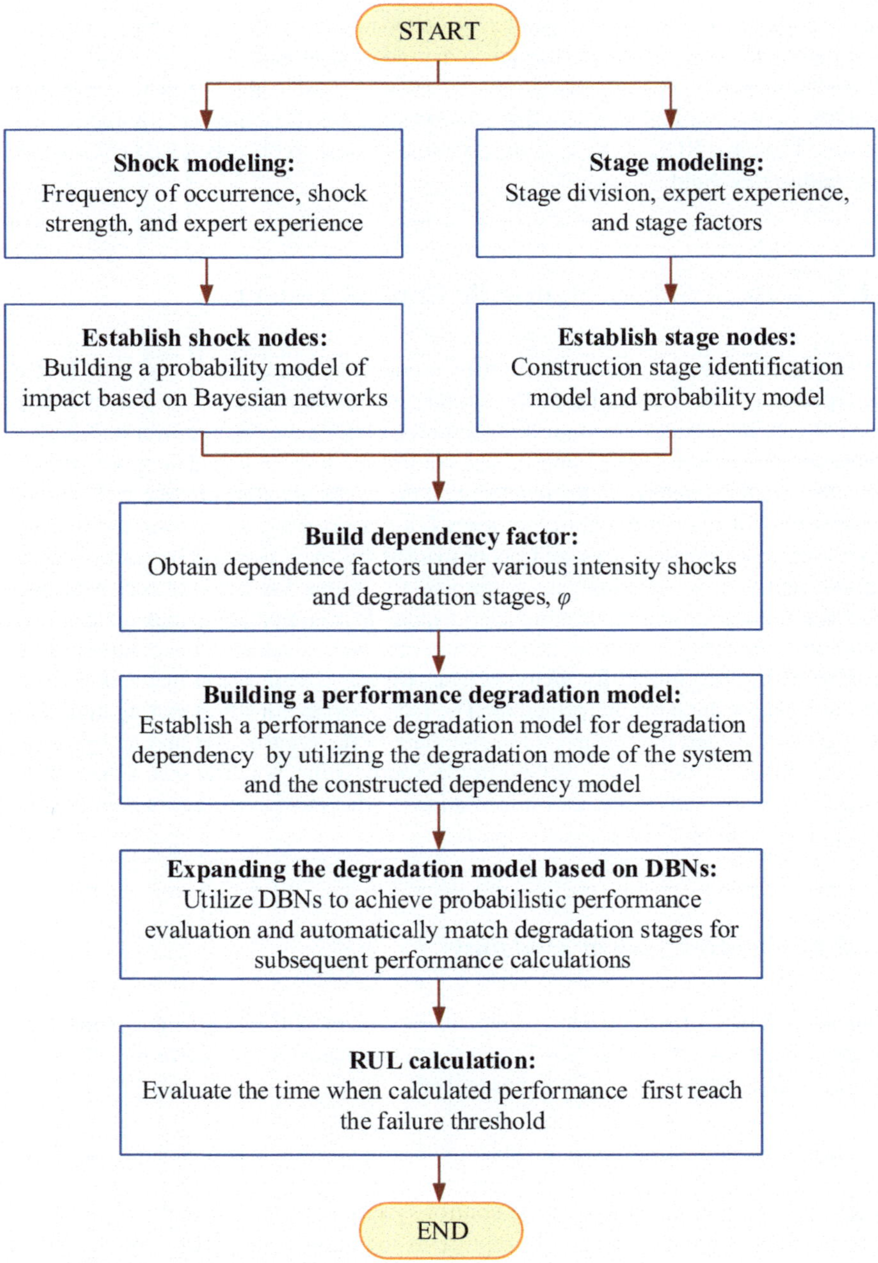

Fig. 11.2 Flowchart of the DBN-based RUL calculation method

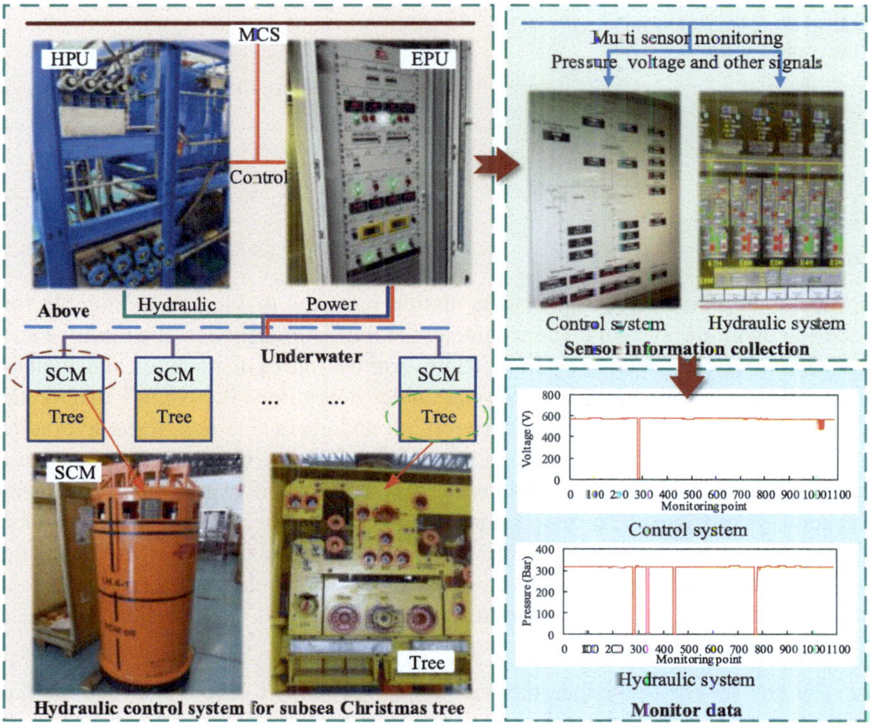

Fig. 11.3 Hydraulic control system for a subsea Christmas tree

Table 11.1 Shock parameters of the control system

Shocks	Strength (I)	Probability (%)
Low	1.1	50
Mid	1.3	40
High	1.6	10

estimated based on historical monitoring data and event data records and corrected by onsite experts. The degradation-related parameters of the control system are shown in Tables 11.1 and 11.2.

11.3.2 Hydraulic System Modeling

The degradation process of hydraulic systems can be effectively described by Gamma processes. Hydraulic systems' degradation is often influenced by multiple factors, including working time, load changes, and temperature variations. Gamma processes

Table 11.2 Stage parameters of the control system

Stages	Performance range	Strength (s)
1	0.8–1	1.01
2	0.6–0.8	1.05
3	0.4–0.6	1.08
4	0.2–0.4	1.1
5	0–0.2	1.2

can accurately simulate complex degradation patterns, making them suitable for modeling hydraulic systems. The parameters of the Gamma process are intuitive and easily interpretable. The shape parameter governs the rate of degradation and dictates how it changes over time. The scale parameter determines the overall magnitude of the degradation rate. The degradation trend and characteristics of the hydraulic system can be inferred by estimating the two parameters. Gamma processes have good mathematical properties, such as additivity and stability. The performance of hydraulic systems can be expressed as ${p_h}^t = \Gamma\,(t, \alpha, \varphi_h\beta)$, where ${p_h}^t$ is the performance of hydraulic systems at time t, α is the shape parameter, β is the scale parameter, and φ_h is the dependency factors of hydraulic systems. Assume that $\{d(t), t \geq 0\}$ is a random Gamma process with the following characteristics:

(a) ${p_h}^0 = 0$;
(b) For any $\tau > t \geq 0$, the increment ${p_h}^\tau - {p_h}^t$ obeys the Gamma distribution, that is,

$$p_h^\tau - p_h^t \sim Ga[\alpha(\tau) - \alpha(t), \beta] \tag{11.14}$$

(c) ${p_h}^t$ is a process with an independent increment, and ${p_h}^{t2} - {p_h}^{t1}, \ldots, {p_h}^{tn} - {p_h}^{tn-1}$ are independent of each other for any $t_1 < t_2 < \cdots < t_n$ $(n > 2)$.

The shocks and degradation stages that hydraulic systems undergo differ from those encountered by control systems, because the two types of systems experience different degradation modes, are in different working environments, withstand different working loads, and cope with different external shocks. The degradation-related parameters of the hydraulic system are shown in Tables 11.3 and 11.4.

Table 11.3 Shock parameters of the hydraulic system

Shocks	Strength (I)	Probability (%)
Low	1.5	40
Mid	1.7	30
High	2	30

Table 11.4 Stage parameters of the hydraulic system

Stages	Performance range	Strength (s)
1	0.8–1	1.01
2	0.6–0.8	1.05
3	0.4–0.6	1.08
4	0.2–0.4	1.1
5	0–0.2	1.2

11.3.3 Modeling of the Hydraulic Control System on the Basis of DBNs

The overall performance of the hydraulic control system is determined by the connection between the control and hydraulic systems, and any failure in either system leads to system failure. The estimated degradation parameters of the two subsystems are shown in Table 11.5. The overall performance of the hydraulic control system can be described as

$$p^t = p_e^t \times p_h^t. \tag{11.15}$$

Subsequently, the DBN model of the hydraulic control system that considers degradation and shock dependence is constructed for system performance evaluation, as shown in Fig. 11.4. l_e is the shock node of the control system, s_e is the stage node of the control system, and φ_e is the dependency factor node calculated based on both. The dependency factor node points to control system degradation parameter λ. The performance of the system based on degradation parameters and time is evaluated. Subsequently, the next stage node identifies the performance distribution at the current time and obtains the distribution probability of the degradation stage for performance evaluation at the next time.

Similarly, l_h is the shock node of the hydraulic system, s_h is the stage node of the hydraulic system, and φ_h is the dependency factor node calculated based on both. The dependency factor node points to degradation rate parameter β. α and β jointly point to the incremental *ga* node. The performance of the hydraulic system is calculated based on the current time, the *ga* node, and the performance at the previous time. The distribution probability of the degradation stage at the next moment is subsequently obtained for dynamic performance evaluation. Overall, the dynamic expansion of DBN is driven by two factors: time and performance nodes. The time

Table 11.5 Degradation parameters of the hydraulic control system

Parameter	Mean	Standard deviation
λ	0.065	0.025
α	1	-
β	0.02	0.014

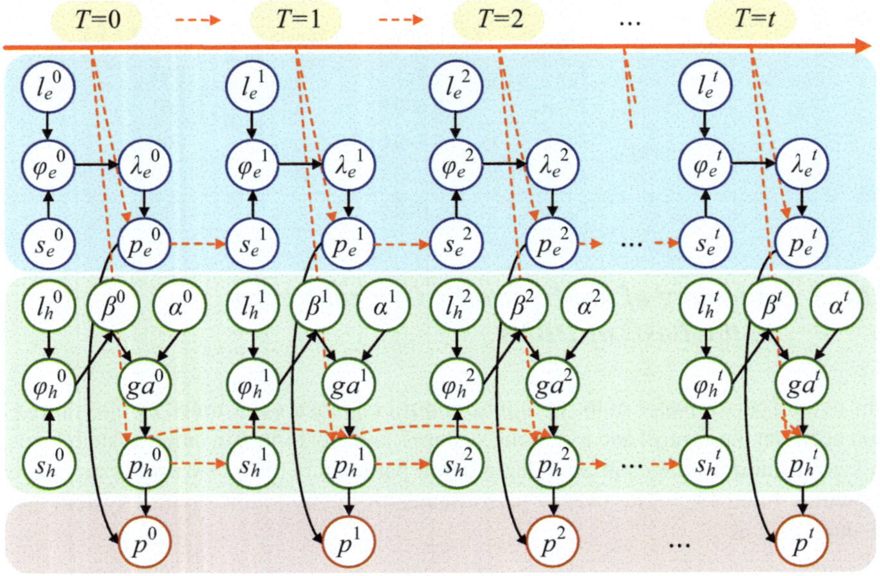

Fig. 11.4 DBN model of the hydraulic control system

node is expanded so that DBN can calculate the performance in different time slices; the performance node provides parameter updates for the stage node at the next moment, thus allowing the conditional probability table between time slices to be updated. Afterward, the performance of the hydraulic control system is calculated based on Eq. (11.12) in the DBNs.

11.3.4 Results and Discussion

11.3.4.1 System Performance Evaluation

The performance evaluation results of the control and hydraulic systems on the basis of the DBNs are shown in Figs. 11.5 and 11.6, which present a the comparison of performance degradation modeling using traditional methods and the proposed method. Traditional methods consider the effect of shocks on degradation acceleration, but disregard the interaction during the degradation stage when the shock occurs. In this case, traditional methods are implemented by assigning the effect at the initial moment while assuming that the effect coefficient is constant at each degradation stage. The degradation of control systems is an exponential model, and compared with the traditional methods, the proposed method calculates performance degradation faster, and its data are closer to the degradation data of engineering monitoring. Monitoring data are obtained by integrating continuous monitoring data at the early stage and interval monitoring data at the late stage into the degradation database. In the evaluation of the hydraulic systems, the degradation process calculated by the proposed method is smooth and shows a remarkable degradation acceleration. The performance comparison at some critical time points is also discussed. The time difference among the proposed method, traditional methods, and monitoring data performance of 0.5 is negligible; However, when the performance reaches 0.2 or 0, the proposed method is close to monitoring data, whereas the traditional methods deviate from such data.

The hydraulic control system performance degradation is shown in Fig. 11.7. The overall performance comparison is calculated based on the performance of two subsystems (that are in a series relationship). Similarly, the performance of traditional method calculations is evaluated for a comparison of the overall degradation. At the early stages of degradation, the degradation trends of the three are similar. However, differences emerge after the sixth year. The performance of the proposed method is close to that of the monitoring data, and the traditional methods exhibit

Fig. 11.5 Performance evaluation of the control system

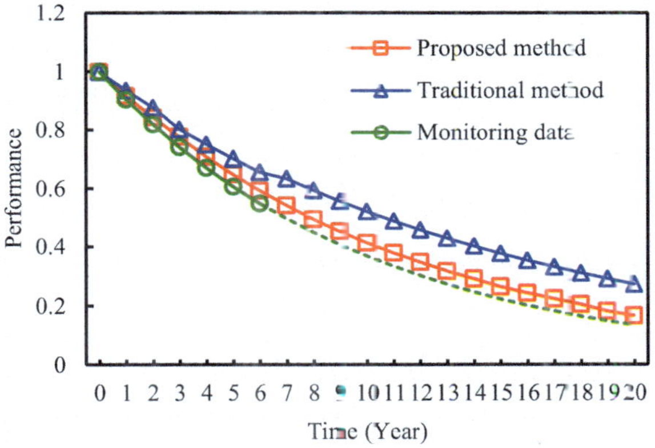

Fig. 11.6 Performance evaluation of the hydraulic system

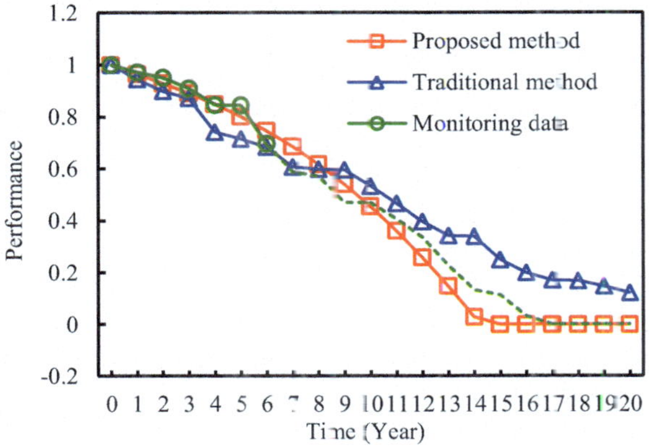

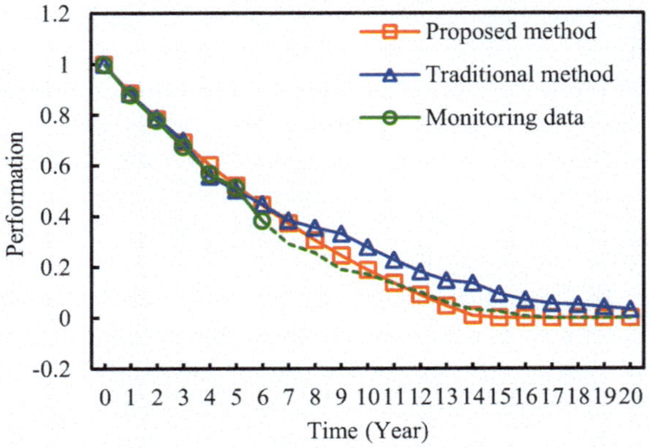

Fig. 11.7 Overall performance evaluation

large computational errors, which make the subsequent RUL calculations inaccurate. When the performance reaches 0.2, a system failure occurs. The system RUL evaluated by monitoring data is 8.8 years. The RUL calculated by the proposed method is 9.7 years, and the RUL evaluated based on the traditional methods is 11.5 years. The proposed method has good prediction accuracy.

11.3.4.2 Degradation Stage Analysis

The degradation stage in the full performance degradation cycle is of concern. The performance degradation process of the control system, hydraulic system, and overall performance from Stage 1 to 5 over time can be obtained based on the predicted mean values, as depicted in Fig. 11.8. The control system quickly degrades from Stage 1–3 in the early phase of degradation and continues to degrade for 8 years at Stage 4, finally reaching Stage 5. The degradation of the hydraulic system is slow at the early stage, lasting for 6 years at Stage 1, but accelerates at the middle stage and quickly deteriorates to Stage 5. The hydraulic control system as a whole deteriorates to Stage 5 in the 10th year, where it reaches its maximum degradation rate.

The probability distribution of performance at some time points of the control and hydraulic systems is shown in Fig. 11.9. Given the randomness of the parameters, the evaluated performance is a probabilistic result. Figure 11.9 shows the probability distribution of the performance of (*a*) the control system and (*b*) the hydraulic system during the degradation process. At the initial moment, the performance distribution of the control system is generally a normal distribution and concentrated at Stage 1. As the degradation progresses, the probability distribution gradually exhibits multiple peaks, and the probability distribution changes, with performance spanning multiple degradation stages. During the degradation cycle of the hydraulic system, the performance follows a normal distribution, and as the degradation progresses, its standard deviation gradually increases, leading to increased degradation stages in the performance range.

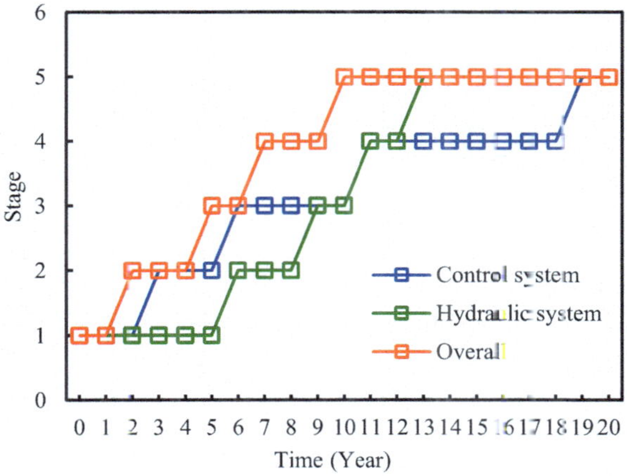

Fig. 11.8 System degradation stage over time

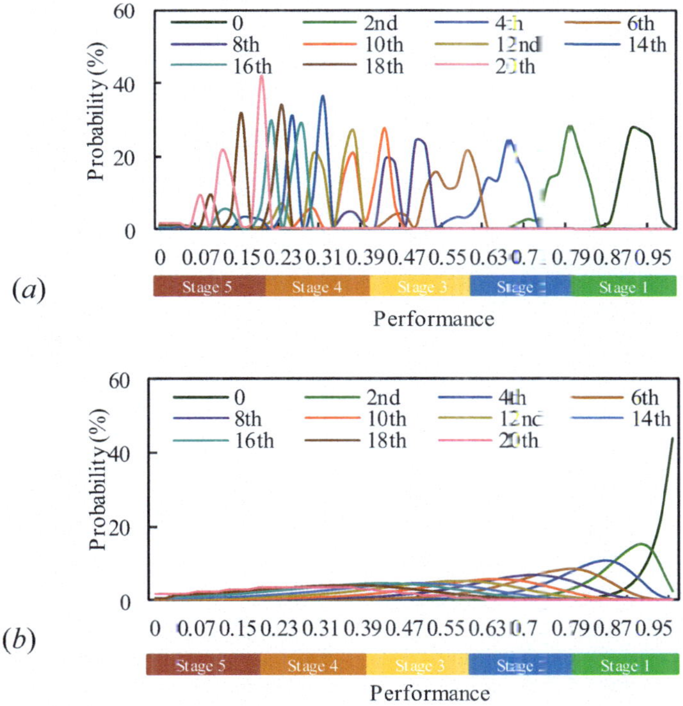

Fig. 11.9 Degradation stage over time of **a** the control system and **b** the hydraulic system

The degradation stage of the control and hydraulic systems undergoes an evolutionary process characterized by a probability distribution, and the stage at any given moment is not static. Figure 11.10 depicts the gradual transition of the degradation stage over time. Specifically, Fig. 11.10 displays the probability distribution of the degradation stage calculated using the performance probability distribution from the initial moment (commencing at year 0) to year 20. During the degradation process, the performance gradually transitions from Stage 1–5. In fact, there is a partial probability that the performance of the control and hydraulic systems will be at Stage 5 by the 8th year. The late probability distribution shows that the probability distribution of control system performance is concentrated mainly at Stages 4 and 5, whereas the probability distribution of hydraulic system performance is relatively loose at Stages 2, 3, 4, and 5.

11.3.4.3 Degradation Parameter Analysis

The degradation parameters of the control and hydraulic systems are illustrated in Fig. 11.11. Traditional prediction methods, which consider the effects of shocks, typically presume that the degradation model's parameters remain constant and do not vary with degradation time. By contrast, the proposed method can dynamically update the degradation parameters in response to the evolving degradation stages. It allows for a precise representation of the system's degradation modes at different stages, thus enhancing the accuracy of the predictions. The degradation parameters of the control system gradually increase, with two obvious fluctuations and an unsteady increase. The degradation parameters of the final control system increase to around 0.09, showing an increase of 12.5% compared with the initial value of 0.08. The degradation parameters of the hydraulic system change smoothly and approximately linearly. The degradation parameters evaluated by the proposed method increased by 13% relative to the initial value of 0.034.

11.3.4.4 Analysis of the Effect of Shocks

Shocks are the main external factors that influence system degradation, and the effect of shocks with different intensities on the degradation rate is shown in Fig. 11.12. The shock strengths (low, mid, and high) in Tables 11.1 and 11.3 are converted to Levels I, II, and III. The difference in the degradation rates of the control and hydraulic systems represents the dependent effects of different intensity combinations. The higher the strength of the shock is, the faster the system performance degrades. After the determined effect information is obtained, it is inputted to the DBNs through the evidence nodes for performance re-evaluation. One of the main advantages of the proposed method is that it can integrate monitoring data into model updates. The mean of the degradation stage is used to represent the effects of different-intensity shocks on stage degradation, as shown in Fig. 11.13. The degradation stage of the control system in response to the shocks still exhibits a jumping-upward trend. The

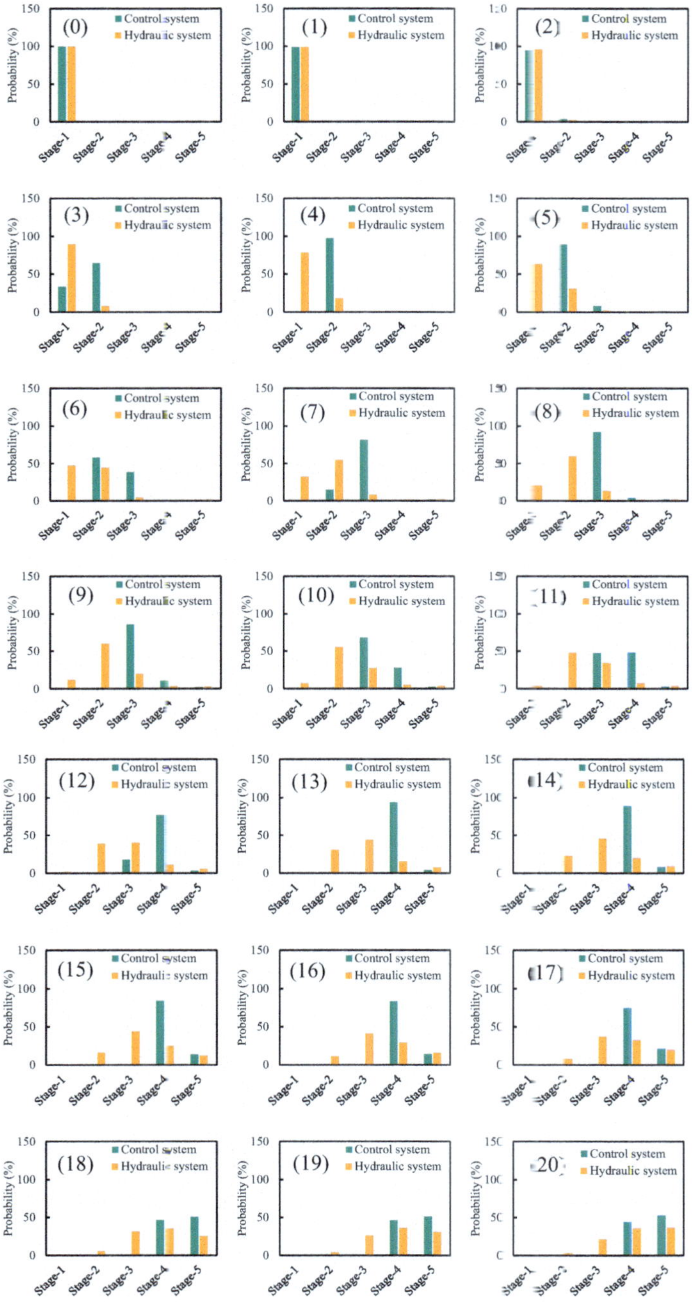

Fig. 11.10 Probability distribution of degradation stage form year 0 to the year 20

Fig. 11.11 Degradation parameter change trend of the **a** control system and **b** hydraulic system

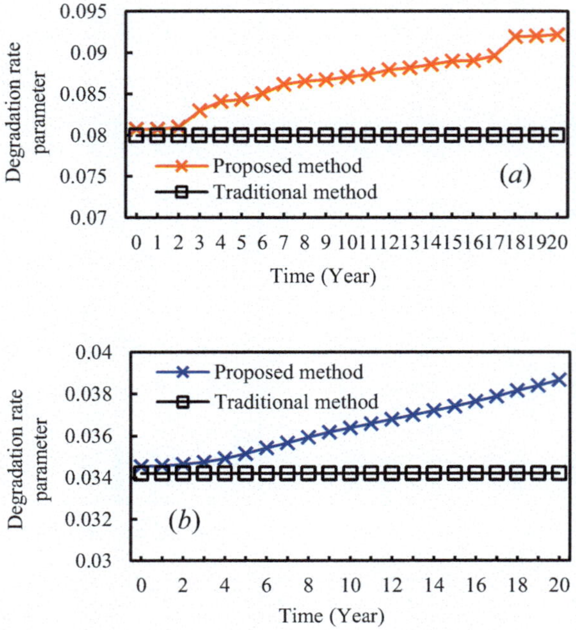

average stage factor of the hydraulic system is approximately linearly increasing. The hydraulic system demonstrates relatively strong stability when facing shocks. The transition speed of the degradation stage of the control system is higher than that of the hydraulic system, and under Level III shocks, the degradation stage of the control system reaches its maximum value in the 16th year, with a corresponding stage factor of 1.2. The control system is more susceptible to shocks than the hydraulic system because electronic systems are more fragile than hydraulic systems.

Fig. 11.12 Performance under different-strength shocks

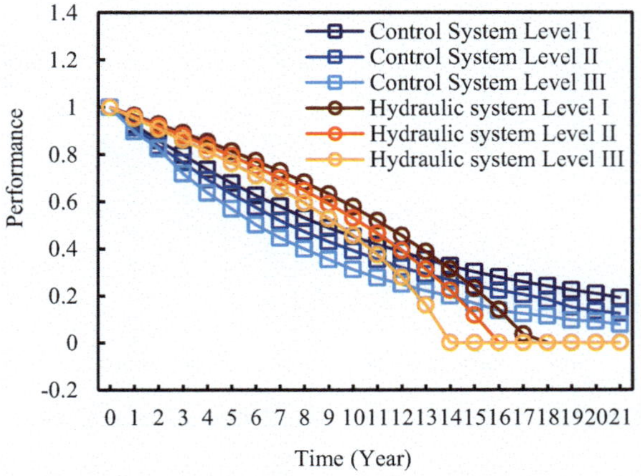

Fig. 11.13 Stage factors under different-strength shocks

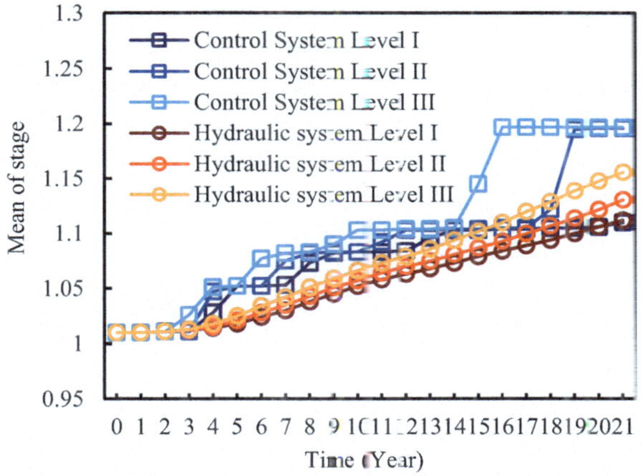

The system RUL under different-intensity shocks is evaluated, and shown in Fig. 11.14. Water shock acts on the control system, and underwater shock acts on the hydraulic system. The RUL under the different shock combinations is used to analyze the sensitivity of the system to shocks, where I-I represents Level I for water shock and Level I for underwater shock. When the combination of water shock Level I and underwater shock Level I is utilized, the lifespan is the longest at 11.3 years. On the contrary, the combination of water shock Level III and underwater shock Level III has the shortest lifespan of 8.7 years. Under the same combination of shocks, the greater the shock of water on the control system is, the shorter the RUL of the system is. For example, under the shock of Levels I-II and II-I, the RUL of the system is 10.9 and 10.5 years, respectively. Under the shock of Levels I-III and III-I, the RUL of the system is 10 and 9.5 years, respectively. Under the shock of Levels II-III and III-II, the RUL of the system is 9.3 and 8.9 years, respectively. Water shock has a considerable effect on the performance degradation of the system.

Fig. 11.14 System RUL under different-intensity shocks

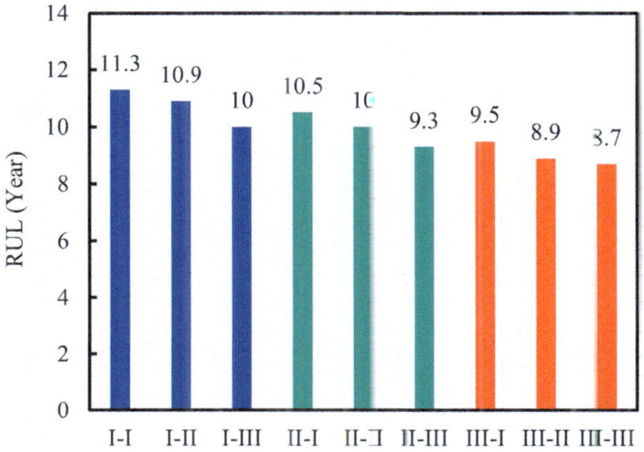

Typically, the water part is susceptible to the effects of typhoons and high temperatures, and the underwater components face potential threats from internal waves, tsunamis, earthquakes, and others. Notably, underwater equipment, predominantly mechanical and hydraulic in nature, undergo multiple protective measures before immersion, to ensure a consistently high level of reliability. Despite the elevated shock risks in the underwater environment, these robust protections measures contribute to the sustained reliability of submerged equipment.

By contrast, the control system has high vulnerability to shock effects, which are often accompanied with the system's own performance degradation. In the hydraulic control system, the control system of the above-water part exhibits high sensitivity to external influences. This susceptibility stems from the intricate interplay of various factors that affect the control system's performance. The comprehensive assessment suggests that, although the underwater components face severe potential effects, the effective protective measures contribute considerably to maintaining the operational integrity of these components. In a broad context, understanding the nuanced vulnerabilities of above-water and underwater elements is crucial for ensuring the overall reliability and resilience of hydraulic control systems.

Chapter 12
RUL Re-prediction

Abstract The traditional Wiener-process-based models often overlook historical data and suffer from missing data issues, thereby resulting in inaccurate RUL prediction. This chapter proposes a novel RUL re-prediction method based on the Wiener process, incorporating both current monitoring data and historical degradation information to improve prediction accuracy. In the initial prediction phase, the Wiener process is used to model system degradation, with the drift and diffusion coefficients estimated using the Expectation Maximization algorithm. To account for uncertainty due to missing data, a DBN model is established. In the re-prediction phase, multiple sets of performance degradation data and historical predictions are combined to refine the degradation stages in the Wiener process, with DBNs used for modeling. The RUL is calculated as the time difference between the detection point and the predicted fault point, determined by a failure threshold. The method is demonstrated on a subsea Christmas tree system, showing its effectiveness in improving RUL prediction by integrating historical data and addressing data uncertainties.

12.1 Introduction

As mentioned in Sect. 9.1, RUL prediction techniques generally fall into model-driven, data-driven, and hybrid approaches. The hybrid approach, combining model-driven and data-driven techniques, has emerged as a promising solution. However, challenges remain in addressing uncertainties, particularly in complex systems with noisy data, and in improving the generalizability of predictive models.

Current RUL prediction methods still face limitations in handling uncertainties inherent in degradation processes and in accounting for incomplete or sparse data (Jardine et al. 2006). To address these challenges, this research proposes a novel hybrid approach that integrates DBNs with stochastic degradation models, such as the Wiener process, to better manage uncertainty and improve prediction accuracy. By utilizing DBNs, the method offers enhanced flexibility in capturing the probabilistic

B. Cai et al., *Intelligent Operation and Maintenance for Subsea Production Systems*,
https://doi.org/10.1007/978-981-96-2677-9_12

relationships between degradation variables, while the Wiener process models non-linear degradation patterns. This combination effectively handles noisy data and small sample sizes, making it well-suited for complex systems.

This chapter focuses on developing a hybrid RUL re-prediction method that leverages both model-based and data-driven approaches to enhance prediction accuracy in complex systems. The integration of DBNs with Wiener process-based degradation models addresses key challenges such as uncertainty management and the incorporation of incomplete data. The proposed method demonstrates improved predictive performance over traditional approaches, providing a more reliable and practical solution for managing system health and ensuring operational safety.

12.2 RUL Prediction and Re-prediction Model

12.2.1 Wiener Process

Wiener process, also known as Brownian motion process, is one of the most important stochastic processes in probability theory. Theoretically, Wiener process is a continuous time stochastic process. The distribution of its increment is only related to the time difference, so it is a homogeneous independent increment process. At the same time, Wiener process is also a normal process, its distribution is completely determined by its mean value function and self-covariance function. If

(1) $X(0) = 0$;
(2) $\{X(t), t \geq 0\}$ have a smooth independent increment;
(3) For any random variable $t > 0$, $X(t)$ is a normal random variable with mean 0 and variance $\sigma^2 t$;

So the random process $\{X(t), t \geq 0\}$ is called the Wiener process. When $\sigma = 1$, this process is called standard Brownian motion and expressed as $\{B(t), t \geq 0\}$. Wiener process with linear drift $\{X(t), t \geq 0\}$ is expressed as follows:

$$X(t) = X(0) + \mu t + \sigma B(t) \tag{12.1}$$

where $X(0)$ represents the initial performance degradation, $X(t)$ represents the current performance degradation, μ represents the drift coefficient, reflecting the degradation rate and determining the degradation path, σ represents the diffusion coefficient, $B(t)$ is the standard Brownian motion.

According to the properties of Wiener process, its increment is independent and follows normal distribution, i.e.

$$\Delta X(t) = X(t + \Delta t) - X(t) \sim N(\mu \Delta t, \sigma^2 \Delta t) \tag{12.2}$$

where $\Delta X(t)$ represents the accumulative amount of the performance degradation of Δt period.

Probability density function of degradation increment is shown as follows

$$f_{\Delta X}(\Delta x|\mu, \sigma) = \frac{1}{\sqrt{2\pi\sigma^2\Delta t}} \exp(-\frac{(\Delta x - \mu\Delta t)^2}{2\sigma^2\Delta t}) \tag{12.3}$$

where $f_{\Delta X}$ represents the probability density function of degradation increment

Probability density function of the RUL is shown as follows

$$f_{Ri}(r_i|X_{0\to i}, \mu, \sigma) = \frac{D - x_i}{\sqrt{2\pi\sigma^2 r_i^3}} \exp(-\frac{(D - x_i - \mu r_i)^2}{2\sigma^2 r_i}) \tag{12.4}$$

where f_{Ri} represents the probability density function of the RUL, D is the failure threshold in terms of reliability, r_i represents the RUL.

12.2.2 RUL Prediction Method Based on Wiener Process

A RUL re-prediction method for complex systems based on Wiener process is proposed in this work. Historical data based on Wiener process and online monitoring data are effectively combined in this method, and the influence of the amount of historical data combined from the time span on the RUL prediction is studied creatively. The RUL re-prediction framework based on Wiener process is shown in Fig. 12.1. In the case of a mechanical or electrical system, historical failure databases, online monitoring data and approximate distributions of degradation trends are obtained. First, an initial model-driven prediction model based on database historical failure data is established. The degradation process represented by the model generally follows exponential distribution or gamma distribution according to system classification. Then, in order to make the prediction result closer to the engineering application, a prediction model based on the Wiener process of data and model driven hybrid is established. The specific application of the Wiener process is mentioned above, EM algorithm is used to analyze the historical data, and the drift parameters and diffusion coefficients of Wiener process are obtained. Then, a DBN is established, Wiener process parameters are combined with performance degradation parameters such as corrosion, fatigue, etc. Then the re-prediction model based on Wiener process is established on this basis. Physical degradation process, online monitoring data and degradation increment of initial predictive data are combined and distributed proportionally according to expert experience. More details are introduced as follows.

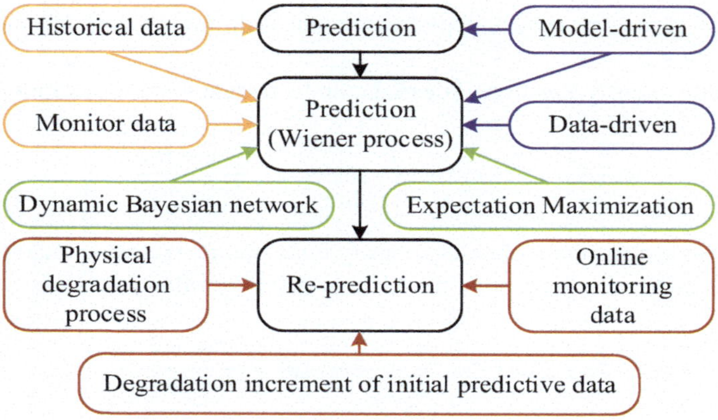

Fig. 12.1 RUL re-prediction framework based on Wiener process

12.2.2.1 Construction of RUL Prediction Model Based on Wiener Process

A system RUL prediction method based on Wiener process is proposed. The ultimate goal of this method is to form a tendency of degradation. Before the trend is modeled, it is necessary to identify the measured variables which carry information on the degradation process. The variables are obtained by EM algorithm, they come from the historical degradation data collected by the sensor. In this chapter, the main variables identified are drift parameters μ and diffusion coefficients σ in the Wiener process, which determine the speed of the degradation process. Some other variables are hard to get directly from historical data, such as fatigue, corrosion, environmental causes, mechanical damage and other factors and their common influence on the system. For these variables, it is necessary to learn from the experience of predecessors to form the corresponding degradation formula according to the degradation of different variables and their mutual influence relations. The initial value of these variables are input into the DBNs through querying the database and combining the experience of experts, and then the trend of degradation is obtained.

This method uses Wiener process with random effect to describe the system's performance degradation process, uses a model based on Bayesian updates and EM algorithm for parameter estimation. In this process, the model drift parameter is to obey the normal distribution of random variables, which is the first passage time under the definition of RUL. That is, the time lag from the current point to reach the failure threshold is the RUL of the system. Among them, EM algorithm adopts the idea of heuristic iteration and divides the solving steps into two steps. Firstly, it assumes the value of hidden data, and then it uses the idea of maximum likelihood function to solve the model parameters. After repeated iteration, the appropriate model parameters are found when the algorithm converges, and finally the parameters related to the RUL of the system are obtained. Furthermore, given that the entire process from installation to failure is threatened by pressure, corrosion, and wear, degradation conditions at each stage are different. In this chapter, the degradation process can be divided into several different stages qualitatively. In each stage, the EM algorithm is used to estimate the initial μ, σ value of Wiener process, and then according to the value and the corresponding formula to predict the degradation process of next time. At the same time, to achieve the objective of the model is more accurate, each stage specific classification, μ, σ of each stage are constantly updated. Finally, the predicted value is compared with the real detected value, and the corresponding relationship between the predicted value and the real value is obtained, so as to make a reasonable prediction for the future degradation of the system and obtain the RUL. The RUL prediction model based on Wiener process is shown in Fig. 12.2.

Firstly, the monitored data is estimated according to the EM algorithm. The EM algorithm is often used as an important step to estimate the degradation performance of Wiener process. Its main function is to generate drift parameters and diffusion coefficients required by Wiener process. Since Wiener process is a basic tool to describe a series of important and complex processes, it can form a complete degradation trend when combined with DBNs. The model parameters are fixed by E-step, and the hidden variables are solved by the mathematical model. The model parameters

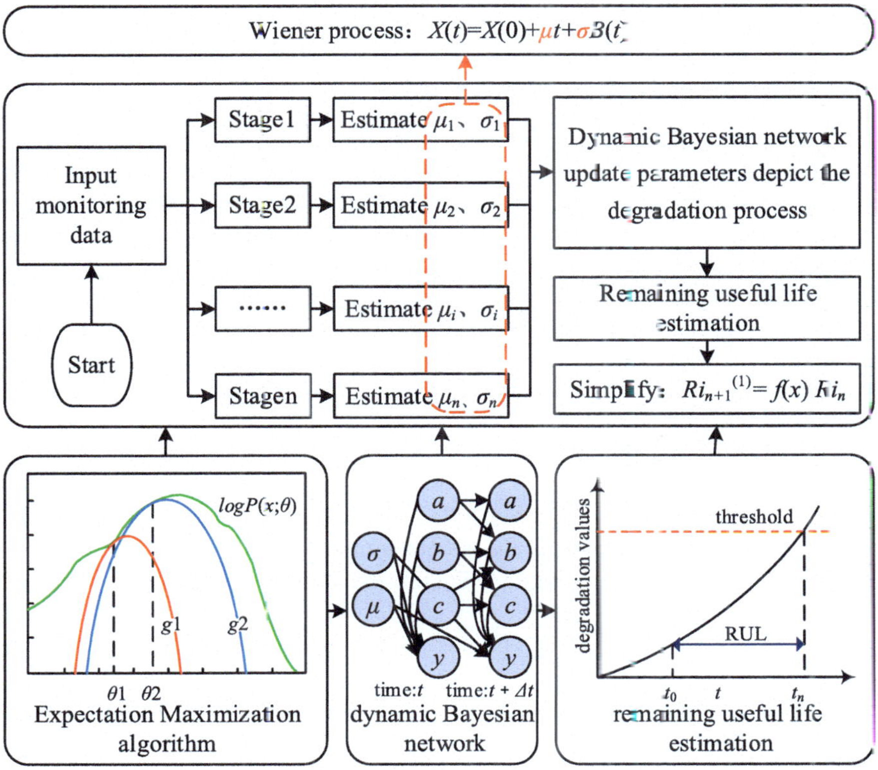

Fig. 12.2 RUL method based on Wiener process

of M-step are obtained according to the implicit variables obtained by E-step and the maximized likelihood function obtained by the maximum likelihood estimation. In this way, the local optimal solution of the model is obtained by mutual iteration. In addition, since the monitoring data comes from sensors, mutations in the system are common, the process can be recorded and simulated when the system degradation occurs suddenly. The specific steps are as follows:

1. Randomly initialize the model parameter with an initial value of $\theta_0, j = 1, 2, \ldots, J$, starts EM algorithm iteration;
2. E-Step: Calculate conditional probability expectation of joint distribution:

$$Q_i(z_i) = p(z_i|x_i, \theta_j) \tag{12.5}$$

$$l(\theta, \theta_j) = \sum_{i=1}^{n} \sum_{zi} Q_i(z_i) \log \frac{p(x_i, z_i, \theta)}{Q_i(z_i)} \tag{12.6}$$

3. M-step: maximize $l(\theta, \theta_j)$, and get θ_{j+1}:

$$\theta_{j+1} = \arg\max l(\theta, \theta_j) \tag{12.7}$$

4. If the results converge, the algorithm ends.

As is shown in Fig. 12.2, in Expectation Maximization algorithm, $P(x; \theta)$ is the maximum log likelihood function, initialization parameters, the posterior probability of the implicit content is estimated at Step E, get the curve $g1$, maximizing in Step M to get curve $g2$, the parameter estimation of EM algorithm can be completed by cycling until convergence.

Then, the DBNs of the system are established, the parameters are substituted into the network, and the degradation of the system at each moment is updated and calculated. The degradation degree and speed of different systems are different in different environments. The system degradation y is defined as

$$y = (y_1, y_2, \cdots, y_m) \tag{12.8}$$

where m indicates time slice, which can be defined as year, month, day, etc., and y_m represents the total degradation of the system at time m.

$\Delta y_i = y_i - y_{i-1}$, Δy_i is the degradation increment of the system from i to (i-1):

$$\Delta y = (\Delta y_2, \Delta y_3, \ldots, \Delta y_m) \tag{12.9}$$

According to the conjugate prior distribution, the posterior joint distribution of μ, σ^2 is as follows:

$$f(\mu, \sigma^2|\Delta y) = \frac{1}{\sqrt{2\pi c\sigma^2}} \exp\left[\frac{(\mu - d)^2}{2c\sigma^2}\right] \cdot \frac{b^a}{\Gamma(a)} \sigma^{-2a-2} \exp\left(-\frac{b}{\sigma}\right) \tag{12.10}$$

According to the structure and functions of the system, the DBN as shown in Fig. 12.3 is established for the system. Here, *a*, *b*, *c* and *d* are unknown hyper-parameters, its initial value can be gained by EM algorithm. Then, combining different degradation indicators of the system, such as environmental cause, fatigue, corrosion, mechanical damage and other factors and their common influence on the system, *a*, *b*, *c* and *d* are updated to obtain new system degradation Δy. This method provides an idea for the RUL prediction of complex systems. Due to the different working environments in the work plans of different systems, the specific ways of application are also slightly different, which is mainly reflected in the determination methods of *a*, *b*, *c*, *d* and other parameters representing the performance indicators and the differences in their interdependence formulas. For example, the main failure modes of different systems are different due to the difference of working environment, which affects the selection of the index to evaluate the RUL of the system. For the power system, the line load, voltage fluctuation and other effects on the system, as well as the existence of cascading failure between them and other internal links need to be considered. On this basis, collecting a large amount of online data and the historical failure database sorted by predecessors, combining with the corresponding formula in this field, deriving a series of performance degradation processes, so as to predict the RUL. In terms of mechanical system, crack, fatigue, corrosion, working conditions and other indicators are the focus of attention. Therefore, in the process of RUL prediction, extract the above indicators from the data, and then establish an appropriate mathematical model according to their importance and interdependence to form the performance degradation trend, so as to obtain the prediction results.

Finally, the RUL is calculated. RUL is determined by the failure threshold A_{th}, which is the critical value. Suppose the performance degradation quantity of the system is A, the system fails when $A \geq A_{th}$, that is, the life of the system is defined as

$$T = \inf\{t : A(t) \geq A_{th} | A(0) > A_{th}\} \tag{12.11}$$

$$A_{th} = 1 - D_i \tag{12.12}$$

where A_{th} is the failure threshold of system performance degradation, D_i is the failure threshold in terms of reliability. When the performance degradation product of the system reaches the A_{th}, the system is judged to have failed.

The calculation process is simplified and corrected, the monitored value and the predicted value are observed, and the predicted value at the latter moment is associated with the monitored value at the previous moment by using function $f(x)$ to obtain the Eq. (12.13). Combined with online monitoring data, near real-time prediction is realized, so once there are abrupt variations in the online monitoring data, there will be corresponding abrupt variations in the prediction process.

$$Ri_{n+1}^{(1)} = f(x) \times Ri_n \tag{12.13}$$

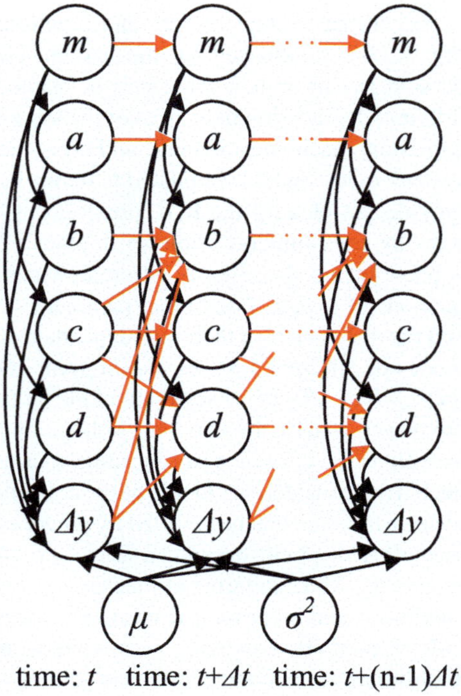

Fig. 12.3 The DBN model of system degradation

where $f(x)$ can be in the form of constant, quadratic function, exponential function, etc. The new RUL value is obtained by online prediction of the degradation amount in the system degradation process with the new formula.

12.2.2.2 Construction of RUL Re-prediction Model Based on Wiener Process

On this basis, from the Wiener process based on the degradation volume Δ at each stage, combined with different number of performance degradation prediction monitoring data and history prediction data, the RUL is estimated, system get more accurate prediction results. The block diagram of RUL re-prediction based on Wiener process is shown in Fig. 12.4. The specific calculation process is as follows:

1. Establish the physical model of system failure efficiency as the initial predicted RUL *Re*;
2. Taking Wiener process as the degradation process in combination with online monitoring data *R*, the drift coefficient μ and diffusion coefficient σ in Wiener process at different stages are estimated by EM algorithm, etc. The predicted value $R^{(1)}$ is obtained by updating the DBN. The EM algorithm and calculation process are shown in Fig. 12.3;

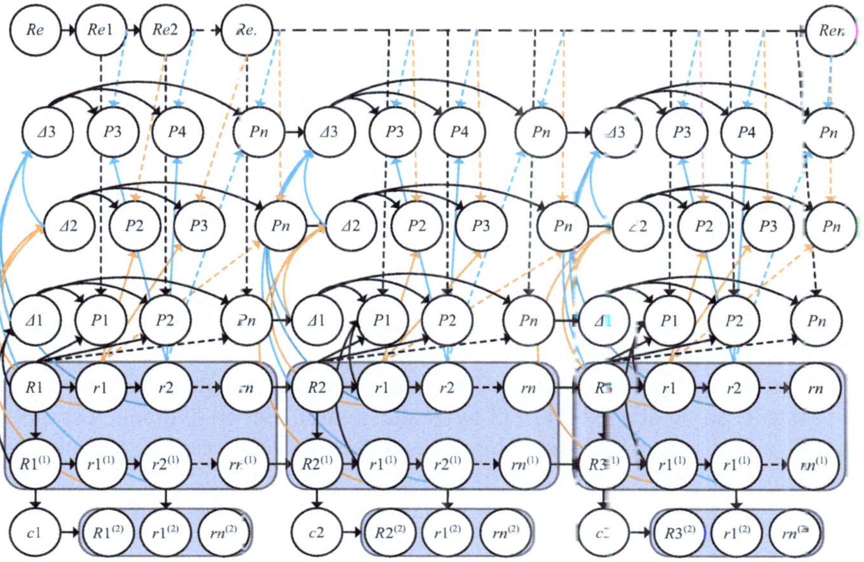

Fig. 12.4 RUL re-prediction model based on Wiener process

3. Establish a DBN re-prediction model based on Wiener process combining 2, 3 and 4 groups of predictive data $R^{(1)}$ respectively. Among them, every moment is relatively a moment the basic amount of degradation of respectively $\Delta 1$, $\Delta 2$ and $\Delta 3$.

$$\Delta n = \frac{\sum_{i=1}^{n} \left(Ri^{(1)} - Ri\right)}{l} \tag{12.14}$$

$$Pm = R1 + pm\Delta i + q \times \text{Ren} \tag{12.15}$$

where Δn *is* degradation increment which combining $(n + 1)$ group historical data, l is a fixed value, namely, when $n = 1$, $l = 2$; When $n = 2$, $l = 3$; When $n = 3$, $l = 6$. m represents time, Pm is the re-prediction data of degradation at the moment of m, p and q are obtained by expert experience, represent the proportion of the initial monitoring data and the degree of influence of the physical degradation process on the degradation respectively, and they add up to 1. Through the effective combination of initial monitoring data, physical degradation process and historical data, whether the system is in a normal state of degradation or mutates, the re-prediction method will get a more reasonable result.

Finally, the RUL of the system is calculated based on Wiener process. RUL is determined by the failure threshold A_{th}, which is the critical value. Suppose the performance degradation of the system is A, that is, the system fails when $A \geq A_{th}$.

Suppose the actual performance degradation of the system is A, that is, the system fails when $A \leq A_{th}$.

12.3 Case of a Subsea Christmas Tree System

12.3.1 Introduction of Subsea Christmas Tree System

This section introduces the subsea Christmas tree as a case study. The description of the subsea Christmas tree refers to Sect. 9.3.1.

According to different control modes (Bai and Bai 2012), the subsea Christmas tree system can be divided into full hydraulic control, electro-hydraulic composite control and full electric control. Among them, the tree valve parts of the fully electric control subsea Christmas tree adopt the full electric control, the electric signal replaces the hydraulic signal, the underwater cable replaces the hydraulic pipeline, has high reliability, is the future development direction. The subsea Christmas tree system is in a complex Marine environment of low temperature, high pressure and high corrosion, etc. The environmental causes, fatigue, corrosion and mechanical damage will all become their main forms of failure. It is necessary to put these indicators into the degradation model of the system. On the other hand, due to the complexity of its internal structure, the evaluation of its reliability only from the physical model will bring a large error. In this chapter, the RUL of the subsea Christmas tree system is predicted by means of physical and data driven model.

Furthermore, because the life prediction method of Wiener process needs certain degradation data parameter iteration to get the optimal model parameters, in engineering, there will always be data missing caused by sensor networks, long life cycle properties of system and so on, leading to unsatisfactory results. In terms of the subsea Christmas tree system, because of its special working environment and working principle, only simple sensors can be installed. As a result, data representing the status of the tree cannot be fully obtained. With the increase of the service life of the subsea Christmas tree system, and the long-term harsh Marine environment and complex working conditions, the trees will inevitably suffer from defects such as rust, corrosion and erosion of the body. The method of combining the monitoring data with database failure data is considered, and the DBN is suitable for this method to solve the uncertainty problem. This chapter combines the two aspects, posterior distribution is derived according to the prior distribution and the test data, the posterior distribution combines prior distribution and sample information, to get reliable Bayesian inference. Taking the subsea Christmas tree control system as an example, the RUL of the system is calculated by combining its overall failure efficiency, historical monitoring data of different stages and multiple moments and the initial prediction value of Wiener process. For other systems or more complex systems, the steps for RUL re-prediction based on Wiener process are completely generic. This chapter only takes the subsea Christmas tree as an example. In fact, it is feasible for any system.

12.3.2 Modeling Process

System data sets are usually obtained by various sensors attached to the system, however, due to the complexity of subsea Christmas tree system structure, the influence of the nature of work and the complicated Marine environment, the number and types of corresponding sensors are less, the data is not easy to get. Furthermore, the working conditions of subsea tree systems are different in different sea areas, which leads to individual differences in the collected data and makes it difficult to achieve unity. A method to combine the traditional database information with the randomly generated simulation data is proposed. The method is based on the analysis of the relevant data of the tree in OREDA database and the random degradation property of Wiener process, the corresponding random number is generated by MATLAB tool, and then the RUL of the system can be analyzed. Since the subsea Christmas tree system is dominated by an electronic control system, its overall degradation performance roughly conforms to the exponential degradation model, and its failure efficiency can be expressed as

$$F = 1 - e^{-\lambda t} \tag{12.16}$$

where λ is the failure efficiency, it can be calculated as 1.663×10^{-4} times/day (DNV 2015), a time slice is modeled every half a year, F is the performance degradation of the system.

12.3.2.1 Construction of Initial Prediction Model Based on Wiener Process

In the Wiener process-based modeling of the subsea Christmas tree system, the following assumptions should be made:

1. Assume that the failure threshold is D_i, and failure occurs when the degradation reaches the failure threshold.
2. The degradation obeys normal distribution and is simulated by Wiener process.
3. The parameter μ_i, σ_i is the random variable of corresponding stage, μ_i, σ_i is described by normal distribution.

$$\mathrm{f}(\mu, \sigma^2) = \frac{1}{\sqrt{2c'\pi\sigma^2}} \exp[\frac{(\mu - d')^2}{2c'\sigma^2}] \cdot \frac{b'^{a'}}{\Gamma(a')} \sigma^{-2a-2} \exp(-\frac{b'}{\sigma^2}) \tag{12.17}$$

According to the whole degradation process of subsea Christmas tree system, it can be divided into three stages artificially, namely, early stage, middle stage and late stage. Because the prediction is based on the assumption that there is a complete system lifetime degradation trend, the stages are divided artificially according to the time points at which the trend changes. When the degradation trend of the system has a more obvious change from the j-th time slice, the prediction of this stage ends and jumps to the next stage for prediction. The predicted results of the j-th time slices are calculated as the initial values of the next stage. The degradation process is expressed as follows:

$$X_i(t) = X_i(0) + \mu_i t + \sigma_i B(t) \tag{12.18}$$

where $X_i(0)$ represents the initial performance degradation in phase i, $X_i(t)$ represents the current performance degradation in phase i, μ_i represents the drift coefficient, reflecting the degradation rate and determining the degradation path in phase i, σ_i represents the diffusion coefficient in phase i.

Different stages of the corresponding DBN model are established, substitute μ_i and σ_i which has been estimated into the formula and BNs of subsea Christmas tree system to implement parameters update (Sun et al. 2018), resulting in subsea Christmas tree system performance degradation Δy of different time.

$$a' = a + \frac{m}{2} \tag{12.19}$$

$$b' = b + \frac{d^2}{2c} + \frac{1}{2}\left[\sum_{i=1}^{m} \frac{\Delta y_i^2}{\Delta t} \cdot \frac{\left(\sum_{i=1}^{m} \Delta y_i + \frac{d}{c}\right)}{\left(\sum_{i=1}^{m} \Delta t + \frac{1}{c}\right)}\right] \tag{12.20}$$

$$c' = \frac{c}{1 + c\sum_{i=1}^{m} \Delta t} \tag{12.21}$$

$$d' = \frac{d + c\sum_{i=1}^{m} \Delta y_i}{1 + c\sum_{i=1}^{m} \Delta t} \tag{12.22}$$

where a represents the initial degree of fatigue, a' represents the current degree of fatigue, increasing steadily; b represents the initial degree of corrosion, b' represents the current degree of corrosion, which depends on changes in the external environment and machinery; c represents the initial effect caused by the environment, c' represents the cumulative effect caused by the environment; d represents the initial mechanical damage of subsea Christmas tree system, d' represents the current mechanical damage of subsea Christmas tree system.

Initial relevant parameters are shown in Table 12.1:

On this basis, the prediction value of performance degradation of subsea Christmas tree system based on the current monitoring value is obtained. In order to simplify this complex process, a simple formula is substituted, namely

Table 12.1 DBNs model parameters of subsea Christmas tree system

Variable name	Probability distribution	Mean value	Standard deviation
$a0$	Normal	0.651	0.15
$b0$	Normal	8.1623×10^{-6}	1×10^{-7}
$c0$	Normal	0.035	0.0031
$d0$	Normal	0.0125	0.0001
μ_1	Normal	0.05	0.0001
$\sigma_1{}^2$	Normal	2.25×10^3	1.25
μ_2	Normal	0.0425	0.0001
$\sigma_2{}^2$	Normal	2.243×10^3	1.25
μ_3	Normal	0.075	0.0001
$\sigma_3{}^2$	Normal	2.272×10^3	1.25

$$Ri_{n+1}^{(1)} = (1 + c) \cdot Ri_n \tag{12.23}$$

where $Ri_{n+1}^{(1)}$ indicates the predictive data obtained through simplification at the $n + 1$ moment, Ri_n indicates the monitoring data at the n moment, and c is the constant value, it can be calculated by

$$c_i = \frac{\sum_{t=1}^{m} (R_t^{(0)} - R_t)}{\sum_{t=1}^{m} R_t} \tag{12.24}$$

where c_i *is the predicted constants of phase i*, m is time slice that has been predicted based on the Wiener process, $R_t^{(0)}$ is the t_{th} predicted data based on the Wiener process, R_t is the t_{th} monitoring data.

12.3.2.2 Construction of Re-prediction Model Based on Wiener Process

In the first prediction of the subsea Christmas tree system RUL based on Wiener process, the degradation process is only related to the current state but not the historical state, which results in the inaccuracy of the prediction. Here, in order to reduce the prediction error, considering every stage of degradation Δ on Wiener process, combining with different number of performance degradation monitoring values, the RUL of subsea Christmas tree system is re-predicted. The re-prediction model is shown in Figs. 12.5, 12.6 and 12.7.

As is shown in Fig. 12.5, the Pn is the n-th time slice of the prediction results in "Re-prediction 1", Δ1 is the basic degradation considering one set of history data, n is the time slice, $R1$ indicates the initial historical monitoring data, Ren indicates the degradation nth time slice to the overall amount of failure rate. The calculation of relevant parameters is shown as follows.

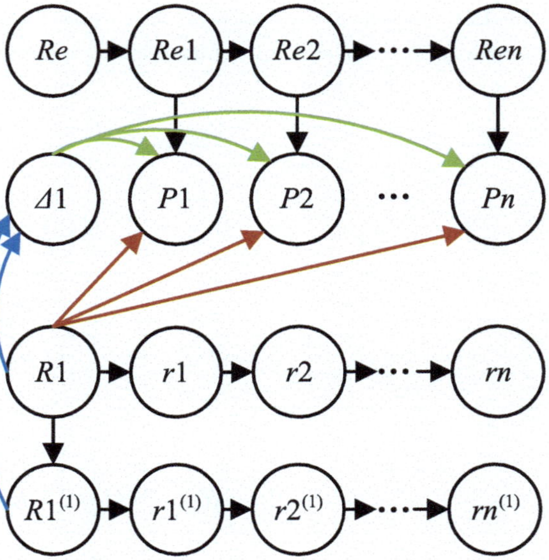

Fig. 12.5 Model construction of "Re-prediction 1"

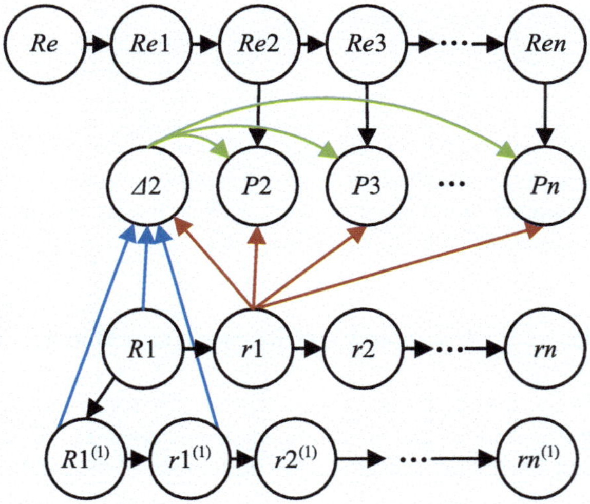

Fig. 12.6 Model construction of "Re-prediction 2"

$$\Delta 1 = \frac{\left(R1^{(1)} - - R1\right)}{2} \tag{12.25}$$

$$P1 = R1 + 0.9 \times \Delta 1 + 0.1 \times \mathrm{Re}1 \tag{12.26}$$

$$P2 = R1 + 0.9 \times 2 \times \Delta 1 + 0.1 \times \mathrm{Re}2 \tag{12.27}$$

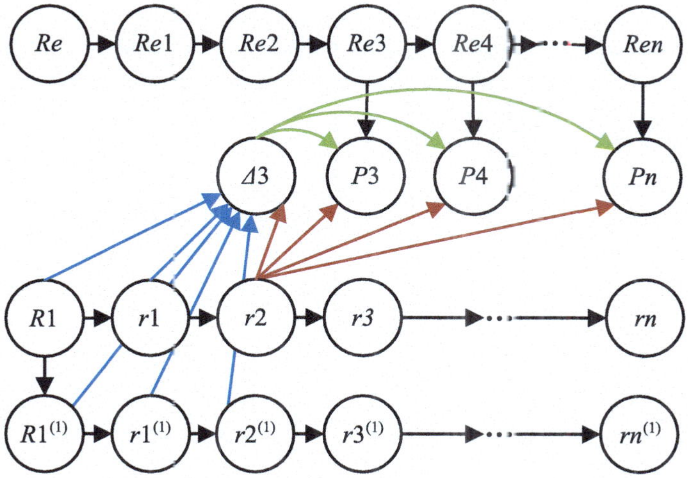

Fig. 12.7 Model construction of "Re-prediction 3"

$$Pn = R1 + 0.9 \times n \times \Delta 1 + 0.1 \times Ren \quad (12.28)$$

Figure 12.6 shows the "re-prediction 2" model, the Pn is the n-th time slice of the prediction results in "Re-prediction 2", Δ2 is the basic degradation considering one set of history data, n is the time slice, $R1$ for initial historical monitoring data, $R1^{(1)}$ is the initial predicted value of the degradation based on the Wiener process, $r1$ is the historical monitoring data under the second time slice, $r1^{(1)}$ is the predicted value under the second time slice of the degradation based on the Wiener process, Ren is the degradation nth time slice to the overall amount of failure rate. The calculation of relevant parameters is shown as follows.

$$\Delta 2 = \frac{\left(R1^{(1)} - R1\right) + \left(r1^{(1)} - r1\right)}{3} \quad (12.29)$$

$$P2 = r1 + 0.9 \times \Delta 2 + 0.1 \times \mathrm{Re}2 \quad (12.30)$$

$$P3 = r1 + 0.9 \times 2 \times \Delta 2 + 0.1 \times \mathrm{Re}3 \quad (12.31)$$

$$Pn = r1 + 0.9 \times n \times \Delta 2 + 0.1 \times \mathrm{Re}n \quad (12.32)$$

The "re-prediction 3" model is shown in Fig. 12.7, the Pn is the n-th time slice of the prediction results in "Re-prediction 3", Δ3 is the basic degradation considering one set of history data, n is the time slice, $R1$ for initial historical monitoring data, $R1^{(1)}$ is the initial predicted value of the degradation based on the Wiener process, $r1$ is the historical monitoring data under the second time slice, $r1^{(1)}$ is the predicted value under the second time slice of the degradation based on the Wiener process, $r2$ is the historical monitoring data under the third time slice, $r2^{(1)}$ is the predicted value of the degradation based on wiener process under the third time slice, *Ren* is the degradation *n*th time slice to the overall amount of failure rate. The calculation of relevant parameters is shown as follows.

$$\Delta 3 = \frac{\left(R1^{(1)} - R1\right) + \left(r1^{(1)} - r1\right) + \left(r2^{(1)} - r2\right)}{6} \tag{12.33}$$

$$P3 = r2 + 0.9 \times \Delta 3 + 0.1 \times \text{Re}3 \tag{12.34}$$

$$P4 = r2 + 0.9 \times 2 \times \Delta 3 + 0.1 \times \text{Re}4 \tag{12.35}$$

$$Pn = r2 + 0.9 \times n \times \Delta 3 + 0.1 \times \text{Re}n \tag{12.36}$$

12.3.3 Results and Discussions

12.3.3.1 Degradation Prediction Based on Physical Models

As is shown in Fig. 12.8, the predicted value of the overall failure efficiency of the system and the online monitoring status are compared. Using Bayesian belief network to construct the regression forecasting model, the degradation process is divided into several phases. The determination of the number of stages should theoretically be learned from the historical data. According to the overall trend of existing data in the whole life cycle of the system database by means of expert experience, the three stages are divided. It can be seen that the degradation trend of the system has a more obvious change from the 5th time slice and the 20th time slice, respectively. Therefore, when the prediction reaches the 5th or 20th time slice, the prediction of this stage ends and jumps to the next stage for prediction. The predicted results of the 5th and 20th time slices are calculated as the initial values of the next stage respectively. $m = 1$–5 is the initial stage of degradation, the degradation rate of this stage is faster; $m = 6$–20 is the middle stage of degradation, in which the degradation rate tends to be flat. $m = 21$–30 is the late stage of degradation, which is faster than the initial stage. Where m stands for time slice, and each m is half a year. The whole degradation process is in accordance with the engineering practice. In the early use of the system, the degradation rate is faster because the components are running in with each other, then the degradation process tends to be smooth, with the long-term use of the system, the loss of each component increases and the degradation rate increases. With the passage of time, the failure rate is only considered modeling method of forecasting results and actual monitoring value gap increases gradually, the degradation value

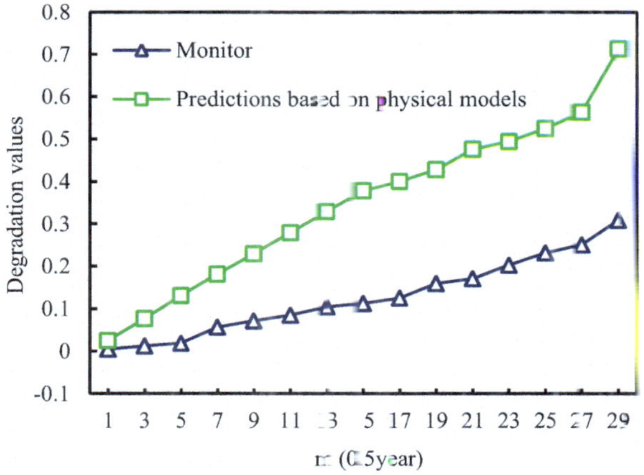

Fig. 12.8 Comparison of the initial predicted value and the monitored value

ended up being more than twice the monitored value. So, considering the modeling method of failure rate is only applicable to the short-term prediction, in the process of long-term prediction, this method needs to be improved.

12.3.3.2 Degradation Prediction Based on Wiener Process

The comparison of incremental degradation predictions values and monitored degradation values in the first stage, second stage and third stage are described in Fig. 12.9. The incremental degradation prediction represents the predicted value of the degradation at each moment from the previous moment, and the cumulative degradation prediction represents the sum of the initial degradation, and the degradation increment at the first n moments.

It can be seen from the figures that the cumulative predicted value at each moment of each stage is higher than the monitored value. Among them, the predicted degradation in the first stage basically showed a linear increase, with the overall increase of about 0.04. Compared with the increase of 0.015 in the monitored value, the difference was relatively obvious. Due to the small number of time slices, the error was relatively large. Incremental predictive value in the second stage degradation growth increases gradually with time, the change trend is relatively flat, the cumulative amount of degradation prediction exponential growth, in the first six time slices are almost consistent with the monitoring value, the predicted results are more accurate, as time changes, prediction and monitoring values gap increases gradually, the final result about monitoring and estimated values of about 1.57 times, the prediction results in failure rate prediction results more accurate. In the third stage, the predicted value of cumulative degradation is growing exponentially, and the final predicted value is about 1.3 times the monitored value, less than the previous two stages. To sum up, in each stage, the accuracy of the cumulative predicted value decreases gradually with the change of time, but in terms of the whole degradation

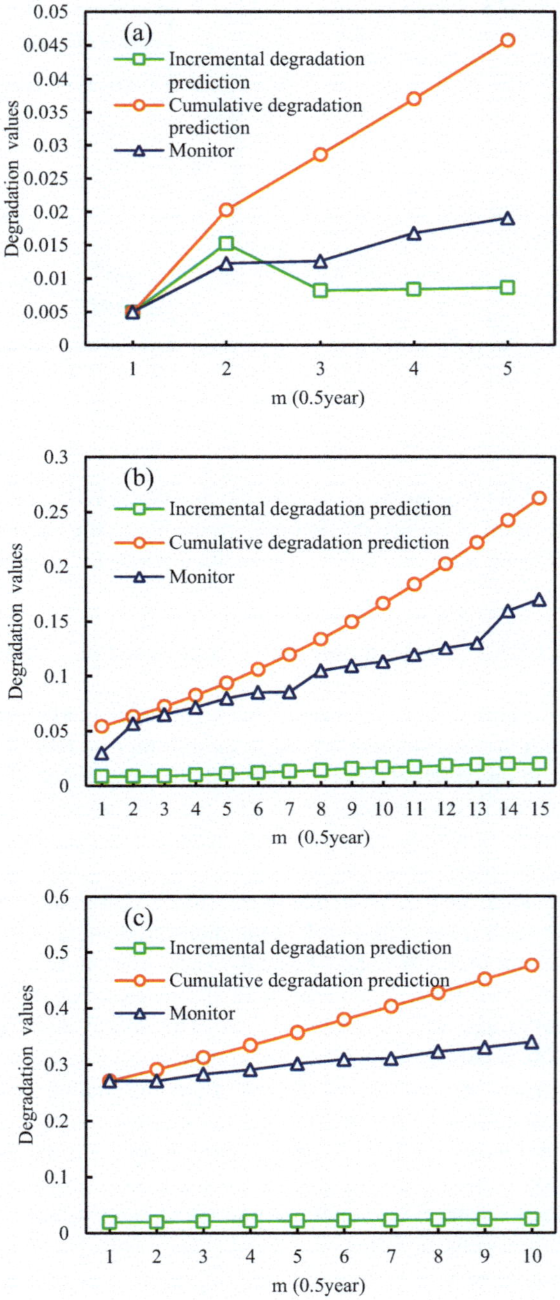

Fig. 12.9 Comparison of predicted and monitored degradation values **a** in first stage **b** in second stage **c** in third stage

process, the accuracy of the predicted value improves. This means that the RUL prediction method based on Wiener process is effective, however, due to the limitations of Wiener process, its prediction results are only relevant to the current state, not the historical degradation data, and there are still some errors between the prediction results and the subsequent monitoring state. In this respect, this method still needs to be improved, which will be introduced in detail in the following part.

Combined with the error of the predicted value compared with the monitored value in each stage, substitute the data into formula $Ri_{n+1}^{(1)} = (1 - c) \times Ri_n$ to simplify and modify, the prediction formula of the performance degradation in the three stages can be obtained as follows. And the new predictive value of performance degradation is called simplified predictive value.

$$R1_{n+1}^{(1)} = (1 + 0.905456383) \times R1_n \tag{12.37}$$

$$R2_{n+1}^{(1)} = (1 + 0.404597775) \times R2_n \tag{12.38}$$

$$R3_{n+1}^{(1)} = (1 + 0.212829063) \times R3_n \tag{12.39}$$

The simplified prediction results are shown in Figs. 12.10, 12.11 and 12.12. Figures 12.10a, 12.11a and 12.12a are the overall prediction results, and Figs. 12.10b, 12.11b and 12.12b are the local enlarged images of corresponding stages, in which, $m = 5$–6, $m = 15$–16 and $m = 10$–11 are respectively partially enlarged. Each stage is estimated backward one time slice in turn. The solid line represents the initial prediction results, and the dotted line represents the simplified prediction results.

As shown in Fig. 12.10 for the first stage of simplified predicted results of the actual situation and local amplification figure. The predicted value by simplified calculation is 0.0088 less than the predicted value based on the Wiener process in the previous time. When $m = 5$, the predicted value of cumulative degradation based on Wiener process is about 2.4 times the monitored value. After simplifying and revising, when m is 6, the predicted value is about 1.9 times of the monitored value, which improves the accuracy of the predicted results to some extent.

The second stage of simplified predicted results of the actual situation and local amplification figure is shown in Fig. 12.11. The predicted value by simplified calculation is 0.02 less than the predicted value based on the Wiener process in the previous time. When m is 15, the predicted value of cumulative degradation based on wiener process is about 1.5 times the monitored value. After simplifying and revising, when m is 16, the predicted value is about 1.4 times of the monitored value, the accuracy of the prediction results is improved and higher than that of the first stage.

As shown in Fig. 12.12 for the third stage of simplified predicted results of the actual situation and local amplification figure, the predicted value by simplified calculation is 0.06 less than the predicted value based on the Wiener process in the previous time. When m is 10, the predicted value of cumulative degradation based on wiener process is about 1.4 times the monitored value. After simplifying and revising, when

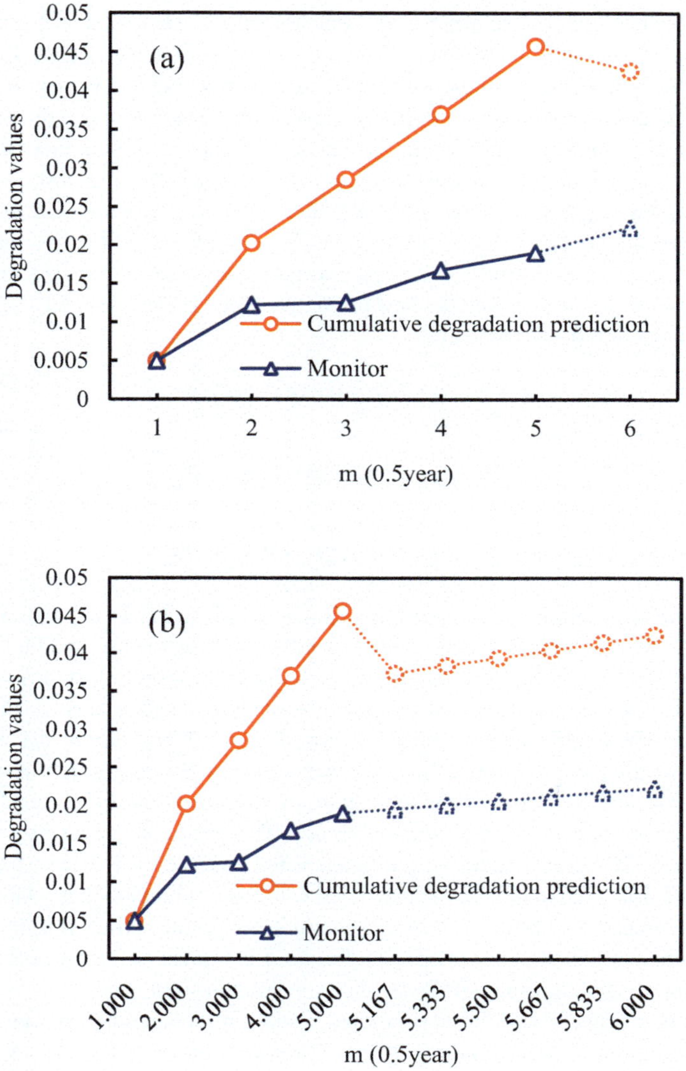

Fig. 12.10 Predicted and monitored values in the first stage **a** normal degradation **b** local enlarged degradation

m is 11, the predicted value is about 1.2 times the monitored value, the accuracy of the prediction results is improved and higher than the previous two stages.

In conclusion, the simplified predicted results at each stage are significantly reduced, which is basically parallel to the monitored value. From the first stage to the third stage, with the increase of the data used, the gap between the simplified

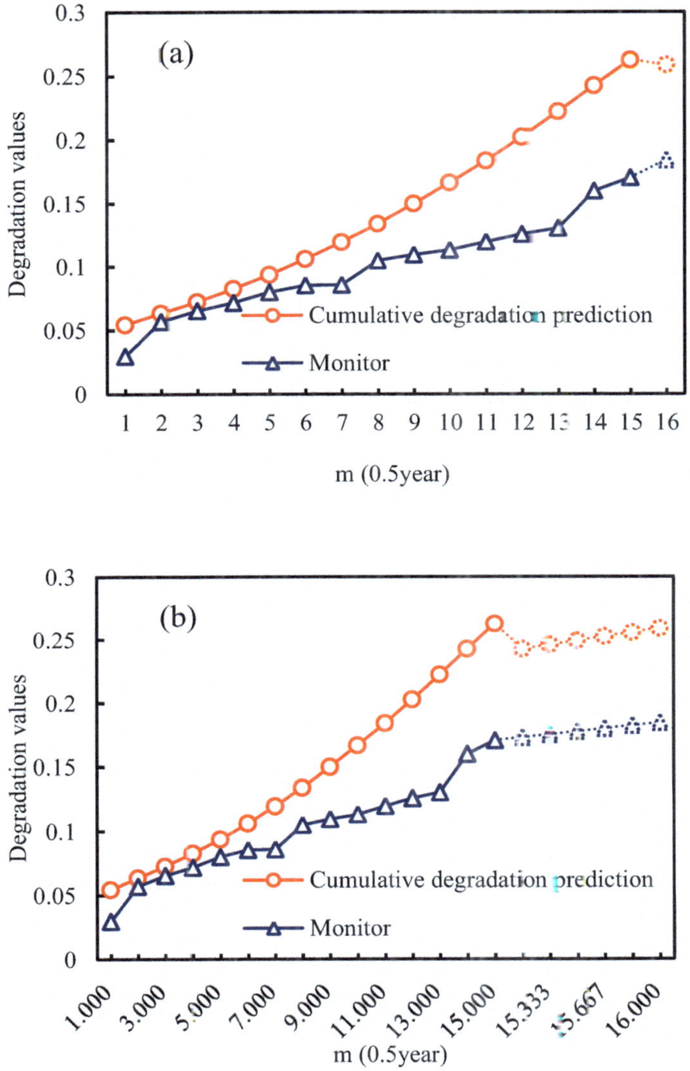

Fig. 12.11 Predicted and monitored values in the second stage **a** normal degradation **b** local enlarged degradation

prediction results and the monitored values decreases. This indicates that the increase of data volume has a great impact on the prediction results.

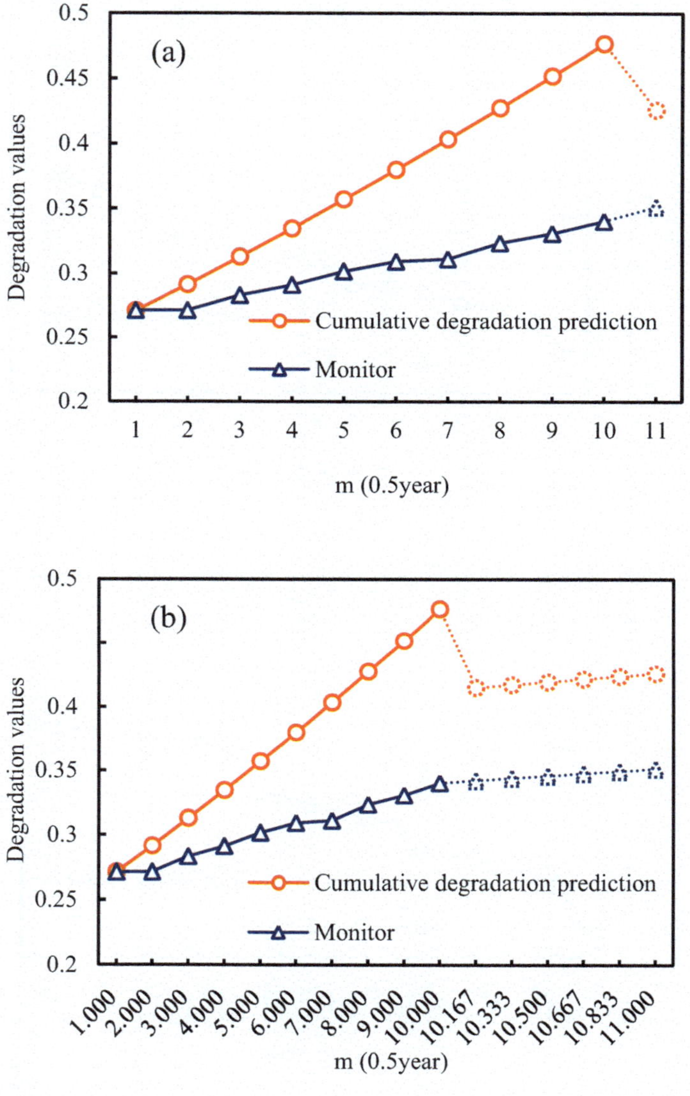

Fig. 12.12 Predicted and monitored values in the third stage **a** normal degradation **b** local enlarged degradation

12.3.3.3 Re-prediction of Degradation Under Different Models

There is still a gap between the predicted performance degradation value based on Wiener process and the actual monitoring value. On this basis, the prediction model proposed in this chapter is adjusted, which is called re-prediction model, and the re-prediction results are improved to some extent. The re-prediction results under different prediction models in the first stage, second stage and third stage are shown in the Fig. 12.13a–c respectively. The RUL re-prediction results of the subsea Christmas tree system are obtained by combining the basic degradation quantity Δ at each stage based on Wiener process with the monitored and predicted performance degradation values of multiple groups. *Here, the multiple groups mean that multiple sets of data collected at different degradation moments corresponding to the same degradation stage.* Among them, combined with a set of data, the result is "re-prediction 1", the result obtained by combining the two sets of data is "re-prediction 2" and the result obtained by combining the three sets of data is called "re-prediction 3". As can be seen from the figure, the re-prediction data are all between the predicted data of initial cumulative degradation and the monitoring data. With the increase of combined data, the predicted results gradually approach the actual monitored value, and with the increase of time, the difference between the re-prediction result and the monitored value gradually decreases, which is far less than 2 times the monitored value, making the predicted results more accurate. A confidence interval of 30% is set. In the first stage, the re-prediction results deviate from the confidence intervals due to the absence of data. But in the second stage and the third stage, both the re-prediction results are within confidence intervals, the results of cumulative degradation prediction are outside the confidence intervals which means that the methodology of re-prediction is excellent.

It can be seen from Fig. 12.13a that in the first stage, the curve "re-prediction 1" is the closest to the monitored value. When m equals to 4, the values of the three re-prediction curves begin to approach, and when m is 5, "re-prediction 2" and "re-prediction 3" basically coincide. As can be seen from Fig. 12.13b, in the second stage, the prediction effect is more obvious due to the increase of monitoring data. At the initial time of the re-prediction, the predicted value increases sharply, all of which are higher than the predicted value of the cumulative degradation. However, from the next moment, the change of the re-predicted value slows down and varies roughly linearly. When m equals to 7, the re-prediction value begins to be lower than the predicted value of the cumulative degradation value. In addition, the gap between curve "re-prediction 1" and curve "re-prediction 2", curve "re-prediction 3" gradually increases, and the final re-predicted value is about 1.12 times as much as the monitored value when m is 6. The curves of "re-prediction 2" and "re-prediction 3" gradually coincide with each other over time since m is 5. It can be seen from Fig. 12.13c that the curves "re-prediction 1" and "re-prediction 2" are more consistent, and the difference between the final re-predicted value and the monitored value is less than 0.005. When m equals to 5, the three re-prediction curves begin to approach. This also provides the guiding significance for engineering practice. In the first 2.5 years, each component of the system is in the working running-in period, with a rapid degradation rate. In this stage, attention should be paid to the maintenance of the system. In addition, due to the unstable work in this stage, which is prone to abrupt changes, close attention should be paid to the operation status, and the sensor data collection should be relatively intensive. Between 2.5 and 10 years, the degradation of the system flattens out, and attention to this phase can be modestly reduced. After the 10th year, components tend to age and the degradation of the system enters into an accelerated state, facing

Fig. 12.13 Prediction results under different prediction models **a** in the first stage **b** in the second stage **c** in the third stage

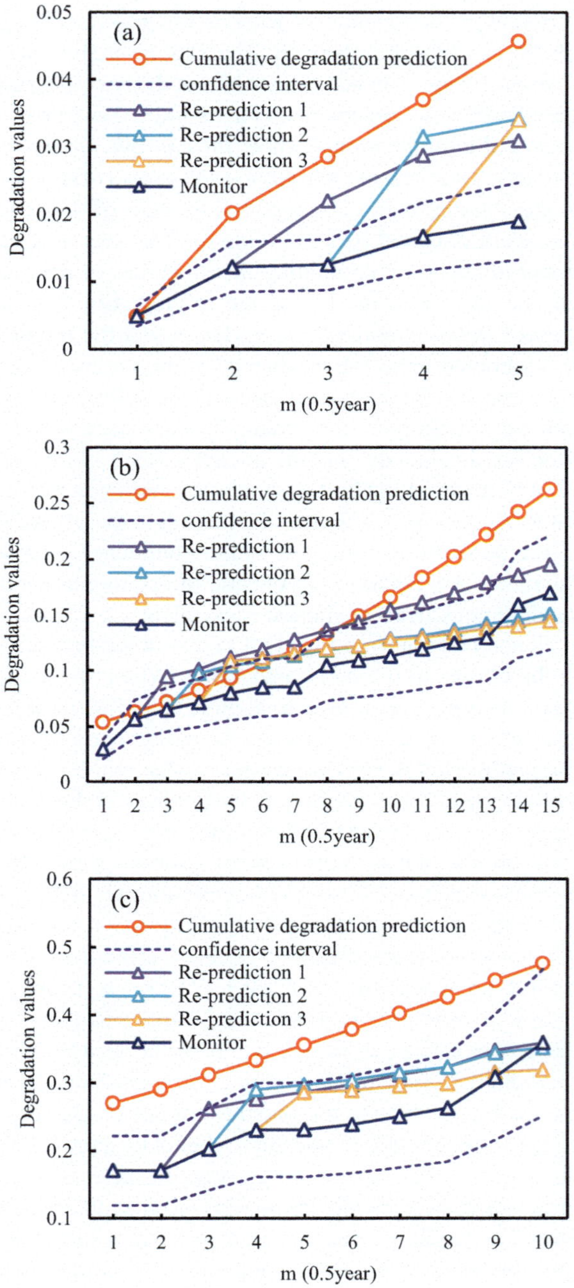

the risk of failure. Therefore, it is necessary to always pay attention to the relevant data in this stage to avoid the occurrence of catastrophic stocks.

To sum up, the effect of re-prediction is obviously better than the traditional degradation prediction based on Wiener process. The re-prediction result is basically stable when the combined historical degradation data reaches three or more groups.

12.3.3.4 Calculation of RUL

RUL of system can be obtained by failure threshold. There are two ways to define the failure threshold: reliability and performance degradation. In terms of performance degradation, the failure threshold is 0.3, which is determined by historical data and expert experience. The RUL of the subsea Christmas tree system can be obtained as shown in the Fig. 12.14. The RUL under the monitoring condition (Monitor) is 15 years. In the case of Predictions based on physical models (P-phy), the RUL suddenly changes to 5 years, because it only considers the historical failure efficiency data and ignores the online status, the prediction is far from the reality. The predicted RUL of the system based on Wiener process considering only the cumulative performance degradation (P-cumu) is 11.25 years. Combined with a set of initial predicted values and monitored values from the previous moment, the re-predicted value of RUL based on Wiener process (Re-1) is 13 years. Combined with the two sets of initial predicted values and the monitored values data at the previous two moments, the RUL predicted value (Re-2) based on the Wiener process is 12.5 years. Combined with the first three moments initial prediction values and monitored values, the RUL prediction based on Wiener process of "Re-3" is 14 years, which is closest to the RUL value of the online monitoring status. Combining as many sets of historical degradation data as possible, the prediction results will be more accurate.

The degenerate track of RUL over time is shown in Fig. 12.15. The RUL of the system decreases linearly over time. The RUL of "P-phy" drops to 0 in the 9th year;

Fig. 12.14 RUL of the subsea Christmas tree system

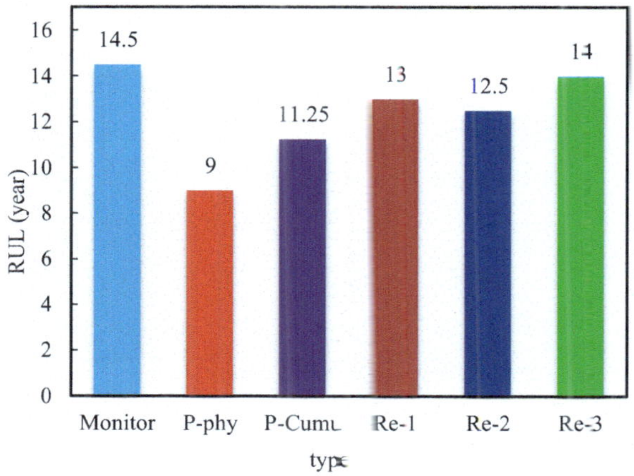

The RUL of "P-cumu" drops to 0 in the 11th year, while the RUL of "Monitor" and "Re-3" drops to 0 in the 14th year. This also illustrates that the RUL re-prediction method based on the Wiener process makes up for the defect that the degradation process of the traditional Wiener process is only related to the current state and has nothing to do with historical data. Furthermore, for complex systems, the method based on physical models is difficult to predict accurately.

In addition, the change of failure threshold will also affect the prediction results. The threshold is set to 0.24, 0.26, 0.28, 0.3(default) and 0.32 in this chapter. The corresponding RUL is calculated as shown in Fig. 12.16, and RUL are 12.2 years, 12.25 years, 12.4 years, 14 years, and 14.95 years respectively. This shows that the determination of failure threshold has a certain influence on RUL. Therefore, how to accurately determine the failure threshold is still a challenge.

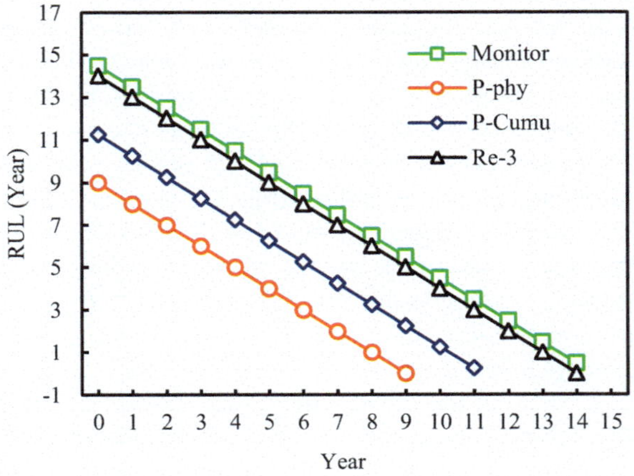

Fig. 12.15 The influence of different prediction methods on RUL

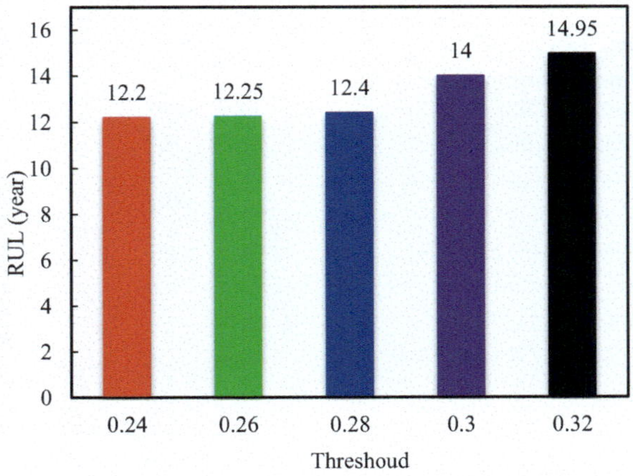

Fig. 12.16 Sensitivity analysis results of threshold

Chapter 13
Fault Prediction with Industrial Incomplete Information

Abstract This chapter presents a novel approach for fault prediction in industrial systems, specifically designed to handle incomplete and missing data, which often undermine the accuracy of predictive models. The method leverages artificial intelligence to address incomplete information by dividing it into incomplete variables and missing variables, each managed with tailored solutions. A DBN is used to integrate complete data with uncertain information, effectively compensating for data integrity issues and reducing prediction errors. The approach also incorporates advanced techniques such as parameter uncertainty analysis, sensitivity analysis, and dynamic range analysis, enhancing the precision and comprehensiveness of predictions. The method's effectiveness is demonstrated through its application to predicting CO_2 corrosion in subsea pipelines, where it significantly improves fault prediction accuracy even in the presence of incomplete data. This approach sets a new benchmark in fault prediction, providing a reliable solution for industrial systems where incomplete data is a common challenge.

13.1 Introduction

The motivation for this research stems from the critical need to address the issues associated with incomplete data in industrial fault prediction. Despite advancements in data collection and sensor technologies, a large proportion of industrial information remains incomplete, either due to missing data or partial degradation parameters that cannot be directly monitored. Traditional approaches to handling missing data, such as data deletion or mean imputation, often lead to significant loss of information integrity, distorting the prediction outcomes. Although machine learning techniques have shown promise in fault prediction by leveraging multivariate data, they struggle when faced with insufficient or incomplete information. In particular, the inability to accurately assess the full scope of equipment degradation due to missing or uncertain data complicates predictive maintenance efforts. Therefore, there is a pressing need

B. Cai et al., *Intelligent Operation and Maintenance for Subsea Production Systems*,
https://doi.org/10.1007/978-981-96-2677-9_13

for a more refined approach that can not only handle missing data but also improve prediction accuracy by incorporating dynamic parameter uncertainty and degradation mechanisms.

The primary contribution of this chapter is the development of a novel hybrid-driven prediction model that effectively handles incomplete data, integrates prior knowledge, and incorporates the uncertainty inherent in industrial systems. Unlike traditional methods that rely solely on statistical models or simplistic data imputation techniques, the proposed approach leverages a combination of data-driven and model-driven strategies. By incorporating DBN and ARIMA algorithms, the model fills missing data gaps and accounts for uncertainty in degradation behavior. Additionally, the method introduces virtual critical variable data based on degradation models to enhance the accuracy of predictions. This hybrid model dynamically updates parameters to provide a more comprehensive and accurate estimation of equipment performance, significantly improving fault prediction and RUL estimation in complex industrial settings.

This chapter introduces the proposed predictive framework for handling incomplete information in industrial fault prediction tasks. It outlines the theoretical background of the hybrid data-model approach, detailing the integration of machine learning techniques with physical degradation models to address missing data and parameter uncertainty. The chapter then discusses the key methodologies employed in the development of the prediction model, including the application of ARIMA for data pre-filling, the construction of virtual critical variables, and the use of LSTM networks for time series forecasting. Finally, the chapter highlights the advantages of the proposed model in terms of prediction accuracy, adaptability, and its ability to handle complex industrial data with missing or uncertain features.

13.2 Proposed Fault Prediction Method

The fault prediction method proposed in this chapter mainly includes industrial information preprocessing, data enhancement with incomplete information, and failure and RUL estimation. The method implementation process, as shown in Fig. 13.1, involves several steps. Firstly, incomplete industrial data is identified and divided into incomplete variables and missing variables through degradation patterns and physical models. Incomplete variables are pre-filled and verified before being output to the BN. Missing variables are constructed as BN nodes with uncertain changes based on prior information such as expert experience, historical data, and databases. An extended DBN over time allows for enhanced performance evaluation, and long-term time series prediction is then implemented using this model. The predicted performance and failure threshold provide critical information for obtaining RUL, which along with system failure information and RUL distribution, can be used for subsequent maintenance decision making.

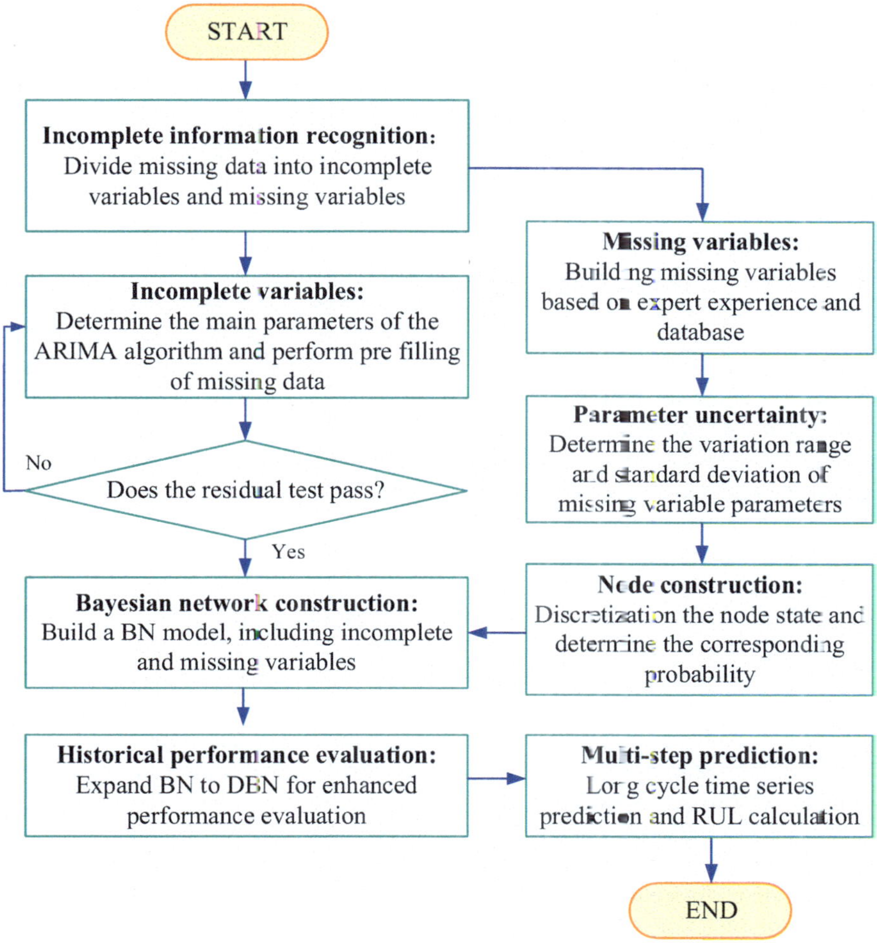

Fig. 13.1 Flow chart for prediction based on incomplete information

13.2.1 Industrial Information Preprocessing

The preprocessing of industrial information includes the collection of preliminary data and the identification of incomplete information. The prior information mainly consists of the degradation model of experience, the approximate range of characteristic parameters obtained from expert experience and some critical evidence information.

Generally, the degradation model of equipment can be expressed as follows:

$$d_t = f(a_t^1, a_t^2, \ldots, a_t^i, t) \tag{13.1}$$

where $a_t{}^1$, $a_t{}^2$,...,$a_t{}^i$ are the critical parameters causing degradation at time t.

The evidence information refers to some basic and available information obtained from detection or estimation. Essential knowledge of some variables, including mean, variance, and probability distribution, is obtained by inputting evidence information:

$$[E, V, P] = g(d_t, e^1, e^2, ..., e^k) \tag{13.2}$$

where E is the mean information, V is the variance information, P is the probability information, and e^1, e^2,...,e^k are the evidence information.

In this chapter, incomplete information is defined into two types: incomplete variables and missing variables. The incomplete variables refer to variables with partial missing data in the degradation model. Incomplete variables are usually variables that can be monitored and have fragments missing. The missing variables refer to variables that are not monitored by the sensor. Compared with incomplete variables, missing variables are the absence of all data. The main task of incomplete information identification is distinguishing incomplete and missing variables. If there are x incomplete variables and y missing variables, the overall degradation can be expressed as:

$$d = z[h_1(a_n^1, \cdots, a_n^x), h_2(a_m^1, \cdots, a_m^y), h_3(a_c^1, \cdots, a_c^{i-x-y})] \tag{13.3}$$

where a_n is the incomplete variable, a_m is the missing variable, and a_c is the normal variable.

Figure 13.2 illustrates the differences between the three types of monitoring variables. Among them, normal variables have complete monitoring data, incomplete variables refer to parameters that can be monitored, such as temperature and flow rate, but have significant amounts of missing data due to sensor limitations. These variables are partially available. Missing variables, on the other hand, refer to parameters that are not monitored at all, such as CO_2 content, which is difficult to measure and typically lacks any available monitoring data.

13.2.2 Data Enhancement with Incomplete Information

Data enhancement refers to making the sample data more comprehensive to contain the degradation characteristics of equipment through data expansion, filling, and optimization methods. The data enhancement method proposed in this chapter for incomplete information includes three parts: incomplete variable filling, missing variable supplement, and time series prediction of degradation and failure.

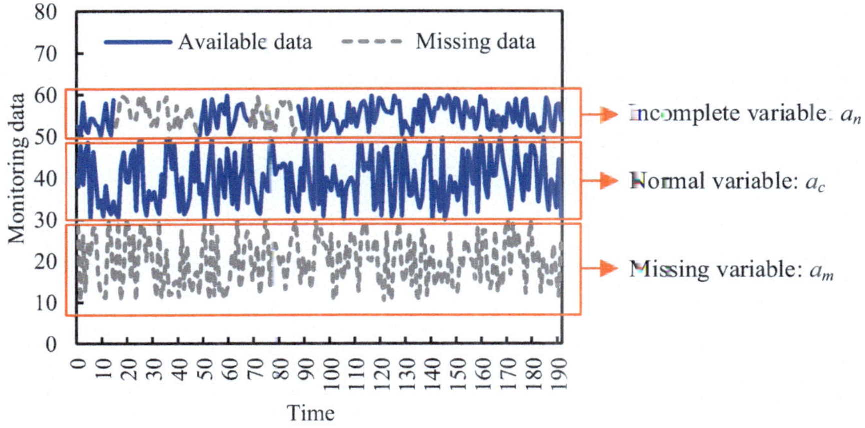

Fig. 13.2 Three types of monitoring information

13.2.2.1 Incomplete Variable Filling Using ARIMA

The ARIMA model is a well-established method for time series prediction and analysis (Fan et al. 2021; Di Mauro et al. 2024). Its selection for data imputation is justified by its strong capabilities in modeling time series data. ARIMA is particularly effective at capturing temporal dependencies and trends, making it an ideal choice for datasets with these characteristics. Its ability to handle non-stationary data through differencing, combined with the intuitive interpretability of its parameters, ensures robust and reliable imputation. By leveraging historical observations, ARIMA excels in scenarios where ample past data is available, thereby enhancing the accuracy of predictions. These strengths collectively make ARIMA a highly suitable and justified choice for our specific imputation needs.

For the p-order AR model, the current value of the time series can be expressed as (Wang et al. 2015):

$$y_t = \Theta_1 y_{t-1} + \Theta_2 y_{t-2} + \cdots + \Theta_p y_{t-p} + \varepsilon_t \tag{13.4}$$

The q-order MA model represents the current value of the time series as the present value of the random error and q previous values, which can be expressed as follows:

$$y_t = \varepsilon_t - \theta_1 \varepsilon_{t-1} - \theta_2 \varepsilon_{t-2} - \cdots - \theta_q \varepsilon_{t-q} \tag{13.5}$$

Therefore, a general expression of the ARMA (p, q) process can be expressed as follows:

$$y_t = \Theta_1 y_{t-1} + \Theta_2 y_{t-2} + \cdots + \Theta_p y_{t-p} + \varepsilon_t - \theta_1 \varepsilon_{t-1} - \theta_2 \varepsilon_{t-2} - \cdots - \theta_q \varepsilon_{t-q} \tag{13.6}$$

where y_t is the predicted value, y_{t-i} are the previous observation, Θ_i are the coefficients associated with each previous observation, θ_i are coefficients associated with previous white noises, ε_t is the normal white noise, and ε_{t-i} are the previous noise.

The ARIMA process can be expressed as follows:

$$\Delta^d y_t = \Theta_1 y_{t-1} + \Theta_2 y_{t-2} + \cdots + \Theta_p y_{t-p} + \varepsilon_t - \theta_1 \varepsilon_{t-1} - \theta_2 \varepsilon_{t-2} - \cdots - \theta_q \varepsilon_{t-q} \tag{13.7}$$

where $\Delta^d y_t$ is a stationary sequence formed by d times difference transformation of non-stationary sequence y_t.

13.2.2.2 Missing Variable Supplement with BN

According to the probability distribution information obtained from the evidence, the missing variable' BN node is established. The average interval and ChiMerge methods are usually used to discretize continuous variables for BN calculation. ChiMerge is a binning algorithm widely used for discretization in data mining and machine learning. It involves merging adjacent intervals based on the Chi-square statistic until a predefined stopping criterion is met. In the ChiMerge method, the chi-square test is used to determine whether there is a significant association between two variables in adjacent intervals.

Suppose there is a contingency table containing observed frequencies and expected frequencies, as shown in Table 13.1:

where O_{ij} represents the observed frequency at level i of variable A and level j of variable B.

The statistic for the chi-square test is:

$$\chi^2 = \sum \frac{(O_{ij} - E_{ij})^2}{Eij} \tag{13.8}$$

where E_{ij} represents the expected frequency of level i of variable A and level j of variable B under independent conditions.

The calculation of expected frequency is usually described as:

Table 13.1 Contingency table for chi-square test

	Level 1 of variable B	Level 2 of variable B	…	Level m of variable B
Level 1 of variable A	O_{11}	O_{12}	…	O_{1m}
Level 2 of variable A	O_{21}	O_{22}	…	O_{2m}
…	…	…	…	…
Level n of variable A	O_{n1}	O_{n2}	…	O_{nm}

$$E_{ij} = \frac{O_i \cdot O_j}{N} \tag{13.9}$$

where O_i represents the sum of observation frequencies in the i-th row, O_j represents the sum of the observation frequencies in the j-th column, and N represents the total observation frequency. Determine whether to merge adjacent intervals based on the set threshold. The detailed process is shown in Algorithm 1

Algorithm 1: ChiMerge method for discretization scheme
Inputs: Continuous Variable
Outputs: Discretization scheme
1: Divide continuous variables into initial intervals
2: Perform a chi-square test on adjacent intervals
3: **if** Chi-square statistic less than threshold?
4: Merge adjacent intervals
5: **else** Do not merge, continue to the next step
6: **if** Stopping criterion reached (it may be that the number of merges reaches a predetermined value or the set minimum number of intervals is reached)
7: End process
8: **else** Go back to the Chi-square test step
9: Based on the results of chi-square merging, the final discretized intervals are obtained
10: Use the discretized intervals to establish BN nodes related to missing variables

13.2.2.3 Time Series Prediction with DBN and LSTM

The DBN is particularly advantageous for this application because it allows for the modeling of complex dependencies between variables and can accommodate uncertainty in the input data. This is critical in scenarios where some degradation parameters cannot be directly measured, leading to completely missing data, while other parameters suffer from partial data gaps. Traditional approaches, such as standard regression models or neural networks, may fail to account for these complexities adequately. In contrast, the DBN framework not only integrates available data but also updates the probabilistic estimates of the unknown variables based on the observed evidence, leading to more accurate predictions (Zeng et al. 2023).

Several studies (Loughney and Wang 2018; Xu et al. 2023) have demonstrated the effectiveness of DBNs in similar applications, particularly in handling uncertainty and incomplete information in industrial settings. These studies provide a strong foundation for the use of DBNs in the proposed method. Furthermore, compared to other techniques like Markov models or simple machine learning algorithms, DBNs offer a more comprehensive approach to managing the intricacies of incomplete data, making them particularly suitable for this chapter.

The construction of a DBN primarily involves structure modeling and parameter modeling, which are key steps in DBN application. These steps determine how the network represents the dependencies between variables and how data is used for inference and prediction. Structural modeling defines the dependencies between variables (nodes) in the network, represented by directed edges in the graph. This begins with identifying all key variables related to the research objectives, such as equipment degradation factors, environmental conditions, and operating parameters, which are modeled as nodes in the DBN. Time is represented through multiple time slices, each indicating the system state at a specific moment, with nodes reflecting the state of variables at those moments. These nodes can be observed (known data) or hidden (data to be inferred), and directed edges connect nodes to depict causal or dependency relationships. For example, a variable at a given time may depend on the same variable or other related variables from the previous time slice. The initial structure can be established using expert knowledge, prior information, or data-driven methods such as structural learning algorithms. Besides dependencies within the same time slice, connections between adjacent time slices are defined to capture the system's temporal dynamics, such as a degradation variable at time $t + 1$ depending on its state at time t.

Parameter modeling specifies probability distributions and conditional probabilities for each node and edge in the model. Each node's conditional probability distribution is defined based on its own state and the states of its parent nodes. Discrete variables use a conditional probability table (CPT), while continuous variables may use Gaussian distributions or other suitable probabilities. CPTs between time slices are derived from the physical model by first discretizing the child nodes and incorporating previously discretized parent performance variables. Relationships between all performance nodes and variable nodes are determined based on the physical model, with parent nodes sampled using methods like Monte Carlo simulation. Finally, CPTs are obtained using the physical relationships and corresponding probabilities, ensuring sufficient sampled values.

The data enhancement of samples is realized by filling incomplete variables and constructing missing variables. The enhanced data is input into the DBN to evaluate the overall degradation process. The initial condition of DBN is built as follows:

$$P(x_0) = \prod_{i=1}^{n} P(x_0^i | Pa(x_0^i)) \tag{13.10}$$

The conditional probability transition of adjacent time slices is built as follows:

$$P(x_t | x_{t-1}) = \prod_{i=1}^{n} P(x_t^i | Pa(x_t^i)) \tag{13.11}$$

The posterior probability of the DBN node is built as follows:

$$P(x_{0:t}) = P_{B_0}(x_0)\prod_{i=1}^{t} P_{B\rightarrow}(x_t|x_{t-1}) \quad (13.12)$$

where $x_{0:t} = \{x_0, x_1,\ldots,x_t\}$is the i-th node of $x_t{}^i$ at time t; $P_{B\rightarrow}$ is the conditional probability transition between two adjacent nodes of DBN; $Pa(x_t{}^i)$is the parent node of $x_t{}^i$.

The enhanced degradation information evaluated with complete information is more suitable for time series prediction of the degradation state because uncertain information and enhanced data are fully considered. The LSTM network is a kind of time-recurrent neural network. LSTM network can detect short-term and long-term dependence, which is one of the most effective methods in the field of time series prediction (Aydemir and Paynabar 2020). In the LSTM network, there are three control gates to control information update: forget gate $\boldsymbol{f}_t$, input gate $\boldsymbol{i}_t$, and output gate $\boldsymbol{o}_t$. The forget gate is a layer of the neural network, which is written as:

$$\mathbf{f}_t = \sigma(\mathbf{W}_f \cdot [\mathbf{h}_{t-1}, \mathbf{x}_t] + \mathbf{b}_f) \quad (13.13)$$

where σ is the Sigmod activation function, $\boldsymbol{W}_f$ is the forget gate matrix, $\boldsymbol{h}_{t\text{-}1}$ is the hidden layer output of the previous time, $\boldsymbol{x}_t$ is the input of time t, and $\boldsymbol{b}_f$ is the corresponding biases vector.

The input gate determines the throughput of the information to be updated, which can be expressed as:

$$\mathbf{i}_t = \sigma(\mathbf{W}_i \cdot [\mathbf{h}_{t-1}, \mathbf{x}_t] + \mathbf{b}_i) \quad (13.14)$$

where $\boldsymbol{W}_i$ is the input gate connection matrix, and $\boldsymbol{b}_i$ is the corresponding biases vector.

The $\underset{t}{\tilde{\mathbf{c}}}$ is generated by inputting $\boldsymbol{x}_t$ and $\boldsymbol{h}_{t\text{-}1}$ through a tanh layer and can be written as:

$$\underset{t}{\tilde{\mathbf{c}}} = \tanh(\mathbf{W}_c \cdot [\mathbf{h}_{t-1}, \mathbf{x}_t] + \mathbf{b}_c) \quad (13.15)$$

where $\boldsymbol{W}_c$ is the connection matrix, and $\boldsymbol{b}_c$ is corresponding biases vector.

The $\boldsymbol{c}_t$ is a new memory unit state, which is composed of the last time memory unit state determined by the forget gate and the new unit value determined by the input gate, and can be expressed as:

$$\mathbf{c}_t = \mathbf{f}_t * \mathbf{c}_{t-1} + \mathbf{i}_t * \underset{t}{\tilde{\mathbf{c}}} \quad (13.16)$$

Then the value of the output gate $\boldsymbol{o}_t$ is written as:

$$\mathbf{o}_t = \sigma(\mathbf{W}_o \cdot [\mathbf{h}_{t-1}, \mathbf{x}_t] + \mathbf{b}_o) \quad (13.17)$$

where $\boldsymbol{W}_o$ is the output gate connection matrix, and $\boldsymbol{b}_o$ is the corresponding biases vector.

The hidden layer output $\boldsymbol{h}_t$ at the current time is calculated as:

$$\mathbf{h}_t = \mathbf{o}_t * \tanh(\mathbf{c}_t) \tag{13.18}$$

13.2.3 Failure and RUL Estimation

The equipment degradation can be predicted through the time sequence prediction of the LSTM network so that the failure state is predicted. The RUL of equipment is expressed as:

$$\text{RUL} = \inf\{t : X(t) \geq th | X(0) < th\} \tag{13.19}$$

where X is the cumulative degradation, and th is the fault threshold.

The overall framework of the proposed fault prediction method for incomplete information is shown in Fig. 13.3. Firstly, prior information and incomplete information are input into the developed model. Among them, pre filling is performed by solving d, p, and q, and the feasibility of data filling is verified based on residuals. ARIMA model and DBN are solutions for insufficient information and realize variable data enhancement. Missing variables and incomplete variables are integrated into DBN to enhance degraded data. By dividing the training and testing sets, the LSTM network is trained for continuous degradation time series prediction. Then, RUL is calculated based on the predicted degradation model and failure threshold. A more complete failure analysis is further developed.

13.3 Case Study: Corrosion of Subsea Pipelines

Internal corrosion of subsea pipelines is a critical threat to the safety and integrity of pipelines in service, which has caused numerous subsea pipeline failures in recent years, especially for pipelines that are not intelligently detected. Because of the significant changes in temperature and pressure at the inlet and outlet and the corrosive substances such as CO_2, H_2S, O_2, and Cl^- contained in the transport medium, the subsea pipelines are subject to substantial internal corrosion risks. Full-scale corrosion is the most common form of corrosion. For example, CO_2 causes full-scale corrosion, resulting in relatively uniform thinning of pipe wall thickness. CO_2 gas dissolves in an aqueous solution to form carbonic acid, which is highly corrosive to steel materials. There are many factors affecting the CO_2 corrosion characteristics of steel. The main factors are materials, CO_2 partial pressure, temperature, medium composition, flow rate, steel surface facial mask, load, etc. These factors lead to

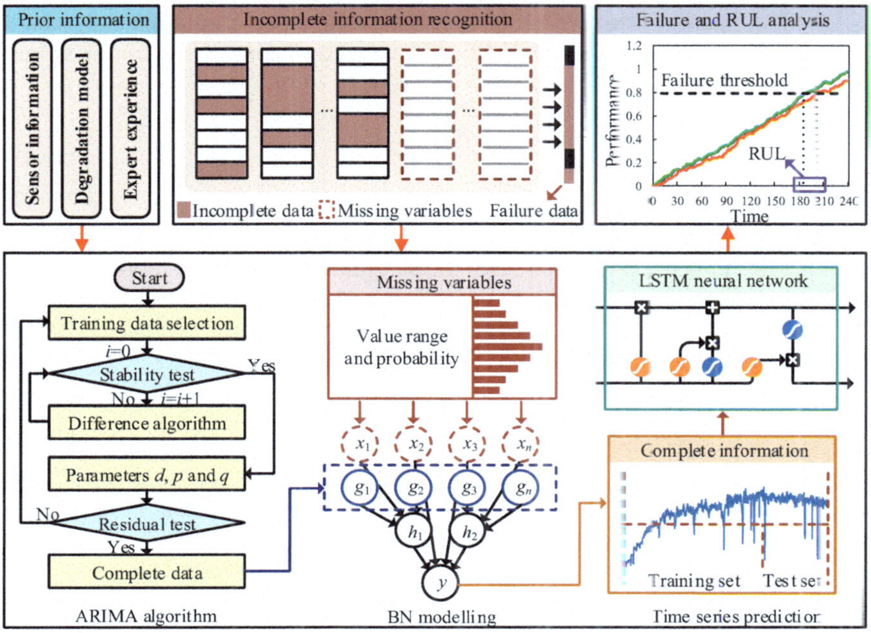

Fig. 13.3 The framework of proposed prediction method

various corrosion damage, high corrosion rate, severe local corrosion, perforation, and even stress corrosion cracking of steel (Zheng, Matin et al. 2016). The proposed method has been validated using extensive, real-world monitoring data from an oil field in the South China Sea, covering the period from 2012 to 2015. This data reflects the changes in key parameters of the target pipeline over several years of operation, ensuring a high level of credibility and relevance for the validation process. The inherent parameters of the target pipeline are shown in Table 13.2.

Table 13.2 Parameters of the pipeline in this case

Parameters	Numerical value
Steel grade	API 5L PSL2 X65
Conveying medium	Oil, gas and water
Diameter	18"/22"
Average thickness	15.9 mm
Design pressure	8100 Kpa
Design temperature	65 °C
Inhibition efficiency	50%
Protection efficiency	50%
Corrosion allowance	5 mm

The CO_2 corrosion rate is usually composed of two parts: fluid-related part and fluid-unrelated part (Bai and Bai 2012). The CO_2 corrosion rate V_c can be written as:

$$V_c = \frac{R_{in}(1 - R_c)}{\frac{1}{V_r} + \frac{1}{V_t}} \tag{13.20}$$

where V_r is the corrosion rate of the part unrelated to the fluid, representing the reaction rate; V_t is the corrosion rate of the part related to the fluid, representing the mass transfer rate; R_{in} is the inhibition efficiency; R_c is the protection efficiency.

The CO_2 corrosion assessment of reaction rate is based on the improved de Waard empirical model mentioned in Chap. 8, which is written as follows:

$$\log(V_r) = 4.93 - \frac{1119}{T + 273} + 0.58 \log(n_{CO2}P) \tag{13.21}$$

where T is the fluid temperature in the pipeline, n_{CO2} is the content of CO_2 in the pipeline, and P is the pressure in the pipeline.

The CO_2 corrosion assessment of mass transfer rate is expressed as:

$$V_t = 2.45 \frac{v^{0.8}}{d_p^{0.2}} n_{CO2}P \tag{13.22}$$

where v is the liquid flow rate, and d_p is the inner diameter of the pipeline.

The liquid flow rate v is usually obtained from the liquid flow:

$$v = \frac{Y}{S_p \cdot t} \tag{13.23}$$

where Y is the liquid yield, and S_p is the cross-sectional area of the pipeline.

The temperature and pressure of the fluid in the pipe are challenging to monitor, and the outlet parameters and inlet parameters are usually used for estimation:

$$T = \frac{T_{in} + T_{out}}{2} \tag{13.24}$$

$$P = \frac{P_{in} + P_{out}}{2} \tag{13.25}$$

where T_{in} and T_{out} are inlet and outlet temperature, respectively, and P_{in} and P_{out} are inlet and outlet pressure, respectively.

Equations above represent physical models for assessing corrosion rates, which rely on the integrity of input parameters. However, it is noted that solely using these models is insufficient for accurately evaluating corrosion rates over time due to two primary reasons. First, the data for various parameters exhibit different extents of incompleteness, leading to significant gaps in the dataset that hinder accurate corrosion assessment. Additionally, the CO_2 content, a critical parameter, is not directly monitored and would be treated as a constant in traditional physical models, which does not reflect the actual fluctuating conditions. To address this, a BN is employed to model CO_2 content as a stochastic variable, capturing its variability. Therefore, a hybrid data-model-driven approach is proposed, where AI is utilized to compensate for incomplete data, enhancing the accuracy of the model's assessments and enabling more reliable long-term predictions.

The monitoring data of the target pipeline contains a lot of incomplete information, which is very suitable for verifying the proposed method. The primary process of CO_2 corrosion prediction is shown in Fig. 13.4. In addition to the monitoring data, the target pipeline is detected every six months, which can be used as evidence for model evaluation.

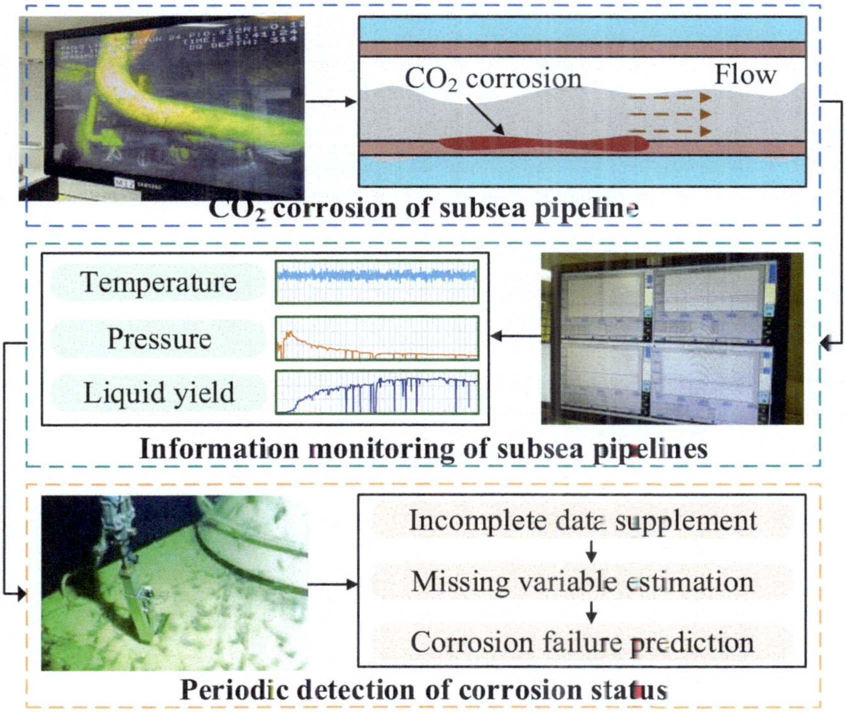

Fig. 13.4 The CO_2 corrosion assessment of target pipeline

13.3.1 Incomplete Information of Monitoring Data

The temperature, pressure, and liquid level data are collected by sensors within the subsea production system's control system. Useful features in the monitoring data include inlet temperature, outlet temperature, inlet pressure, outlet pressure, and daily liquid yield. Through the analysis of the monitoring data, it is found that the above five features have missing data. Incomplete information can arise from factors such as sensor malfunctions, communication interruptions, data storage issues, environmental interference, and delays in maintenance or calibration. The detailed location of missing data for each variable is shown in Fig. 13.5. The available monitoring data of five elements are marked with blue dots, and missing data is marked with red dots. The missing data is summarized in Table 13.3, and each missing part is labeled. In addition, CO_2 content is a missing variable due to a lack of monitoring means.

13.3.2 Data Enhancement of Monitoring Data

The corrosion prediction scheme for subsea pipelines is shown in Algorithm 2. The ARIMA algorithm fills in the missing data of each incomplete variable. The adjacent available data is extracted according to the location and length of the missing data. Then the stability of the available data is verified to determine the number of differences. The optimal p and q are determined according to the Bayesian information criterion criteria. The model parameters of ARIMA are shown in Table 13.4.

Algorithm 2: Data enhancement based on incomplete information
Inputs: Incomplete monitoring data of subsea pipelines
Outputs: Prediction sequence of daily average corrosion
for $i = 0$ to n **do**
Identify incomplete information
end for
Determine the number of incomplete variables l and missing variables k
for $i = 0$ to l do
Determine training data and ARIMA parameters
Using (4)–(6) to supplement incomplete variables
Qualitative and quantitative residual testing
if Test failed
Re select parameters and training data
else output complete data
end for
for $i = 0$ to k do
Using prior information to supplement missing variables

(continued)

(continued)

15:	Convert missing variables to BN nodes
16: **end for**	
17: Building a BN model of corrosion degradation based on supplementary information and (21)–(25)	
18: Expand BN to DBN using (10)–(12) to calculate enhanced	
corrosion data	
19: Using (13)–(18) to predict long-term daily average corrosion	

The verification of the ARIMA model is mainly to verify the fitting effect. Suppose the model thoroughly explains the correlation of data. In that case, the noise sequence of the model is a white noise sequence, and the model's verification is also the noise sequence's independence test. The Quantile–Quantile diagram of the white noise test is shown in Fig. 13.6. The quantity of the sample data (blue points) is roughly distributed near the normal (red straight line) quantity, indicating that the residual error roughly conforms to the normal distribution. The Durbin-Watson (DW) test is used to test the correlation of residuals quantitatively, and the results are shown in Fig. 13.7. When the DW test result is 2, the residual follows the standard normal distribution. The results are valid within the range of 1–3. It is seen that the inspection results of all models are in the red area, indicating that all models passed the inspection.

After passing the residual test, the missing data of incomplete variables are predicted according to the available data, and the missing data is filled. Figure 13.8 shows the filling result of missing data, where the green mark indicates the supplied data.

The liquid flow velocity can be obtained through (13.23), as shown in Fig. 13.9. Through the detection information, it is estimated that the CO_2 content follows the normal distribution of $N(0.3,0.01)$. The known information of temperature, pressure, velocity, and CO_2 content is input into BN, and the estimation model of CO_2 corrosion is constructed first. Then DBN is built through time expansion, and the average daily corrosion is calculated using the enhanced data. The built DBN model is shown in Fig. 13.10. Where the T_1 node is the inlet temperature, the T_2 node is the outlet temperature, the T node is the fluid temperature in the pipeline, the P_1 node is the inlet pressure, the P_2 node is the outlet pressure, the P node is the pressure in the pipeline, the Y node is the output, the v node is the velocity, and the p_{co2} node is the content. The Co node is the daily average corrosion amount.

13.3.3 CO_2 Corrosion Prediction with Enhanced Data

The data filling of incomplete variables and missing variables enables data enhancement. The average daily corrosion can then be implemented using enhanced data, as shown in Fig. 13.11. A total of 10 daily average corrosion data are obtained

Fig. 13.5 Incomplete data for degradation parameters

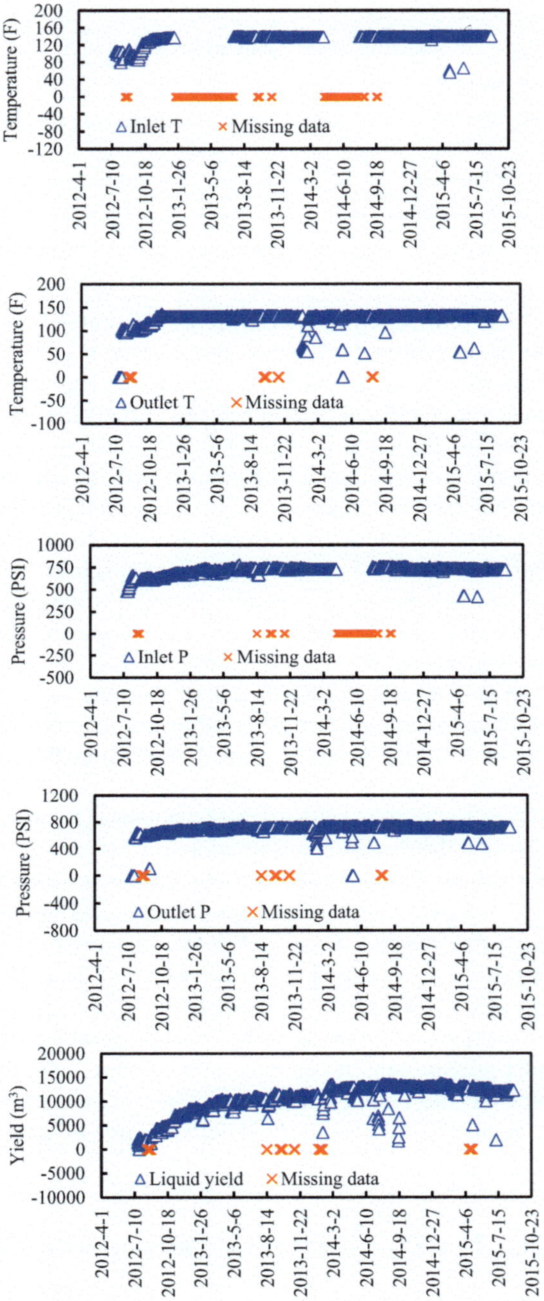

Table 13.3 Incomplete data of monitoring data

Time	Inlet T	Outlet T	Inlet P	Outlet P	Liquid yield
2012/8	1-1	1-2	1-3	1-4	1-5
2013/1-7	2-1				
2013/8			3-3	3-4	3-5
2013/9	4-1	4-2	4-3	4-4	4-5
2013/11	5-1	5-2	5-3	5-4	5-5
2014/1					6-5
2014/4-7	7-1		7-3		
2014/8	8-1	8-2	8-3	8-4	
2014/9	9-1		9-3		
2015/4		10-2			10-5

Table 13.4 Parameters of ARIMA algorithm

No	1-1	1-2	1-3	1-4	1-5	2-1	3-3	3-4	3-5	4-1
p	1	1	1	1	1	2	1	1	1	1
q	1	1	1	3	1	2	1	1	1	1
d	1	1	1	1	1	1	1	1	1	1
No	4-2	4-3	4-4	4-5	5-1	5-2	5-3	5-4	5-5	6-5
p	1	1	1	2	1	1	1	1	1	1
q	1	1	1	1	1	1	1	1	1	1
d	1	1	1	1	1	1	1	1	2	1
No	7-1	7-3	8-1	8-2	8-3	8-4	9-1	9-3	10-2	10-5
p	1	2	1	2	1	1	1	1	1	1
q	3	2	1	1	1	1	1	1	1	1
d	1	1	1	1	1	1	1	1	1	1

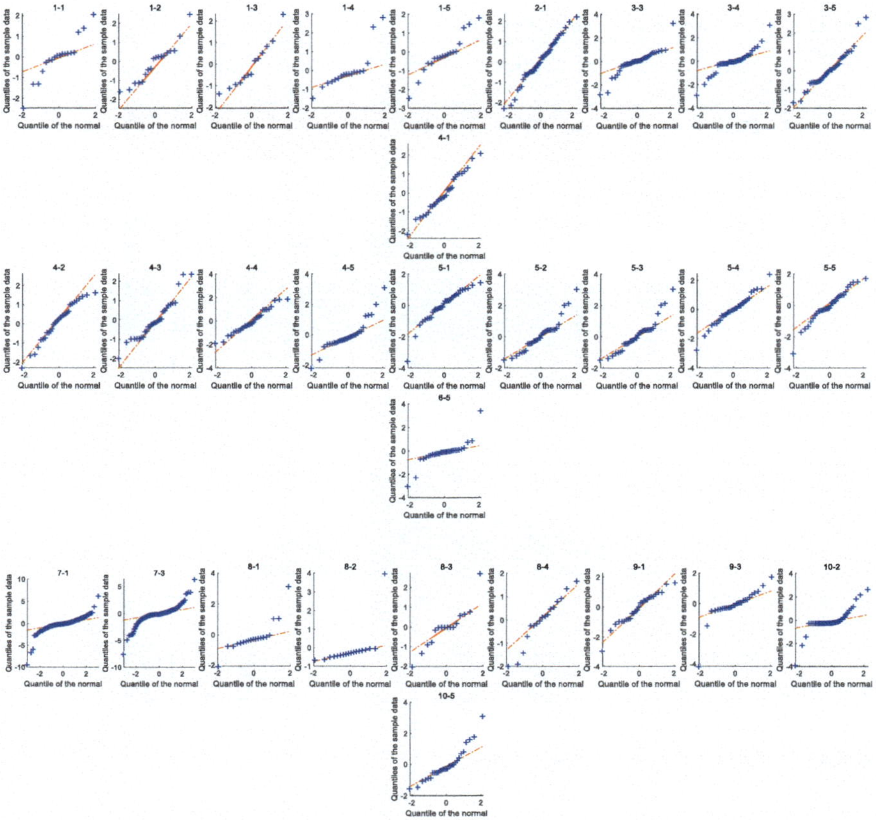

Fig. 13.6 Test of normal distribution of residuals

using incomplete information. The 10 marked areas of different colors represent the calculated corrosion depth after enhancing the data using the proposed method. The evaluated daily average corrosion data is used to calculate the CO_2 corrosion of pipelines. The cumulative CO_2 corrosion amount of the pipeline is shown in Fig. 13.12. The red area is the 95% confidence interval of the estimated cumulative corrosion amount. The proposed data enhancement method is verified by comparing the confidence interval with the detection data. All the detection points are within the confidence interval, indicating that the proposed method can handle the incomplete information in the monitoring data.

The assessed daily average corrosion data is used to implement prediction and to illustrate the confidence of the LSTM network in the time series prediction. 80% of the corrosion data is used as the training set, and the remaining 20% is used as the test set. The single-layer LSTM network is selected. The number of input layer nodes is 1, the number of hidden layer nodes is 200, the number of output layer nodes is 1, the number of training times is 400, the threshold of the gradient is set to 1, the

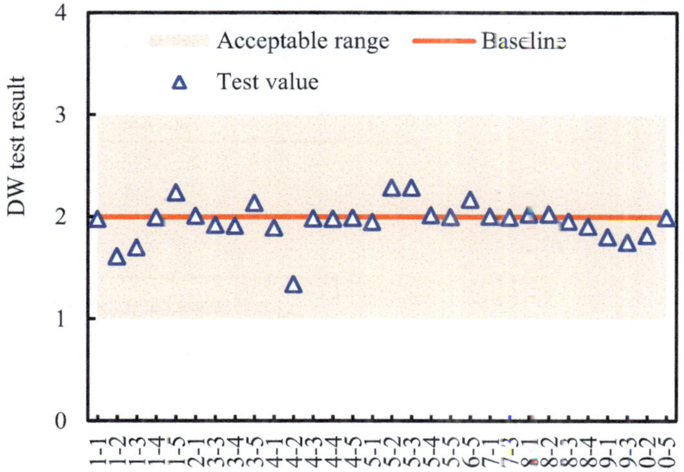

Fig. 13.7 DW test of residuals

initial learning rate specified is 0.005. The learning rate is reduced by multiplying the factor 0.2 after 125 training cycles. The prediction results of the LSTM network are shown in Fig. 13.13. The trend of average daily corrosion predicted by enhanced data is consistent with that of actual daily average corrosion. To better illustrate the advantages of the proposed method, the ARIMA model with enhanced data, LSTM with average filled data, and LSTM with missing data deletion method are used to compare the results with the proposed method. Four evaluation metrics are used to compare the performance of the proposed model: (1) Mean Absolute Error (MAE), which quantifies the average magnitude of prediction errors by computing the average of the absolute differences between predicted and actual values; (2) Mean Square Error (MSE), which calculates the average of the squared differences between predicted and actual values, giving more weight to larger errors; (3) Root Mean Square Error (RMSE), which measures the average magnitude of errors in the same units as the target variable by taking the square root of MSE, thus being sensitive to larger errors; and (4) Mean Absolute Percentage Error (MAPE), which evaluates forecasting accuracy as a percentage by averaging the absolute percentage errors between predicted and actual values, providing a normalized measure of prediction accuracy. These metrics are computed according to the following equations:

$$\mathrm{MAE} = \frac{1}{n}\sum_{i=1}^{n} |\hat{y}_i - y_i| \tag{13.26}$$

$$\mathrm{MSE} = \frac{1}{n}\sum_{i=1}^{n} (\hat{y}_i - y_i)^2 \tag{13.27}$$

Fig. 13.8 Data filling of degradation parameters

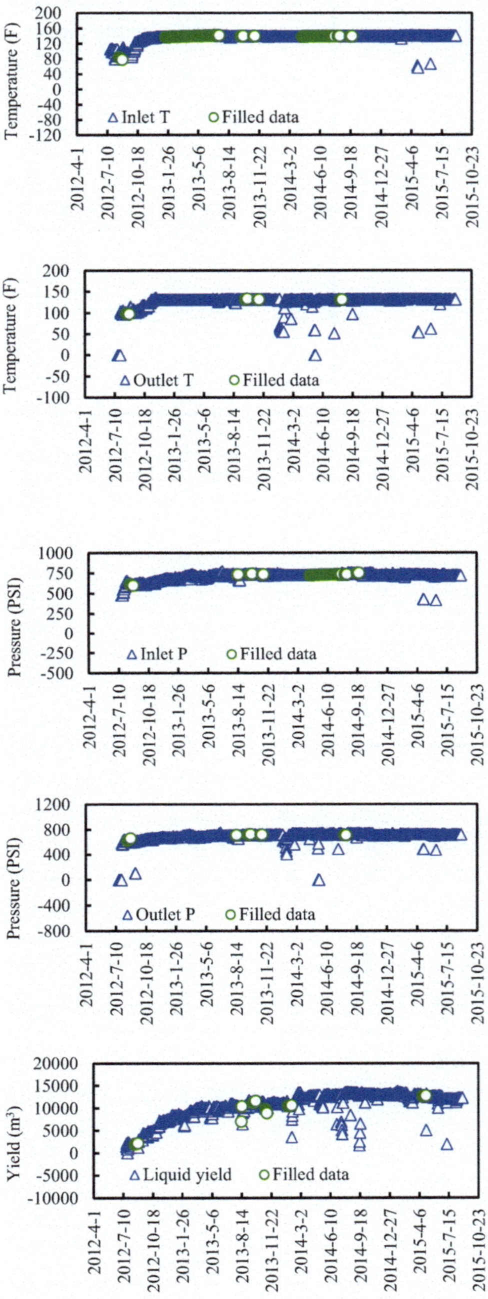

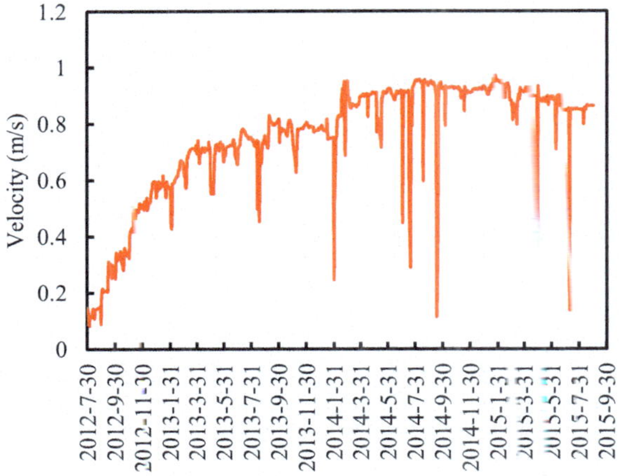

Fig. 13.9 Liquid flow velocity of mixture in pipelines

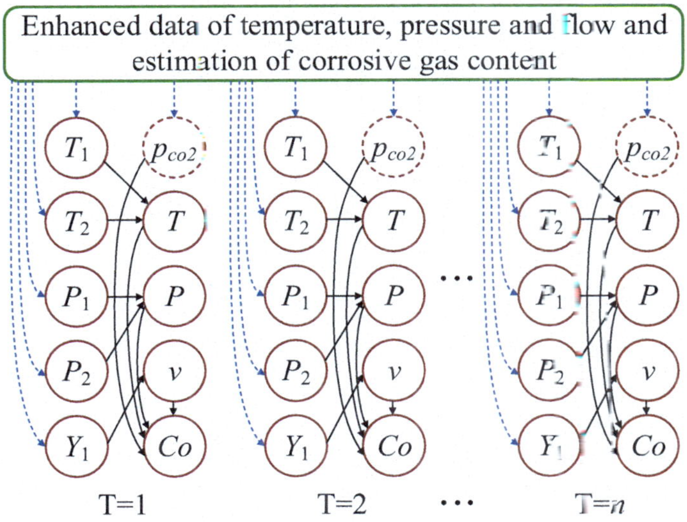

Fig. 13.10 DBN model of daily average corrosion

$$\text{RMSE} = \sqrt{\frac{1}{n}\sum_{i=1}^{n}(\hat{y}_i - y_i)^2} \tag{13.28}$$

$$\text{MAPE} = \frac{100\%}{n}\sum_{i=1}^{n}\left|\frac{\hat{y}_i - y_i}{y_i}\right| \tag{13.29}$$

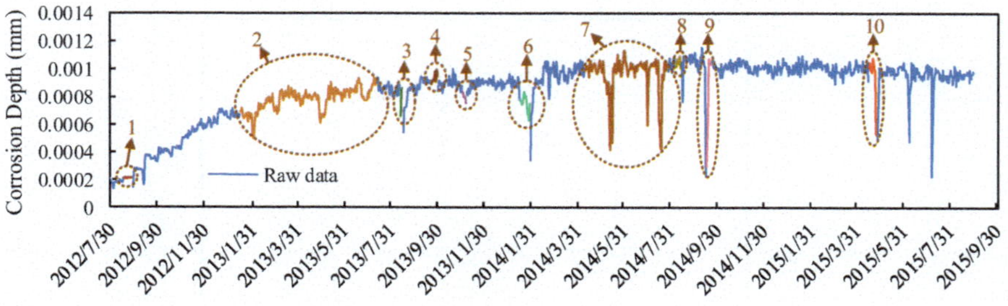

Fig. 13.11 Daily average corrosion after data enhancement

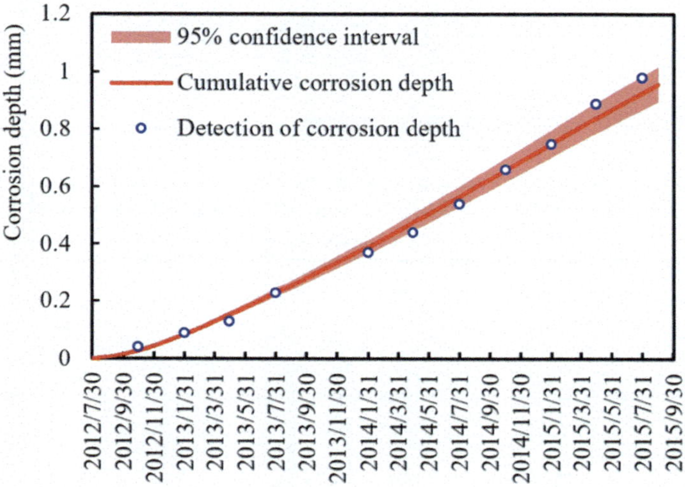

Fig. 13.12 Comparison between prediction and detection data

where y is the predicted values in the model, and $\hat{y}$ is the target value.

The comparison of the proposed method with other prediction models is crucial and is beneficial to verify the adaptability and superiority of the proposed model. A combination of different prediction methods and different data processing methods is used for validation. The error analysis results are listed in Table 13.5. Among them, ARIMA with enhanced data is a long-term prediction model using ARIMA based on enhanced data. LSTM with average filled data is a long-term prediction model based on average filled data combined with DBN. LSTM with missing deletion is a model that first deletes missing data, combines it with DBN, and then uses LSTM to make long-term predictions. LSTM with ARIMA filling first uses ARIMA for data filling, but does not consider uncertainty (without DBN), and then uses LSTM for long-term prediction model. The proposed method's MAE, MSE, RMSE and MAPE are 0.2307, 0.1967, 0.4435, and 0.5736, respectively. It shows that the prediction error of the proposed method is minimal. The predicted cumulative depth

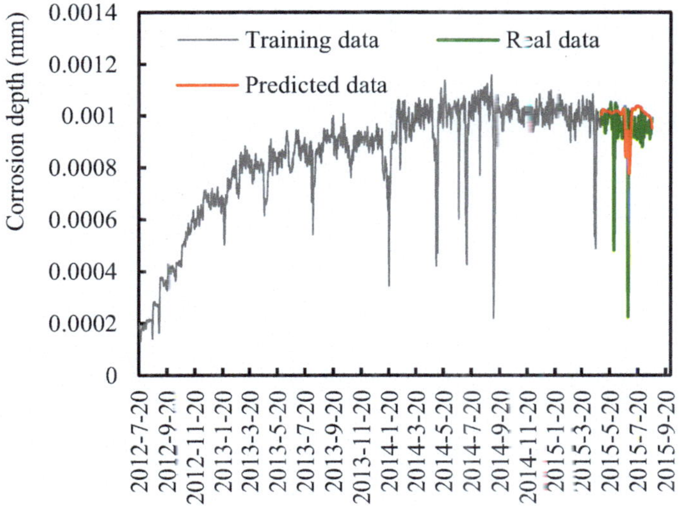

Fig. 13.13 Prediction verification of LSTM

of different methods is shown in Fig. 13.14. The cumulative corrosion depth of the proposed method is also closer to the actual depth. Compared with other methods, the proposed method has minor error, indicating that more accurate prediction can be achieved using the proposed method for data enhancement. Running efficiency is also an important indicator for method comparison. In this case, running time is used to compare time consumption associated with the method. Since the average filling and actual deletion take longer to run, the LSTM with average filled data and LSTM with missing deletion methods have the longest running time, reaching 1124 and 1026 s. The running efficiency of the proposed method is not the fastest because the uncertainty of missing variables is processed, but the running time is only extended by about 10–40 s, which is very close to the ARIMA with enhanced data and LSTM with ARIMA filling methods. The proposed method sacrifices a very small amount of running efficiency but greatly improves the prediction accuracy, and also brings huge benefits to the safety assessment and maintenance of equipment.

Table 13.5 The MAE, MSE, RMSE, MAPE and runtime of different methods

Model	MAE	MSE	RMSE	MAPE	Runtime (s)
Proposed method	**0.2307**	**0.1967**	**0.4435**	**0.5736**	**798**
ARIMA with enhanced data	0.4168	0.3556	0.5963	0.9754	743
LSTM with average filled data	0.2818	0.2408	0.4907	0.6943	1124
LSTM with missing deletion	0.2715	0.2326	0.4823	0.6695	1026
LSTM with ARIMA filling	0.2601	0.2174	0.4722	0.5926	784

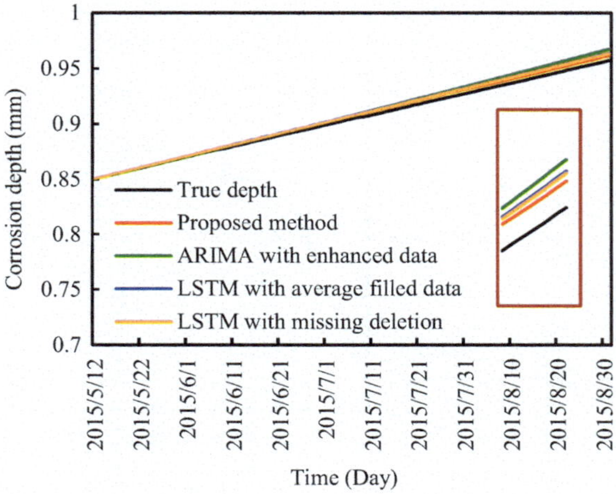

Fig. 13.14 Comparison of predicted cumulative depth of different methods

The predicted daily average corrosion amount fluctuates around 0.001. The predicted cumulative corrosion depth is shown in Fig. 13.15. According to the confidence interval range, the maximum depth and minimum depth are also obtained. The pipeline's corrosion allowance is set at 5 mm, and according to the predictions, the pipeline is expected to reach this threshold between 3930 and 4523 days from the start of monitoring, with a high likelihood of failure around December 2026. This range of RUL underscores the variability in corrosion progression and emphasizes the importance of continuous monitoring and maintenance. The effective service life of approximately 15 years, from commissioning to failure, aligns with the expected operational lifespan, reinforcing the accuracy of the prediction model. The predicted failure timeline suggests that proactive measures should be taken in the period leading up to this point to ensure the pipeline's integrity and safety. The observed fluctuations in the corrosion rates can be attributed to several factors, including variations in environmental conditions, operational parameters, and inherent uncertainties in the prediction model. These fluctuations highlight the necessity for a dynamic and responsive maintenance strategy, capable of adapting to changing conditions and improving the reliability of the pipeline.

13.3.4 Sensitivity Analysis of Degradation Parameters

Sensitivity analysis can evaluate the stability and reliability of the results and whether the combined results are affected by a single study and have significant changes. The variance of beliefs and variance reduction are two critical indicators of sensitivity analysis, indicating the influence of parameters on the overall degradation. The sensitivity analysis results of crucial parameters of the corrosion assessment model are shown in Fig. 13.16. The analysis of parameters at three distinct stages (early,

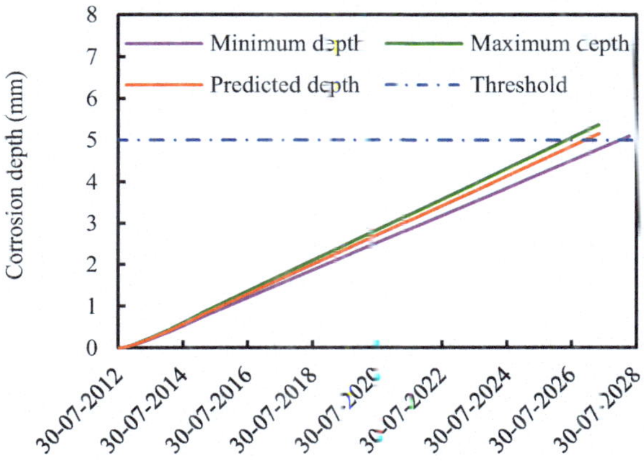

Fig. 13.15 RUL range calculated according to predicted depth

middle, and stable) reveals that temperature and pressure have minimal impact on overall degradation. In contrast, CO_2 content and flow rate emerge as the most critical parameters affecting degradation significantly. CO_2 content plays a substantial role in overall degradation as it influences the reaction rates both in fluid-related and fluid-independent parts of the system. This impact is particularly pronounced because CO_2 accelerates corrosion processes and contributes to the deterioration of pipeline integrity. During the middle stage of production, the flow rate becomes the dominant factor impacting degradation. This period is characterized by a rapid increase in liquid velocity, which represents a transitional phase in the production process. The increased flow rate enhances the abrasive and erosive effects on the pipeline, thus exacerbating degradation. In summary, while temperature and pressure have limited effects, CO_2 content and flow rate are crucial parameters that should be closely monitored. Their significant influence on degradation underscores the need for targeted strategies to manage CO_2 levels and control flow rates to mitigate corrosion and extend the pipeline's service life.

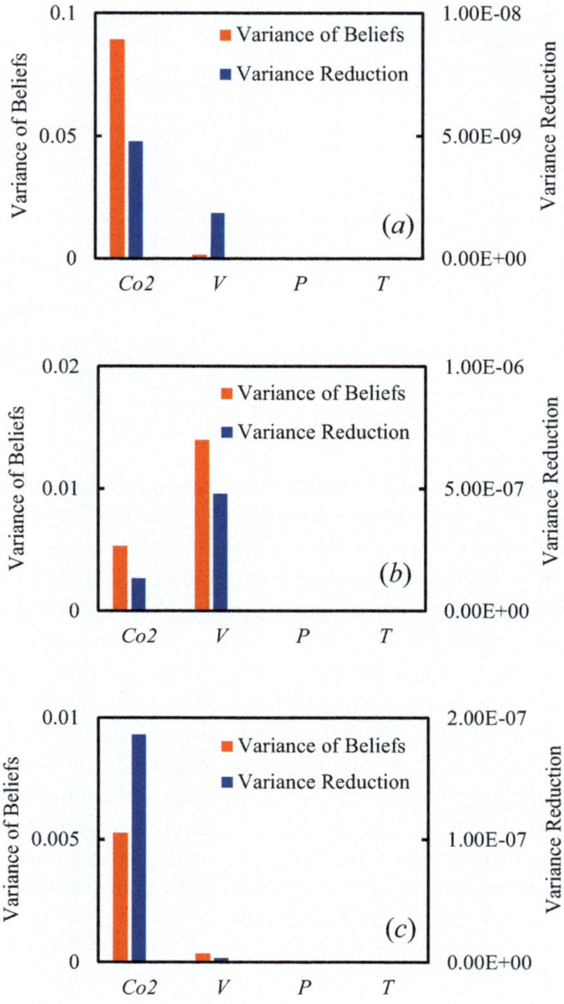

Fig. 13.16 Parameter sensitivity analysis of the early (**a**), middle (**b**) and stable stages (**c**)

Part II
Concluding Remarks

Part II *Fault prognosis* has provided a detailed exploration of various RUL prediction models tailored for subsea production systems, each designed to address the specific challenges inherent in the complex environments. By introducing methods that function with small and multi-sample data, handling multiple failure causes, dealing with failure dependences, and accounting for interactions between degradation and shocks, this part of the book offers a robust backbone for accurate fault prognosis. Additionally, the concept of RUL re-prediction underscores the importance of adaptability, allowing for continual updates as system conditions and data change. The integration of industrial incomplete information into RUL models further enhances the predictive capability. Through practical case studies, this part has shown how theoretical models can be applied in real-world scenarios, offering valuable insights for both researchers and industry professionals. The methodologies presented here form the foundation for predictive maintenance strategies, essential for reducing downtime, optimizing maintenance resource usage, and enhancing the overall performance of subsea production systems.

Part III
Maintenance

For subsea production systems, effective maintenance strategies are critical for ensuring long-term operational efficiency, safety, and reliability. Part III *Maintenance* focuses on the advanced maintenance methodologies, offering a comprehensive exploration of various models and approaches tailored to the unique challenges of subsea environments. These systems operate under highly demanding conditions, where component degradation, maintenance delays, and failure dependences can significantly impact the system performance. Traditional maintenance models often fall short in addressing these complexities, necessitating more intelligent strategies. This part provides a holistic view of CBM, starting with the integration of spare parts management and imperfect maintenance, which acknowledges the real-world scenario of non-ideal repair actions. The exploration then extends into the challenges posed by maintenance delays, a common issue in remote subsea operations. The discussion of CBM optimization under discrete state conditions further addresses how varying operational states affect maintenance decisions. A critical aspect of subsea systems is the interdependence of failures across multiple components, which is addressed through establishment of dependent multi-component degradation models. The part also recognizes the need to balance emergency maintenance with CBM, ensuring that unforeseen failures can be managed alongside scheduled maintenance to avoid operational disruptions. Finally, the chapter emphasizes sustainability as a core element of modern maintenance strategies. As subsea production occurs in environmentally sensitive regions, maintenance strategies must not only ensure system reliability but also align with sustainability goals. By presenting a range of case studies that apply these CBM models to real-world subsea systems, such as subsea trees, production systems, and transmission systems, this chapter offers both theoretical insights and practical applications. The integrated approach ensures that maintenance strategies are not only technically sound but also adaptable to the evolving demands of subsea operations.

Chapter 14
CBM with Spare Parts Management and Imperfect Maintenance

Abstract CBM is the last phase of diagnosis-prognosis-maintenance, which has arouse wide attention in recent years. A CBM method based on RUL prediction for the multi-component system is proposed with considering imperfect maintenance and spare part management in this chapter. Due to limitations of the special working environment or maintenance technology for some multi-component systems, imperfect maintenance method is more frequently applied in practical engineering application. At the same time, proper spare part management is needed to be scheduled to reduce maintenance time and maintenance cost. The imperfect maintenance model is constructed based on the virtual age rule. Spare parts are managed based on practical maintenance requirements to reduce the total numbers of spare parts and maintenance preparation cost. Maintenance preparation threshold, spare part ordering threshold and spare part number threshold are optimized based on genetic algorithm.

14.1 Introduction

RUL prediction involves predicting the degradation state of components according to degradation data collected through sensors. Predictive maintenance necessitates integrating the current state and considering the future state of the system. Incorporating RUL into maintenance decisions represents a more advanced maintenance method and a new research direction for maintenance decision-making. CBM is a predictive maintenance approach, whose decisions are determined based on the RUL prediction information provided by condition monitoring or on-line condition monitoring system (Zhou and Yin 2019). It has been proved that CBM can effectively minimize maintenance costs while ensuring system reliability. During the CBM decision process, maintenance option and spare part management need to be considered.

Maintenance can be categorized into three types based on their impact: minimal maintenance, perfect maintenance, and imperfect maintenance. Imperfect maintenance falls between perfect and minimal maintenance in terms of its effectiveness. Given the limitations of the unique working environment, maintenance technology, or technical level of maintenance personnel, perfect maintenance is hard to be realized.

B. Cai et al., *Intelligent Operation and Maintenance for Subsea Production Systems*,
https://doi.org/10.1007/978-981-96-2677-9_14

Generally, imperfect maintenance options are considered and applied in practical engineering applications, which offers the potential to further reduce maintenance costs. The imperfect maintenance model has good applicability in practical because it is more consistent with the aging characteristics and early failure condition of equipment (Ding and Tian 2012). When combined with RUL prediction technology, the effects of imperfect maintenance can be more intuitively reflected through changes in the operating time and RUL of components, based on the concept of virtual age, providing a more micro and intuitive perspective.

Effective spare part management for multi-component systems is crucial to minimizing maintenance preparation time and costs. However, managing spare parts poses significant challenges in practice. Overstocking spare parts can lead to unnecessary inventory costs, while a shortage can result in system downtime and subsequent production losses. Achieving a balance between these two extremes is required based on practical maintenance requirement, which is scheduled based on the optimal maintenance strategy. For the multi-component system with heterogeneous components, the diverse degradation mechanisms and life distributions further complicate spare part management. The advancement of RUL prediction technology has the potential to rationalize CBM strategies including spare part strategies, making it a worthwhile area of study.

Given the previous considerations, this chapter introduces a RUL-based CBM method for the multi-component system with considering imperfect maintenance and spare part management.

14.2 Imperfect CBM Model with Spare Part Strategy

The CBM approach suggested for the multi-component system encompasses primarily RUL prediction, spare part management and maintenance optimization, all while taking into account the fact that maintenance may not always be perfect, as shown in Fig. 14.1. RUL prediction result is the basis of spare part management and maintenance optimization. In the spare part management stage, starting time of maintenance preparation and spare part ordering time are determined based on RUL prediction results of each component and the multi-component system. The optimal maintenance option of the multi-component system is optimized based on RUL prediction results by genetic algorithm.

14.2.1 RUL Prediction Model

The life distribution of the mechanical system generally follows Weibull distribution, and the density function of Weibull distribution is,

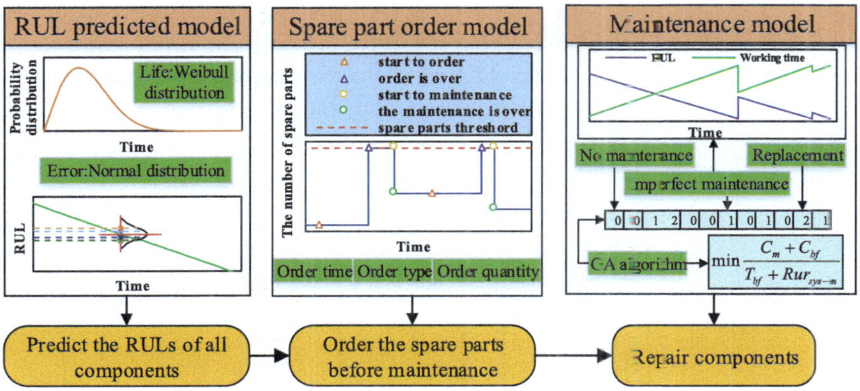

Fig. 14.1 Frame diagram of the maintenance method

$$f(t;\lambda,k)=\begin{cases}\frac{k}{\lambda}(\frac{t}{\lambda})^{k-1}e^{-({t}/{\lambda})^k}, t\geq 0\\ 0, t<0\end{cases} \tag{14.1}$$

where t is the component life, λ is the scale parameter, k is the shape parameter. The mean and variance of Weibull distribution are,

$$E=\lambda\Gamma(1+\frac{1}{k}) \tag{14.2}$$

$$Var=\lambda^2\left[\Gamma(1+\frac{2}{k})-\Gamma(1+\frac{1}{k})^2\right] \tag{14.3}$$

where Γ is the gamma function, which is

$$\Gamma(x)=\int_0^{+\infty} z^{x-1}e^{-z}dz(x>0) \tag{14.4}$$

In the proposed CBM method for the multi-component system, RUL of each component needs to be obtained first. It is assumed that RUL error t_e between the real value and the predicted result follows Normal distribution, where mean value is 0 and variance is σ^2, as shown below,

$$f(t_e)=\frac{1}{\sqrt{2\pi}\sigma}\exp\left(-\frac{t_e^2}{2\sigma^2}\right) \tag{14.5}$$

$$\sigma=T\times v \tag{14.6}$$

where T is the real RUL of the component, and v is the prediction error coefficient.

During the early stages of degradation, a component exhibits a high RUL, and there is limited data available for predicting its RUL, which can result in imprecise predictions. Therefore, the variance of prediction errors in the earlier stage is large. In the later stage of the degradation process, the component has low RUL. Large amount of diagnostic data is available, so the prediction error of RUL is small. The predicted RUL of the component can be calculated as,

$$Rul = T + t_e \tag{14.7}$$

where *Rul* is the RUL prediction result, which follows Normal distribution with mean T and variance σ^2.

14.2.2 Imperfect Maintenance Model

In comparison to perfect maintenance, imperfect maintenance has a lower cost. Additionally, imperfect maintenance can enhance the RUL of a component without using spare parts. The maintenance effect of imperfect maintenance is better than no maintenance and worse than replacement, which will revert to a new stable state. Therefore, imperfect maintenance is widely adopted in practical maintenance. Working time of the component will be shortened after imperfect maintenance, thus leading to an improvement of the RUL.

The number of imperfect maintenance actions performed on a component is limited because the effectiveness of such maintenance diminishes significantly after multiple interventions. However, despite this limitation, imperfect maintenance can still lead to a reduction in the working time and an increase in its RUL. However, RUL of the component will not be higher than RUL after the last imperfect maintenance. That is, the working time will not be lower than the last imperfect maintenance. Suppose that the working time WT_j follows Uniform distribution, then

$$f(WT_j) = \begin{cases} \frac{1}{(\gamma_2-\gamma_1)(WT-WT_{j-1})}, & WT_j \in [WT_{j-1} + \gamma_1(WT - WT_{j-1}), WT_{j-1} + \gamma_2(WT - WT_{j-1})] \\ 0, & else \end{cases} \tag{14.8}$$

where γ_1 and γ_2 are imperfect maintenance coefficients. Based on the historical data, γ_1 is set as 0.6, and γ_2 is 0.8.

14.2.3 Spare Part Models of Multi-component System

After conducting multiple imperfect maintenance operations on a component, the effectiveness of these repairs becomes significantly limited. As a result, replacing the component entirely becomes a more effective maintenance strategy. However, this method necessitates the availability of a corresponding spare part to facilitate the replacement process. However, if all spare parts are purchased and stored at the same time, high maintenance preparation costs will be incurred. In order to meet the need for component replacement and reduce the preparation cost, a reasonable spare part management model of the multi-component system needs to be scheduled. Three different spare part management strategies are proposed

(1) Spare part management strategy 1

During the maintenance process, it is not necessary to replace all components simultaneously. As a result, the threshold of total spare parts, which is represented by S is set as the sum of spare parts of all components. Only one spare part is required for the component when it is replaced. The total number of spare parts will decrease if corresponding spare parts are used in the replacement activity. To ensure that replacement requirements are met during the next maintenance, spare parts must be ordered prior to the upcoming maintenance event.

The ordering time of the spare part is closely related to the system maintenance time. If the necessary spare parts are not available, the replacement process must be discontinued until the ordered spare parts are received. As a result, the maintenance activity will be delayed and the risk of system failure will increase, which may cause additional downtime loss or even serious accidents If spare parts are ordered ahead of schedule, storage costs will incur after completing the ordering activity. Although storage cost is smaller than the downtime loss, the cost cannot be ignored if the storage time is too long. Therefore, it is necessary to set a spare part ordering threshold Tbj to optimize the ordering time. When the RUL of the system is lower than the spare part ordering threshold, spare parts are needed to be ordered.

Components with low RULs require replacement, therefore RUL is used as a metric when ordering spare parts to decide which spare parts need to be ordered. At the time point of ordering spare parts, RULs of all components are sorted from low to high. For components without spare parts, corresponding spare parts are ordered based on the sequence of RUL until the total number of spare parts reach S. In the ordering period TBJ, the number of the ordered spare parts is,

$$Nbjgm = S - Nbj \tag{14.9}$$

where $Nbjgm$ is the number of the ordered spare parts, S is the threshold of total spare parts, Nbj is the number of spare parts which already existed.

(2) Spare part management strategy 2

Spare part management strategy 2 is a special case of spare part management strategy 1. In strategy 2, the threshold of total spare parts is equal to the number of all components N, that is, $S = N$.

(3) Spare part management strategy 3

Spare part management strategy 3 is a specific variation of strategy 1. If spare part for some components is used in the replacement activity, corresponding spare parts are ordered immediately after the maintenance activity.

In summary, the three spare part management strategies vary solely in their approach to managing spare parts. In strategy 1, spare parts are ordered based on the RUL degradation condition. For components without spare parts, corresponding spare parts are ordered based on the sequence of RUL until the total number of spare parts reach the threshold S. In strategy 2, the threshold of total spare parts is equal to the number of all components N. That is, all components will have their corresponding spare part. In strategy 3, corresponding spare parts are ordered immediately after the maintenance activity if the spare part for some components are used in the replacement activity.

14.2.4 Maintenance Optimization Model

Monte Carlo simulation is used to simulate real RULs of all components. According to the with considering the prediction error t_e. RUL of the multi-component system is defined as the minimum RUL of all components, which is,

$$Rul_{sys} = \min(Rul_1, Rul_2, ..., Rul_N) \tag{14.10}$$

where Rul_{sys} is the RUL of the multi-component system, Rul_l is the RUL of the 1st component, Rul_2 is the RUL of the 2nd component, Rul_N is the RUL of the N-th component.

The multi-component system is considered safe if its RUL exceeds the maintenance preparation threshold ST, the system is considered to be in a safe state. Maintenance activity is not required in the safe state. If RUL of the multi-component system is lower than the maintenance preparation threshold, the system may be failure after the maintenance preparation time Lt. Therefore, it is necessary to start the maintenance preparation work and repair the multi-component system after completing the maintenance preparation work. The multi-component system remains operational during the maintenance preparation period in order to minimize downtime losses.

After completing the maintenance preparation activity, it is needed to determine maintenance options for all components. For components which have spare parts, there are three maintenance options, no maintenance, imperfect maintenance and replacement. For components which don't have spare parts, there are two options, no maintenance and imperfect maintenance. It will be more difficult to carry out imperfect maintenance or replacement if the component fails, and costs of imperfect maintenance and replacement will be higher. Maintenance operation cost caused mainly contains four parts, preventive imperfect maintenance cost C_{ipm}, preventive replacement cost C_{pr}, corrective imperfect maintenance cost C_{icm} and corrective replacement cost C_{cr}.

Total maintenance cost C_m of the multi-component system is calculated as the sum of maintenance preparation cost and maintenance operation cost for all components, which is,

$$C_m = c_s + \sum_{i=1}^{N} c_g^i \tag{14.11}$$

$$c_g^i = q^i \cdot c_{ipm}^i + y^i \cdot c_{icm}^i + r^i \cdot c_{pr}^i + p^i \cdot c_{cr}^i, q^i + y^i - r + p^i \leq 1 \tag{14.12}$$

where c_s is the maintenance preparation cost, i represents the i-th component, N indicates the number of all components in the multi-component system, $c_g{}^i$ is the maintenance operation cost of the i-th component, $c^i{}_{ipm}$ is the preventive imperfect maintenance cost of the i-th component, $c^i{}_{icm}$ is the corrective imperfect maintenance cost of the i-th component, $c^i{}_{pr}$ is the preventive replacement cost of the i-th component, $c^i{}_{cr}$ is the corrective replacement cost of the i-th component, q^i is the preventive imperfect maintenance coefficient of the i-th component, y^i is the corrective imperfect maintenance coefficient of the i-th component, r^i is the preventive replacement coefficient of the i-th component, p^i is the corrective replacement coefficient of the i-th component. When the maintenance coefficient is 1, it means that i-th component adopts the corresponding maintenance option, and each component can only choose one maintenance option. If the maintenance coefficient is set as 0, it means that no maintenance operation is adopted.

Genetic algorithm (GA) is used to select and optimize maintenance options for all components in the multi-component system. A chromosome includes N codes, and the one code has 3 states, which represents a group of maintenance options for all components, respectively. For the maintenance options, 0 means no maintenance, 1 means imperfect maintenance, and 2 means replacement. Each chromosome needs to be evaluated to select the most suitable chromosome. Two factors are considered to evaluate the chromosome, which are the maintenance cost and RUL of the multi-component system after maintenance. However, only considering the maintenance cost and RUL after maintenance for a single maintenance is not appropriate. For example, in order to reduce maintenance cost, only a few components with low RULs are repaired. Although the maintenance cost is decreased, RUL of the overall multi-component system is limited. Finally, the maintenance interval for the next maintenance activity will be shortened, which will lead to higher maintenance preparation cost as well as higher maintenance cost. To solve the problem, maintenance cost and working time before maintenance are combined into the optimization objective. The genetic algorithm only makes use of the fitness function of each individual in the population to search. Selection of the fitness function can directly affect the convergence speed of genetic algorithm and determine whether the optimal solution can be found. The fitness function is transformed from the objective function. Maintenance cost and working time before maintenance are combined into the optimization objective. The fitness value of each chromosome V is,

$$W = \frac{C_m + C_{bf}}{T_{bf} + Rul_{sys-m}} \tag{14.13}$$

where C_{bf} is the total maintenance cost before maintenance, C_m is the maintenance cost generated by the current maintenance activity according to maintenance options represented by the chromosome, T_{bf} is the working time before maintenance, $Rul_{sys\text{-}m}$ is the predicted RUL of the multi-component system after taking all maintenance activity for all components as represented by the chromosome.

Based on the reduction effect of imperfect maintenance on the working time, the working time of the i-th component after imperfect maintenance is,

$$WT_j^{i-m} = \frac{(\gamma 1 + \gamma 2)}{2}(WT^i - WT_{j-1}^i) + WT_{j-1}^i \tag{14.14}$$

where $WT_j^{i\text{-}m}$ is the estimated working time after the imperfect maintenance of the i-th component, WT^i is the working time before the maintenance of i-th component, and $WT_{j\text{-}1}^{\ i}$ is the working time after the j-1th imperfect maintenance of i-th component. The estimated RUL of i-th component after imperfect maintenance is,

$$Rul_{i-m}^1 = Rul_i + WT^i - WT_j^{i-m} \tag{14.15}$$

where Rul_i is the predicted RUL of i-th component before maintenance, and $Rul_{i\text{-}m}{}^1$ is the estimated RUL of i-th component after imperfect maintenance.

If the i-th component is replaced, the estimated RUL after replacement is calculated as the mean of the Weibull distribution, that is,

$$Rul_{i-m}^2 = E = \lambda\Gamma\left(1 + \frac{1}{k}\right) \tag{14.16}$$

where $Rul_{i\text{-}m}{}^2$ is the estimated RUL of i-th component after replacement. The estimated RUL of the multi-component system after maintenance is,

$$Rul_{sys-m} = \min(Rul_{1-m}^{k_1}, Rul_{2-m}^{k_2}, \ldots, Rul_{N-m}^{k_N}) \tag{14.17}$$

where k_1, k_2 and k_N refer to maintenance options of the 1st component, 2nd component and N-th component respectively. $Rul_{1\text{-}m}{}^{k1}$, $Rul_{2\text{-}m}{}^{k2}$ and $Rul_{N\text{-}m}{}^{kN}$ are estimated RULs of the 1st component, 2nd component and N-th component after maintenance, respectively. The optimization objective is,

$$\min \frac{C_m + C_{bf}}{T_{bf} + Rul_{sys-m}} \tag{14.18}$$

The optimization process of genetic algorithm is shown in Fig. 14.2. Firstly, the population is initialized. Multiple sets of chromosomes are randomly generated. To avoid the possible maintenance options that may be inconformity with the optimal solution, it is necessary to filter out the maintenance options to meet certain requirements. For example, when there is no spare part for the component, imperfect maintenance should be carried out if the predicted RUL is lower than the maintenance preparation threshold *ST*; When the component has corresponding spare part, imperfect maintenance or replacement can be chosen if the predicted RUL is lower than the maintenance preparation threshold *ST*; When the component has corresponding spare part, it is needed to be replacement if the component is failure or the number of imperfect maintenance are larger than twice. Based on the above requirements, the fitness value for each chromosome is calculated.

Secondly, crossover operation, mutation operation and selection operation need to be performed to select better chromosomes for the next iteration. The crossover operation and mutation operation are shown in Fig. 14.3. Crossover points are randomly generated, and chromosome fragments are exchanged after the crossover point between adjacent chromosomes in the crossover operation. The code value of a certain position of the chromosome is randomly changed in the mutation operation. A new maintenance option of the component obtained by the mutation operation must obey the requirements of the spare part status, that is, the component without spare part cannot be mutated into replacement. The method of roulette is adopted in the selection operation. Finally, it is needed to judge whether the termination condition is met. If the termination condition is met, the optimal maintenance option for

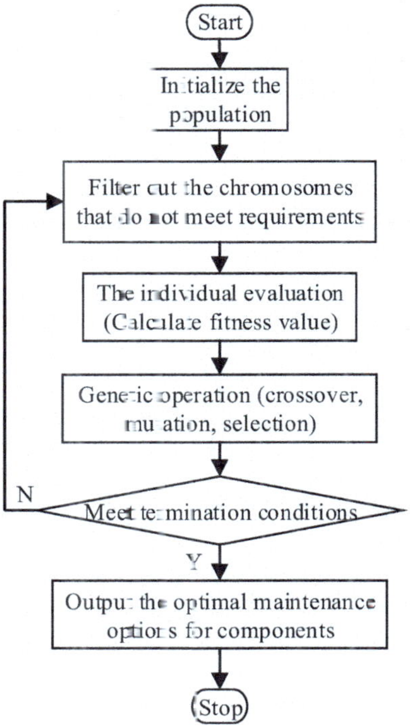

Fig. 14.2 Flow chart of genetic algorithm

all components is output, otherwise executing iteration process is continued until realizing the termination condition.

The total maintenance cost C_z in the whole life cycle is,

$$C_z = c_{cc} + c_{bj} + c_D \cdot N_f + \sum_{l=1}^{M} C_m^l \tag{14.19}$$

where c_{cc} is the storage and maintain cost of spare parts, c_{bj} is the purchase cost of spare parts, c_D is the downtime loss for per unit time, N_f is the downtime, ${C_m}^l$ is

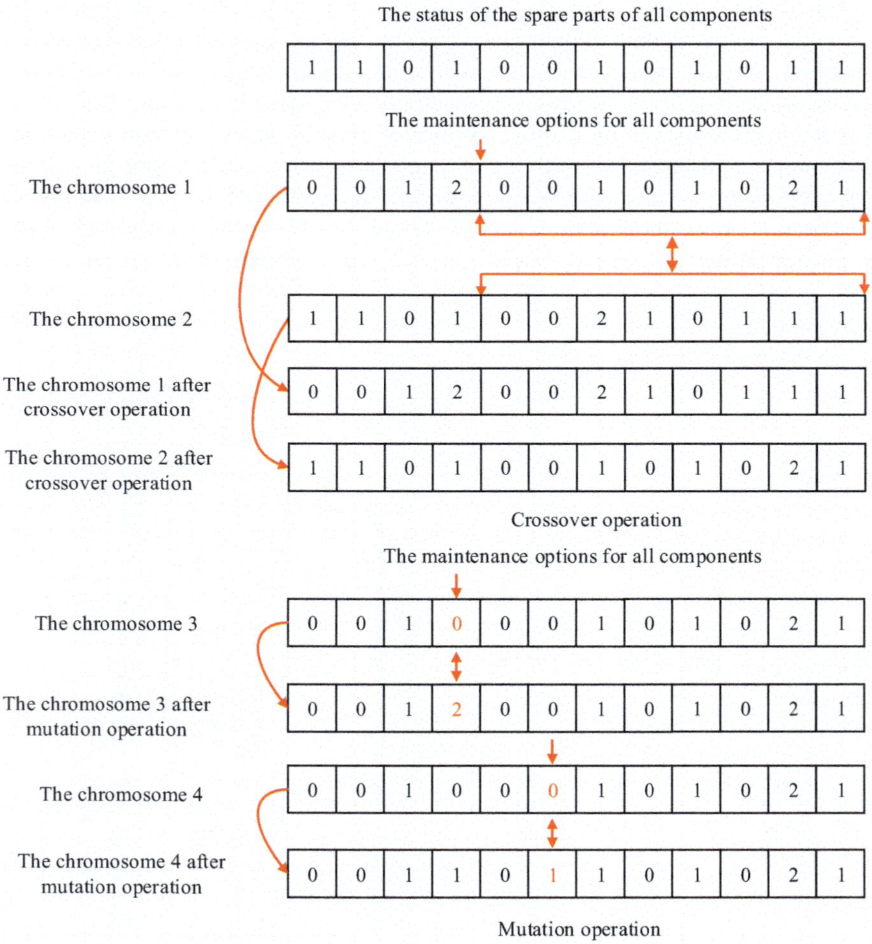

Fig. 14.3 The crossover operation and mutation operation of genetic algorithm

the l-th maintenance cost of the multi-component system, and M is the maintenance times in the whole life of the multi-component system.

The first maintenance strategy, which incorporates spare part management strategy 1, is referred to as maintenance strategy 1, which is shown as Fig. 14.4a. First, RULs for each component at per inspection time are obtained, and then the RUL of the multi-component system can be calculated. If the spare part ordering requirement is met, the spare part ordering type is determined based on the predicted RULs and corresponding spare parts are ordered. Then, after finishing the preparation work, if the maintenance requirement is met, components are maintained according to the maintenance options, which are determined by the optimization algorithm. The maintenance strategy combined with spare part management strategy 2 is named maintenance strategy 2, which is shown as Fig. 14.4b. Compared with the maintenance strategy 1, only the spare part ordering strategy is different. Spare parts are ordered for components without spare parts. The maintenance strategy combined with spare part management strategy 3 is named maintenance strategy 3, which is shown as Fig. 14.4c. Comparing with the maintenance strategy 1 and the maintenance strategy 2, spare parts are ordered after the maintenance activity. The total number of spare parts maintain the optimal value.

Maintenance decision variables of maintenance strategy 1 include maintenance preparation threshold ST, spare part ordering threshold Tbj, and total spare parts threshold S. The optimization objective is,

$$\min C_{t_z \to \infty}(ST, Tbj, S) = \frac{C_z}{t_z} \tag{14.20}$$

where t_z is the total working time.

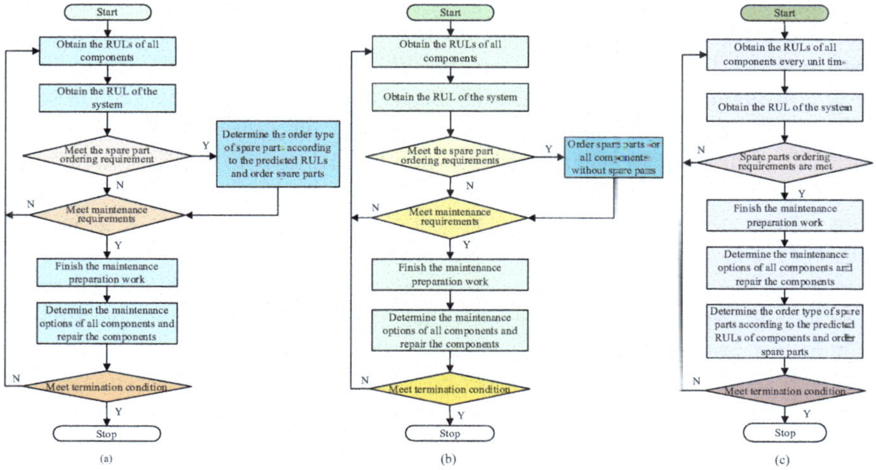

Fig. 14.4 Maintenance flows of the three maintenance strategies

Maintenance decision variables of maintenance strategy 2 include maintenance preparation threshold ST and spare part ordering threshold Tbj. The optimization objective is,

$$\min C_{t_z \to \infty}(ST, Tbj) = \frac{C_z}{t_z} \tag{14.21}$$

Maintenance decision variables of maintenance strategy 3 include maintenance preparation threshold ST and total spare parts threshold S. The optimization objective is,

$$\min C_{t_z \to \infty}(ST, S) = \frac{C_z}{t_z} \tag{14.22}$$

14.3 Case of Subsea Tree System

14.3.1 Introduction of the Subsea Tree System

Subsea tree is composed of a set of valves, such as methanol injection valve (MIV, component 1), chemical injection valve (CIV, component 2), annulus access valve (AAV, component 3), production throttle valve (PCV, component 4), crossover valve (XOV, component 5), annular wing valve (AWV, component 6), production isolation valve (PIV, component 7), annulus master valve (AMV, component 8), production wing valve (PWV, component 9), surface control subsurface safety valve (SCSSV, component 10) and production master valve (PMV, component 11) (Cai et al. 2021). The subsea tree system is a multi-component system, as shown in Fig. 14.5. Oil production can be controlled by switching PMV and adjusting PCV. In order to prevent rusting, freezing and blocking of the oil path, chemical agents such as scale inhibitor, ethylene glycol and methanol are injected into the subsea tree through CIV to keep the oil path smooth and maintain normal production. Subsea tree can also perform routine maintenance, such as pressure measurement, well cleaning and so on. Failure of the subsea tree system will cause huge economic loss, as well as endanger production safety and personnel safety.

Maintenance preparation includes leasing maintenance vessels, preparing maintenance tools and hiring maintenance personnel. The cost of maintenance preparation includes the cost of leasing maintenance vessels, maintenance tools and maintenance personnel. During the maintenance preparation process, the subsea tree system continues to operate while its performance gradually degrades. Significant downtime losses will be caused if the subsea tree system is failure. According to the practical data, the mean time to repair a component in the subsea tree system only needs 0.5 days, but the preparation work may need a week or even longer time. Compared with the maintenance preparation time, the mean time to repair a component in

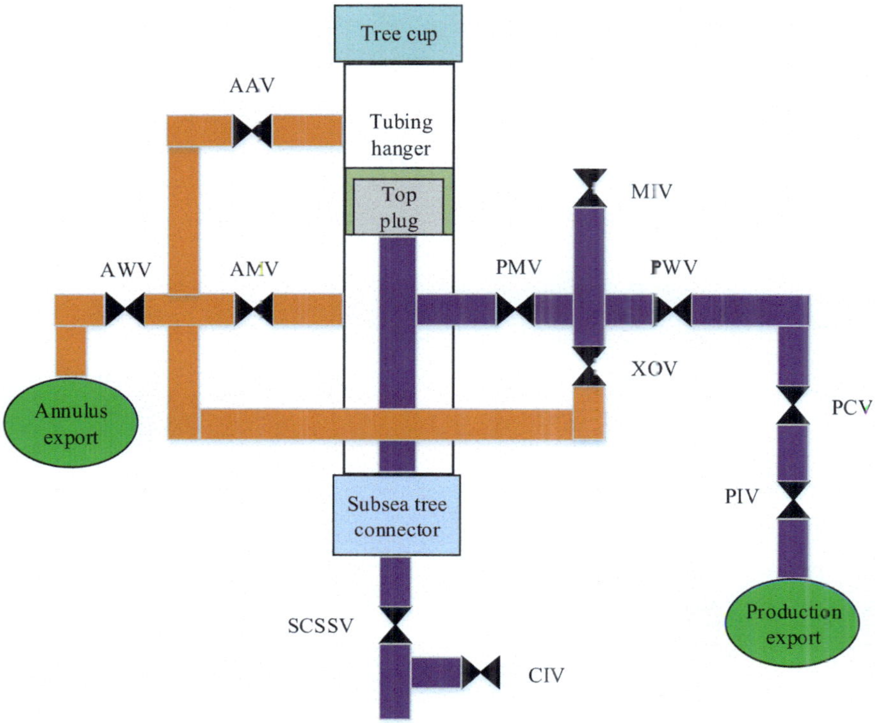

Fig. 14.5 Structure diagram of subsea tree system

the subsea tree system is negligible. The life distribution parameters for all components are shown in Table 14.1. Maintenance costs and spare part costs are shown in Table 14.2.

Table 14.3 lists basic parameters about the maintenance preparation, including maintenance preparation cost, the downtime loss per unit time maintenance preparation time, the spare part ordering period, spare part storage and maintain cost per unit time. Unit time in the case is set as 10d. The prediction error coefficient is set as 0.05, the crossover probability of the genetic algorithm is set as 0.9, and the mutation probability is set as 0.08.

Table 14.1 Weibull distribution parameters for components

Parameters	PMV	PWV	SCSSV	PCV	PIV	AMV	AWV	XOV	AAV	MIV	CIV
λ	301	261	281	161	221	241	201	181	141	100	121
k	121	105	113	84	99	108	90	81	73	56	63

Table 14.2 Maintenance costs and spare part costs

Components	$C_{pim}/\times 10^3\$$	$C_{cim}/\times 10^3\$$	$C_{pr}/\times 10^3\$$	$C_{cr}/\times 10^3\$$	$c_{bj}/\times 10^3\$$
PMV	213	255	152	182	300
PWV	185	221	132	158	260
SCSSV	200	241	143	172	280
PCV	109	132	78	94	160
PIV	153	183	109	131	220
AMV	169	203	121	145	240
AWV	140	168	100	120	200
XOV	125	150	89	107	180
AAV	91	109	65	78	140
MIV	62	74	44	53	100
CIV	78	94	56	67	120

Table 14.3 Basic parameters about the maintenance preparation

$c_s/(\times 10^3\$)$	$c_D/(\times 10^3\$/10\text{d})$	$Lt/(\times 10\text{d})$	$TBJ/(\times 10\text{d})$	$c_{cc}/(\times 10^3\$/10\text{d})$
1000	200	4	10	0.05

14.3.2 Results and Discussions

(1) Results of three maintenance strategies

Figure 14.6 illustrates the varying trends in the RUL of the subsea tree system, as influenced by various prediction error coefficients. It can be seen that when the RUL value is high and the prediction error coefficient is smal, the predicted RUL is close with the real RUL. When the RUL is in a lower level, differences between the predicted RUL and the real RUL are small under the three prediction error coefficients. This is because the prediction error is calculated as the product of the prediction error coefficient and the real RUL. When the real RUL is very low, the prediction error coefficient has little influence on the prediction error.

The optimization process of the genetic algorithm is shown in Fig. 14.7. It can be seen that the optimal solution is no longer updated along with the increasing iteration times, which indicates that convergence is eventually achieved. The optimal threshold, maintenance cost per unit time and maintenance interval time of the three maintenance strategies are shown in Table 14.4. As can be seen from the table, the same optimal maintenance decision variables in the three maintenance strategies are the same. The maintenance cost per unit time is the lowest in maintenance strategy 1, and maintenance interval is the longest in maintenance strategy 2. However, differences between the three maintenance strategies are not too significant.

It is assumed that the total life cycle of the subsea tree system is $1000 \times 10\text{d}$. In the period, the maintenance options and spare parts status of components in maintenance

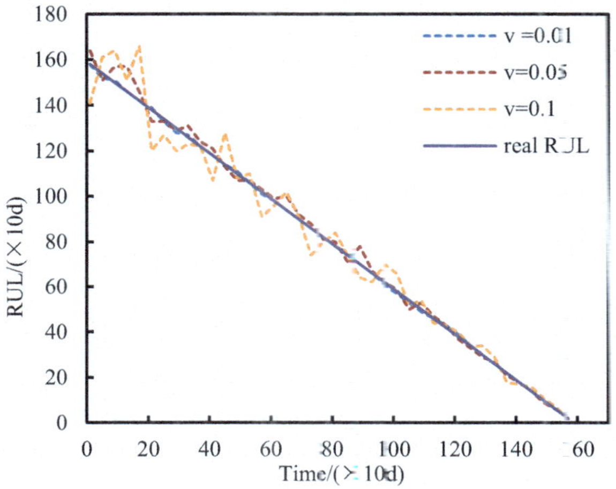

Fig. 14.6 Effects of prediction error coefficients on predicted RUL

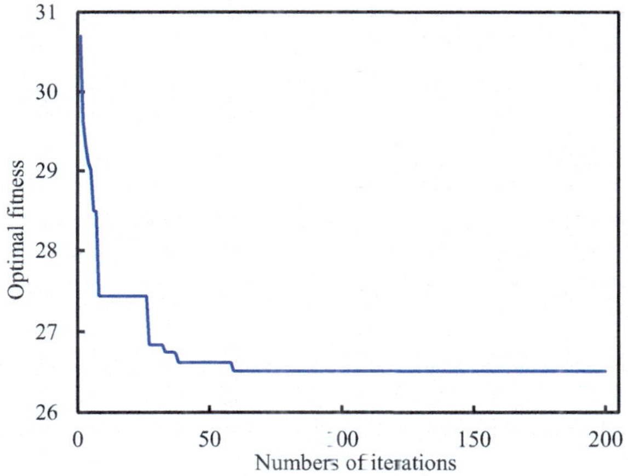

Fig. 14.7 Optimization process of genetic algorithm

Table 14.4 Optimal parameter in 3 maintenance strategies

Maintenance strategy	*ST*/(×10d)	*Tbj*	*S*	Maintenance cost per unit time/(×10^3\$/10d)	Maintenance interval time/(×10d)
1	4	11	7	31.18	95.69
2	4	11	N	32.67	98.03
3	4	N	7	32.14	92.77

strategy 1 are shown in Fig. 14.8. It can be seen that there are 11 maintenance activities in total. More components are chosen to be replaced instead of imperfect maintenance. For components whose using life is low, such as MIV, CIV, AAV and PCV, they are replaced almost every time. Imperfect maintenance will be adopted

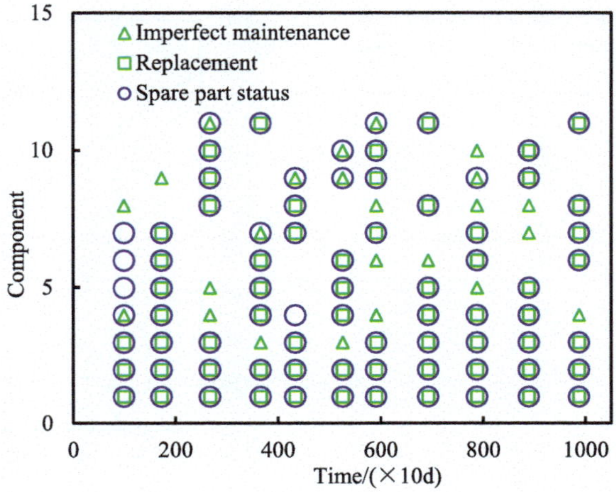

Fig. 14.8 Maintenance options and spare part status in maintenance strategy 1

for components that have longer lives. During each maintenance activity, comparing with components which have lower RULs, components which have spare parts and high RULs are more likely to choose imperfect maintenance.

Figure 14.9 presents the varying trends of the predicted RUL, real RUL, and total number of spare parts for the subsea tree system under maintenance strategy 1. With the decrease of the real RUL, the gap between the predicted RUL and the real RUL also decreases. Before adopting the maintenance activity, the system RUL decreases monotonously with the increase of time. It can be seen that the system has been maintained several times, and the RUL is significantly improved to varying degrees according to the corresponding different maintenance options. The total number of the spare parts is 7. The average number of spare parts is only 1. If components are replaced, the total number of spare parts decreases. If the system RUL falls below the spare part ordering threshold, spare parts are ordered. This ensures that the necessary spare parts are available in time for the next maintenance activity.

The maintenance options of all components when implementing maintenance strategy 2 is shown in Fig. 14.10. In the maintenance strategy 2, spare parts are available for each component. The number of components to be replaced in each maintenance activity is more than that in maintenance strategy 1, because each component has corresponding spare part. The maintenance interval becomes longer because the RUL is greatly improved after the replacement activity. The total number of maintenance activities in maintenance strategy 2 is 10, which is one less than maintenance strategy 1.

Variation trends of the predicted RUL, the real RUL and the total spare parts of the system in maintenance strategy 2 are shown in Fig. 14.11. The RUL variation trend is similar with the maintenance strategy 1. It can be seen that the total number of spare parts are 11. And the average number of spare parts is about 4, which is more than that in the maintenance strategy 1. After the replacement activity, the number

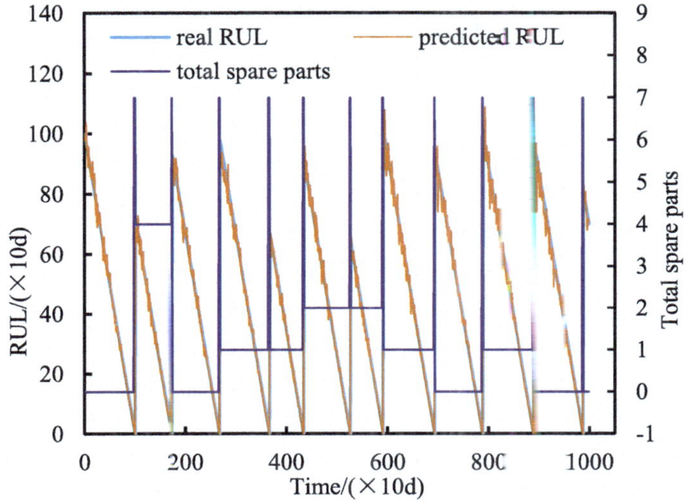

Fig. 14.9 RULs and total spare parts in maintenance strategy 1

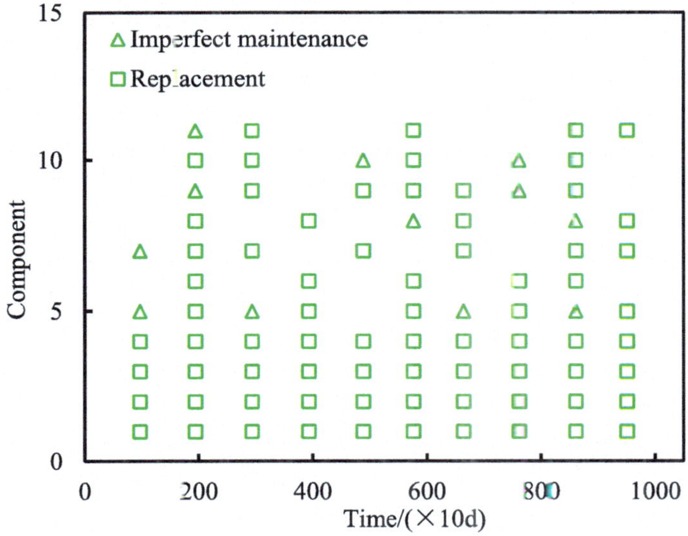

Fig. 14.10 Maintenance options in maintenance strategy 2

of spare parts decreases temporarily, then spare parts are ordered for the components without spare parts, so the total number of spare parts are increasing gradually.

The maintenance options and spare parts status of all components when implementing maintenance strategy 3 are shown in Fig. 14.12. The ordering time of spare parts in maintenance strategy 3 is earlier than maintenance strategy 1, the prediction

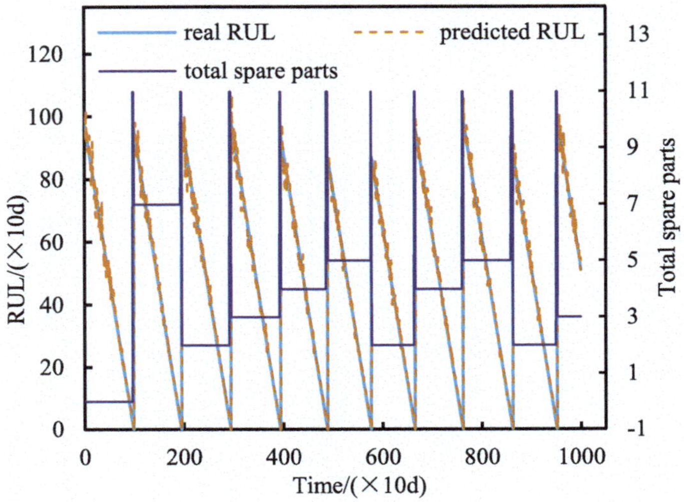

Fig. 14.11 RULs and total spare parts in maintenance strategy 2

error is larger. However, there is little difference of spare parts usage results between maintenance strategy 1 and 3.

Variation trends of the predicted RUL, the real RUL and the total spare parts of the system over time in maintenance strategy 3 are shown in Fig. 14.13. In the maintenance strategy 3, spare parts are ordered after the maintenance activity. In the case, the optimal number spare parts is 7. That is, the total number of spare parts should be 7. It can be seen that the number of spare parts stays maintains in 7 for a long time. After the replacement activity, the number of spare parts decreases temporarily, and then spare parts are ordered immediately. After the ordering period,

Fig. 14.12 Maintenance options and spare part status in maintenance strategy 3

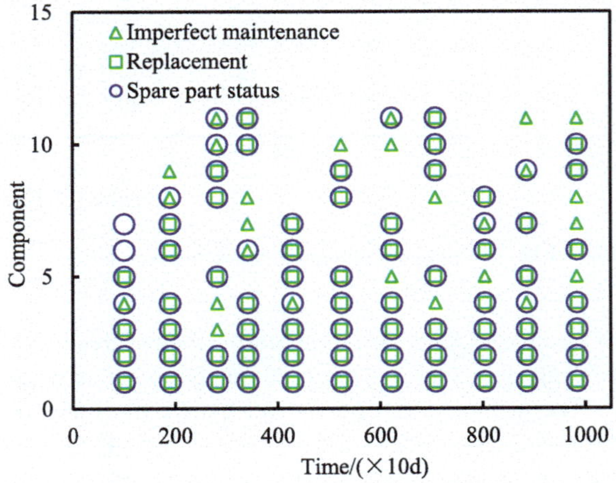

the number of spare parts returns to 7. Maintenance strategy 3 requires the storage and maintenance of a larger number of spare parts for an extended period, resulting in increased storage and maintenance costs. Subsequently, the maintenance preparation cost for this strategy surpasses that of the others.

Frequencies of maintenance activities in the overall simulation period under the three maintenance strategies are shown in Fig. 14.14. As can be seen that the maintenance time of maintenance strategy 1 and 3 is relatively close. The maintenance time of maintenance strategy 2 is slightly higher than the other two. The maintenance frequencies of maintenance strategy 2 is one less than maintenance strategy 1 and 3.

The various maintenance costs under three maintenance strategies are shown in Fig. 14.15. In the overall simulation period, since the maintenance number in strategy 3 is minimal, maintenance preparation cost and reparation cost are 10,000 × $10^3$$ and 8327 × $10^3$$, which are less than maintenance strategy 1 and 3. The purchase cost of spare parts in maintenance strategy 2 is the highest because the maximum spare parts are consumed. The total costs of maintenance strategy 1 and 3 are 31,276.55 × $10^3$$ and 31,797.5 × $10^3$$, which are slightly lower than the total cost of the maintenance strategy 2.

(2) Effect of spare part storage and maintenance cost

Figure 14.16 depicts the impact of spare part storage and maintenance costs on the overall maintenance cost of three different maintenance strategies, per unit time. With the increasing of spare part storage and maintain cost, all maintenance costs increase. But the maintenance cost per unit time of maintenance strategy 3 is the highest, and the maintenance cost of maintenance strategy 1 is the lowest. Therefore, if the spare part storage and maintain cost per unit time is high, the maintenance cost

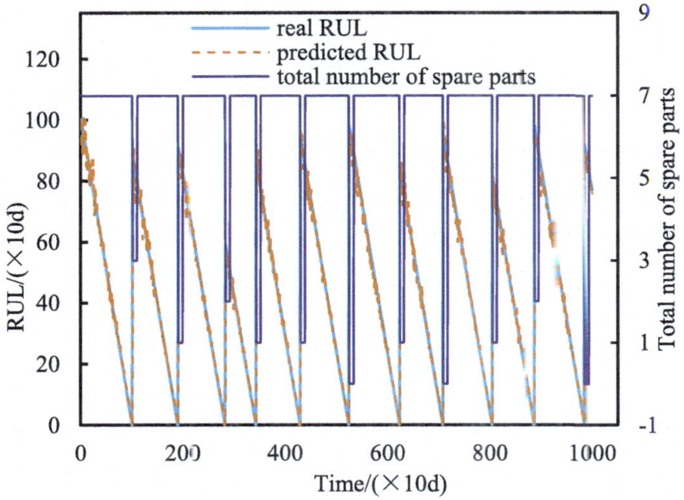

Fig. 14.13 RULs and total spare parts in maintenance strategy 3

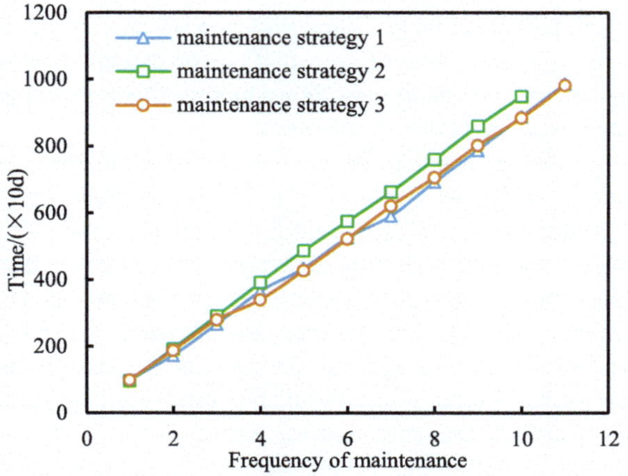

Fig. 14.14 Maintenance freqiencies under three maintenance strategies

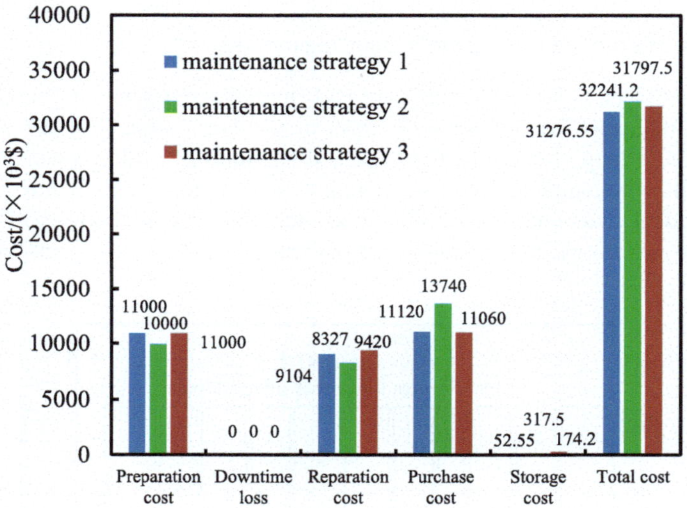

Fig. 14.15 Various maintenance cost under three maintenance strategies

can be significantly reduced by reducing the total number of spare parts and setting an appropriate ordering time for spare parts.

(3) Effect of downtime loss cost

Figure 14.17 illustrates the relationship between maintenance cost and maintenance preparation threshold, taking into account the varying impacts of downtime losses. With the increase of the maintenance preparation threshold, the maintenance cost per unit time decreases rapidly, and then maintenance cost increases slowly under the

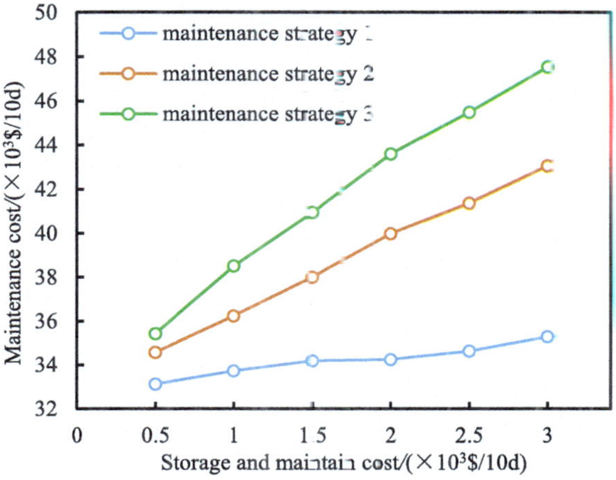

Fig. 14.16 Effects of spare part storage and maintenance cost on the maintenance cost

influence of downtime loss. When the maintenance preparation threshold is equal to the maintenance preparation time, the maintenance cost per unit time reaches to the minimum value. This is because when the system maintenance preparation threshold is less than the maintenance preparation time, the subsea tree system is prone to failure before maintenance, which will lead to extra downtime loss. If the maintenance preparation threshold exceeds the maintenance preparation time, the system can undertake the maintenance activities in a timely manner. However, an excessively large maintenance preparation threshold can lead to a shortened maintenance interval and an increase in the number of maintenance activities, ultimately resulting in a higher maintenance preparation cost. If the RUL of components is high, effect of the maintenance activity on improving the system life is not significant, which is a kind of over-maintenance. The optimal maintenance preparation thresholds under different downtime losses are the same. However, maintenance costs in the three maintenance strategies have little differences. It can be seen that when the prediction error coefficient is 0.05 and the maintenance preparation time is 40d, the prediction error will be small enough to predict the accurate RUL, which can avoid system down and high downtime loss.

(4) Effect of maintenance preparation cost

The relationships between maintenance cost and the total spare part threshold under different maintenance preparation costs are shown in Fig. 14.18. With the increasing of the total spare parts threshold, the maintenance cost decreases significantly at first, and then it remains at a stable level. When the total spare parts threshold is small, there are few components with spare parts, and most components can only be maintained imperfectly. After multiple imperfect maintenance, the improvement in RUL becomes limited. Furthermore, the rising total number of maintenance activities results in an increase in the overall maintenance cost. When the total spare parts threshold is large, more components can be replaced. RUL of the subsea tree system can be greatly improved, and the maintenance interval can be appropriately

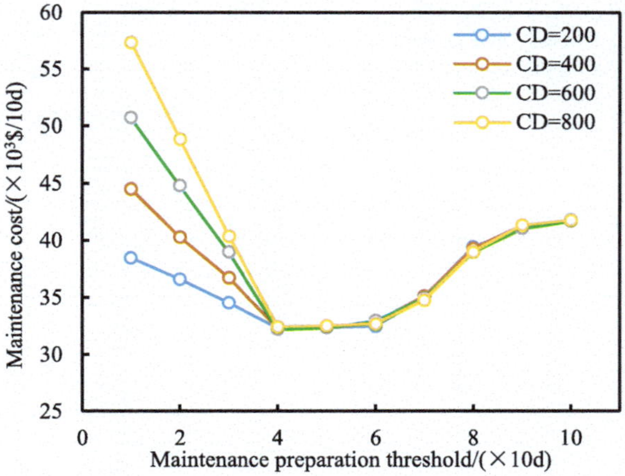

Fig. 14.17 Effect of downtime loss on the maintenance preparation threshold

prolonged. As a result, the number of maintenance activities is reduced, and the expensive maintenance preparation cost can be saved. It can also be seen that the optimal threshold of the total spare parts increases slightly with the increase of the maintenance preparation cost. In other words, to enhance the RUL of the subsea tree system, a greater number of components need to be replaced, which can also contribute to reducing the frequency of maintenance activities.

(5) Effect of prediction error coefficient

Figure 14.19 demonstrates the relationship between maintenance cost and spare part ordering threshold, considering the influence of various prediction error coefficients. Notably, as the spare part ordering threshold increases, the maintenance cost decreases significantly. When the spare part ordering threshold closes to the spare part ordering period, the maintenance cost per unit time reaches the minimum value and then keeps in a relatively stable level. If the spare part ordering threshold falls

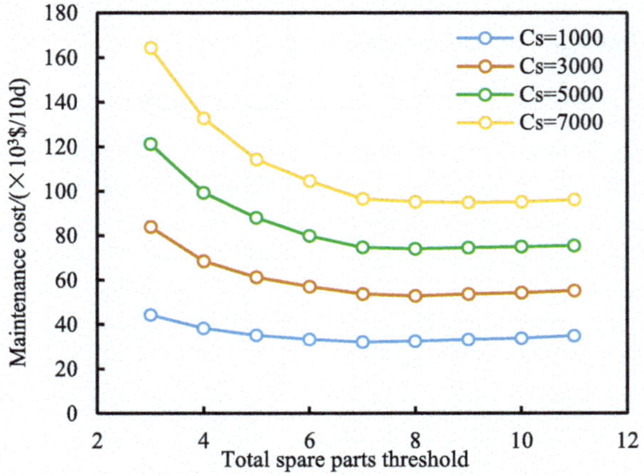

Fig. 14.18 Effects of maintenance preparation cost on the total spare part threshold

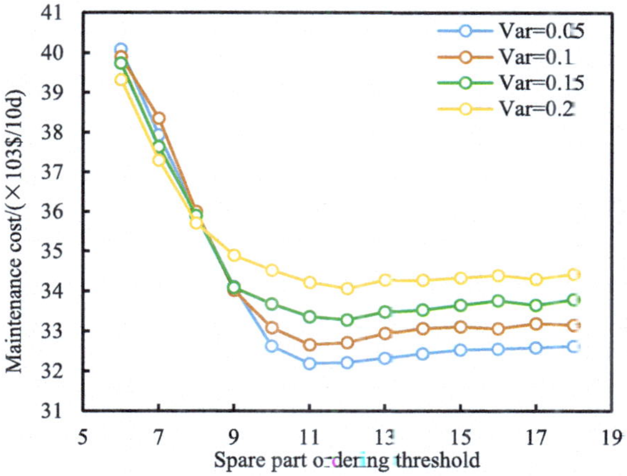

Fig. 14.19 Effects of prediction error coefficient on the spare part ordering threshold

below the ordering period, the system will require a shutdown period until the spare part ordering is completed. In this way, system failure can be prevented, but a certain amount of downtime can be caused. When the spare part ordering threshold is greater than the spare part ordering period, spare parts are ordered in advance and storage and maintenance cost will be generated, which is relatively lower than the downtime loss. It is evident that a larger prediction error coefficient results in higher maintenance costs. Additionally, the optimal threshold for spare part ordering increases, suggesting that advancing the ordering time of spare parts can mitigate the effects of prediction errors.

Chapter 15
CBM with Maintenance Delay

Abstract For multi-component systems, harsh working environments can pose challenges in terms of accessibility for maintenance. Additionally, even after formulating a maintenance strategy, there are delays in the arrival of necessary maintenance tools and personnel. A CBM method considering maintenance preparation delay is proposed. Maintenance preparation activities for the planned maintenance and emergency maintenance are both considered. Inspection time is dynamically determined based on the maintenance preparation time and RUL prediction result. RUL for the multi-component system is predicted by combining Gamma process and Poisson process. Gamma process is used to model the internal degradation process. The homogeneous Poisson process is used to simulate the arrival process of external shocks. A threshold based on RUL prediction results is set as the maintenance standard to evaluate the maintenance effectiveness. Inspection interval, safety threshold, and maintenance threshold are optimized with the goal of minimizing the maintenance cost.

15.1 Introduction

CBM relies on degradation state information gathered from either online or offline monitoring technologies. Due to limitations of resources, scheduling problems, online detection technology, and the cost of continuous monitoring, some systems are unable to achieve real-time monitoring capabilities. Thus, it is necessary to devise an inspection strategy that incorporates offline monitoring of the system to ensure effective oversight and timely detection of degradation states (Nguyen et al. 2019; Zhang et al. 2021). Offline inspection, while unable to provide real-time status updates or immediately address potential emergencies, can be optimized through strategic adjustment of inspection intervals (Tlili et al. 2015; Hu et al. 2021). This approach ensures that repairs are carried out at the most opportune time, maximizing system efficiency and minimizing downtime. Scheduled periodic inspections offer the advantage of facilitating proactive maintenance planning and advance preparations. However, as the system approaches the later stages of degradation, when the rate

B. Cai et al., *Intelligent Operation and Maintenance for Subsea Production Systems*,
https://doi.org/10.1007/978-981-96-2677-9_15

of degradation tends to accelerate, it becomes necessary to consider implementing aperiodic inspections to promptly shorten the inspection intervals (Cao 2021), thereby ensuring timely intervention and minimizing the risk of unforeseen breakdowns.

The optimal approach involves conducting real-time reliability assessments and maintenance practices. However, the working environment and limited accessibility to systems may hinder the timely arrival of maintenance resources and personnel despite having a maintenance strategy in place. Consequently, maintenance delay after scheduling may arise, necessitating careful consideration when determining the next inspection time. Thus, it is crucial to consider delay in the maintenance preparations under normal and emergency conditions, as these can significantly impact the offline inspection duration of the maintenance activity. To address this, a CBM approach is introduced that incorporates maintenance preparation delays into the RUL prediction. This method dynamically adjusts the inspection time based on both the estimated maintenance preparation time and the RUL prediction results.

15.2 CBM Model Based on Maintenance Delay

To simplify the analysis of the complex multicomponent system, assume that:

(1) The degradation status of each component is unknown and can only be determined through inspections, which are conducted offline.
(2) In the absence of maintenance, the degradation trajectory of system components is assumed to be monotonic and non-decreasing, starting from an initial state of 0.
(3) The degradation of one component is assumed to have no impact on the degradation of other components, meaning that the components are treated as independent from each other.

15.2.1 Component Degradation Model

The component performance degradation process is generally an integrated result of both internal degradation and external shock. The Gamma stochastic process is widely used in modeling internal degradation of components, where degradation is viewed as a gradual occurrence through the accumulation of small, positive increments over time. The degradation increment at any t time follows the Gamma distribution as follows:

$$f(x, \alpha, \beta) = \frac{\beta^{\alpha t}}{\Gamma(\alpha t)} x^{\alpha t - 1} e^{-\beta x}, x > 0 \tag{15.1}$$

$$\Gamma(x) = \int_0^{+\infty} t^{x-1} e^{-t} dt, \quad x > 0 \tag{15.2}$$

where α is the shape parameter, β is the inverse scale parameter, x is the degradation quantity at the time interval t, and Γ is the Gamma function. The mean and variance of x are

$$EX = \frac{\alpha t}{\beta} \tag{15.3}$$

$$\mathrm{Var} = \frac{\alpha t}{\beta^2} \tag{15.4}$$

The internal degradation quantity X_1 at time T_1 is

$$X_1 = \sum_{i=1}^{T_1} x_1^i \tag{15.5}$$

where ${x_1}^i$ is the degradation degree at per unit of time. The degradation process of different components can be simulated by controlling parameters α and β.

The homogeneous Poisson process is used to describe the arrival process of an external shock. In any time interval t, the number of external shocks follows Poisson distribution. For any $s \geq 0, t \geq 0$,

$$P\{N(t+s) - N(s) = n\} = e^{-\lambda t} \frac{(\lambda t)^n}{n!}, n = 0, 1, \ldots \tag{15.6}$$

The time interval T_n of successive shocks follows exponential distribution with an average value of $1/\lambda$.

$$f_{T_n}(t) = \begin{cases} \lambda e^{-\lambda t}, & t \geq 0 \\ 0, & t < 0 \end{cases} \tag{15.7}$$

The degradation degree x_2 caused by shocks follows normal distribution with a mean value of μ_1 and a variance of ${\sigma_1}^2$, as shown below:

$$f(x_2) = \frac{1}{\sqrt{2\pi}\sigma_1} \exp\left(-\frac{(x_2 - \mu_1)^2}{2\sigma_1^2}\right) \tag{15.8}$$

In time T_1, when the number of a certain shock is n_1, the degradation degree caused by the external shock is

$$X_2 = \sum_{j=1}^{n_1} x_2^j \tag{15.9}$$

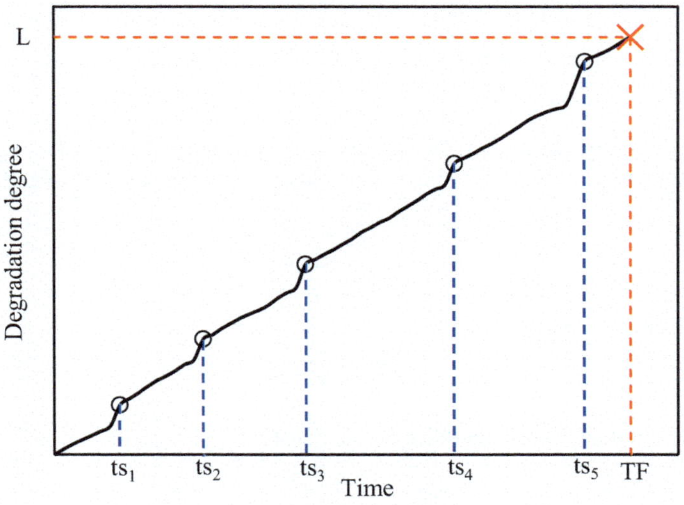

Fig. 15.1 Total degradation process combining internal degradation and external shock

where ${x_2}^j$ is the degradation degree caused by the j-th external shock.

The overall level of degradation experienced by a component is the cumulative result of both internal degradation and external shocks. Specifically, the total degradation degree X exceeds the fault threshold L, the component has failed.

$$X = X_1 + X_2 \tag{15.10}$$

The degradation process of the component is shown in Fig. 15.1. TF is the fault time, and ts_k $(k = 1, 2, 3, \cdots)$ represents the time of the external shock. The degradation degree suddenly increases when a shock occurs, and the amplitude rises depending on the intensity of the external shock.

Offline inspection policy is adopted, and the inspection error follows normal distribution with a mean value of 0 and a variance of ${\sigma_2}^2$. The degradation value obtained through the offline inspection X_z is

$$X_z \sim N\left(X, \sigma_2^2\right) \tag{15.11}$$

15.2.2 RUL Prediction Model

Under the given assumptions, the total degradation inspection value measured at time T_1 is denoted as X_Z and the RUL value of the multicomponent system is set as *rul*. According to the degradation process of components, the mean of external shock times n_s in the *rul* period is

$$n_s = \frac{rul}{\lambda} \tag{15.12}$$

Given that the mean value of degradation caused by the shock is μ_1, the mean value of degradation caused by shock $X_2{}^h$ in the *rul* period is

$$X_2^h = n_s \times \mu_1 \tag{15.13}$$

The mean value of internal degradation at per unit time E is

$$E = \frac{\alpha}{\beta} \tag{15.14}$$

When the total degree of degradation is X_Z, the RUL value of the system is

$$rul = \frac{L - X_Z - X_2^h}{E} \tag{15.15}$$

The estimated RUL value can be further expressed as

$$rul = (L - X_Z)/\left(\frac{\mu_1}{\lambda} + \frac{\alpha}{\beta}\right) \tag{15.16}$$

15.2.3 Maintenance Model

(1) Maintenance strategy 1

The aperiodic inspection strategy is adopted to reduce the frequency of maintenance preparation and thus the cost of maintenance preparation. The basic inspection period is set as T. If the system does not need to be repaired, then the periodic inspection is adopted to inspect each component every T time. When the RUL of the system meets the maintenance requirement, the next inspection interval is set based on the maintenance preparation time and the predicted RUL.

The cost of each inspection is CI. The degradation state of each component is obtained based on the offline inspection. RUL of each component can be predicted, and then, RUL of the multicomponent system is the minimum RUL value of all components,

$$S_{sys} = \min(rul_1, rul_2, \ldots rul_N) \tag{15.17}$$

Prior to undertaking maintenance activities, essential preparatory steps include scheduling maintenance personnel, preparing necessary tools and equipment, ordering spare parts for components, and arranging for the leasing of maintenance vessels. The transportation of both equipment and personnel to the target location via the maintenance vessel is a crucial step in executing maintenance tasks, and the preparation time for maintenance work should not be ignored. The normal maintenance preparation time is *LT*, and the maintenance cost under normal maintenance preparation time is *CS*. Upon the occurrence of an emergency fault, it is imperative to mitigate the resulting production losses and implement preventive measures to avert the severe consequences from the deterioration of the fault. Furthermore, the maintenance preparation work should be completed as soon as possible to realize an emergency maintenance of the multicomponent system. *LTmin* is set as the emergency maintenance preparation time, and *CSmax* is the corresponding emergency maintenance preparation cost.

Errors may exist in the degradation state inspection and RUL prediction of the multicomponent system. To reduce the fault caused by prediction errors, *DT* is set as the safety time threshold, and the safe RUL S_{st} of the multicomponent system is

$$S_{st} = S_{sys} - DT \tag{15.18}$$

The following four different schedules of maintenance preparation are proposed according to S_{st}.

1. As shown in Fig. 15.2a, ○ indicates the inspection time. When $S_{st} \geq T + LT$, the safe RUL of the system is sufficiently long. After the basic inspection period *T*, the normal maintenance preparation work can be completed before the system failure without production loss. Therefore, the next inspection interval is set as the basic period *T*. In this case, no maintenance preparation is required.

As shown in Fig. 15.2b, × indicates failure moment, Δ indicates the start time of maintenance preparation, □ indicates maintenance time, and the time period *Dt* indicates downtime. The maintenance preparation work is not started when a fault occurs within the basic inspection period *T*. The maintenance preparation time needs to be shortened to reduce the production loss. The maintenance preparation time is

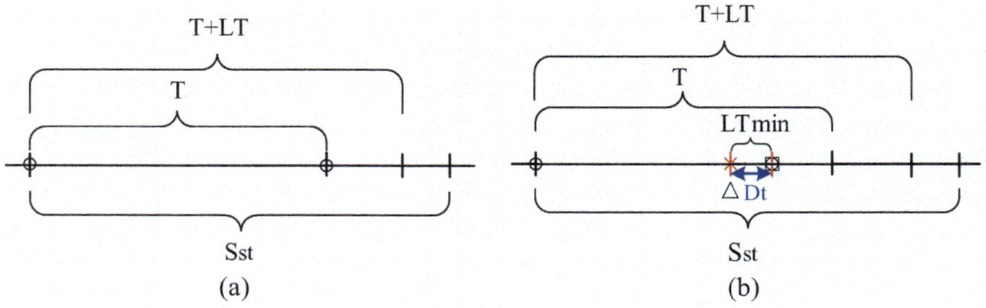

Fig. 15.2 Schedule of maintenance preparation when $S_{st} \geq T + LT$

set to *LTmin*, and the sum of maintenance preparation cost and production loss is

$$C_{S1} = CSmax + LTmin \times CD \tag{15.19}$$

where *CD* represents the production loss per unit time.

2. As shown in Fig. 15.3a, when $LT \leq S_{st} < T + LT$, the RUL of the system meets the maintenance requirements, and the next inspection interval is set to S_{st}. With the maintenance preparation time set as *LT*, the maintenance preparation is completed when the next inspection time arrives. The maintenance preparation cost is

$$C_{S21} = CS \tag{15.20}$$

As shown in Fig. 15.3b, if a fault occurs within the S_{st} interval and the remaining maintenance preparation time is less than the emergency maintenance preparation time, then the maintenance preparation work is taken according to the normal maintenance preparation time. In this case, the sum of maintenance preparation cost and production loss is

$$C_{S22} = CS + CD \times (S_{st} - T_g) \tag{15.21}$$

where T_g indicates the working time from the beginning of the inspection period to the fault time.

As shown in Fig. 15.3c, if the remaining maintenance preparation time is greater than the emergency maintenance preparation time, then the maintenance preparation time is set to *LTmin* to reduce the shutdown loss. In this case, the sum of maintenance preparation cost and production loss is

$$C_{S23} = CSmax + CD \times LTmin \tag{15.22}$$

3. Figure 15.4a shows that, when $0 < S_{st} < LT$, the next inspection interval is set as S_{st}, and the maintenance preparation time is set as *LTmin*. If no failure occurs before completion of maintenance preparation and $S_{st} \geq LTmin$, then no shutdown

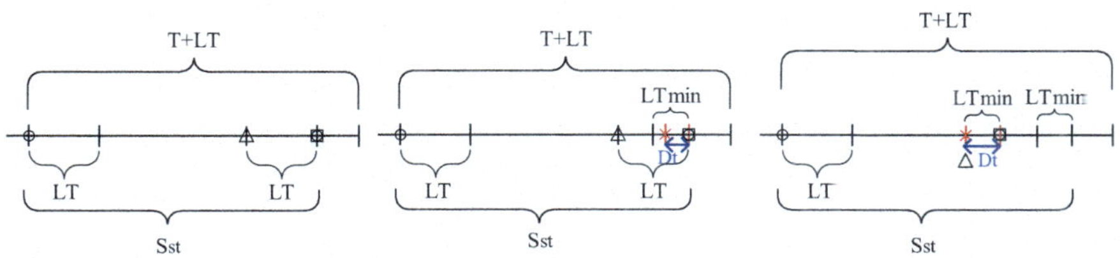

Fig. 15.3 Schedule of maintenance preparation when $LT \leq S_{st} < T + LT$

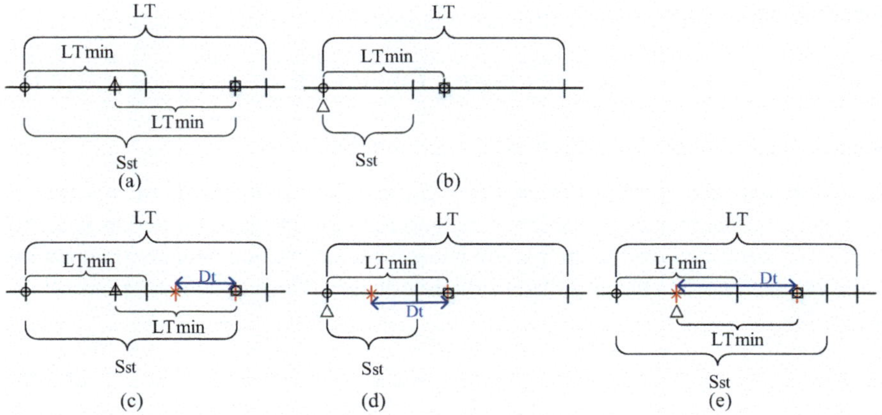

Fig. 15.4 Schedule of maintenance preparation when $0 < S_{st} < LT$

loss will be incurred. Maintenance preparation cost is

$$C_{S31} = CSmax \tag{15.23}$$

As shown in Fig. 15.4b, if no failure occurs before completion of maintenance preparation and $S_{st} < LTmin$, then shutdown loss will be generated. The sum of maintenance preparation cost and production loss is

$$C_{S32} = CSmax + CD \times (LTmin - S_{st}) \tag{15.24}$$

As shown in Fig. 15.4c, if a fault occurs before the completion of maintenance preparation, maintenance preparation has been started at the point of failure, and $S_{st} \geq LTmin$, then the sum of maintenance preparation cost and production loss is

$$C_{S33} = CSmax + CD \times (S_{st} - T_g) \tag{15.25}$$

As shown in Fig. 15.4d, if a fault occurs before the completion of maintenance preparation, maintenance preparation has been started at the point of failure, and $S_{st} < LTmin$, then the sum of maintenance preparation cost and production loss is

$$C_{S34} = CSmax + CD \times (LTmin - T_g) \tag{15.26}$$

As shown in Fig. 15.4e, if a failure occurs before the completion of maintenance preparation and the maintenance preparation is not started at the point of failure, then the sum of maintenance preparation cost and production loss is

$$C_{S35} = CSmax + CD \times LTmin \tag{15.27}$$

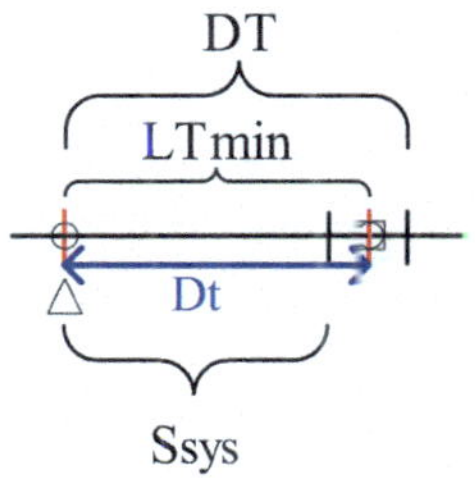

Fig. 15.5 Schedule of maintenance preparation when $S_{sys} \leq DT$

4. Figure 15.5 shows that, when $S_{sys} \leq DT$, the system is stopped to ensure security. When the maintenance preparation time is the shortest, the sum of maintenance preparation cost and production loss is

$$C_{S4} = CSmax + CD \times LTmin \tag{15.28}$$

Performing group maintenance for the multicomponent system can lead to a notable decrease in maintenance cost. If the component fails, then it needs repair. If the RUL of a component is less than the maintenance threshold T_m, then preventive maintenance is performed. Corrective maintenance and preventive maintenance are both perfect maintenance methods. The degradation degree of the repaired component is 0 after the perfect maintenance. Corrective maintenance is more complex than preventive maintenance. Thus, the cost of corrective maintenance surpasses that of preventive maintenance. The cumulative maintenance cost of all components in a maintenance activity is

$$C_M = \sum_{i=1}^{N} (C_{cmi} \times I_{cmi} + C_{pmi} \times I_{pmi}) \tag{15.29}$$

where C_{cmi} is the corrective maintenance cost of the i-th component. I_{cmi} is 1 if the i-th component has failed; otherwise, I_{cmi} is 0. C_{pmi} is the preventive maintenance cost of the i-th component. When the i-th component meets the preventive maintenance requirement, I_{pmi} is 1; otherwise, I_{pmi} is 0.

Each maintenance strategy is optimized based on the average maintenance cost per unit time of the multicomponent system, which is

$$C = \frac{CI \times N_I + C_{SZ} + C_{MZ}}{T_N + T_D} \tag{15.30}$$

where T_N is the working time, T_D is the downtime, CI is the cost for once inspection, N_I is the sum of inspection times, C_{SZ} is the maintenance preparation cost and production loss incurred in maintenance preparation, and C_{MZ} is the maintenance cost of all components.

The flowchart of maintenance strategy 1 is shown in Fig. 15.6. Initially, a fundamental inspection interval is established, drawing upon expert experience or historical data. The next inspection interval is set based on the maintenance preparation time and the predicted RUL. Then, four different schedules of maintenance preparation are proposed according to the safe RUL S_{st}. A suitable schedule of maintenance preparation will be chosen to complete the maintenance activity. The decision variables are basic inspection period T, safety time threshold DT, and maintenance threshold T_m.

(2) Maintenance strategy 2

Maintenance strategy 2 is a fixed period maintenance strategy where the system needs to be maintained once every inspection period T_2. The maintenance preparation work is required to be performed before the inspection time. A fault is supposed to occur before system maintenance and the remaining maintenance preparation time *LTs* exceeds the emergency maintenance preparation time *LTmin*. If the remaining maintenance preparation time *LTs* is shorter than the emergency maintenance preparation time *LTmin*, then the maintenance preparation time will remain the same. After the maintenance preparation is completed, corrective maintenance is taken for failure components, and preventive maintenance is taken for the component with RUL less than T_2. Figure 15.7 shows the flowchart of maintenance strategy 2. The decision variable is the inspection period T_2.

(3) Maintenance strategy 3

Maintenance strategy 3 is a failure-based strategy, where emergency maintenance is taken if the multicomponent system has failed, and the maintenance preparation time is *LTmin*. A maintenance threshold *Tx* is set. Preventive maintenance is conducted for the component with RUL less than *Tx*. The flowchart of maintenance strategy 3 is shown in Fig. 15.8. The maintenance threshold *Tx* is the decision variable.

15.3 Case of Subsea Tree System

Figure 14.5 shows the structure diagram of the subsea tree system. In this chapter, one unit time is set as 10 days. The fault threshold L of each component of the subsea tree is set to 500 × 10 days. The variance σ_2 of inspection error is 4. According to the practical data, mean time to repair a component in the subsea tree system only needs 0.5 day. However, the preparation work may need a week or even longer time. Compared with the maintenance preparation time, the mean time to repair a component in the subsea tree system is negligible. The Gamma process degradation parameters of the components in the subsea tree system are shown in Table 15.1, and the Poisson process degradation parameters are given in Table 15.2. The preventive and corrective maintenance cost parameters of all components are shown in Table 15.3, and the inspection cost, maintenance preparation cost, and production loss are shown in Table 15.4.

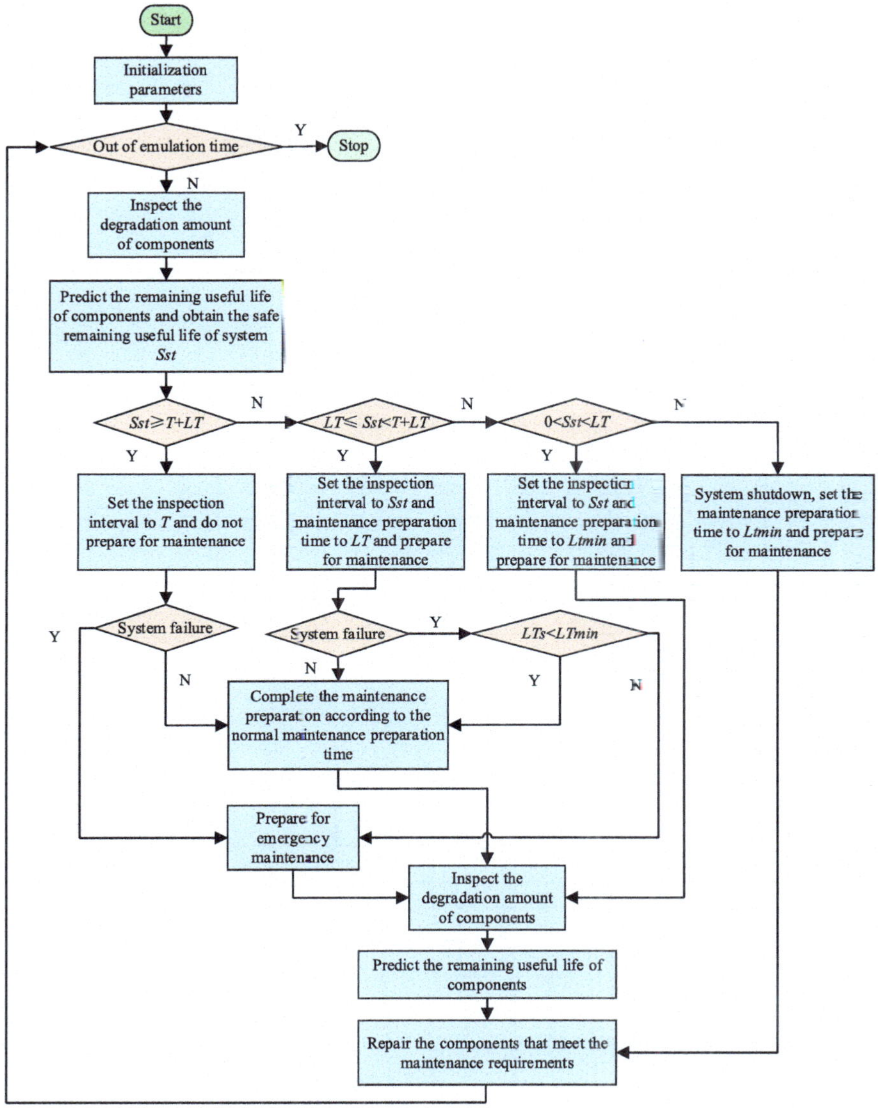

Fig. 15.6 Flowchart of maintenance strategy 1

15.3.1 *Results of Three Maintenance Strategies*

Table 15.5 showcases the optimal thresholds and the lowest maintenance costs associated with each of the three maintenance strategies. Maintenance strategy 1 boasts a lower cost and a lower probability of failure during maintenance in comparison to strategies 2 and 3, making it the superior choice among the three.

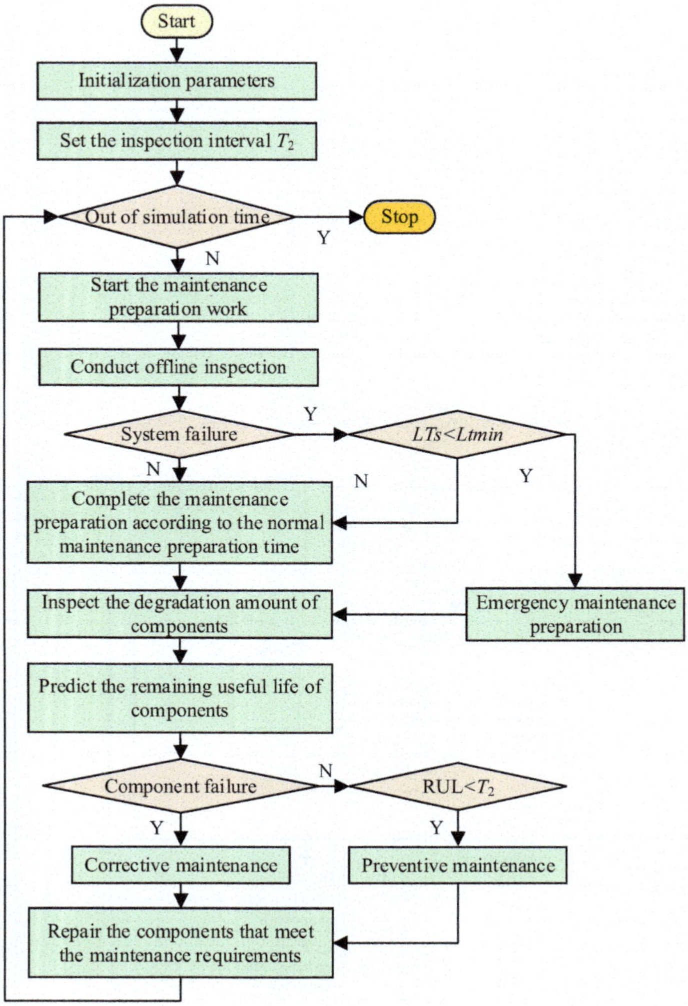

Fig. 15.7 Flowchart of maintenance strategy 2

(1) Result discussions of maintenance strategy 1

The working time is set as 1000 × 10 days. Maintenance results of each component of the subsea tree system under the maintenance strategy 1 are shown in Fig. 15.9. The ordinate represents components, corresponding to the orders shown in Table 15.1. MIV, CIV, AAV, PCV, XOV, and AWV are replaced every time, and AMV, PWV, SCSSV, and PMV are replaced every two times.

Figure 15.10 shows the degradation and repair process of MIV under maintenance strategy 1. The graph depicts the RUL prediction curve for the subsea tree system overlapping with MIV, suggesting that the RUL of MIV is the lowest during

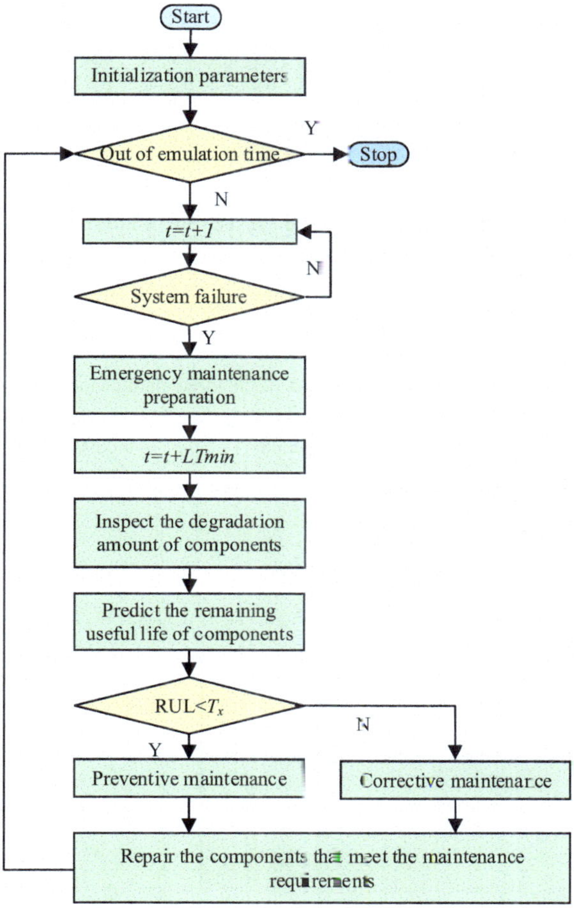

Fig. 15.8 Flowchart of maintenance strategy 3

Table 15.1 Gamma process degradation parameters

	1	2	3	4	5	6	7	8	9	10	11
	MIV	CIV	AAV	PCV	XOV	AWV	PIV	AMV	PWV	SCSSV	PMV
α	99.51	99.49	97.06	99.12	97.38	98.46	97.07	97.62	93.56	97.69	96.08
$1/\beta$	0.044	0.036	0.031	0.026	0.023	0.020	0.018	0.016	0.015	0.013	0.012

Table 15.2 Poisson process degradation parameters

λ	μ_1	σ_1
10	10	0.1

Table 15.3 Maintenance cost parameters of subsea tree components

	MIV	CIV	AAV	PCV	XOV	AWV	PIV	AMV	PWV	SCSSV	PMV
$C_{pm}/\times 10^3\$$	44	56	65	78	89	100	109	121	132	143	152
$C_{cm}/\times 10^3\$$	53	67	78	94	107	120	131	145	158	172	182

Table 15.4 Inspection cost, maintenance preparation cost, and production loss

$CI/\times 10^3\$$	$CS/\times 10^3\$$	$CSmax/\times 10^3\$$	LT/10 days	$LTmin$/10 days	$CD/\times 10^3\$$
50	1000	1400	10	5	200

Table 15.5 Optimal parameters of the three maintenance strategies

	Optimal maintenance thresholds	Optimal maintenance cost/($\times 10^3\$$/10 days)	Failure probability (%)
Maintenance strategy 1	$DT = 3, T = 70, T_m = 95$	21.34	12.48
Maintenance strategy 2	$T_2 = 89$	22.10	21.80
Maintenance strategy 3	$Tx = 93$	34.42	100

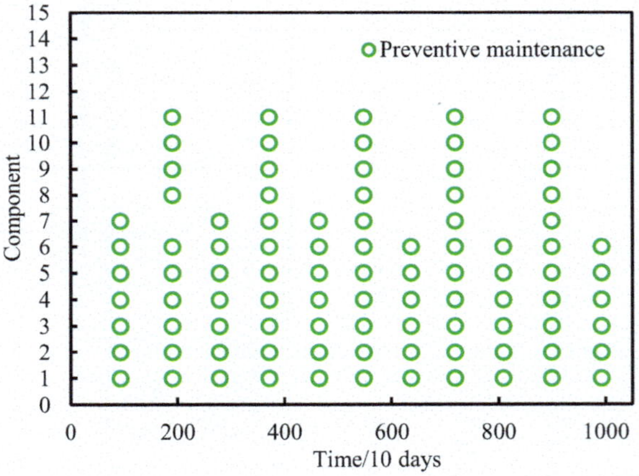

Fig. 15.9 Maintenance process of all components under maintenance strategy 1

each inspection and maintenance. The system undergoes a single inspection between maintenance intervals, followed by a scheduled inspection and maintenance preparation. MIV exhibits high life efficiency due to its replacement at a point when its RUL is relatively low.

The degradation and repair process of PMV under maintenance strategy 1 are shown in Fig. 15.11. The predicted RUL of PMV at the maintenance time point is significantly greater than the expected RUL.

(2) Results of maintenance strategy 2

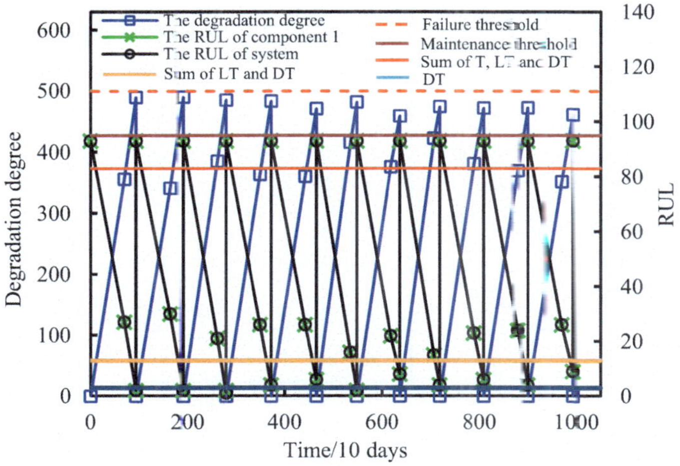

Fig. 15.10 Degradation and maintenance process of MIV

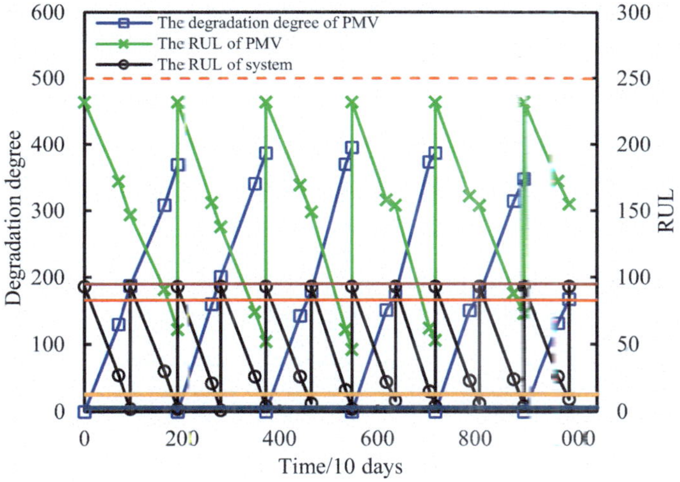

Fig. 15.11 Degradation and maintenance process of PMV

The maintenance results of each component of the subsea tree system under maintenance strategy 2 are shown in Fig. 15.12. MIV, CIV, AAV, PCV, and XOV are repaired every time, PIV, AMV, PWV, SCSSV and PMV are repaired twice, and AWV is repaired occasionally. Throughout the entire operational period, there were a total of three failures recorded: one each for AWV, MIV, and PIV. Maintenance strategy 2 has a higher probability of failure than maintenance strategy 1.

Fig. 15.12 Maintenance process of all components under maintenance strategy 2

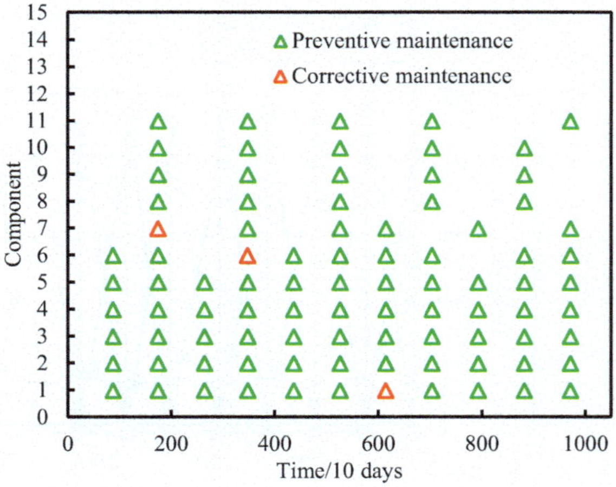

(3) Results of maintenance strategy 3

The maintenance results of each component of the subsea tree system under maintenance strategy 3 are shown in Fig. 15.13. The figure shows that each maintenance is due to the failure of a component. MIV has a total of nine failures, and PIV has one failure. Thus, MIV, CIV, AAV, PCV, XOV, AWV, and PIV are repaired every time, and AMV, PWV, SCSSV, and PMV are repaired twice.

The maintenance times for each component under the three proposed maintenance strategies are shown in Fig. 15.14. The maintenance time of maintenance strategy 1 is slightly longer than that of maintenance strategy 2. The maintenance time of maintenance strategy 3 is longer than that of maintenance strategies 1 and 2. The reason is that maintenance strategy 3 only repairs when the component has failed.

Fig. 15.13 Maintenance process of all components under maintenance strategy 3

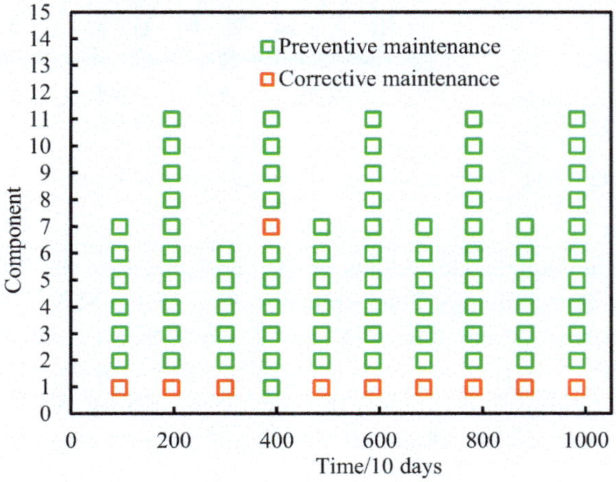

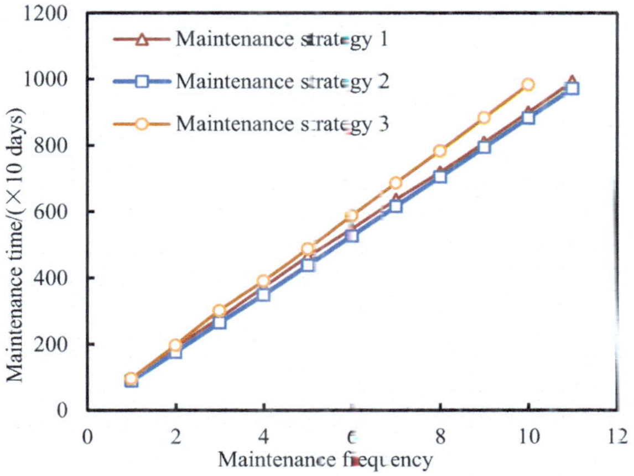

Fig. 15.14 Relationships between maintenance time and frequencies

However, maintenance strategies 1 and 2 schedule the maintenance activity before the failure, which will increase maintenance times.

Figure 15.15 displays the maintenance costs of the different components under the three maintenance strategies. In strategies 1 and 3, AWV has the highest maintenance cost, whereas in strategy 2, PIV incurs the highest maintenance cost. Notably, while PMV has the highest single maintenance cost among all components, its overall maintenance cost remains relatively low due to its lower frequency of maintenance, which is attributed to its higher RUL.

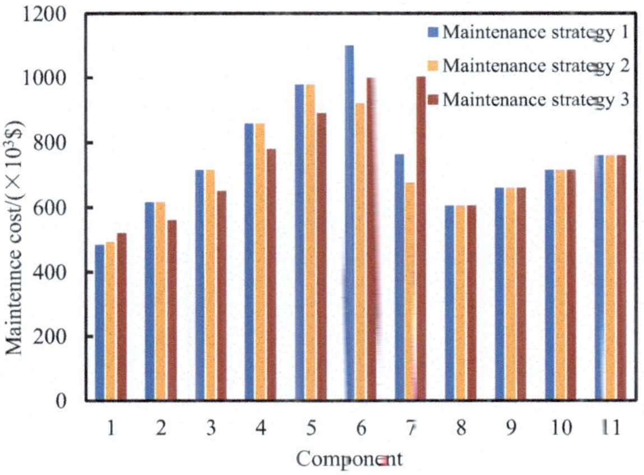

Fig. 15.15 Maintenance cost of different components under three maintenance strategies

15.3.2 Influence of Safety Time Threshold on Maintenance Cost

Figure 15.16 illustrates how the safety time threshold affects maintenance costs under varying degrees of inspection errors. When the inspection error increases from 4 to 28, the maintenance cost increases by 11.7% when optimal safety time threshold is selected. The maintenance cost first decreases and then increases as the safety time threshold increases. At a safety time threshold of 140 days, the maintenance cost initially experiences a significant decline followed by a gradual increase. A low safety time threshold increases the possibility of system failure, potentially leading to production losses. Conversely, a high safety time threshold shortens the maintenance intervals, resulting in more frequent maintenance operations. This, in turn, increases maintenance preparation costs and ultimately leads to a sharp increase in overall maintenance costs.

As the safety time threshold increases, both the second inspection interval and the maintenance preparation time become shorter. When the safety time threshold surpasses a certain point, emergency maintenance preparations become imperative, leading to a steep rise in maintenance preparation costs. When the safety time threshold increases to 140 days, the predicted RUL of the system after maintenance is less than the sum of T, DT, and LT, and maintenance is required at the first inspection. On the one hand, it saves the cost of maintenance inspection. On the other hand, it provides sufficient time for maintenance preparation such that the maintenance cost dramatically decreases. The possibility of failure is reduced because the safety time threshold is high. The maintenance cost also varies slightly with different inspection errors.

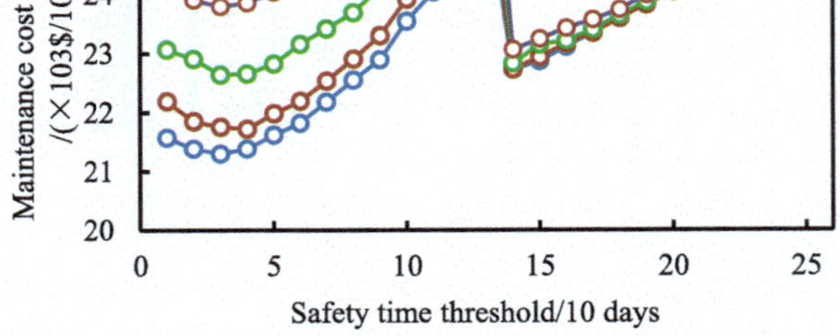

Fig. 15.16 Influence of safety time threshold on maintenance cost

15.3.3 Influence of Basic Inspection Period on Maintenance Cost

The influences of the basic inspection period on the maintenance cost under different production losses are shown in Fig. 15.17. As production loss per unit time increases from 200 × 10^3\$/10 days to 800 × 10^3\$/10 days, the maintenance cost increases by 6.1% when the basic inspection period is the optimal. With the increase in the basic inspection period, the maintenance cost initially declines and reaches its lowest point when the period is set at 700 days. Subsequently, the maintenance cost begins to rise. It reaches its maximum value when the basic inspection period is around 800 days, then begins to decline, and finally remains at a relatively stable level.

When the basic inspection period is short, the degradation detected during the first inspection is minimal, amplifying the impact of inspection errors and subsequently leading to a larger prediction error in the RUL. The probability of failure before the second inspection and maintenance is higher and the system will remain shut down in a longer period when the prediction error during the first inspection is greater. Consequently, as the basic inspection period lengthens, the maintenance cost per unit time initially decreases. However, if the basic inspection period becomes excessively long, the time for the second inspection shortens, necessitating emergency maintenance preparations that increase both the maintenance preparation cost and the maintenance cost per unit time.

As the basic inspection period further extends, the system may require maintenance activities during the first inspection. At this time, the maintenance preparation time is sufficient, the maintenance preparation cost decreases, and the maintenance cost per unit time decreases accordingly. When the basic inspection period is greater than 850 days, it has no effect on the formulation of inspection and maintenance strategy. As a result, the maintenance cost remains stable.

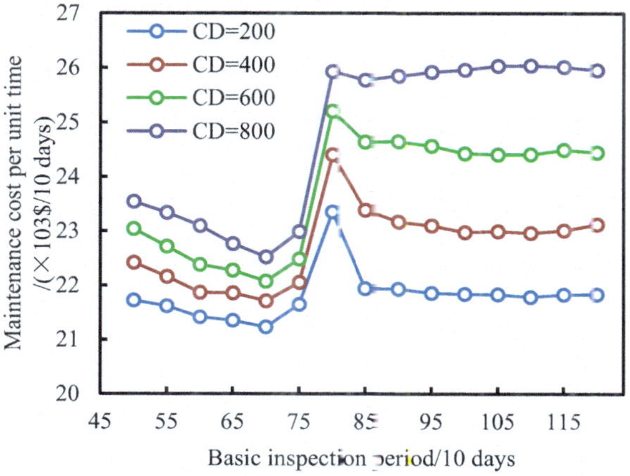

Fig. 15.17 Influence of basic inspection period on maintenance cost

15.3.4 Influence of Maintenance Threshold on Maintenance Cost

The influences of maintenance threshold on maintenance cost under different maintenance preparation costs are shown in Fig. 15.18. As the maintenance threshold increases, the maintenance cost per unit time decreases first and then increases. Fewer components meet the maintenance requirements during maintenance when the maintenance threshold is smaller. Shorter RUL of the system after maintenance reduces the maintenance interval time and increases the number of maintenances. Ultimately, the maintenance cost increases. Additionally, when the maintenance threshold is set higher, more components meet the maintenance requirements. When most components undergo repair during maintenance, the RUL of the system after maintenance is equal to that of MIV. As a result, the variation in maintenance intervals remains minimal, ultimately leading to an increase in the maintenance cost per unit time.

When the maintenance preparation cost increases from 500×10^3\$ to 2000×10^3\$ and the maintenance threshold increases from 700 days to their optimal values of 970 and 940 days, the maintenance cost per unit time decreases by 12.57% and 15.27%, respectively. The effect of the increase in the maintenance threshold on the decrease in the maintenance cost per unit time is more significant when the maintenance preparation cost is higher. Therefore, increasing the maintenance threshold becomes crucial in reducing the maintenance cost per unit time, especially when the maintenance preparation cost increases.

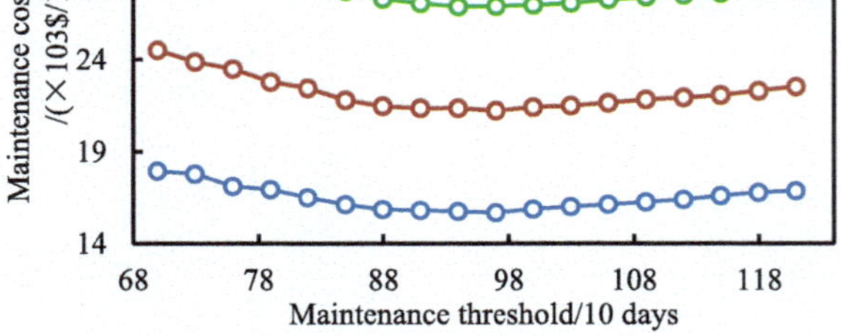

Fig. 15.18 Influence of maintenance threshold on maintenance cost

15.3.5 Discussion

The proposed CBM methodology is exemplified by its application to a subsea tree system. In maintenance strategy 2, the component exhibits a higher failure probability compared to strategy 1. The maintenance time of maintenance strategy 3 is the longest, which is due to that the component will only be repaired if it has failed. The maintenance cost of PMV is the highest of all components. However, the maintenance number of PMV is much lower than those of others. Specifically, the average maintenance cost of PMV in the overall simulation period is relatively low. Proper safety time threshold needs to be determined, which should be about 140 days. Furthermore, during the maintenance preparation phase, selecting a higher maintenance threshold in scenarios where maintenance costs are already high can help mitigate maintenance preparation cost.

Chapter 16
CBM with Discrete States

Abstract Currently, the CBM methods are determined based on expert experience to obtain failure thresholds. The simplistic perfect maintenance measure is utilized to achieve a perfect state. To address these limitations, a CBM method under discrete-state conditions is proposed for multi-component systems. A maintenance list is established based on the historical maintenance data, which contains various maintenance measures. Markov chain is employed to construct a single-component state transition model, which can calculate the component state transition matrix, and obtain the component state recovery matrix. A single-step model is established to predict the state transition probability for each component under different states and different times. The maintenance optimization model is applied to determine the overall maintenance plan of the multi-component system. The proposed method is illustrated using an example involving critical equipment of a subsea production system.

16.1 Introduction

CBM generally encompasses three basic steps: collecting the operating data to determine states, predicting the RUL, and proposing maintenance strategies. Specifically speaking, the effectiveness of a CBM plan is significantly influenced by the quality of condition monitoring (Karatu and Arslanoglu 2022), which requires complex inspections are necessary, and subsequently leads to higher costs. By implementing a partially observable Markov decision process, it becomes feasible to refine the quality of condition monitoring, ultimately contributing to a reduction in the overall maintenance cost (Nguyen et al. 2019). Besides, RUL is a crucial bridge between data acquisition and maintenance decision making. By using the observed degradation signals containing some health-related information of the system, and by constructing a degradation modeling framework grounded in maximum likelihood estimation, it is possible to examine these signals and make accurate predictions about the future degradation status of the system. For equipment operating in complex environments with few sensors, it is necessary to establish an equipment degradation model through

B. Cai et al., *Intelligent Operation and Maintenance for Subsea Production Systems*,
https://doi.org/10.1007/978-981-96-2677-9_16

historical data or accelerated degradation experiments to effectively compensate for data shortages. Based on the RUL prediction of system, maintenance strategies could be developed, including selecting the maintenance threshold. The issue of optimizing maintenance can be reframed within the context of a Markov decision process. By modeling the degradation states of each component by a discrete-time Markov chain, a finite state space for the entire system is obtained, which enables dynamic programming to determine optimal maintenance plans for a multi-component system.

For multi-component systems, CBM planning becomes more challenging owing to the varying degradation states among individual components. Most current CBM models for these systems rely solely on current state or degradation information. In the context of maintenance decision-making, failure distributions are commonly presupposed to be known constants, remaining unchanged throughout the process, which can compromise prediction accuracy (Shi et al. 2020, Wang et al. 2024). Moreover, certain CBM strategies prioritize optimizing replacement policies for individual components to maintain their degradation below a safety threshold, overlooking the variation in degradation rates across components. Consequently, these strategies may fail to accurately reflect the complex and dynamic maintenance requirements of the overall system (Wang et al. 2022).

To address the differences in degradation states among individual components, this chapter established maintenance model to outline the multi-level maintenance actions tailored to each component. A CBM framework for multi-component systems operating in discrete states is then constructed, emphasizing the analysis of component transition probability variations. This degradation model integrates Markov chain theory with the Wiener process. An optimization model for maintenance planning is formulated, incorporating the maintenance schedule, with the aim of maximizing the RUL while minimizing associated maintenance costs.

16.2 CBM Optimization Model Under Discrete-State Condition

To enhance the accuracy of system failure detection and facilitate more targeted repairs, decision-makers typically divide the system into distinct components, considering them as the fundamental units for maintenance operations. At the component level, a degradation prediction model and a state transition matrix are formulated. This approach can provide more precise maintenance measures and speed up the repair process by quickly locating problems when failures occur. However, due to factors such as complex environments, spatial constraints, and the need to control production costs, it is often impractical to accurately monitor the specific parts of the system. Consequently, comprehensive troubleshooting and maintenance operations can be generally conducted on the entire functional component, with maintenance strategies formulated at the system-wide level. To analyze the probability of state transitions of the components and the maintenance strategies of the system in a discrete state, the following assumptions or explanations are applied to a multi-component system:

(1) It is presumed that, in the absence of maintenance, the degradation trajectory of system components is monotonous and non-decreasing. The initial state of the component is 0.
(2) It is assumed that the degradation of one component is isolated and does not influence the degradation of other components. Furthermore, the maintenance of each component necessitates the system shutdown, implying a lack of interdependence among components.
(3) Each component has several maintenance measures, each with distinct varying levels of recovery effects. While the maintenance time and inspection time of the components are ignored, the inspection cost is taken into account.
(4) The degradation process of components is characterized by discrete states, and the state transition matrix is used to represent the transitions between each state.

A CBM method tailored for multi-component systems under discrete-state conditions is proposed. According to the historical operation data, the parameters for Wiener process are estimated to develop a single-step prediction model. This model predicts component performance degradation data based on Wiener process, allowing for real-time adjustments to the prediction data based on monitoring inputs. Once the amount of prediction data meets the calculation accuracy of the state transition matrix, the state degradation matrix could be derived by counting the performance degradation data. Based on historical maintenance data, the state recovery matrix is obtained by Monte Carlo simulation. To maximize the RUL of a system and minimize the maintenance costs during its service life, an optimization model for maintenance planning is established incorporating decision variables such as component maintenance time and maintenance measures. The flowchart of the model is illustrated in Fig. 16.1.

16.2.1 Establishment of the Model List

Creating a maintenance list (ML) based on historical maintenance information can facilitate the identification of appropriate maintenance measures for each component and formulate an effective strategy for CBM of multi-component systems under the current status. Each component has multiple maintenance measures, and maintenance measures $a1$, $a2$, ..., an are obtained based on the historical maintenance data of the component. Feature values are extracted, and the measures are classified based on the different maintenance duration, cost, and effects, as shown in Fig. 16.2. t is the maintenance time, c is the maintenance cost, and r is the performance recovery coefficient after the maintenance activity. L is the maintenance lever, which is decided by integrating the maintenance time, maintenance cost and performance recovery coefficient. Each measure of component is described by a parameter set $G = \{g^1, g^2, \ldots, g^m\}$. To classify the maintenance measure, the K-nearest neighbor (KNN)

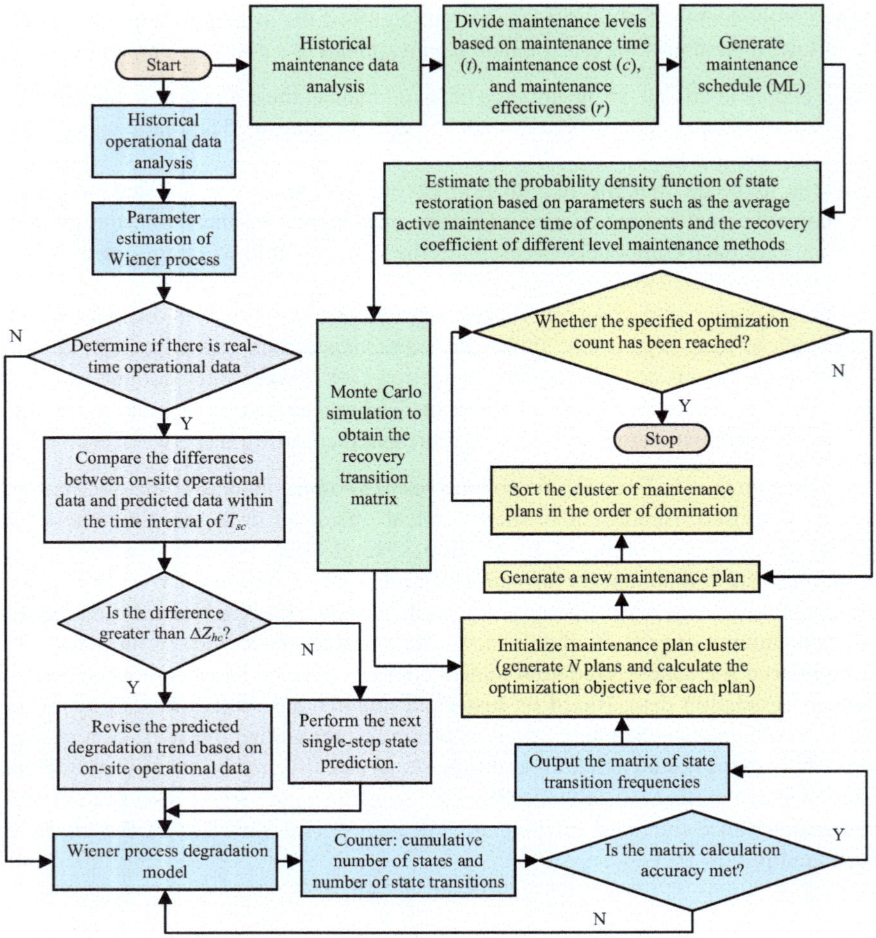

Fig. 16.1 Flowchart of maintenance plan optimization model

algorithm is used. The classification is based on the distance D between the test sample and known samples, instead of using a fixed K value.

The operating state of a component can typically be categorized into four types: normal state, attention state, abnormal state, and severe state. The component state initiates from the normal state and goes through all the above four states in order. Accordingly, the whole degradation process of the component is divided into four states, namely, the initial state, the stable degradation state, the waiting-for-repair state and the maintenance state. Correspondingly, the maintenance method is divided into four levels, represented by $L1 - L4$.

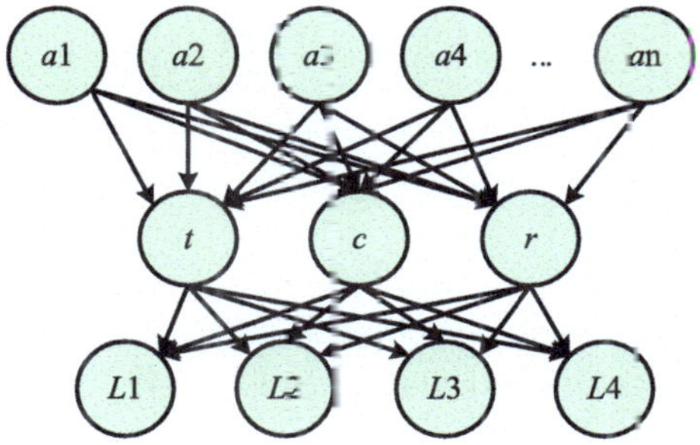

Fig. 16.2 Maintenance measures classification

16.2.2 Markov Chain

A Markov process is a stochastic process that exhibits the Markov property, which states that the future state of the system at any given time is solely determined by its current state and is independent of its past states. In the context, the system state transition process is modeled using a Markov model, which incorporates the following assumptions:

(1) The states of components are revealed upon periodic inspections.
(2) The maintenance policies are based on the detected state of system.
(3) At inspection, a maintenance action can begin without any delay.
(4) The inspection and repair time could be ignored compared to its long service lifetime.

To make the degradation simulation and maintenance optimization process tractable by using Markov chain, the continuous degradation level of each component is discretized into discrete degradation states according to the time point. In the Markov process $x(t)$, if the current value is specified, the past values will not affect the future values, i.e. if $t_{n-1} < t_n$, then:

$$P\{x(t_n) \leq x_n | x(t), t \leq t_{n-1}\} = P\{x(t_n) \leq x_n | x(t_{n-1})\} \tag{16.1}$$

Markov chain is a special type of Markov process that has a finite or countable infinite number of states: e_1, e_2, …, e_n. In a Markov chain, the future evolution is independent of the past, given the current state of the system. Markov chains can be either discrete-time or continuous-time processes, depending on whether the time index is discrete or continuous. The deterioration of the system is modeled by a discrete time Markov chain with a state space $E = \{e_1, e_2, \ldots, e_n\}$, and a time set $H = \{t_n = nT \mid T > 0, n \in N\}$ (Zhang et al. 2024). N is the set of non-negative integers and T is the time period for state transitions. For any positive integers l, n, and s, if $t_s < t_n < t_l$, then the following equation holds:

$$\begin{aligned}P\{x(t_l) = e_j|x(t_n) = e_i, x(t_s) = e_k\} \\ = P\{x(t_l) = e_j|x(t_n) = e_i\} \\ = p_{ij}(n, l)\end{aligned} \tag{16.2}$$

which represents the probability of the system in state e_i at time t_n and in state e_j at time t_l. In particular, when $l = n+1$,

$$P\{x(t_{n+1}) = e_j|x(t_n) = e_i\} = p_{ij}(n) \tag{16.3}$$

which represents the one-step transition probability of $x(t)$ at time t_n. Since starting from state e_i will always result in a single subsequent state, the transition probability $p_{ij}(n)$ satisfies the condition:

$$\sum_{e_j \in E} p_{ij}(n) = 1, 0 \le p_{ij}(n) \le 1 \tag{16.4}$$

According to the time-stationarity property for the homogeneous Markov chain (Sahin 2023), $p_{ij}(n, l)$ only depends on the time difference l-n. So $p_{ij}(n, l)$ can be described as:

$$p_{ij}(n, l) = p_{ij}(n)^{l-n} = p_{ij}^{l-n} \tag{16.5}$$

where $p_{ij} = \{x(t_{n+1}) = e_j|x(t_n) = e_i\}$ represents the probability of the system transitioning from state e_i to state e_j by one step at any time t_n.

The corresponding one-step state transition probability matrix is:

$$U = \begin{bmatrix} p_{00} & p_{01} & \cdots & p_{0m} \\ p_{10} & p_{11} & \cdots & p_{1m} \\ \vdots & \vdots & \ddots & \vdots \\ p_{m0} & p_{m1} & \cdots & p_{mm} \end{bmatrix}, p_{ij} \ge 0, \sum_{j=0}^{m} p_{ij} = 1 \tag{16.6}$$

where m refers to the total number of states.

The k-step transition probability of a homogeneous Markov chain can be expressed as:

$$U(n, n+k) = U^k \tag{16.7}$$

Given that the initial state is b_0 and the initial state vector is B_0, the probability distribution of the state after k-steps can be expressed as:

$$Q(k) = B_0 U^k \tag{16.8}$$

16.2.3 Establishment of the Degradation State Transition Matrix

(1) Establishment and modification of degradation prediction model

To accurately predict the states of various components using Markov chains, it is necessary to estimate the state transition matrix of the Markov chain based on the measured data. However, the operational data of the components may contain outlier values due to environmental factors or sensor inaccuracies, which can compromise the assessment and prediction of component performance. Therefore, a method is needed to eliminate the influence of these outlier values and obtain accurate component performance data. In this case, the Wiener process is used to fit the historical operating data of the component, effectively smoothing out the outlier values. This approach enables the extraction of more accurate component performance data at each discrete time point, thereby facilitating a more reliable assessment and prediction of the health status of the component based on the refined data.

The Wiener process, also known as Brownian motion, is more suitable for describing no monotone degradation process due to its increment adhering to a normal distribution centered at zero. In other words, it has not only matured theoretical system, but also good physical characteristics. Therefore, Wiener process has stronger applicability in practical engineering applications. As a key stochastic process in probability theory, the Wiener process offers robustness in describing complex phenomena. It is theoretically a continuous-time stochastic process. The expression for a Wiener process $\{Z(t), t \geq 0\}$ with a linear drift is as follows:

$$Z(t) = Z(0) + \lambda t + \gamma D(t) \tag{16.9}$$

where $Z(0)$ is the initial degradation amount; $Z(t)$ is the current degradation amount; λ is the drift coefficient, which reflects the degradation rate; γ is the diffusion coefficient; $D(t)$ refers to the standard Brownian motion.

Expectation maximization (EM) algorithm is used to analyze the degradation data, and the drift parameter and diffusion parameter of Wiener process are obtained. The specific steps of EM algorithm are as follows:

(1) Randomly initialize the model parameter θ ($\theta = [\mu, \lambda]^{\mathrm{T}}$) with an initial value θ_0.
(2) E-Step: Calculate conditional probability expectation of joint distribution:

$$Q_i(w_i) = p(w_i|z_i, \theta_j) \tag{16.10}$$

$$l(\theta, \theta_j) = \sum_{i=1}^{n} \sum_{zi} Q_i(w_i) \log \frac{p(z_i, w_i, \theta)}{Q_i(w_i)} \tag{16.11}$$

(3) M-step: maximize $l(\theta, \theta_j)$, and get θ_{j+1}:

$$\theta_{j+1} = \arg\max l(\theta, \theta_j) \tag{16.12}$$

(4) If the results are convergent, the algorithm ends.

According to the properties of the Wiener process, its increments are independent and follow a normal distribution. If $\Delta Z(t)$ represents the degradation amount of the state in the time interval Δt, then:

$$\Delta Z(t) = Z(t + \Delta t) - Z(t) \sim N(\lambda \Delta t, \gamma^2 \Delta t) \tag{16.13}$$

The probability density function of the degradation increment is expressed as:

$$f_{\Delta Z}(\Delta z | \lambda, \gamma) = \frac{1}{\sqrt{2\pi\gamma^2 \Delta t}} \exp(-\frac{(\Delta z - \lambda \Delta t)^2}{2\gamma^2 \Delta t}) \tag{16.14}$$

In the real-world operation of a component, there will be varying degrees of differences between on-site sensor monitoring data and historical statistical data. Additionally, the operating load and output adjustments of the production system can significantly influence the estimation of the remaining service life of the component, thereby affecting the accuracy of fault time estimation and leading to problems of excessive or insufficient maintenance. Therefore, the degradation model of the component obtained using limited on-site sensor data collected over a short period of time is revised, as shown in Fig. 16.3.

In Fig. 16.3, curve B represents the predicted data, while curve A represents the actual monitoring data. ΔZ_r and ΔZ_p respectively indicates the state difference between the on-site monitoring data and the predicted data within the time interval

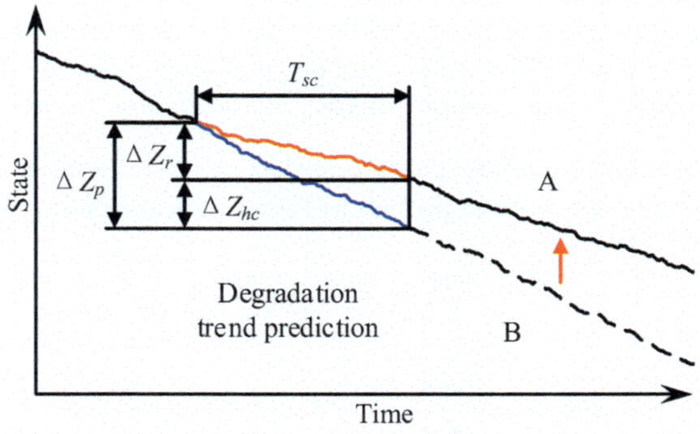

Fig. 16.3 Forecast data correction

T_{sc}. If the difference between the on-site component degradation amount and the predicted value is greater than a specified value Δz_{hc} during a limited time interval T_{sc}, the predicted data in the limited time interval T_{sc} will be replaced by the monitoring data.

(2) Discrete degradation state transition model based on Markov chain

Markov processes are used to study transfer relationships between states. If the variables describe the system change from one specific value or specific area to another, the system undergoes a state transition. If there exists a time point $t_n \geq 0$ such that $p_{ij}(n) > 0$, it implies a transition from state e_i to state e_j and could be denoted as $e_i \rightarrow e_j$. Conversely, if $p_{ij}(n) = 0$ holds for any time point $t_n \geq 0$, state e_i is deemed incapable of transition to state e_j.

Based on the performance degradation process of the component, the characteristic points are identified in the degradation process as a state of the component, and a single-component state transition model is formulated as shown in Fig. 16.4. In this model, the operation process of the component is divided into several states, and the state space is $E=\{e_0, e_1,\ldots, e_n\}$, where e_0 represents the perfect state of the component, i.e., the initial state. The model is divided into two processes: the state degradation process and the recovery process. During normal operation of the system, the component gradually degrades according to the state degradation model until it meets the maintenance conditions. At this point, the component state is corrected according to the recovery model. After repair, the component continues to degrade according to the state degradation model, and the process repeats.

By analyzing component performance degradation data, we can obtain the frequency of component state transitions noted as $f_{ij}{}^d$. Subsequently, the state transition probability $p_{ij}{}^d$ could be estimated as follows (Holmes and Hassini 2021):

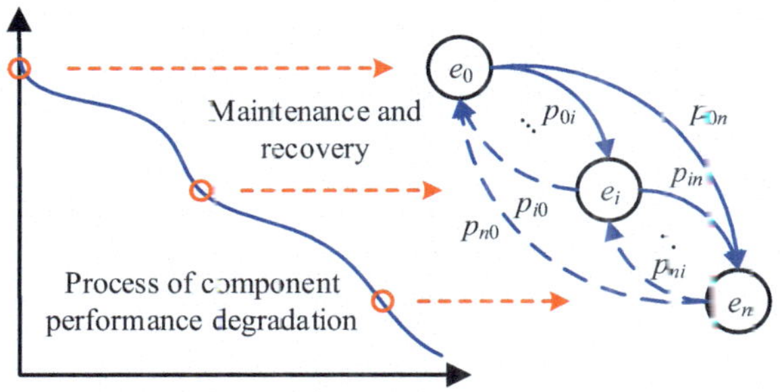

Fig. 16.4 State transition model for a single component

$$p_{ij}^{d} = \frac{f_{ij}^{d}}{\sum_{j=0}^{m} f_{ij}^{d}} \tag{16.15}$$

The state degradation matrix is

$$U_d = \begin{bmatrix} p_{00}^{d} & p_{01}^{d} & p_{02}^{d} & p_{03}^{d} \\ p_{10}^{d} & p_{11}^{d} & p_{12}^{d} & p_{13}^{d} \\ p_{20}^{d} & p_{21}^{d} & p_{22}^{d} & p_{23}^{d} \\ p_{30}^{d} & p_{31}^{d} & p_{32}^{d} & p_{33}^{d} \end{bmatrix}, p_{ij}^{d} \geq 0, \sum_{j=0}^{3} p_{ij}^{d} = 1 \tag{16.16}$$

16.2.4 Establishment of the Recovery State Transition Matrix

Upon reaching a specific state, it becomes imperative to promptly initiate a maintenance program for a component. Given variations in operational environment and functionality among different components, diverse maintenance procedures are required. These procedures for a single component are categorized into four distinct levels, each tailored to address variations in maintenance time, cost, and effectiveness. Following the maintenance activity, the component exhibits a different probability of returning to any previous state. The impact of the maintenance actions can be described as a normal distribution (Zhou et al. 2012, Hu et al. 2018, Ma et al. 2023). The probability of component recovery is shown as below:

$$P^{r} \sim N(\mu, \sigma) \tag{16.17}$$

There is a small probability that the component state will transfer to a better state when it degenerates, and the component state will not transfer to a worse state when it is maintained. The normal distribution is truncated within a suitable range [0,1].

The probability density function for the value $p^{r} \in [0, 1]$ of the state recovery probability is:

$$f_{pr}(pr|\mu, \sigma) = \frac{1}{K} \frac{1}{\sqrt{2\pi}\sigma} \exp\left(-\frac{(pr-\mu)^2}{2\sigma^2}\right) \tag{16.18}$$

$$K = \int_{0}^{1} \frac{1}{\sqrt{2\pi}\sigma} \exp\left(-\frac{(x-\mu)^2}{2\sigma^2}\right) dx \tag{16.19}$$

Calculate the mean and variance of the normal distribution based on the component's average active maintenance time t_{ar} and the recovery coefficient c_{ml} for different maintenance levels:

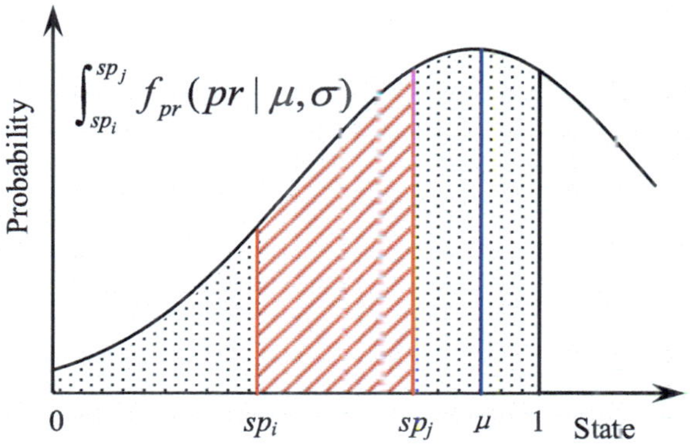

Fig. 16.5 Monte Carlo simulation

$$\sigma = \frac{1}{t_{ar}} \tag{16.20}$$

$$\mu = sp_n + (1 - sp_n) \cdot c_{ml} \tag{16.21}$$

where sp_n represents the current operating state of the component at runtime.

Using the Monte Carlo simulation method, as shown in Fig. 16.5, the state recovery probability $p_{ij}{}^r$ is obtained, and the state recovery matrix is calculated as follows:

$$U_r = \begin{bmatrix} p_{00}^r & p_{01}^r & p_{02}^r & p_{03}^r \\ p_{10}^r & p_{11}^r & p_{12}^r & p_{13}^r \\ p_{20}^r & p_{21}^r & p_{22}^r & p_{23}^r \\ p_{30}^r & p_{31}^r & p_{32}^r & p_{33}^r \end{bmatrix}, p_{ij}^r \geq 0, \sum_{j=0}^{3} p_{ij}^r = 1 \tag{16.22}$$

16.2.5 Optimization Goals for Maintenance Plans

When using a Markov chain to calculate the component state, the future state of the component at the next moment solely depends on the current state and is independent of its previous states, due to the nature of a Markov process. This single-step state prediction is achieved through the integration of online real-time monitoring data with state prediction data. Since the maintenance of complex systems entails scheduling of maintenance personnel, preparation of maintenance tools, spare parts management, etc., it becomes imperative to establish a maintenance schedule in advance. In order to ensure higher reliability at lower costs during system operation, a reasonable maintenance strategy is necessary. By continuously predicting the component state in a step-by-step manner, the state transition probability for each component at the subsequent moment is determined, as depict in Fig. 16.6.

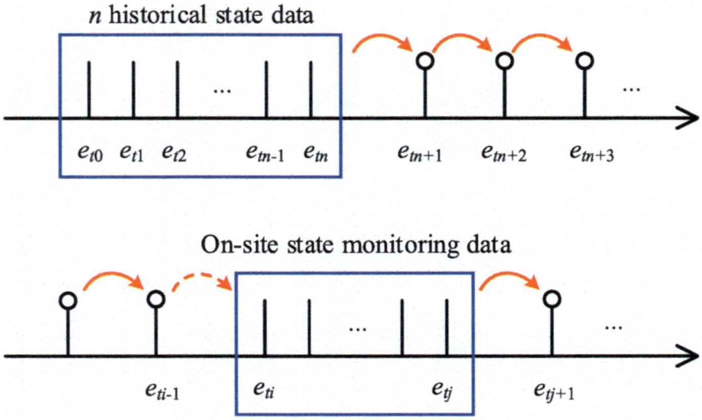

Fig. 16.6 Continuous single step prediction of components

At the onset of the state prediction process, the initial probabilities of state transitions are predicted based on the historical state data. The first state prediction, e_{tn+1}, is then generated, followed by the prediction of the next state and the determination of transition probability based on the established Markov state transition model. This process continues iteratively. Since on-site conditions vary, the equipment state predicted from historical data may differ from the actual state. Therefore, when there are on-site state monitoring data that do not match the predicted data, adjustments are made timely to ensure the accuracy of the prediction. During the continuous degradation process of a component, the time interval $Rul_{ij}(t_n)$ is defined for any two states e_i, $e_j \in E$. It represents the duration from state e_i to state e_j that the system reaches for the first time at time t_n.

$$Rul_{ij}(t_n) = \min\{\tau : x(t_n) = e_i, x(t_{n+\tau}) = e_j, \tau \geq 1\} \tag{16.23}$$

When $j = 3$, indicating that the component is in state e_3 during its degradation process, $Rul^k(t_n) = Rul_{i3}(t_n)$ is defined as the RUL of component k at time t_n. During the entire service life of the component, multiple maintenance activities are required. The average of the remaining life at all state transition moments is taken as the remaining life of the component, i.e.,

$$Rul^k = \frac{1}{c} \sum_{t_n \in H} Rul^k(t_n) \tag{16.24}$$

In multi-component systems, components are interconnected through logical series-parallel relationships. The failure of a single component does not necessarily cause a system shutdown but does affect the overall operational efficiency of the system. Therefore, the remaining life of the system is determined by the minimum value of remaining life among all components, i.e.,

$$R_s = \min(Rul^1, Rul^2, ..., Rul^N) \tag{16.25}$$

Once the predicted state reaches e_3, the component needs immediate maintenance. Maintenance operations require advance preparation, and due to the substantial costs associated with preparation work for a single maintenance operation, it is common to repair multiple components simultaneously. The maintenance cost is categorized into system maintenance cost and component maintenance cost. System maintenance cost can be understood as maintenance preparation cost, which includes inspection cost, monitoring cost and so on. The component maintenance cost includes equipment rental cost and spare part cost for the individual component. Therefore, the maintenance cost of a maintenance activity can be expressed as:

$$C_M = C_s + \sum_{i=1}^{N} \varepsilon C_e^i \tag{16.26}$$

where C_M is the cost of a maintenance activity, C_S is the system maintenance cost, ${C_e}^i$ is the maintenance cost of the i-th component, and ε is the maintenance operation flag. If the component meets the maintenance conditions, $\varepsilon = 1$; otherwise, $\varepsilon = 0$.

An optimization model for maintenance planning is formulated, incorporating variables such as component maintenance duration and measures, with the aim of maximizing the system RUL and minimizing maintenance expenses throughout its service life. In the optimization process, a maintenance scheme cluster is firstly initialized. Each maintenance scheme involves maintenance level and maintenance time for each component, and the optimization goals are calculated based on the initialized maintenance scheme. The maintenance scheme cluster is organized based on the primary order of importance. In the event that the optimization objectives are not met, new maintenance schemes are developed to repeat the optimization process.

16.3 Case of Subsea Production System

16.3.1 Introduction of Subsea Production System

This section introduces the subsea production system as a case study. The description of the subsea production system could refer to Sect. 1.4.1. A CBM model is developed for seven crucial components of the subsea production system, namely the master control station (MCS), electrical power unit (EPU), hydraulic power unit (HPU), umbilical, subsea distribution unit (SDU), subsea control module (SCM), and subsea Christmas tree (Tree). Due to limited operational data for the system, the component degradation model is established based on the failure rate data of critical equipment in the subsea production system, as shown in Table 16.1. The OREDA database covers reliability data from a wide range of equipment used in oil and gas exploration and production. According to the statistics in the OREDA database, the overall degradation performance of each component of the subsea production system in the service stage roughly conforms to the exponential degradation model.

The degradation time for a single component is 300 days, and the total service time of the system is 3000 days. The optimal number of individuals for maintenance optimization is 50. According to the data in Table 16.1, degradation data can be generated for the parameter simulation of Wiener process. The predicted data can be furtherly used for calculating the state transition matrix based on Eqs. (16.15) and (16.16). In the process of component degradation, the difference between on-site operational data and predicted data in the time interval T_{sc} is compared. If it is greater than the specified value, prediction model will be modified according to on-site operational data.

The operational process of the components is divided into four states, with the state space $E = \{e_0, e_1, e_2, e_3\}$, where e_0 is the perfect state, i.e., the initial state of the component; e_1 is the stable degradation state; e_2 is the waiting-for-repair state, where the component can still operate stably for some time, but there is a high probability of failure; e_3 is the repair state where the component must be repaired immediately to avoid failure. When a component enters the repair state, it is imperative to stop operating immediately and carry out repairs together with some components in the previous state to reduce maintenance costs. The state transition model is shown in Fig. 16.7.

The maintenance methodology is categorized into four distinct levels. Levels 1, 2, and 3 correspond to imperfect maintenance, whereas Level 4 signifies perfect maintenance. Each maintenance level yields varying degrees of maintenance effectiveness. The maintenance costs and times for different maintenance levels are determined based on the maintenance records about the subsea production system in the South China Sea. Maintenance costs for different maintenance levels are shown in Table 16.2, and maintenance times for different maintenance levels are shown in Table 16.3. For maintenance level 1, the performance recovery coefficient is set as 0.5, the performance recovery coefficient in the maintenance level 2 is set as 0.65, and the performance recovery coefficient in the maintenance level 3 is set as 0.8. The

Table 16.1 Failure and maintenance data for subsea production system critical equipment

	Average failure rate/(10^{-6}/h)	Variance of failure rate	Active repair time/h
MCS	123.90	145.84	8.80
HPU	83.45	167.89	9.80
SCM	25.11	41.90	11.30
Tree	17.79	25.41	20.50
EPU	17.68	24.05	8.30
Umbilical	16.46	32.29	33.75
SDU	10.31	11.17	26.90

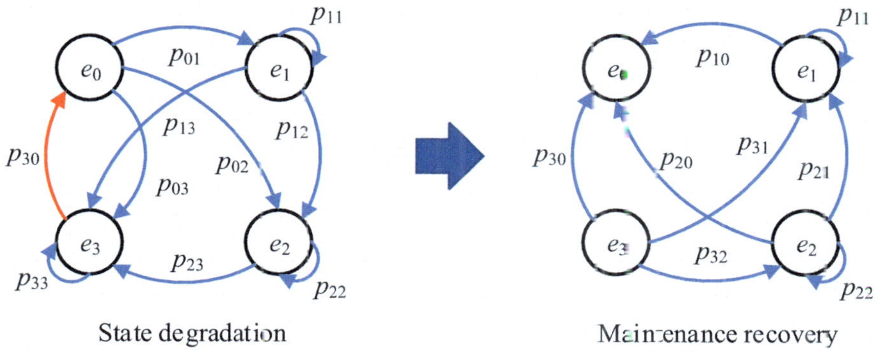

Fig. 16.7 Single component state transition model of subsea production system

performance recovery coefficient in the maintenance level 4 is 1, that is, the perfect maintenance can restore the component state to the initial state.

Taking the subsea distribution unit as an example with an average active repair time of $t_{ar} = 8.8$ and a repair measure recovery coefficient of c_{m_i}=0.6, the probability

Table 16.2 Maintenance costs for different maintenance levels

	Level 1 c/$\times 10^4$CNY	Level 2 c/$\times 10^4$CNY	Level 3 c/$\times 10^4$CNY	Level 4 c/$\times 10^4$CNY
MCS	1.00	2.00	3.00	4.00
HPU	0.80	1.60	2.40	3.20
SCM	0.40	0.80	1.20	1.60
Tree	4.00	8.00	12.00	16.00
EPU	0.60	1.20	1.80	2.40
Umbilical	1.20	2.40	3.60	4.80
SDU	0.70	1.40	2.10	2.80

Table 16.3 Maintenance times for different maintenance levels

	Level 1 t/Day	Level 2 t/Day	Level 3 t/Day	Level 4 t/Day
MCS	1	1	2	2
HPU	1	1	2	2
SCM	2	2	3	4
Tree	2	4	6	8
EPU	1	2	2	3
Umbilical	3	5	7	10
SDU	2	3	5	7

density function of state recovery is calculated as follows:

$$f_{pr}(pr|\mu,\sigma) = \frac{1}{K} \cdot \exp(-38.72 \cdot (pr - 0.4sp_n - 0.6)) \tag{16.27}$$

$$K = \int_0^1 \exp(-38.72 \cdot (x - 0.4sp_n - 0.6)^2)dx \tag{16.28}$$

The probability matrix of state recovery is obtained through Monte Carlo simulation. The resulting degradation matrix and the recovery matrix under maintenance level 1 are as follows:

$$U_d = \begin{bmatrix} 0.855 & 0.002 & 0 & 0 \\ 0.145 & 0.994 & 0.007 & 0 \\ 0 & 0.004 & 0.975 & 0.001 \\ 0 & 0 & 0.018 & 0.999 \end{bmatrix}, U_r = \begin{bmatrix} 1.000 & 0.016 & 0.009 & 0 \\ 0 & 0.984 & 0.954 & 0.499 \\ 0 & 0 & 0.037 & 0.311 \\ 0 & 0 & 0 & 0.190 \end{bmatrix} \tag{16.29}$$

During the maintenance optimization phase, given the derived probability matrix for system component state transitions, three strategies are presented to determine the optimal maintenance time and component maintenance plan.

Strategy 1: The probability of component transition to each state at the next time point is calculated based on the proposed state transition model. The state with the highest transition probability is selected as the current state. When the state is in e_3, the maintenance operation is adopted.

Strategy 2: The next state of the component is simulated by using random probability. When the component is in state e_2, it enters the waiting for maintenance stage (the maintenance time and plan are selected based on the system maintenance plan). When the component is in state e_3, it requires immediate maintenance, and the system enters the maintenance state.

Strategy 3: The probability of state transitions is analyzed during the degradation process of the component and the characteristics of the component degradation is identified. The optimal maintenance threshold is determined when the state transition probability exceeds this threshold, perform maintenance operation.

16.3.2 Results and Discussion

(1) Analysis for component state transition probability

Throughout the operational lifespan of the equipment, degradation models are formulated for each individual component, accompanied by a component state transition

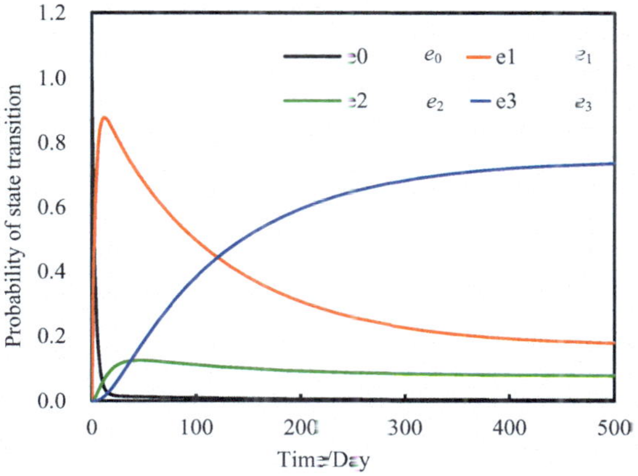

Fig. 16.8 State transition probability of a component

model and a maintenance table, both of which are derived from historical maintenance data. When the initial state is $B_0 = [1, 0, 0, 0]$, the probability of component state transition during a 500-day period is shown in Fig. 16.8.

It shows that the state transition probability of state e_3 intersects with states e_1 and e_2 at one point each. Before 40 days, the state transition probability of e_2 surpasses that of e_3, indicating that the component has a very low probability of entering a maintenance state. At this time, the reliability of the component is high, and the system can be operated stably. From 40 to 120 days, the state transition probability of e_3 exceeds that of e_2 but is still less than e_1, and the component has a relatively low probability of entering a maintenance state. At this time, the reliability of the component is lower, but it can still operate stably. After 120 days, the state transition probability of e_3 exceeds that of e_1 and e_2, indicating that the component has a high probability of entering a maintenance state, and the reliability of the component is very low with a high probability of failure.

Based on the transition trends of the status probability for key equipment, the component operation process can be divided into four distinct stages: initial operating stage, stable degradation stage, maintenance waiting stage, and shutdown maintenance stage. The status probability transition trends are shown in Fig. 16.9. When there is no external intervention, multiple components of the system go through these four stages independently.

Once the system starts up, its components enter the initial operating stage. During the short period following the system activation, the probability of transition to state e_0 decreases rapidly, while the probability of transition to state e_1 increases rapidly, reaching its peak. This indicates that the component will quickly transition from its initial perfect state to the stable degradation stage, maintaining a relatively high transition probability for a long time. The component will not return to its perfect state unless a perfect maintenance operation is carried out. In the subsequent long duration, the probability of transition to state e_1 gradually decreases, while the probability of transition to state e_2 decreases slowly and remains stable, and the probability of

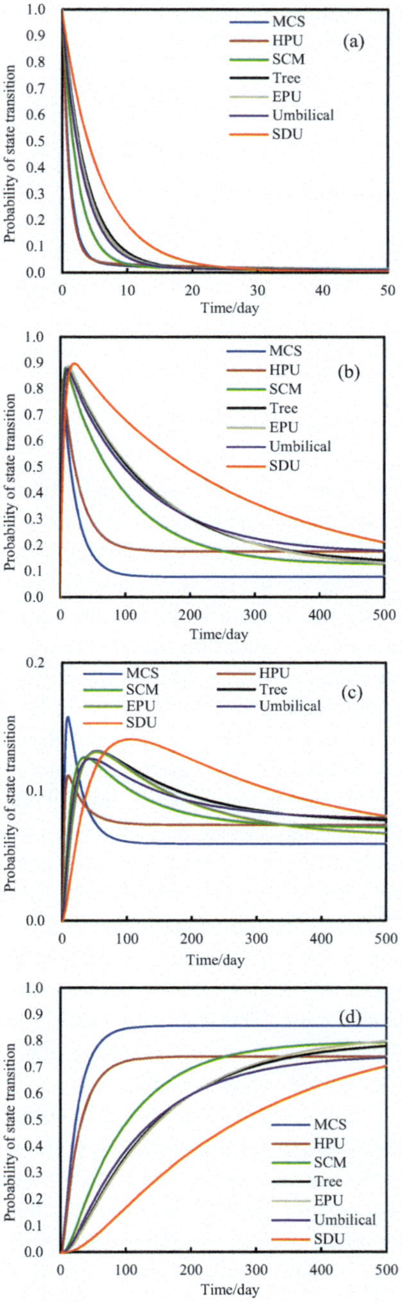

Fig. 16.9 Probability changes of state transition for critical equipment in different stages, **a** initial operating stage, **b** stable degradation stage, **c** maintenance waiting stage, **d** shutdown maintenance stage

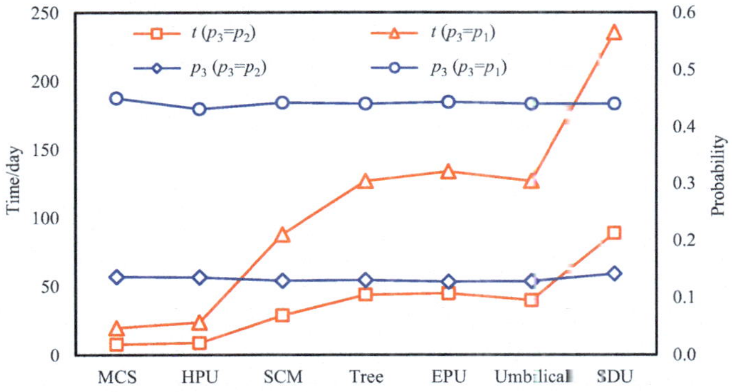

Fig. 16.10 Time and probability of state intersection

transition to state e_3 gradually increases. This indicates that as the system operates, the component's state gradually transitions from the stable degradation stage to the maintenance waiting stage and ultimately to the maintenance stage.

The intersection times and probabilities of the state transition probability curves for components e_1, e_2, and e_3 are shown in Fig. 16.10, where p_1, p_2, and p_3 represent the transition probabilities for states e_1, e_2, and e_3 respectively. It is evident that the intersection times among the states of different components exhibit notable variations, primarily attributed to the distinct fault rates of different components resulting from the unique structures and working environments. Components with higher fault rates have a faster increase in the probability of transition to the state e_3. The probabilities of intersection among various component states are basically consistent, serving as a foundation for assessing the component state and determining maintenance threshold. Once the probability reaches the maintenance threshold or is identified as being in the maintenance stage, the system immediately shut down for maintenance operations.

In the maintenance decision-making phase, simulations are conducted for the three proposed maintenance strategies to determine the component state and the maintenance time based on the state transition probability of the component. The proposed maintenance strategy is optimized using an optimization algorithm to obtain the optimal maintenance interval and maintenance measures for each component of the subsea production system under this strategy.

(2) Results of three maintenance strategies

1. Strategy 1

The current state is determined by considering the maximum probability of the state transition at the current time, whereas the subsequent state is computed by applying the one-step state transition rule. The selection of current state may be affected by the calculation method of the probability of state transition. This strategy simulates the method of determining maintenance time with fixed threshold. The maintenance time and measures for each component within 200 days are shown in Fig. 16.11.

Fig. 16.11 Maintenance plan for each component within 200 days in strategy 1

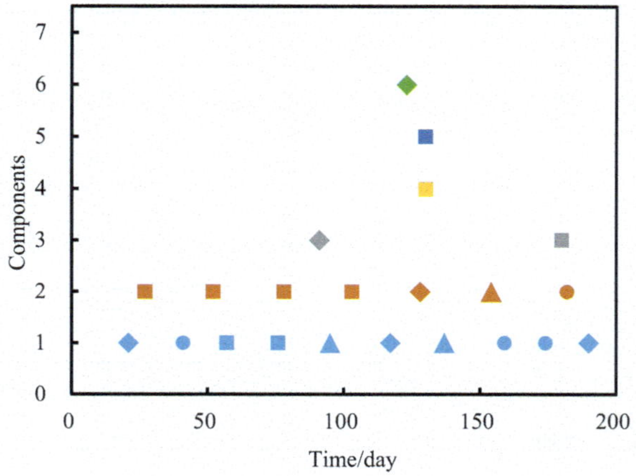

In the figure, "○" represents maintenance level 1, "□" represents maintenance level 2, "◊" represents maintenance level 3, and "Δ" represents maintenance level 4. Components 1 to 7 represent the master control station, hydraulic power unit, subsea control module, subsea Christmas tree, electrical power unit, umbilical, and subsea distribution unit, respectively. Maintenance activities were performed on the hydraulic power unit at maintenance level 2 around 27, 52, 78, and 103 days, with maintenance intervals that remained consistent and closely aligned with the scheduled maintenance.

The state probability changes of the subsea distribution unit within 550 days are shown in Fig. 16.12. On the 246th and 493rd days, the transition probabilities of the e_1 and e_3 states were equal. If no maintenance operation was performed, the transition probability of the e_3 state would exceed that of the other states. At this point, it was determined that maintenance was needed, and maintenance at level 4 was performed to restore the transition probabilities to the initial level. The subsea distribution unit underwent a total of 12 maintenance operations during its operation, including 8 maintenance operations at level 4. It can be seen that for components with a slower degradation rate, this strategy tends to favor the use of high-level maintenance measures.

For strategy 1, the optimization results about the average cost and the average RUL for all maintenance plans are shown in Fig. 16.13. It can be shown that the maintenance cost and system RUL are centered in a certain area. The strategy 1 simulates the method of determining maintenance time with fixed threshold. The optimal RUL is 34.95 days, with an average maintenance cost is 119.2×10^3 CNY.

2. Strategy 2

Throughout system operation, components exhibit varying probabilities of transitioning to any given state. Based on the current state transition probability of the

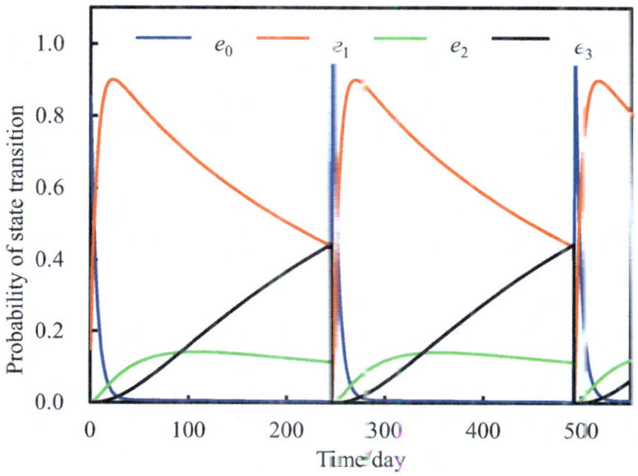

Fig. 16.12 Change of state transition probability of subsea distribution unit within 500 days

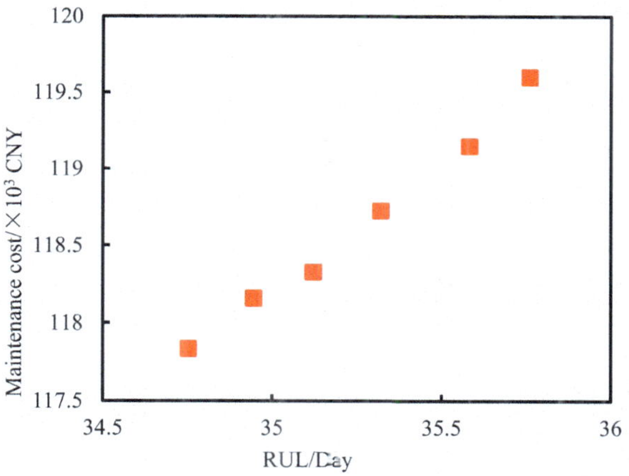

Fig. 16.13 The Pareto front solutions of the strategy 1

component at a given time, the component's state is determined through random simulation. The decisions regarding whether or not to conduct maintenance on the component and which maintenance measure to employ are made based on the obtained state. This strategy simulates the assessment method used by field operators. Due to the lack of accurate calculation methods and assessment criteria, operators often rely on experience to evaluate the running state of equipment based on on-site sensor data. The maintenance plan for different components over a 200-day period are shown in Fig. 16.14. The repair frequencies for components of different levels are shown in Table 16.4.

Under this plan, the average number of repairs per component is 204.42, indicating a higher frequency of maintenance compared to policy one. All components in the system undergo more maintenance operations than under policy one. The longest average maintenance interval among all components is 26.55 days, while the shortest

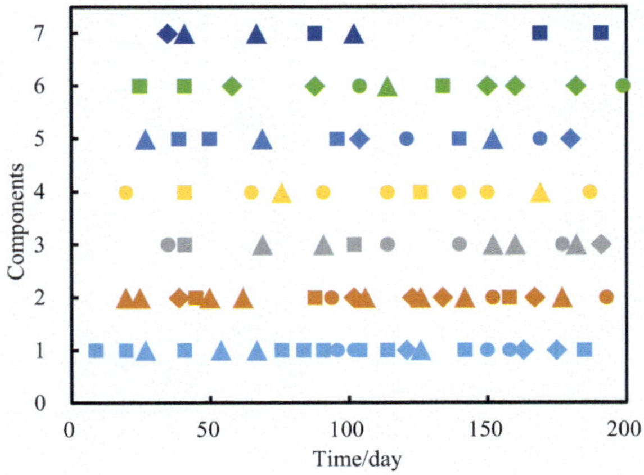

Fig. 16.14 Maintenance plan for each component within 200 days in strategy 2

Table 16.4 Maintenance times for each component

Level	MCS	HPU	SCM	Tree	EPU	Umbilical	SDU
1	87	65	35	51	33	50	28
2	93	67	53	44	37	39	28
3	79	84	50	34	40	39	35
4	94	91	47	27	42	37	22
Total	353	307	185	156	152	165	113
Maintenance intervals/day	8.50	9.77	16.22	19.23	19.74	18.18	26.55

is 8.50 days. There is no clear pattern in component selection of different level maintenance measures, unlike in policy one. The maintenance time chart shows that lower-level maintenance has shorter maintenance intervals.

With strategy 2, the state changes of the subsea distribution unit and subsea control module are shown in Fig. 16.15. As components degrade, their states gradually worsen until a maintenance operation is performed. Lower-level maintenance operations result in minor recovery of the component state, while higher level maintenance operations result in greater recovery, sometimes even approaching the initial operating stage. Different level maintenance operations allow components to remain in the stable degradation stage of state e_1 for as long as possible during operation. The state e_2 is the waiting-for-maintenance state. As shown in the figure, the subsea distribution unit enters this state at 23rd days and remains the state for a long time. This methodology is adopted to minimize downtime costs and prioritize the repair of as many components in lower states as possible during each downtime. When the time reaches 34 days, this component enters state e_3, and the subsea control module runs in state e_2. Therefore, both components undergo maintenance simultaneously, reducing the maintenance cost to some extent.

For the strategy 2, the optimization results about the average cost and the average RUL for all maintenance plans are shown in Fig. 16.16. Strategy 2 can be understood

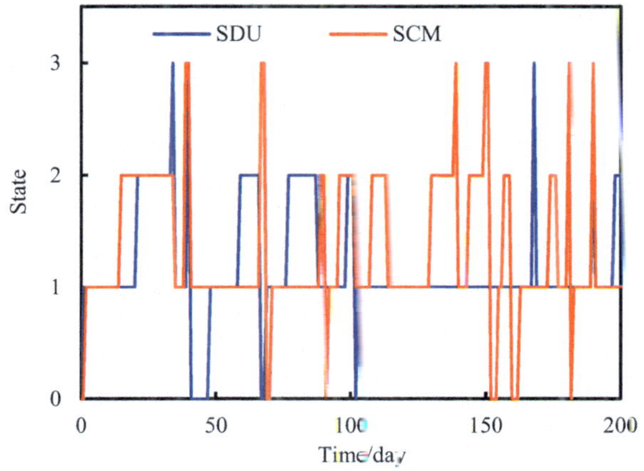

Fig. 16.15 State change of subsea distribution unit and subsea control module within 200 days

as a state optimal scheme. For the complex running environment, the strategy 2 should be chosen to ensure a better system running status to cope with the impact of environment. It can be seen that the maintenance cost remains in a stable value, and the average maintenance cost is 7.6×10^3 CNY. The optimal RUL is 104.65 days, which is obviously larger than that of the strategy 1.

3. Strategy 3

Based on the characteristics of the state transition probability curve, the intersection points between state e_1 and other curves is taken as the key point, i.e., $sp_0 = 0.99$, $sp_1 = 0.13$, $sp_2 = 0.44$. The optimized maintenance plans for each component in the system are obtained, and the maintenance plans for each component in a 200-day period are shown in Fig. 16.17.

During system operation, the MCS conducted 171 maintenance operations of different levels, while the SDU underwent 30 maintenance operations of different

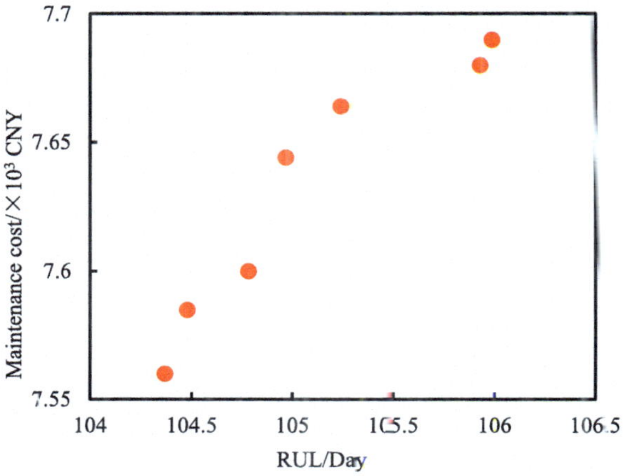

Fig. 16.16 The Pareto front solutions of the strategy 2

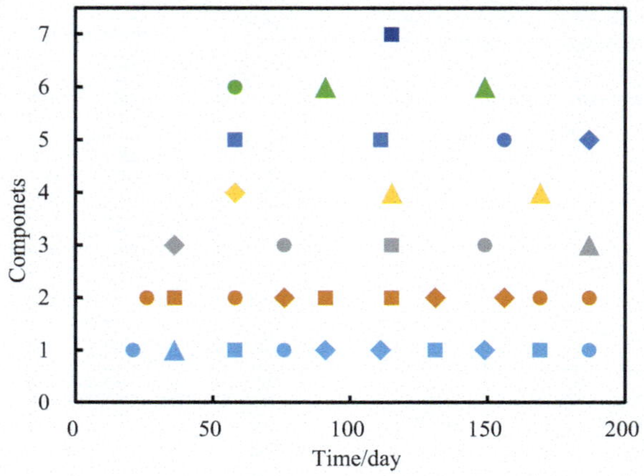

Fig. 16.17 Maintenance plan for each component within 200 days in strategy 3

levels. On average, each component underwent 81.14 maintenance operations. Among the four maintenance levels, components with lower failure rates tend to undergo level 1 simple maintenance and level 4 replacement operations. The state transition probability curve for the SDU over a 500-day period is shown in Fig. 16.18. It can be seen that after 291 days, this component adopted level 4 maintenance, and the transition probability for state e_0 became 1. When other maintenance measures were used, there were different probabilities of returning to the previous state. For example, level 2 maintenance was carried out on the 115th and 473rd days, while level 3 maintenance was carried out on the 202nd and 390th days. During normal system operation, the transition probabilities for states e_0 and e_1 of each component remained at a high level.

Regarding the strategy 3, Fig. 16.19 presents the optimization results about the average cost and the average RUL for all maintenance plans. The optimal RUL is 57.17 days, and the corresponding maintenance cost is 6.69 × 10^3 CNY. In the

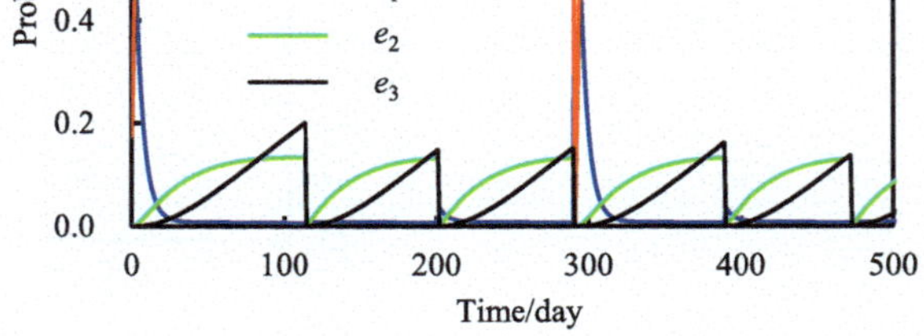

Fig. 16.18 Change of state transition probability of umbilical within 500 days

Fig. 16.19 The Pareto front solutions of the strategy 3

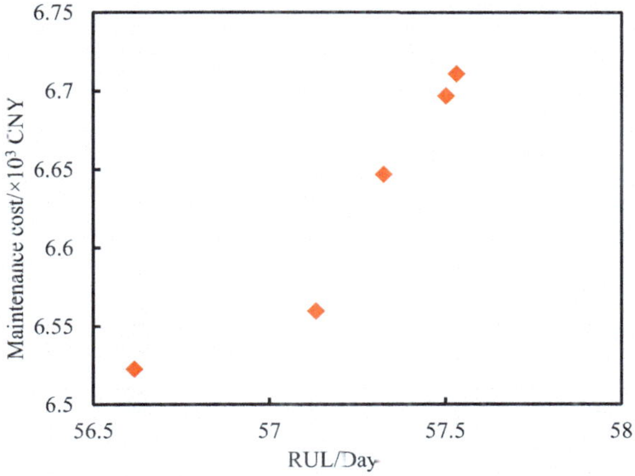

strategy 3, the average maintenance cost is lower than that in the strategy 2. On the premise of ensuring the system reliability, the maintenance cost is reduced as much as possible. In order to ensure a stable operational environment while also minimizing maintenance costs and conserving resources, Strategy 3 should be chosen as the preferred option.

(3) Discussions of three maintenance strategies

The three proposed maintenance strategies cater to distinct application scenarios, offering decision-makers a range of options to tailor a suitable maintenance plan based on factors such as maintenance budget constraints or system reliability requirements. The comparison of the average RUL and average maintenance cost under the three strategies is shown in Fig. 16.20. The maintenance frequencies of each component under different strategies are shown in Table 15.5. It is observed that the maintenance frequency of a component is closely related to its degradation rate. All three strategies demonstrate that components exhibiting faster degradation rates necessitate more frequent maintenance interventions to maintain stable operational performance.

The comparison of the average RUL and total maintenance costs under the three strategies is shown in Fig. 16.20. More frequent maintenance operations ensure high system RUL but require higher maintenance costs in the strategy 2, the RUL and the maintenance cost are the highest. But the total maintenance cost under strategy 3 is 79.01% of that under strategy 1, but the average remaining service life of components is increased by 63.60% under strategy 1. This highlights the importance of accurate assessment of component degradation status, which can lead to enhanced system reliability and reduced maintenance costs.

The average maintenance costs for each maintenance strategy are calculated. The average maintenance costs under each strategy are 119.4 × 10^3 CNY, 76.7 × 10^3 CNY, and 65.6 × 10^3 CNY, respectively. Although the maintenance frequency of components under strategy 1 is much lower than that under the other two strategies,

Fig. 16.20 Comparison of RUL and total maintenance cost for different strategies

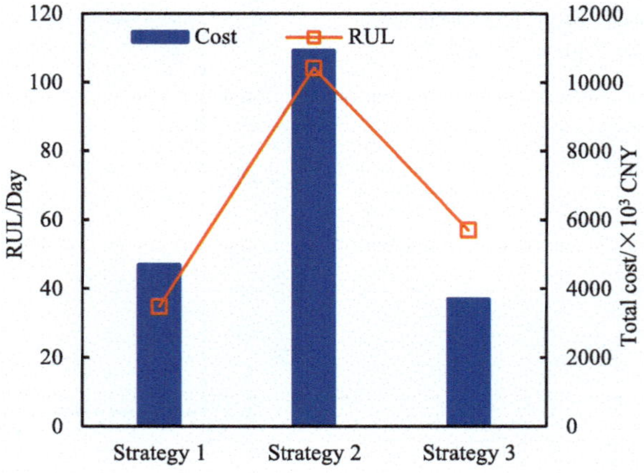

Table 16.5 Number of maintenance times for each component in the three strategies

Strategy	MCS	HPU	SCM	Tree	EPU	Umbilical	SDU
1	153	121	35	24	25	25	12
2	353	307	185	156	152	165	113
3	171	144	68	49	51	55	30

the average maintenance cost is the highest. In strategy 1, the component is maintained when it is about to fail, or when there is an omen of failure. The system will be frequently shut down for maintenance, resulting in high maintenance cost. According to the characteristics of the state transition probability curves of components, in accordance with the state assessment method of strategy 1, components do not enter the waiting-for-maintenance state. Upon entering state e_3, the component undergoes immediate repairs, leading to system downtime for each component's maintenance and significantly increasing maintenance costs. In systems where downtime losses are significant, the opportunity maintenance method should be considered to mitigate these losses and improve overall system efficiency.

Chapter 17
CBM with Heterogeneous Failure Dependence

Abstract Considering complexity of subsea production systems, interactions among components are often characterized by failure dependences that may accelerate their degradation. Besides, not all components share the same failure dependences due to system design and functional interactions. Such heterogeneous failure dependences further complicate maintenance strategies and activities in the dependent multi-component subsea production systems. This chapter introduces a comprehensive approach to assess heterogeneous failure dependences and a maintenance optimization model using Markov processes tailored for multi-component systems. The method is demonstrated through a practical example of the subsea transmission system, highlighting the impact of heterogeneous failure dependences. The results indicate that the proposed framework and model can optimize maintenance strategies, enhancing system availability while reducing maintenance costs.

17.1 Introduction

The complexity of many technical systems is escalating, with multiple components are becoming more vulnerable to failure dependence. Within such complex systems with failure dependences, the failure of a component or some components may accelerate the degradation of other components, usually increasing their failure probabilities and may lead to a series of failures. This phenomenon is referred to as a cascading failure (CAF). The degradation or failure of the first component is defined as the triggering event of the cascading process. The failed component is defined as the triggering component. The components among which CAFs occur, are identified as dependent components or coupled components. Failure dependence, as implied by its terminology, indicates the relationship between components that the failures of the components are dependent. In this context, the scope of failure dependence could be extensive. Firstly, with regard to failure mechanisms, failure dependence identifies failures caused by shock or damage, as well as those that stem from loading dependence. Secondly, concerning failure manifestation, failure dependence induces changes in the states or failure rates of the components affected. Finally, in the context

B. Cai et al., *Intelligent Operation and Maintenance for Subsea Production Systems*,
https://doi.org/10.1007/978-981-96-2677-9_17

of consequences, failure dependence has the potential to result in not only the immediate end of component operation, but rather the degradation of components. Thus, the failure dependences can be classified into two types (Zhao et al. 2023) to encompass the failure and degradation interactions more comprehensively and accurately: (1) Type I failure dependence: the direct damage caused by the initial failure can trigger other failures. Specifically, a component subject to type I failure dependence may fail due to the combined impact of its inherent degradation and the shock from failures in other components. (2) Type II failure dependence: the working load is redistributed in the overall system and the load redistribution leads to other failures. Specifically, a component suffered type II failure dependence may fail due to the combined impact of its inherent degradation and the accelerated degradation caused by failures in other components.

Due to diverse system layout and functional interactions, not all components have identical failure dependence. In complex systems, failure dependences are generally not single and identical but rather multiple and heterogeneous, introducing further complexities in the exploration of failure dependence in operation and maintenance. To the best of our knowledge, most current studies consider a system with two (kinds of) components with identical failure dependence. However, it is inconsistent with reality as the multi-component system has more than two components and their failure dependence is complex and heterogeneous (Kong et al. 2022). Heterogeneous failure dependences refer to such situations where there are at least two types of non-identical failure dependence in a complex multi-component system. Therefore, a flexible framework to model the heterogeneous failure dependences in the context of engineering is desired. In this chapter, we focus on modeling the heterogeneous failure dependence within a complex multi-component system. Compared to modeling methodologies applied only in a two-component system, or in a multi-component system with identical components, this modelling approach may also apply in a multi-component system with non-identical components. Compared to the current failure dependence modelling approaches, heterogeneous failure dependences modelling accounts better for the interaction and dependence among nonidentical components in a complex multi-component system.

Exploration to identify the most effective maintenance policies is essential for achieving a balance between system performance, asset lifespan, resource allocation, and cost-effectiveness within complex systems with failure dependence. Based on the model to characterize the degradation state and framework to evaluate heterogeneous failure dependences, a general maintenance model for systems with heterogeneous failure dependences could be constructed. Given that the degradation status can be observed or measured, CBM emerges as one of the most widely applied maintenance strategies to keep system reliability while reducing maintenance cost. Among the maintenance activities, CBM is recognized as a proactive approach preceding system failures, and is proven to be more cost-effective, compared to traditional maintenance solutions. Therefore, there is a need for a general framework that can capture the heterogeneous failure dependences in the context of maintenance management.

Given the above motivations and knowledge basis, this chapter aims to quantify the failure dependences and construct a general maintenance model with heterogeneous failure dependences. This chapter is expected to derive some suggestions concerning maintenance management in complex systems with heterogeneous failure dependence. The proposed approach is discussed in the following sections.

17.2 Dependent Multi-component Degradation Model

17.2.1 Independent General Degradation Model

To explain the inherent independent degradation of components in the dependent system, the independent general degradation model is performed by applying a general degradation path, as shown in Fig. 17.1. This model serves as the foundation of the degradation model for dependent multi-component system (DMDM) when degradation dependences are considered in a dependent multi-component system.

Starting with a fully functioning state $x = 0$ at time $t = 0$, the component is observed until failure. State $x = k$ represents the failed state of component i as an absorbing state. Between $x = 0$ and $x = k$ there are $k - 1$ intermediate states. Let $P_x(t)$ be the probability that the component is in state x at time t. A time dependent probability vector $\boldsymbol{P}(t) = [P_0(t), P_1(t), \ldots, P_k(t)]$, could be obtained, denoting the sojourn probability of the Markov process at time t. The initial state probability $P(0) = [1, 0, \ldots, 0]$, and the sum of state probabilities is equal to 1 at any time.

Let $\mathbb{A}$ be a $k \times k$ matrix where the element $a_{x,y}$ denotes the transition rates from state x to state y for all $x \neq y$ and $x, y \in \{0, 1, 2, \ldots, k\}$. State 0 is the brand-new state. For this simple independent degradation model, assume that the degradation process proceeds all states chronologically from 0 to k. The degradation rate could be represented by λ_x from state x to state $x + 1$. Then the state equation may be written according to Kolmogorov forward equations in matrix terms as

$$\boldsymbol{P}(t) \cdot \mathbb{A} = \dot{\boldsymbol{P}}(t) \tag{17.1}$$

from which it follows

$$\dot{P}_y(t) = \sum_{x=0}^{k} a_{x,y} P_x(t) \tag{17.2}$$

Fig. 17.1 State transition diagram of individual component

If $P_x(t)$ tends to a constant value when $t \to \infty$, then

$$\lim_{t \to \infty} \dot{P}_y(t) = 0 \tag{17.3}$$

The steady state probabilities $\boldsymbol{P} = [P_0, P_1, \ldots, P_k]$ must therefore satisfy the matrix equation

$$\boldsymbol{P} \cdot \mathbb{A} = 0 \tag{17.4}$$

17.2.2 Failure Dependence Model

When there exists influence on the degradation rates of the component from other degrading or failed components, the state transition situation is supposed to shown in Fig. 17.2. The state of two-component system is expressed as $X = (x_1, x_2)^T$, and so is the state of the n components system $X = (x_1, x_2, \ldots x_n)^T$, where $x_i \in \{0, 1, 2, \ldots, k\}$ could characterize the degradation state of component i in this system. Each component has $k + 1$ states, and the state in which the component is depends on how degraded it is in comparison to the failed state. As a result, components in the same state may exhibit varying degrees of degradation. This means that although there may be some components in the same degradation state, they could have distinct levels of degradation. The degradation of the two-component system could be illustrated by $\{X_0, X_1, \ldots X_{(k+1)^2}\}$ since there are $k + 1$ states for each component and $(k + 1)^2$ states for the two-component system. Similarly, the degradation of the n components system is governed by the state space $\boldsymbol{\chi}$, which is taken to be $\{X_0, X_1, \ldots X_{(k+1)^n}\}$ since there are $k + 1$ states for each component and $(k + 1)^n$ states for the n components system in total. State $X_0 = (0, 0, \ldots, 0)^T$ is the brand-new state. State $X_{(k+1)^n} = (k, k, \ldots, k)^T$ is an absorbing state.

As shown in Fig. 17.2, the process starts with $X = (0, 0)^T$ at time $t = 0$. State $X = (k, k)^T$ represents the failed state of the two-component system as an absorbing state. The transition rates in this state transition diagram are illustrated as follows. The $(k + 1)^2 \times (k + 1)^2$ matrix $\mathbb{A}$ now represents the transition rates from state (x_1, x_2) to next state. $\lambda^1_{x_1, x_2}$ is the degradation rate of component 1 from state x_1 to state $x_1 + 1$ when there exists failure dependence between it and another component 2 whose state is x_2. Similarly, in an n components system, the $(k + 1)^n \times (k + 1)^n$ matrix $\mathbb{A}$ represents the transition rates from state $(x_1, x_2, \ldots, x_i, \ldots, x_n)$ to next state. $\lambda^i_{x_1, x_2, \ldots, x_i, \ldots, x_n}$ is the degradation rate of component i from state x_i to state $x_i + 1$ when there exists failure dependence between it and other components whose states are $(x_1, x_2, \ldots, x_n)$. Observe that events of multiple transitions are not included in the state transition diagram, such as a transition between state (x_i, x_j) and $(x_i + 1, x_j + 1)$, since it is assumed to be impossible for all the components in a system to degrade simultaneously during a short time interval from the point of practical.

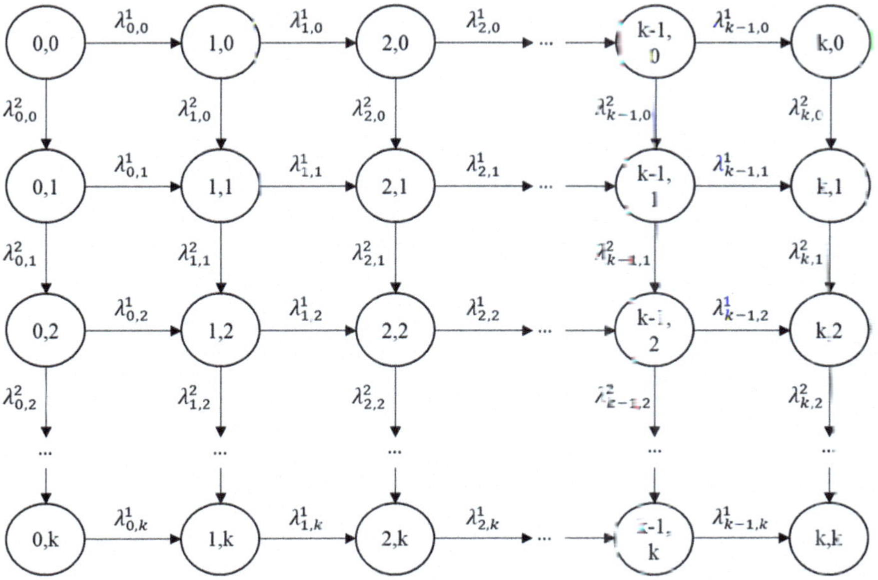

Fig. 17.2 State transition diagram of a two-component system with failure dependence

The failure dependence model is established from examination of two random components i and j, and then expanded to conclude n components. For failure dependence between two components i and j, when component j degrades, the degradation rate of component i increased, and the procedure to calculate the new degradation rate of component i is demonstrated through the flowchart in Fig. 17.3. The new degradation rate for component i from state x_i to state $x_i + 1$ and influenced by degradation of component j is expressed by

$$\lambda^i_{x_i,x_j} = \left(1 + \gamma_{ij}\phi_{x_j}\right)\lambda_{x_i}, \quad \forall i \neq j \tag{17.5}$$

where γ_{ij} is the cascading intensity between components i and j, representing the possibility that failures or degradations are cascaded to components i from component j; ϕ_{x_j} is the influencing level from component j, representing the degree of influence from components j, whose value is determined by the degradation degree of component j compared to its failed state. Detailed explanations about the parameters are reported in the following:

(1) Cascading intensity γ_{ij}

The cascading intensity between components is determined considering various practical constraints such as system layout, material backup, safety redundancy, and other practical constraints. According to industrial standards, expert experience, and practical scenarios, it is possible to obtain γ_{ij}, which characterizes the influence of component j on component i in a probabilistic manner, and the same goes for the

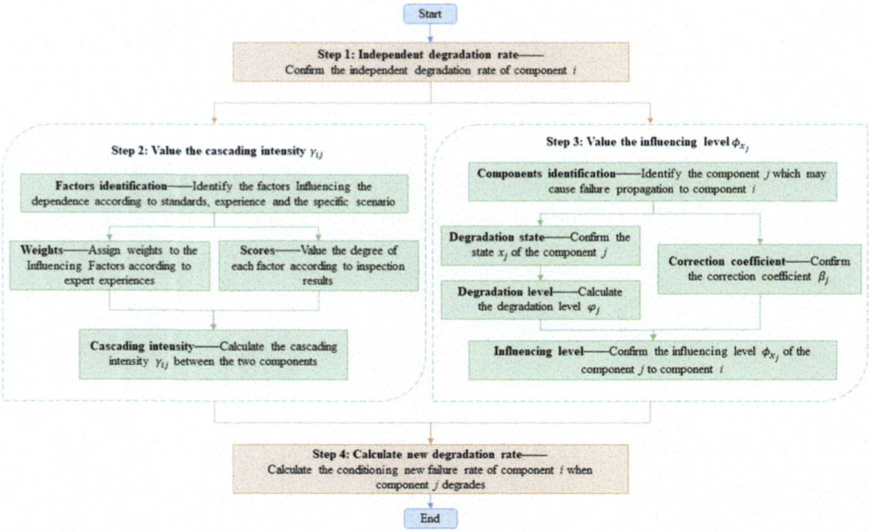

Fig. 17.3 Flowchart of new degradation rate identification considering failure dependence

influence of component i on component j. Furthermore, the value of cascading intensity is supposed to be between 0 and 1. When the value of γ_{ij} is closer to 0, it indicates that the degradation of component j has little influence on the degradation of component i. When the value of γ_{ij} is closer to 1, it indicates that the failure dependence between components is quite strong. The value of cascading intensity depends upon the importance of influencing factors and the situation of each factor in the specified context.

(2) Influencing level ϕ_{x_j}

The influencing level ϕ_{x_j} denotes the loading increment or shock from degrading or failed inducing component j to the induced component i. It is considered to be determined by the degradation degree of component j:

$$\phi_{x_j} = \varphi_j \beta_j \tag{17.6}$$

where φ_j is the degradation degree of inducing component j, which depends solely on its degradation level compared to the failed state. Since the components degrade overtime and may be repaired after maintenance actions, the value of φ_j varies over time.

$$\varphi_j = x_j / k \tag{17.7}$$

Engineering data can be adversely affected by technical errors or environmental factors, leading to inaccuracies that may not accurately reflect the actual conditions. To minimize such errors, a correction coefficient β_j is introduced considering that the effect of component degradation degree on other components is not strictly distributed due to unstable factors. Introducing of correction coefficients enables adjustments of the influencing level model to better reflect the actual situation based on historical or experimental data. The range of correction coefficient should be [0, 1] empirically. Normally linear regression could be used to estimate the value of β_j based on historical data or experimental data. Assume that the correction coefficient is uniformly distributed in [0, 1].

17.2.3 Dependent Multi-Component Degradation Model

For the degradation process in a n-component system, the state transition influenced by degradation and failures of other components could be expressed by a Markov model. Since there are n^2 correlations denoting the cascading intensity in the n components system, the cascading intensity among components could be expressed by a matrix $\boldsymbol{\gamma} = \{\gamma_{ij}\}_{n\times n}$. In addition, $\phi = (\varphi_1\beta_1, \varphi_2\beta_2, \ldots, \varphi_k\beta_k)^T$ is a vector of influencing level for all the components $j \neq i$. Use an $n \times n$ matrix $\mathbb{D}$ to denote the failure dependences among components

$$\mathbb{D} = \begin{pmatrix} \mathbb{D}_1 \\ \mathbb{D}_2 \\ \vdots \\ \mathbb{D}_n \end{pmatrix} = \begin{pmatrix} 0 & \boldsymbol{D}_{1,2} & \cdots & \boldsymbol{D}_{1,n-1} & \boldsymbol{D}_{1,n} \\ \boldsymbol{D}_{2,1} & 0 & & \boldsymbol{D}_{2,n-1} & \boldsymbol{D}_{2,n} \\ \vdots & & \ddots & \vdots & \\ \boldsymbol{D}_{n,1} & \boldsymbol{D}_{n,2} & \cdots & \boldsymbol{D}_{n,n-1} & 0 \end{pmatrix} \tag{17.8}$$

where $\mathbb{D}_i$ is the matrix of failure dependences from other components to component i following the selection of component i as the target component, which is **component 1**. For simplifying the issue, component 1 is presumed to have been decided. 0 is the null matrix whose order is corresponding by the dimensions of blocks $\boldsymbol{D}_{i,j}$, indicating that the component suffers no failure dependence from itself. This also means that $D_{i,x_i} = 0$ for any component i. $\boldsymbol{D}_{i,j} = \left(D_{i,x_j=0}, D_{i,x_j=1}, \ldots, D_{i,x_j=k}\right)$ denoting the failure dependence from component j on component i. Furthermore, the element in the submatrix $D_{i,x_j} = \gamma_{ij}\phi_{x_j}$ is vector to represent the failure dependence from component j on component i when component j is in state x_j. After the qualification of the correlations among components, the new transition rates of the component could be updated as below.

$$\lambda^i_{x_i,x_j} = \left(1 + D_{i,x_j}\right)\lambda_{x_i} \tag{17.9}$$

$$\lambda^j_{x_i,x_j} = \left(1 + D_{j,x_i}\right)\lambda_{x_j} \tag{17.10}$$

where $\lambda^i_{x_i,x_j}$ is the degradation rate of component i from state x_i to state $x_i + 1$ when there exists failure dependence between it and another component j whose state is

x_j, and $\lambda^j_{x_i,x_j}$ is the degradation rate of component j from state x_j to state x_j+1 when there exists failure dependence between it and another component i whose state is x_i.

The states of the system $(x_1, x_2, \ldots x_n)$, from (0,…,0) to (k,…,k), are divided into $(k+1)^{(n-1)}$ subsets, where only the component 1 (any component could be chosen as component 1) degrades and other components remain the constant states in every subset $\{(0, x_2, \ldots, x_n), (1, x_2, \ldots, x_n), \ldots (k, x_2, \ldots, x_n)\}$. Consequently, the matrix $\mathbb{A}$ could be expressed as a $(k+1)^{(n-1)} \times (k+1)^{(n-1)}$ matrix as follows.

$$\mathbb{A} = \begin{pmatrix} \mathbb{A}_{\boldsymbol{n}}^0 & \boldsymbol{A}_n^0 & & & & \\ & \mathbb{A}_{\boldsymbol{n}}^1 & \boldsymbol{A}_n^1 & & & \\ & & \mathbb{A}_{\boldsymbol{n}}^2 & & & \\ & & & \ddots & & \\ & & & & \mathbb{A}_{\boldsymbol{n}}^{\mathbb{k}-1} & \boldsymbol{A}_n^{\mathbb{k}-1} \\ & & & & & \mathbb{A}_{\boldsymbol{n}}^{\mathbb{k}} \end{pmatrix} \tag{17.11}$$

The blocks $\mathbb{A}_{\boldsymbol{n}}^{x_n}$ are the sub matrixes of the whole matrix $\mathbb{A}$, denoting the transition rates of the system where the state of component n keeps constant x_n; The blocks $\boldsymbol{A}_{\boldsymbol{n}}^{x_n}$ are also the sub matrixes of the whole matrix $\mathbb{A}$, denoting the transition rates of the system where only the component n degrades from state x_n to state x_n+1. The other blocks are null, and their orders are corresponding by the dimensions of blocks $\mathbb{A}_{\boldsymbol{n}}^{x_n}$ and $\boldsymbol{A}_{\boldsymbol{n}}^{x_n}$. The blocks $\mathbb{A}_{\boldsymbol{n}}^{x_n}$ could also be further represented by smaller sub matrixes $\mathbb{A}_{\boldsymbol{n}-1}^{x_{n-1}}$ and $\boldsymbol{A}_{\boldsymbol{n}-1}^{x_{n-1}}$ using the same recursive way from $\mathbb{A}$ to $\mathbb{A}_{\boldsymbol{n}}^{x_n}$ and $\boldsymbol{A}_{\boldsymbol{n}}^{x_n}$. The same applies to the general sub matrixes $\mathbb{A}_{\boldsymbol{i}}^{x_i}$. With the recursive method, the general blocks $\mathbb{A}_{\boldsymbol{i}}^{x_i}$ could also be further represented by smaller sub matrixes $\mathbb{A}_{\boldsymbol{i}-1}^{x_{i-1}}$ and $\boldsymbol{A}_{\boldsymbol{j}}^{x_j}$ as below. The blocks $\mathbb{A}_{\boldsymbol{i}}^{x_i}$ are the sub matrixes denoting the transition rates of the system where the state of component i keeps constant x_i; The blocks $\boldsymbol{A}_{\boldsymbol{j}}^{x_j}$ are the sub matrixes denoting the transition rates of the system where only the component j degrades from state x_j to state x_j+1. The other blocks are null, and their orders are corresponding by the dimensions of blocks $\mathbb{A}_{\boldsymbol{i}}^{x_i}$ and $\boldsymbol{A}_{\boldsymbol{j}}^{x_j}$.

$$\mathbb{A}_{\boldsymbol{i}}^{x_i} = \begin{pmatrix} \mathbb{A}_{\boldsymbol{i}-1}^0 & \boldsymbol{A}_{\boldsymbol{i}-1}^0 & & & & \\ & \mathbb{A}_{\boldsymbol{i}-1}^1 & \boldsymbol{A}_{\boldsymbol{i}-1}^1 & & & \\ & & \mathbb{A}_{\boldsymbol{i}-1}^2 & & & \\ & & & \ddots & & \\ & & & & \mathbb{A}_{\boldsymbol{i}-1}^{\mathbb{k}-1} & \boldsymbol{A}_{\boldsymbol{i}-1}^{\mathbb{k}-1} \\ & & & & & \mathbb{A}_{\boldsymbol{i}-1}^{\mathbb{k}} \end{pmatrix} \tag{17.12}$$

$$A_j^{x_j} = \begin{pmatrix} A^j_{x_2,\dots,0,x_{j+1},\dots,x_n} & & & \\ & A^j_{x_2,\dots,1,x_{j+1},\dots,x_n} & & \\ & & \ddots & \\ & & & A^j_{x_2,\dots,k,x_{j+1},\dots,x_n} \end{pmatrix} \text{ for } j = 2, 3, \dots n \tag{17.13}$$

where the elements $A^j_{x_2,\dots,x_j,\dots x_n}$ in $A_j^{x_j}$ are also sub matrixes, as shown in Eq. (17.14), denoting the transition rates of the system where only the component j degrades from state x_j to state $x_j + 1$ when other components are at state $(x_1, \dots, x_n)$ for $j \neq 1$. For example, $A^j_{x_2,\dots,0,x_{j+1},\dots,x_n}$ refers to the transition rates of the system where only the component j degrades from state 0 to state 1 when other components are at state $(x_1, \dots, x_n)$ for $j \neq 1$.

$$A^j_{x_2,\dots,x_j,\dots x_n} = \begin{pmatrix} \lambda^j_{0,x_2,\dots,x_j,\dots x_n} & & & \\ & \lambda^j_{1,x_2,\dots,x_j,\dots x_n} & & \\ & & \ddots & \\ & & & \lambda^j_{k,x_2,\dots,x_j,\dots x_n} \end{pmatrix} \tag{17.14}$$

As mentioned before, the whole matrix $\mathbb{A}$ is recursed to the blocks $\mathbb{A}_n^{x_n}$, and further recursed to the blocks $\mathbb{A}_i^{x_i}$ for $i = 3, \dots n-1$. The recursive process stops when i equals 3, and at this point we have

$$\mathbb{A}_3^{x_3} = \begin{pmatrix} \mathbb{A}_{0,x_3,\dots,x_n} & A^2_{0,x_3,\dots,x_n} & & & & \\ & \mathbb{A}_{1,x_3,\dots,x_n} & A^2_{1,x_3,\dots,x_n} & & & \\ & & \mathbb{A}_{2,x_3,\dots,x_n} & & & \\ & & & \ddots & & \\ & & & & \mathbb{A}_{k-1,x_3,\dots,x_n} & A^2_{k-1,x_3,\dots,x_n} \\ & & & & & \mathbb{A}_{k,x_3,\dots,x_n} \end{pmatrix} \tag{17.15}$$

The block $A^j_{x_2,\dots,x_j,\dots x_n}$ could be obtained by Eq. (17.14). The block $\mathbb{A}_{x_2,\dots,x_n}$ is the submatrix denoting the transition rates of components 1 when other components are at state $(x_2, \dots, x_n)$.

$$\mathbb{A}_{x_2,\ldots,x_n} = \begin{pmatrix} -\sum_{j=1}^{n} \lambda^{j}_{0,x_2,\ldots,x_n} & \lambda^{1}_{0,x_2,\ldots,x_n} & & & \\ & -\sum_{j=1}^{n} \lambda^{j}_{1,x_2,\ldots,x_n} & & & \\ & & \ddots & & \\ & & & -\sum_{j=1}^{n} \lambda^{j}_{k-1,x_2,\ldots,x_n} & \lambda^{1}_{k-1,x_2,\ldots,x_n} \\ & & & & -\sum_{j=2}^{n} \lambda^{j}_{k,x_2,\ldots,x_n} \end{pmatrix} \quad (17.16)$$

where $\lambda^{1}_{x_1,x_2,\ldots,x_n}$ is transition rate of the component 1 from state from state x_1 to state $x_1 + 1$ when there exist failure dependences between it and other components whose state are $(x_2, \ldots, x_n)$.

$$\lambda^{1}_{x_1,x_2,\ldots,x_n} = \lambda_{x_1} \cdot \prod_{j=1}^{n} (1 + D_{1,x_j}) \quad (17.17)$$

and $\lambda^{j}_{x_1,\ldots,x_n}$ is transition rate of the component j from state from state x_j to state $x_j + 1$ when there exist failure dependences between it and other components whose state are $(x_1, \ldots, x_n)$ for $j \neq 1$.

$$\lambda^{j}_{x_1,\ldots,x_n} = \lambda_{x_j} \cdot \prod_{i=1}^{n} \left(1 + D_{j,x_i}\right) \quad (17.18)$$

In this situation, let $P_{x_i}(t)$ be the probability that the component i is in state x_i at time t and $P_{X_i}(t)$ be the probability that the entire system is in state X_i at time t. The vector $\boldsymbol{P}(t) = \left[P_{X_0}(t), P_{X_1}(t), \ldots, P_{X_{k^n-1}}(t)\right]$ denotes the time dependent state probability, and the initial state probability $P(0) = [1, 0, \ldots, 0]$.

17.3 CBM Model Based on Failure Dependence

The general maintenance policies for multi-component systems when there exist heterogeneous failure dependences are discussed in this section. The state transition process of the system is modelled with a Markov model. In the model, assume that:

- The states of components are revealed upon periodic inspections.
- Every inspection could reveal the states of all components.
- The maintenance policies are based on the current state of system.
- Upon inspection, maintenance activities can begin without any delay.
- Compared to long service life, the inspection and repair time are ignored.

17.3.1 Inspections and Maintenances

The inspection-based monitoring is the case for many passive items in practical engineering. As assumed above, the inspection interval is $(s-1)\tau \leq t \leq s\tau$ for $s = 1, 2, \ldots, N_{in}$, where τ is a constant value independent of the component state and the time. The inspection intervals are recounted after each inspection, maintenance, or repair in the overall lifecycle of the system, and could be modeled as $[0, T_1], [T_1, T_2], \ldots, \left[T_{N_{in}-1}, T_{N_{in}}\right]$. If the states of components are revealed to reach the thresholds of maintenance measures, then a corresponding maintenance task will be carried out timely. The time immediately before inspection is denoted by t^- and the time right after inspections, maintenances and repairs (IMRs) is denoted by t^+. When the state of the system immediately before $t = T_s$, namely the state of the system when $t = T_s^-$, is given, the maintenance activities for the system could be then decided. Note that CBM is a maintenance strategy that involves monitoring the actual condition of systems to determine the appropriate maintenance activities. Based upon the maintenance policy, the possible maintenance actions and the state of the system just after IMRs are assumed to depend on the state of the system when $t = T_s^-$, but independent of all transitions of the system before T_s. The effect of IMRs at time $t = T_s$ could be illustrated by

$$\Pr\left(X\left(T_s^+\right) = X_j \middle| X\left(T_s^-\right) = X_i\right) = b_{X_i,X_j} \text{ for all } X_i, X_j \in \chi \tag{17.19}$$

where b_{X_i,X_j} is the probability that the system is in state X_j after IMRs, given that it was in state X_i before inspection.

Based on the previously mentioned inspection strategies, several maintenance policies are proposed. Minor preventive maintenance (PM) and major corrective maintenance (CM) are implemented according to the inspection results. For an independent component, the maintenance policy is classified into three phases, as shown in Fig. 17.4.

In phase I ($x \leq a$), the component is in an acceptable state, and no maintenance activities (NM) are required

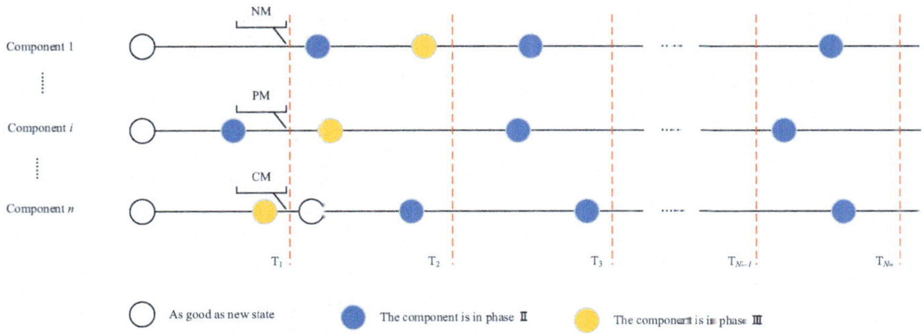

Fig. 17.4 Illustration of maintenance policies

In phase II ($a + 1 \leq x \leq b$), the component is operating in a degrading state, and PM will be performed to take the component back to last state.

In phase III ($b + 1 \leq x \leq k$), It is time to do the CM to restore the component to an as good as new state when the component degrades to a bad state.

Following the implementation of maintenance actions, the state transition model could be represented by degradation transition model, repair transition model, and combinations of those, as shown in Fig. 17.5. Assume that the components have constant degradation rate between two states.

Let $\mathbb{B}$ describes the corresponding maintenance transition matrix of the system, then

$$\boldsymbol{P}\left(T_s^+\right) = \boldsymbol{P}\left(T_s^-\right) \cdot \mathbb{B} \tag{17.20}$$

The corresponding maintenance matrix $\mathbb{B}$ is expressed by the sub matrixes

$$\mathbb{B} = \begin{pmatrix} \mathbb{B}_n^{\mathbf{I}} & 0 & 0\,0 \\ 0 & \mathbb{B}_n^{\mathbf{a}} & 0\,0 \\ 0 & \mathbb{B}_n^{\mathbf{a}+1} & 0\;\;0 \\ 0 & 0 & \mathbb{B}_n^{\mathbf{II}}\;0 \\ \mathbb{B}_n^{\mathbf{III}} & 0 & 0\;\;0 \end{pmatrix} \tag{17.21}$$

The blocks $\mathbb{B}_n^{\mathrm{I}}$ and $\mathbb{B}_n^{\mathbf{a}}$ are the sub matrixes, separately denote the maintenance transition of the system where the component n is in phase I or in state a. The explanation for other sub matrixes in Eq. (17.21) can be obtained similarly. These sub matrixes could be generalized and recursively defined using Eq. (17.22).

$$\mathbb{B}_i^{x_i} = \begin{pmatrix} \mathbb{B}_{i-1}^{I} & 0 & 0 & 0 \\ 0 & \mathbb{B}_{i-1}^{a} & 0 & 0 \\ 0 & \mathbb{B}_{i-1}^{a+1} & 0 & 0 \\ 0 & 0 & \mathbb{B}_{i-1}^{II} & 0 \\ \mathbb{B}_{i-1}^{III} & 0 & 0 & 0 \end{pmatrix} \tag{17.22}$$

The blocks 0 are null, and their orders are corresponding by the dimensions of sub matrixes.

Recurse the matrix until reaching the matrix of component 2.

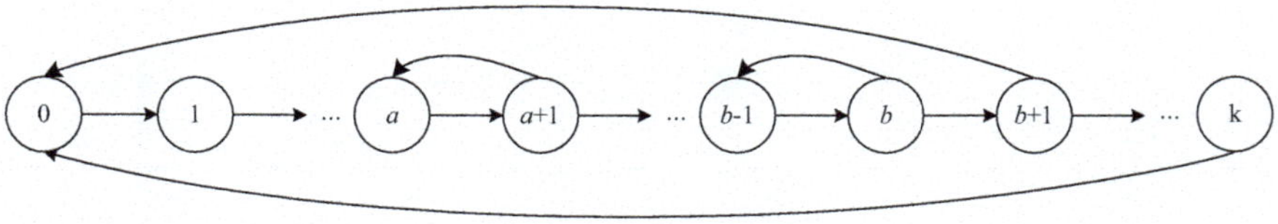

Fig. 17.5 Markov model of an individual component

$$\mathbb{B}_2^{x_2} = \begin{pmatrix} 1 & & & & & & & & & & \\ & \ddots & & & & & & & & & \\ & & 1 & & & & & & & & \\ & & & 1 & & & & & & & \\ & & & 1 & 0 & & & & & & \\ & & & & & \ddots & & & & & \\ & & & & & & 0 & & & & \\ & & & & & & 1 & 0 & & & \\ & 1 & & & & & & & 0 & & \\ & \vdots & & & & & & & & \ddots & \\ & 1 & & & & & & & & & 0 \end{pmatrix} \tag{17.23}$$

Since the process has no steady state and could be divided into several time units on a given finite time horizon, the time-dependent state probability vector $\boldsymbol{P}(t)$ at time t could be then given by

$$\boldsymbol{P}(t) = \boldsymbol{P}(0) \cdot \left(\prod_{s=1}^{s=N_{in}} \exp(\mathbb{A}(T_s - T_{s-1})) \cdot \mathbb{B} \right) \cdot \exp(\mathbb{A}(t - T_{N_{in}})) \tag{17.24}$$

17.3.2 System Availability Analysis

The time-dependent state probabilities equations are used to quantify the system availability, which signifies the proportion of time that the system remains operational under normal conditions to perform its intended function. Suppose that the system is not available only when it fails, the mean value of the system failure probability over a period could then be calculated to represent the unavailability of the system

$$\overline{A_s} = \frac{1}{T} \int_0^T P_{X_F}(t)dt \tag{17.25}$$

where X_F denotes that the component or the entire system is in the failed states immediately at time t. $P_{X_F}(t)$ denotes the probability that the entire system is in the failed states at time t. Based on the identification of all the failed states and the probabilities that the system is in various states at time t included in vector $\boldsymbol{P}(t)$, $P_{X_F}(t)$ could be calculated by summing up all the probabilities that the entire system is in the failed state at time t.

The system availability is the probability of being operational given by

$$A_S = 1 - \overline{A_s} \tag{17.26}$$

The model in this subsection seeks for the optimal maintenance threshold to increase the system availability to an acceptable level.

17.3.3 Maintenance Cost

The maintenance cost consists of the inspection cost, the downtime cost, and the maintenance cost. Suppose that the inspection cost is c_{in} for each time. The downtime cost contains the planned downtime cost c_{p} caused by the scheduled maintenance activities and the unplanned downtime cost c_{u} induced by unexpected failures.

The cumulative maintenance cost between two inspections in the time interval $(T_{s-1}, T_s]$ accounts for the maintenance cost at time $t = T_s$. The maintenance cost is supposed to includes $c_{\mathrm{m1},i}$ and $c_{\mathrm{m2},i}$ respectively for maintenance activities PM and CM to component i. Therefore, the cumulative maintenance cost for the system in $(T_{s-1}, T_s]$ is

$$\begin{aligned} C((T_{s-1}, T_s]) &= \sum_{i=1}^{n} \left[c_{m1,i} \Pr(a+1 \le x_i(T_s) \le b) + c_{m2,i} \Pr(b+1 \le x_i(T_s) \le k) \right] \\ &= \sum_{i=1}^{n} \left[c_{m1,i} P_{a+1 \le x_i \le b}(T_s) + c_{m2,i} P_{b+1 \le x_i \le k}(T_s) \right] \end{aligned} \tag{17.27}$$

where $x_i(t)$ is the degradation state of component i at time t.

The average life-time cost during the period T could be given by

$$C_S = \left[c_{in} N_{in} + c_p N_{in} + \sum_{s=1}^{N_{in}} C((T_{s-1}, T_s]) \right] / T + c_u \overline{A_s} \tag{17.28}$$

The model in this subsection seeks for the optimal maintenance threshold to minimize the maintenance cost.

17.4 Case of Subsea Transmission System

For illustration, this subsection provides the detailed description of a part of subsea separation system. Figure 17.6 shows the scheme of the system. Three pipes transporting gas, oil and water after the outlet of separator are directed to the pump station and compression station, which are located nearby the separation station. The compression station and pump station are operating dependently for the purpose of transmission, which could be regarded as the subsea transmission system for the sake of brevity. One compressor and two pumps are installed in parallel in the simplified subsea transmission system model. Wet gas is compressed by a compressor routed

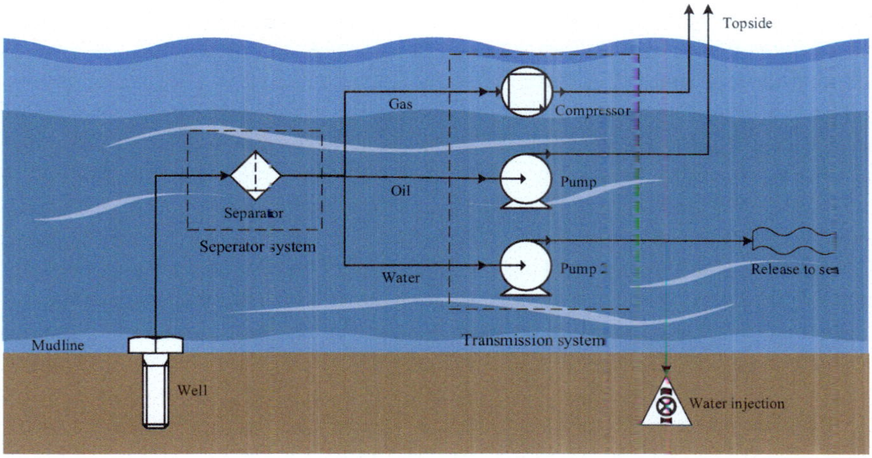

Fig. 17.6 Scheme of the subsea transmission system

to the topside platform. Then, the separated oil and water are respectively pumped following the topside direction or reinjected into a reservoir via the water injection.

The compressor and pump are typically engineered to serve for 5–10 years without any intervention and expected to serve for 30–50 years with IMRs. During their long service lifetime, they degrade stochastically, and the degradation process could also be accelerated by the failures or degradation of other components. The compressor and the two pumps normally transport different substances at the desired power under ideal conditions. In practice, however, devices degrade naturally, resulting in a variety of failure modes such as low output, leakage, vibration, overheating, spurious stop, etc. It has been revealed that some failures of the devices affect not only their own production efficiency, but also the degradation rates of other devices in the system. In the case study, the load redistribution and failure-induced damage mainly lead to failure and degradation dependences. For example, the separator cannot separate the three substances completely, and the mutual doping of substances will aggravate the degradation of compressor and pumps. Similarly, if a component such as compressor malfunctions, but somehow the system cannot be inspected and repaired timely, and it still needs to continue working, gas will enter the pipeline that transports oil or water, and the doping of the gas will compound the damage to the pumps. This type of failure dependence is classified as type II failure dependence. Another example to illustrate type I failure dependence is that the vibration and overheating of one pump may have a direct impact on the operation and aging process of another pump within a certain distance in the pump station. This kind of failure dependence is related to the safety distance, and safety barrier measures. Hence, the compressor and pumps may experience gradual degradation failure and heterogenous failure dependences.

Periodic inspections are conducted to evaluate the condition of the subsea transmission system. PM and CM are implemented according to the inspection results. PM could lower the accumulated damage to a certain level, such as anti-corrosion coating, de-rusting and cleaning. CM include overhaul and preventive replacement that components are perfectly overhauled or replaced, and their states are reset to "as-good-as-new" state.

The case of a subsea transmission system is explored to validate the application of the proposed maintenance policies. Inspections are performed regularly to examine the system to confirm consistency with the performance requirements. Table 17.1 lists the parameter setting of the degradation, inspection and maintenance. The failure rates values are obtained from the existing literature (Zhao et al. 2021; Bhardwaj et al. 2022) and from the application of the Cox model, using the data derived from OREDA database. The service life and repair cost were obtained from the article and thesis.

Assume that there are only four states for each component: normally operating, moderately degraded, severely degraded, and failed. The initial state probability $P(0) = [1, 0, \ldots, 0]$. The states of the system $X = (x_1, x_2, x_3)$ from (0, 0, 0) to (3, 3, 3), are divided into 4^2 subsets: (0,0,0), (1,0,0), (2,0,0), (3,0,0); (0,1,0), (1,1,0), (2,1,0) (3,1,0); …; (0,3,0), (1,3,0), (2,3,0), (3,3,0); ……; (0,0,3), (1,0,3), (2,0,3), (3,0,3); (0,1,3), (1,1,3), (2,1,3), (3,1,3); …; (0,3,3), (1,3,3), (2,3,3), (3,3,3).

As illustrated before, three key factors are generally considered to impact on the failure dependences between components: load redistribution, distance, and safety barrier. In this system, weights could be assigned according to experts' experience: distance (2), load redistribution (5), and safety barrier (3). Here load redistribution denotes the material transfer and doping. After expert experience, the parameters of the failure dependences could be evaluated as Table 17.2 based on the method proposed previously. Since the states of all the components are expressed as $x_i \in \{0, 1, 2, 3\}$, the degradation level of the components could be estimated as $\varphi_j \in \{0, 1/3, 2/3, 1\}$. It is reasonable to deduce the vector of influencing level for all the components when correction coefficient β_j is assumed to be 1, denoted as $\phi = (\phi_0, \phi_1, \phi_1, \phi_3)^T = (0, 1/3, 2/3, 1)^T$. To highlight the necessity of considering failure dependence, set all the parameters in Table 17.2 as 0 or other values to imitate the scenario when failure dependence is neglected or varied in this example. By adjusting the values in the table after evaluating the varying degrees of failure dependences, the proposed model could be applied to compute the system performance under various conditions.

Table 17.1 Parameter setting of the subsea system considered in the case-study

Parameter	Value (*/year*)		Parameter	Value (€)		Parameter	Value (€)
	Compressor	Pumps		Compressor	Pumps		
λ_0	0.046	0.104	c_{m1}	1.93×10^6	2.41×10^6	c_{in}	1.21×10^6
λ_1	0.021	0.105	c_{m2}	2.89×10^6	3.86×10^6	c_p	7.23×10^5
λ_2	0.041	0.056				c_u	6.51×10^7

Table 17.2 Parameter setting of the failure dependences

Parameter	Value	Parameter	Value
γ_{12}	0.34	ϕ_0	0
γ_{13}	0.24	ϕ_1	1/3
γ_{23}	0.66	ϕ_2	2/3
γ_{21}	0.44	ϕ_3	1
γ_{31}	0.34		
γ_{32}	0.56		

Based on the data from Table 17.2, a 3×3 matrix $\mathbb{D}$ could be obtained to denote the failure dependences among components

$$\mathbb{D} = \begin{pmatrix} \boldsymbol{E} & \boldsymbol{D}_{1,2} & \boldsymbol{D}_{1,3} \\ \boldsymbol{D}_{2,1} & \boldsymbol{E} & \boldsymbol{D}_{2,3} \\ \boldsymbol{D}_{3,1} & \boldsymbol{D}_{3,2} & \boldsymbol{E} \end{pmatrix} \tag{17.29}$$

Using the failure dependence of component 2 on component 1 as an example to illustrate the quantification of failure dependences, $\boldsymbol{D}_{1,2} = \left(D_{1,x_2=0}, D_{1,x_2=1}, \ldots, D_{1,x_2=3}\right) = (0, 0.103, 0.207, 0.31)$ could be obtained if the correction coefficient β_2 is supposed to be 1. The same calculation approach can be extended to other submatrices of failure dependences. With the confirmation of failure dependences among components, the DMDM could be denoted by the transition matrix $\mathbb{A}$. The transition matrix $\mathbb{A}$ could be expressed as a $4^2 \times 4^2$ matrix

$$\mathbb{A} = \begin{pmatrix} \mathbb{A}_3^0 & \boldsymbol{A}_3^0 & & \\ & \mathbb{A}_3^1 & \boldsymbol{A}_3^1 & \\ & & \mathbb{A}_3^2 & \boldsymbol{A}_3^2 \\ & & & \mathbb{A}_3^3 \end{pmatrix} \tag{17.30}$$

$$\mathbb{A}_3^{x_3} = \begin{pmatrix} \mathbb{A}_{0,x_3} & \boldsymbol{A}_{0,x_3}^2 & & \\ & \mathbb{A}_{1,x_3} & \boldsymbol{A}_{1,x_3}^2 & \\ & & \mathbb{A}_{2,x_3} & \boldsymbol{A}_{2,x_3}^2 \\ & & & \mathbb{A}_{3,x_3} \end{pmatrix} \text{ for } x_3 = 0, 1, 2, 3. \tag{17.31}$$

$$\boldsymbol{A}_3^{x_3} = \begin{pmatrix} \boldsymbol{A}_{x_2,x_3}^3 & & & \\ & \boldsymbol{A}_{x_2,x_3}^3 & & \\ & & \boldsymbol{A}_{x_2,x_3}^3 & \\ & & & \boldsymbol{A}_{x_2,x_3}^3 \end{pmatrix} \text{ for } x_3 = 0, 1, 2. \tag{17.32}$$

$$\boldsymbol{A}_2^{x_2} = \begin{pmatrix} \boldsymbol{A}_{x_2,x_3}^2 & & & \\ & \boldsymbol{A}_{x_2,x_3}^2 & & \\ & & \boldsymbol{A}_{x_2,x_3}^2 & \\ & & & \boldsymbol{A}_{x_2,x_3}^2 \end{pmatrix} \text{ for } x_2 = 0, 1, 2. \tag{17.33}$$

where the blocks $\mathbb{A}_{x_2,x_3}$ and $\boldsymbol{A}^j_{x_2,x_3}$ are as follows

$$\mathbb{A}_{x_2,x_3} = \begin{pmatrix} -\sum_{i=1}^{3}\lambda^i_{0,x_2,x_3} & \lambda^1_{0,x_2,x_3} & & \\ & -\sum_{i=1}^{3}\lambda^i_{1,x_2,x_3} & \lambda^1_{1,x_2,x_3} & \\ & & -\sum_{i=1}^{3}\lambda^i_{2,x_2,x_3} & \lambda^1_{2,x_2,x_3} \\ & & & -\sum_{i=2}^{3}\lambda^i_{3,x_2,x_3} \end{pmatrix} \tag{17.34}$$

$$\boldsymbol{A}^j_{x_2,x_3} = \begin{pmatrix} \lambda^j_{0,x_2,x_3} & & & \\ & \lambda^j_{1,x_2,x_3} & & \\ & & \lambda^j_{2,x_2,x_3} & \\ & & & \lambda^j_{3,x_2,x_3} \end{pmatrix} \text{ for } j = 2, 3. \tag{17.35}$$

The matrix $\mathbb{B}$ could be expressed as a $4^2 \times 4^2$ matrix.

$$\mathbb{B} = \begin{pmatrix} \mathbb{B}^0_3 & 0 & 0 & 0 \\ 0 & \mathbb{B}^1_3 & 0 & 0 \\ 0 & \mathbb{B}^2_3 & 0 & 0 \\ \mathbb{B}^3_3 & 0 & 0 & 0 \end{pmatrix} \tag{17.36}$$

$$\mathbb{B}^{x_3}_3 = \begin{pmatrix} \mathbb{B}^0_2 & 0 & 0 & 0 \\ 0 & \mathbb{B}^1_2 & 0 & 0 \\ 0 & \mathbb{B}^2_2 & 0 & 0 \\ \mathbb{B}^3_2 & 0 & 0 & 0 \end{pmatrix}, \mathbb{B}^{x_2}_2 = \begin{pmatrix} 1 & 0 & 0 & 0 \\ 0 & 1 & 0 & 0 \\ 0 & 1 & 0 & 0 \\ 1 & 0 & 0 & 0 \end{pmatrix} \tag{17.37}$$

$\mathbb{B}^{x_3}_3$ and $\mathbb{B}^{x_2}_2$ represents the maintenance matrix when components 3 or 2 are respectively in the state x_3 or x_2.

The whole degradation transition matrix $\mathbb{A}$ and the whole maintenance transition matrix $\mathbb{B}$ could be obtained after splicing the matrices. The state probability vector $\boldsymbol{P}(t)$ could then be calculated by Eq. (17.24). The time dependent failure probabilities $P_{X_F}(t)$ are obtained by summing the probabilities of different failure states of the system.

The maintenance cost for an independent component in $(T_{s-1}, T_s]$ is

$$C((T_{s-1}, T_s]) = c_{m1,comp}P_{x_1=2}(T_s) + c_{m2,comp}P_{x_1=3}(T_s) + \sum_{i=2}^{3}\left[c_{m1,pump}P_{x_i=2}(T_s) + c_{m2,pump}P_{x_i=3}(T_s)\right] \tag{17.38}$$

where the $c_{m1,comp}$ and $c_{m2,comp}$ are respectively the PM and CM for a compressor, $c_{m1,pump}$ and $c_{m2,comp}$ are respectively the PM and CM for a pump.

For this kind of system, the availability of the system and its average life-time cost could be obtained by Eqs. (17.28) and (17.38).

17.4.1 Failure Probabilities

To evaluate the effect of mean time between inspections (MTBI) on the failure probabilities, observe the difference of $P_{X_F}(t)$ curves in the log-run horizon by testing various MTBI ($MTBI$ = 1 year; 2 years; 5 years). The results are shown in Fig. 17.7.

The failure probability is observed to increase with time but experiences a sudden decrease at the IMRs timepoints under varying MTBI. The continuous degradation of the system and its components over time results in a steadily increasing failure probability for the system. As the system approaches the IMRs timepoints, the failure probability peaks in this specific time interval. However, following the IMRs timepoints, the status of the system and its components could be noticeably improved, and the system failure probability is close to zero, indicating a peak value at the IMRs timepoints. Another finding can be obtained by comparing the failure probability curves under different MTBI. It is obvious that with smaller MTBI, the maximum values of failure probabilities are expected to be lower. On the contrary, the maximum values of failure probabilities tend to be higher when the MTBI increases, which means that the system tends to be less reliable. In this regard, the reliability

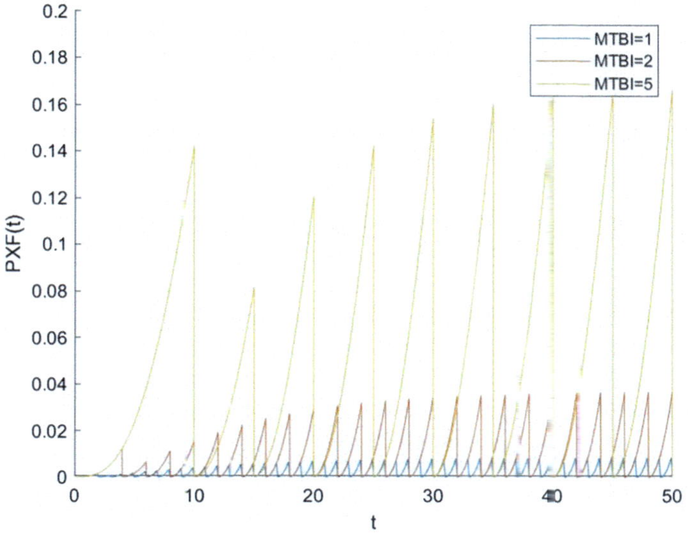

Fig. 17.7 Failure probabilities for different MTBI

and availability of the system can be enhanced by reducing the value of MTBI, i.e., shortening the IMRs interval. However, a lower MTBI is not always preferable. In addition to system reliability and availability, maintenance cost is also key factor that must be considered when performing maintenance activities. The following subsections will go through how to determine the optimum MTBI value, balancing the system performance and maintenance cost.

17.4.2 Maintenance Strategies with Various Failure Dependences

The availability and average life-time cost of the subsea transmission system under the condition of with various failure dependence are respectively shown in Fig. 17.8. The actual failure dependence in Table 17.2 is denoted as normal dependence; the strong dependence is set when all the γ take the maximum value (0.66) in Table 17.2; the case that all the γ take the minimum value (0.24) in Table 17.2 is weak dependence; there is no dependence when all the γ take the value of 0.

The figures show that the availability of the system decreases with the increase of MTBI. One interesting observation is that these curves are not smooth, but rather contain distinct breaking lines. After examining these fold points, it is found that they are always located at certain MTBI values that enable the IMRs number to be an integer. For the A_s-MTBI curves, the small the MTBI is, a larger number of inspections and maintenance activities are needed, the higher the availability reached, and vice versa. This trend is consistent with the conclusion of the previous subsection. At each fold point, the IMRs frequency drops by one, which leads to a sudden increase in system failure probability and steady state probability of failure, resulting in a sudden decrease in system availability. Besides, the curves C_s-MTBI show the similar trend that the average life-time cost falls initially and subsequently climbs as MTBI grows, indicating that there is a point to minimize the cost. A reasonable explanation is that when the MTBI is relatively small, more inspections and maintenance are undertaken, which may lower the failure probability of system and the unexpected downtime cost, also may impose considerable IMRs costs. However, when the MTBI is greatly increased, the IMRs costs can be accordingly decreased; but the system unavailability rises, inevitably leading to more production loss due to unplanned downtime. Similarly, before the cost reaches the lowest value, the variation of IMRs cost dominates the trend of C_s-MTBI curves. As MTBI increases, the amount of IMRs may drop by one, causing the immediate drop of total IMRs cost and the average life-time cost. After the lowest value, the variation of unexpected downtime cost dominates the trend of C_s-MTBI curves. Hence the effect of drop amount of IMRs on the unexpected downtime cost is stronger than the effect on the IMRs cost. As the amount of IMRs drops, the failure probability increases suddenly, as well as the downtime cost, which is strongly proportional to it.

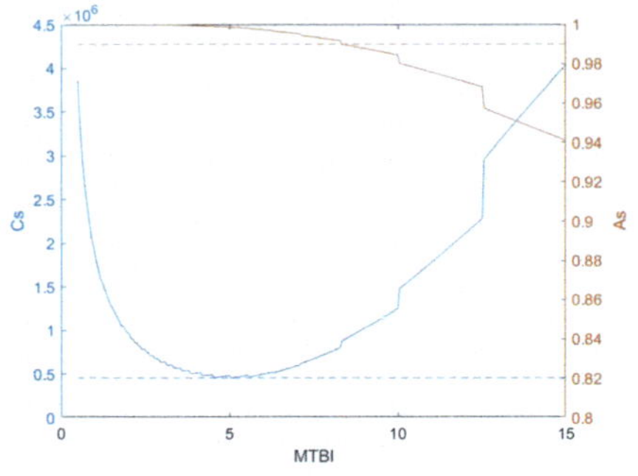

(a) With strong dependence

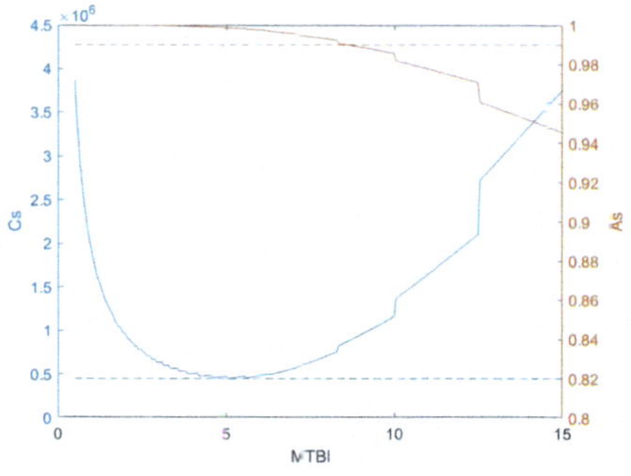

(b) With normal dependence

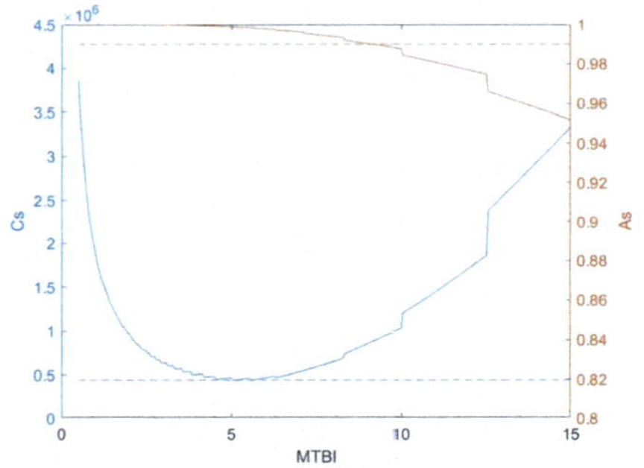

(c) With weak dependence

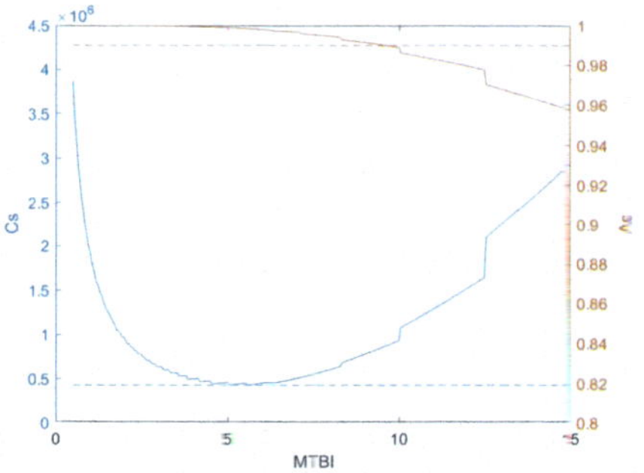

(d) Without dependence

Fig. 17.8 Availability and average life-time cost of the subsea transmission system under different MTBI

It is generally too costly to pursue extensive system availability in practical engineering applications. An acceptable availability threshold is therefore required to be determined. In this case the average life-time cost should be minimized while ensuring system availability over the acceptable threshold 0.99. The optimum of the maintenance policy could be achieved by adjusting the parameter MTBI. From the figures of Fig. 17.8, the minimal cost appears within the range of system availability exceeding 0.99, thus this point representing the minimal cost could be considered as the optimal option of the maintenance strategy.

Distinct differences can be also observed between the findings with various failure dependence. Table 17.3 displays the comparison of the results. In terms of the impact of MTBI on system availability, the availability of the system with stronger failure dependence is generally lower than that of the system with weaker failure dependence and that of the system without failure dependence. The thresholds of MTBI for system availability under 0.99 increases as system failure dependence weakens: 8.35 (strong), 8.65 (normal), 9.15 (weak), 9.65 (without). This means that in scenarios where failure dependence is stronger, more frequent inspections and maintenance are required to improve its functionality. From a financial standpoint, the minimal average life-time cost considering strong, normal, and weak failure dependence are

Table 17.3 Results for the subsea transmission system with and without failure dependence

	Availability		Average life-time cost	
	When As is 0.99	When MTBI = 15	When Cs is minimized	When MTBI = 15
With strong dependence	(8.35, 0.99)	(15, 0.9408)	(4.55, 454525)	(15, 4047150)
With normal dependence	(8.65, 0.99)	(15, 0.9452)	(5.05, 448474)	(15, 3755460)
With weak dependence	(9.15, 0.99)	(15, 0.9516)	(5.6, 438202)	(15, 3333170)
Without dependence	(9.65, 0.99)	(15, 0.9574)	(5.6, 428870)	(15, 2948640)

respectively 454,525€, 448,474€, and 438,202€, higher than the minimal average life-time cost without failure dependence (428,870€). This also supports a similar result that a higher investment is required with stronger failure dependence. The comparison and analysis reveal the necessity of addressing the failure dependence of complex systems.

17.4.3 Maintenance Strategies for Various Initial Costs Input

In this subsection, the impact of some cost parameters on the average life-time cost is explored, while the remaining parameters remain constant. By resetting the inspection cost $c_{in} = \left[1.21 \times 10^5, 1.21 \times 10^6, 1.21 \times 10^7\right]$, the planned downtime cost $c_p = \left[7.23 \times 10^4, 7.23 \times 10^5, 7.23 \times 10^6\right]$, the unplanned downtime cost $c_u = \left[6.51 \times 10^6, 6.51 \times 10^7, 6.51 \times 10^8\right]$, the influence of costs input on the average life-time cost is depicted in Fig. 17.9. The result shows that the average lifetime cost basically increase as the three kinds of cost increase. However, the effects of inspection cost and planned downtime cost are most significant when the MTBI value is small, whereas the impact of unplanned downtime cost is most pronounced when the MTBI value is high. This finding can provide guidance for adjusting costs in line with the current maintenance strategy. For example, when the MTBI is small and the amount of IMRs is high, the inspection cost can be appropriately decreased to control the average life-time cost. When the value of MTBI is high and the amount of IMRs is low, it is recommended to prioritize reducing the unplanned downtime cost to minimize the average life-time cost.

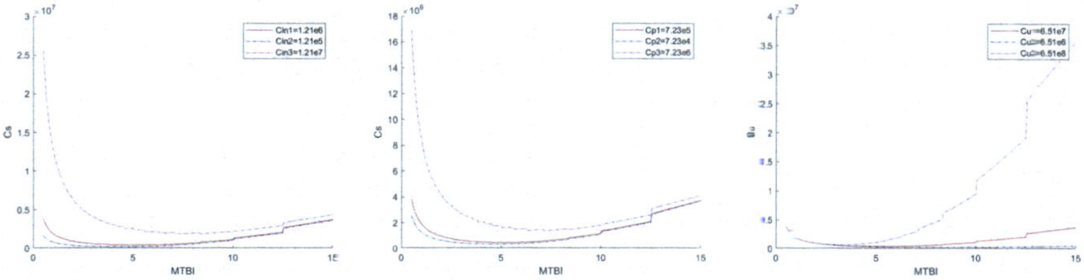

Fig. 17.9 Maintenance cost for different initial costs input

Chapter 18
Joint Optimization for Emergency Maintenance and CBM

Abstract In practical engineering, effects of random external shocks on the engineering system are unavoidable. The resistant ability of an engineering system can be improved by taking CBM operation. Meanwhile, further deterioration of system performance can be avoided with adopting emergency maintenance (EM) operation. A joint multi-objective optimization method for EM and CBM throughout the service life cycle is proposed in this chapter. A cumulative degradation prediction model is established with considering the competitive failure mechanism. In the multi-objective optimization model, the objectives are the performance loss ratio and maintenance cost in the overall service life cycle, and the priority of optimization objectives are determined according to the practical application. The joint maintenance optimization process is realized using the multi-objective particle swarm optimization algorithm to obtain the optimal maintenance thresholds of EM and CBM. Based on the optimal maintenance thresholds, reasonable maintenance management plan, such as maintenance sequence, can be further formulated.

18.1 Introduction

A significant challenge in CBM is the creation of health factors or quantitative indicators that accurately portray the state of a system related to a specific maintenance concern. Furthermore, relationships between these indicators and the associated maintenance cost need to be determined (Wang et al. 2024). Accurate degradation state prediction of a system is crucial for making correct maintenance decisions and serves as a crucial benchmark for overall service life cycle management. In practical engineering, the engineering system often encounters random shocks (Wang et al. 2020; Sun et al. 2021) in addition to the typical degradation it undergoes. These random shocks can vary in intensity and significantly impact the system performance, leading to dramatic changes. To predict the system state accurately, it is necessary to consider both normal degradation process and the impact of random

B. Cai et al., *Intelligent Operation and Maintenance for Subsea Production Systems*,
https://doi.org/10.1007/978-981-96-2677-9_18

shocks when modeling the degradation process. Ignoring these shocks poses difficulties in precisely forecasting the degradation status, which may ultimately lead to wrong maintenance strategies.

The random shocks can directly lead to a huge variation in the system degradation level, though it might not immediately lead to a complete failure. High-intensity impacts can severely affect system performance, potentially result in failure at any moment (Zhang et al. 2018). To avoid further deterioration of the system state, it is necessary to take emergency maintenance (EM) activity (Cai et al. 2021; Wang et al. 2023). EM refers to unscheduled and immediate operations taken to fix unexpected failures or breakdowns of a system, which aims to address immediate safety concerns and prevent further deterioration (Levitin et al. 2022). Due to the combined effects of system degradation and external random shocks, the system performance may degrade to an arbitrary state. The decision to execute EM activities significantly influences the future state of the system. Therefore, in the context of CBM decision optimization process throughout the whole service life cycle, incorporating EM activities is crucial to mitigate the risk of system failure and improve the reliability and availability of the system. During the CBM decision optimization process, focusing solely on a single objective for optimization can result in either excessive or inadequate maintenance. To balance this dilemma, it is essential to optimize the maintenance strategy with multiple indexes.

Given the above considerations, this chapter proposed a joint multi-objective optimization method for EM and CBM throughout the service life cycle.

18.2 Joint Optimization Model of Emergency Maintenance and CBM

Given the inherent instability of the working environment, system performance loss caused by unexpected random shocks is unavoidable. To achieve precise prediction of the system performance, the degradation prediction model is established based on the competitive failure mechanism. That is, effects of the normal degradation and external random shocks should be considered when establishing the degradation prediction model.

To ensure that the system can operate normally and stably during the overall service life cycle, it necessitates the implementation of multiple CBM activities. In response to unexpected random shocks, EM is needed to address immediate safety concerns and prevent further deterioration. So, the premise of taking EM is that there is an occurrence of an external random shock, which shows urgency of EM. The system performance can be recovered to a new stable state by the EM activity. In the overall service life cycle for a system, the joint consideration of the EM and CBM is necessary. It can ensure a long-term normal operation of the system and can also realize immediate crisis handling. The selection of the EM threshold for the system and the CBM threshold for each component can influence the system performance

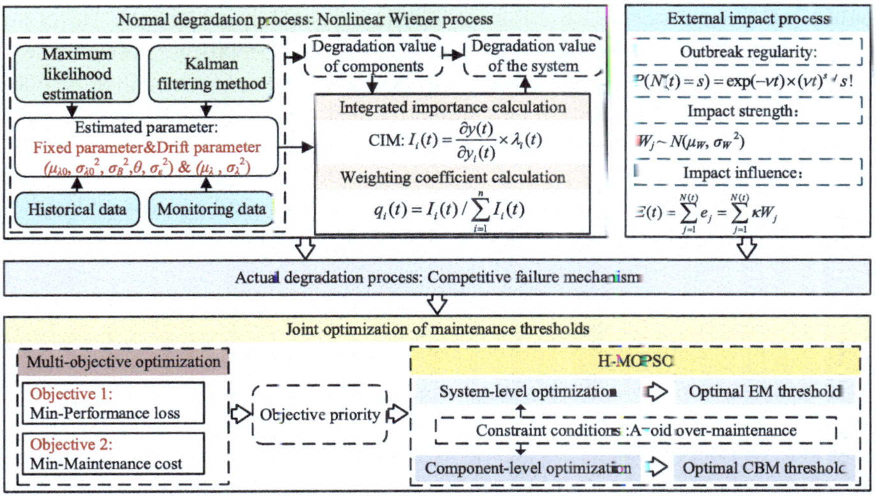

Fig. 18.1 Framework of the joint optimization method

loss and the maintenance cost in the overall service life cycle. The challenge in optimizing these thresholds lies in the inherent conflict between minimizing performance loss and controlling maintenance costs. To achieve a balance, a hierarchical multi-objective particle swarm optimization (H-MOPSO) model is established based on the priority of multiple optimization objectives. A joint optimization of EM threshold in the system level and CBM threshold in the component level can be realized. The framework of the joint optimization method is shown in Fig. 18.1.

18.2.1 Cumulative Degradation Value Prediction Model

Considering the impact of external random shocks, the actual degradation process of the system at time t includes two parts: the normal cumulative degradation result and the cumulative impact of external random shocks. The total degradation value of the system at time t is

$$De(t) = N(t) + E(t) \tag{18.1}$$

where, De(t) is the total system degradation value, N(t) is the predicted cumulative degradation result, and E(t) is the cumulative impact of external random shocks.

(1) Cumulative degradation value prediction in the component level

The process of system performance degradation is complex and subject to various uncertainties, including time variation, individual difference and measurement error uncertainty. To accurately capture the nonlinear nature of this degradation, the nonlinear Wiener process is used as a modeling tool (Han et al. 2023, Hu et al. 2023). Firstly, the state space model of nonlinear degradation process is established, followed by which the model parameters are updated online by Kalman filter method. In this way, real-time prediction of the cumulative degradation values for each component can be realized. The integration importance and weighting values of each component are calculated based on predicted cumulative degradation values. Finally, the cumulative degradation value for each component can be calculated and updated.

Based on the nonlinear model, the component degradation value can be described as

$$X(t) = x_0 + \lambda\Lambda(t;\theta) + \sigma_B B(t) \tag{18.2}$$

where, x_0 is the initial degradation state, which is usually set as 0. λ is a drift parameter, which characterizes the degradation rate of the component. Due to the differences in the production materials and in the production process, the degradation rate of each component in the degradation process is different. To describe individual differences between components, it is assumed that λ follows a normal distribution with a mean of μ_λ and a variance of $\sigma_\lambda{}^2$. $\Lambda(t;\theta)$ denotes an nonlinear function, and θ is a nonlinear coefficient. σ_B is a diffusion coefficient. B(t) represents the standard Brownian motion, which can describe the time-varying uncertainty in the degradation process.

To describe the influence of measurement error on the prediction accuracy, it is assumed that the measurement error ε follows normal distribution a mean of 0 and a variance of $\sigma_\varepsilon{}^2$. The degradation model can be further described as

$$Y(t) = X(t) + \varepsilon_t \tag{18.3}$$

Supposing that there are n sets of degradation data, and the degradation data is measured respectively at $t_1, t_2, \ldots t_j, \ldots, t_m$. Let $y =$ Y(t), and the degradation value of the i-th component at time t_j is

$$y_{i,j} = Y_i(t_j) = \lambda_i\Lambda(t_j;\theta) + \sigma_B B(t_j) + \varepsilon_{i,j} \tag{18.4}$$

Parameters to be estimated for the nonlinear Wiener model are $\Theta = \{\mu_\lambda, \sigma_\lambda{}^2, \sigma_B{}^2, \theta, \sigma_\varepsilon{}^2\}$. The parameter vector is estimated using the degenerate increment of the component. The degradation increment is

$$\Delta y_{i,j} = \lambda_i \Delta v_j + \sigma_B B(\Delta t_j) + (\varepsilon_{i,j} - \varepsilon_{i,j-1}) \tag{18.5}$$

where, $\Delta v_j = \Lambda(t_j;\theta) - \Lambda(t_{j-1};\theta)$, $\Delta t_j = t_j - t_{j-1}$. Let $\Delta v = (\Lambda(t_1;\theta), \Delta v_2, \ldots, \Delta v_m)'$, $\Delta t = (\Delta t_1, \Delta t_2, \ldots, \Delta t_m)'$, $\Delta y_i = (\Delta y_{i,1}, \Delta y_{i,2}, \Delta y_{i,3}, \ldots, \Delta y_{i,m})'$, $Y = (\Delta y_1{}', \Delta y_2{}', \ldots, \Delta y_m{}')'$. Δy_i follows a multidimensional normal distribution, which can be expressed as

$$\Delta y_i \sim MVD(\mu_\lambda \Delta v, \Sigma) \tag{18.6}$$

$$\Sigma_{i,j} = \begin{cases} \sigma_\lambda^2 \Delta v \Delta v' + \sigma_B^2 \Omega + \sigma_\varepsilon^2, i = j = 1 \\ \sigma_\lambda^2 \Delta v \Delta v' + \sigma_B^2 \Omega + 2\sigma_\varepsilon^2, i = j > 1 \\ \sigma_\lambda^2 \Delta v \Delta v' + \sigma_B^2 \Omega - \sigma_\varepsilon^2, i = j + 1 \\ \sigma_\lambda^2 \Delta v \Delta v' + \sigma_B^2 \Omega, i > j + 1 \end{cases} \tag{18.7}$$

where, $\Omega = \text{diag}(\Delta v_1, \Delta v_2, \ldots, \Delta v_m)$.

To simplify the parameter estimation process, let $\Phi = \sigma_3{}^2/\sigma_\varepsilon{}^2$, $\Sigma = \sigma_\lambda{}^2 \Delta v \Delta v' + \sigma_B{}^2 \Omega + \sigma_\varepsilon{}^2 U$. U is a m-order matrix, which is expressed as

$$U = \begin{bmatrix} 1 & -1 & 0 & 0 & \cdots & 0 \\ -1 & 2 & -1 & 0 & \cdots & 0 \\ 0 & -1 & 2 & -1 & \cdots & 0 \\ 0 & 0 & -1 & 2 & \cdots & 0 \\ \vdots & \vdots & \vdots & \vdots & \ddots & -1 \\ 0 & 0 & 0 & \cdots & -1 & 2 \end{bmatrix} \tag{18.8}$$

The unbiased estimation expressions for the estimated parameters in the nonlinear Wiener model are

$$\lambda_i = \frac{\Delta y_i'}{\Delta v'} \tag{18.9}$$

$$\mu_\lambda = \frac{\Omega + \Phi U}{n \Delta v' \Delta v} \sum_{i=1}^{n} \frac{1}{\Omega + \Phi U} \Delta y_i' \Delta v \tag{18.10}$$

$$\sigma_\lambda^2 = \frac{1}{n-1} \sum_{i=1}^{n} (\mu_\lambda - \lambda_i)^2 - \frac{\Omega + \Phi U}{n(m-1) \Delta v \Delta v'} \sum_{i=1}^{n} \frac{(\Delta y_i - \lambda_i \Delta v)'(\Delta y_i - \lambda_i \Delta v)}{\Omega + \Phi U} \tag{18.11}$$

$$\sigma_B^2 = \frac{1}{n(m-1)} \sum_{i=1}^{n} \frac{(\Delta y_i - \lambda_i \Delta v)'(\Delta y_i - \lambda_i \Delta v)}{\Omega + \Phi U} \tag{18.12}$$

The estimated values of the parameters θ and Φ can be obtained through solving the minimum value of the multivariable function. Values of parameters μ_λ, $\sigma_\lambda{}^2$, $\sigma_B{}^2$, and $\sigma_\varepsilon{}^2$ can be further calculated based on the unbiased estimation expressions. The likelihood function of the nonlinear Wiener model is

$$\ln L(\theta, \Phi|Y) = -\frac{mn}{2}(\ln 2\pi + 1) - \frac{n(m-1)}{2}$$
$$\times \ln\left(\frac{1}{n(m-1)}\sum_{i=1}^{n}\frac{(\Delta y_i - \lambda_i \Delta v)'(\Delta y_i - \lambda_i \Delta v)}{\Omega + \Phi U}\right)$$
$$-\frac{n}{2}\left[\ln\left(\frac{\Delta v' \Delta v}{n(\Omega + \Phi U)}\sum_{i=1}^{n}(\mu_\lambda - \lambda_i)^2\right) + \ln|\Omega + \Phi U|\right] \quad (18.13)$$

The drift coefficients are updated according to the monitoring data by Kalman filtering method. The basic equation of Kalman filter is a recursive form in the time domain, and its calculation process is a continuous "prediction-correction" process. Once new data is observed, a new filter value can be calculated, making it suitable for real-time processing. The state space equation at time t can be expressed as

$$\begin{cases} x_t = x_{t-1} + \lambda_{t-1}(\Lambda(t;\theta) - \Lambda(t-1;\theta)) + \sigma_B(B(t) - B(t-1)) \\ y_t = x_t + \xi_t \end{cases} \quad (18.14)$$

Let $\eta_t = \sigma_B(B(t) - B(t-1))$. The state space model can be transformed into

$$\begin{cases} \begin{bmatrix} x_t \\ \lambda_t \end{bmatrix} = \begin{bmatrix} 1 & \Lambda(t;\theta) - \Lambda(t-1;\theta) \\ 0 & 1 \end{bmatrix}\begin{bmatrix} x_{t-1} \\ \lambda_{t-1} \end{bmatrix} + \begin{bmatrix} \eta_t \\ 0 \end{bmatrix} \\ y_t = [1\ 0]\begin{bmatrix} x_{t-1} \\ \lambda_{t-1} \end{bmatrix} + \xi_t \end{cases} \quad (18.15)$$

The state space model can be further simplified as

$$\begin{cases} X_t = F_t X_{t-1} + W_t \\ y_t = HX_{t-1} + \xi_t \end{cases} \quad (18.16)$$

where, $X_t = [x_t, \lambda_t]'$. F_t is the state transition matrix, and $F_t = [1, \Lambda(t;\theta) - \Lambda(t-1;\theta); 0, 1]$. H is the observation matrix, and $H = [1, 0]$. ξ_t is the observed noise, which follows the normal distribution with a mean of 0 and a variance of R_t. R_t is related to the measurement accuracy of the sensor. W_t is the input white noise, and $W_t = [\eta_t, 0]'$. W_t follows the normal distribution with a mean of 0 and a variance of Q_t. Q_t is the predicted noise, which is

$$Q_t = \begin{bmatrix} \sigma_B^2 \Delta\Lambda(t;\theta) & 0 \\ 0 & 0 \end{bmatrix} \quad (18.17)$$

The initial value matrix of the state and the covariance matrix can be described as

$$X_0 = \begin{bmatrix} 0 \\ \mu_{\lambda 0} \end{bmatrix} \quad (18.18)$$

$$P_0 = \begin{bmatrix} 0 & 0 \\ 0 & \sigma_{\lambda 0}^2 \end{bmatrix} \tag{18.19}$$

(2) Calculation of system cumulative degradation value

In order to accurately calculate the cumulative degradation value of the multi-component system, it is necessary to first establish weighting coefficients for each component. These coefficients are derived based on the corresponding integrated importance degree. Integrated importance degree represents the influence degree of component failure in different states on the system cumulative degradation value, and it has state failure superposition property (Si et al. 2019). The influence of component state transition rate on the system cumulative degradation overall the overall service life cycle is considered. The importance degree of the i-th component at time t can be expressed as

$$I_i(t) = \frac{\partial y(t)}{\partial y_i(t)} \times \lambda_i(t) \tag{18.20}$$

where $I_i(t)$ is the importance degree of the i-th component at time t, $y_i(t)$ is the cumulative degradation value for the i-th component at time t, and $y(t)$ is the original system cumulative degradation value at time t, which is calculated based on the logical relationship among all components.

In the overall service life cycle, the integrated importance degree of the i-th component is

$$IM_i = \sum_{t=1}^{T} I_i(t) \tag{18.21}$$

where IM_i is the integrated importance degree of the i-th component, and T is the overall service time.

Based on the integrated importance degree of each component, the weighting coefficient for the i-th component $q_i(t)$ is

$$q_i(t) = I_i(t) / \sum_{i=1}^{n} I_i(t) \tag{18.22}$$

Based on the degradation value and weighting coefficient of each component, the cumulative degradation of the system at time t can be calculated as

$$N(t) = \sum_{i=1}^{n} y_i(t) \times q_i(t) \tag{18.23}$$

(3) System performance degradation estimation under random shocks

In practical, the system fails not only due to the natural performance degradation, but also due to the impact of random shocks to the system (Feng et al. 2023). External random shocks may bring great risks to the system normal operation, or even cause the system failure directly. Since the external shock occurs randomly, the impact on the system remains uncertain. The external random shock is modeled based on a random process (Zheng et al. 2023). The arrival of random shocks follows a homogeneous Poisson process (HPP) with the arrival rate ν_s. In time t, it is assumed that the random shock number $Num(t)$ is s. The distribution probability is

$$P(Num(t) = s) = \exp(-\nu_s t) \times (\nu_s t)^s / s! \tag{18.24}$$

The shock strength follows a normal distribution. The random impact strength W_j is

$$W_j \sim N(\mu_W, \sigma_W^2), j = 1, 2, \cdots, s \tag{18.25}$$

where μ_W is the mean, ${\sigma_W}^2$ is the variance.

At time t, the cumulative degradation E(t) caused by external shocks is

$$E(t) = \sum_{j=1}^{N(t)} e_j = \sum_{j=1}^{N(t)} \kappa W_j \tag{18.26}$$

where, $e_j(t)$ represents the transient degradation increment caused by the j-th external shock, κ is a conversion factor.

18.2.2 Joint Optimization Method of EM and CBM

To enhance the overall usability of the system across its entire service life cycle, it is necessary to prioritize the performance recovery ability. This encompasses two primary modes: EM and CBM. Anti-interference ability is the capability of a system to function correctly in the presence of external disturbances, which reflects the robustness of the system (Dui et al. 2023, Dui et al. 2023). The system performance gradually declines due to the natural degradation process, resulting in a decline in the anti-interference ability to future external shocks. The system performance can be restored to the initial state through CBM, and the anti-interference ability can be improved. Recovery capability refers to the ability of a system to return to normal operations or functionality after experiencing a disruption, failure, or adverse event. When the system encounters external shocks that may lead to failure, reasonable EM can quickly restore the system performance. The recovery ability of the system can be significantly enhanced through the implementation of reasonable EM activities.

EM represents an imperfect maintenance (IPM) approach, which cannot restore system performance to its initial state. After IPM, the degradation state and degradation rate are affected. c_k is used to represent the cumulative degradation discount factor of the component after the k-th EM, and λ_{ik} represents the effect of the k-th EM on the component degradation rate (Guo et al. 2013). The probability density function of the discount factor is

$$f(c_k) = \frac{1/(1 - \exp(-ab^{k-1})) * ab^{k-1}}{T_{EM}/T_{total}} e^{\left(\frac{-b^{k-1}a(T_{EM}/T_{total} - c_k)}{T_{EM}/T_{total}}\right)} \tag{18.27}$$

where, a and b are hyper-parameters, k is EM frequency, T_{EM} is the EM threshold and T_{total} is the whole service life.

λ_{ik} represents draft parameter of the i-th component, which describe the effect of the k-th EM on the degradation rate. The draft parameter follows the normal distribution. That is, $\lambda_{ik}(t) \sim N(\mu_{\lambda ik}(t), \sigma_\lambda{}^2(t))$. The mean value is

$$\mu_{ik}(t) = (\exp(k) + 1)\mu_{\lambda ik}(t) \tag{18.28}$$

(1) Indexes for the joint optimization

To improve the comprehensiveness and adaptability of joint maintenance optimization decision, multi-objective optimization is adopted. The cumulative performance loss ratio and maintenance cost of the system are taken as the joint optimization indexes.

Inspired by the concept of system resilience (Dui, Liu et al. 2023), the system performance loss ratio is proposed as one of the maintenance optimization indexes. Based on the cumulative degradation value of the system, the performance loss during the whole life cycle can be directly measured, providing a direct reflection of the impact of maintenance activities. Compared with the traditional indexes, such as reliability or availability, the proposed performance loss ratio is more intuitive. The cumulative performance loss is

$$Me_1 = \int_0^T De(t)dt/T \tag{18.29}$$

where $De(t)$ represents the dynamic cumulative degradation value as time t, T represents the length of the overall service life cycle.

Maintenance cost includes maintenance resource cost and downtime loss. The maintenance resource cost mainly includes cost of renting vessels, cost of renting or purchasing maintenance tools and related personnel cost. The downtime loss is related to the time spent on maintenance activities. The total maintenance cost is

$$Me_2 = C = \sum (C_m + C_p) \tag{18.30}$$

where, C is the total maintenance cost, C_m is the maintenance resource cost and C_p is the downtime loss.

The goals of the multiple-objective optimization model are to minimize the system performance loss and the maintenance cost, and the decision variables are the EM threshold for the system and the CBM thresholds for all components. Constraint conditions need to be set to ensure that the optimization results are within a reasonable range. The system performance loss and the maintenance cost should be positive values. The maintenance threshold should be limited in the service lie cycle. The multi-objective function and the constraint conditions are expressed as

$$\begin{cases} \min Me_1 = \int_0^T De(t)dt/T \\ \min Me_2 = \sum (C_m + C_p) \\ \text{s.t.} \quad 0 \leq Me_1 \leq 1 \\ \quad Me_2 \geq 0 \\ \quad 0 \leq T_{EM} \leq T \\ \quad 0 \leq T_{CBM}(i) \leq T \end{cases} \tag{18.31}$$

where T_{EM} represents the EM threshold for the system, and $T_{CBM}(i)$ represents the CBM threshold for the i-th component.

(2) Joint optimization method for EM and CBM based on H-MOPSO model

Within the realm of multi-objective maintenance optimization, a common approach involves objective weighting coefficients to consolidate multiple objectives into a singular objective. The assigned weighting coefficients can reflect the relative importance of various optimization goals. In this way, the single objective optimization algorithm can be applied to find the optimal solution. Nevertheless, the determination of weighting coefficients may be influenced by the subjective judgment of decision-makers. Furthermore, the utilization of multi-objective optimization algorithm entails a direct exploration of Pareto-optimal solutions within the multi-objective space. During the optimization process, here may be a potential trade-off where enhancements in certain objectives might require sacrificing the optimization of others.

To avoid the above problems, a H-MOPSO algorithm is introduced. The optimization flow of H-MOPSO is shown in Fig. 18.2. H-MOPSO is an extension of the multi-objective particle swarm optimization algorithm, which can deal with the multi-objective problem according to the objective priority. First, the initial position and speed of the particle swarm need to be determined. The Latin hypercube sampling method is used to determine the initial position of each particle. The space position x_i of the i-th particle is

$$x_i = (x_{i1}, x_{i2}, \ldots, x_{ij}, \ldots, x_{iD}), i = 1, 2, \ldots, N; j = 1, 2, \ldots, D \tag{18.32}$$

The velocity v_i of the i-th particle is also a D-dimensional vector, which is

$$v_i = (v_{i1}, v_{i2}, \ldots, v_{ij}, \ldots, v_{iD}), i = 1, 2, \ldots, N; j = 1, 2, \ldots, D \tag{18.33}$$

The spatial position of the particle is seen as the solution of the objective optimization problem. The fitness value can be calculated through substituting the solution into the fitness function, that is, the objective function. According to the fitness value, the quality of the particle can be measured. Record P_{ibest} and G_{best}, and then update the position and speed of particles. After a-th iteration, the speed and position for the i-th particle are updated to

$$v_{ij}^{a+1} = w_p v_{ij}^a + c_1 r_1 (P_{ibest} - x_{ij}^a) + c_2 r_2 (G_{best} - x_{ij}^a) \tag{18.34}$$

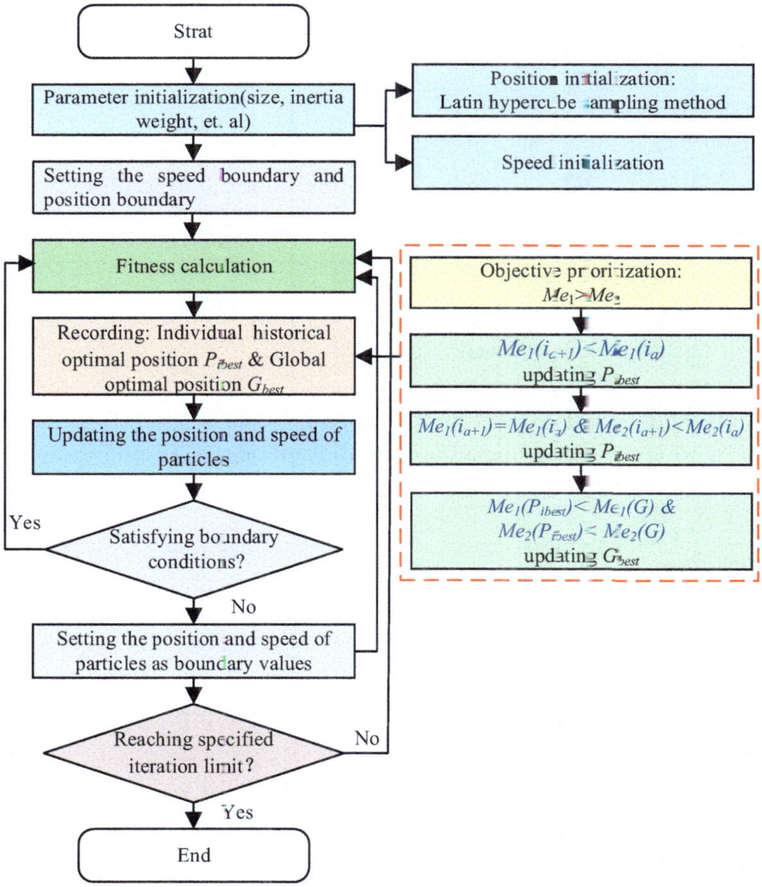

Fig. 18.2 Optimization flow of H-MOPSO

$$x_{ij}^{a+1} = x_{ij}^{a} + v_{ij}^{a+1} \tag{18.35}$$

where, w_p is the inertia weighting coefficient, a is the current iteration number, c_1 and c_2 are scale factors, r_1 and r_2 are random numbers that are independent of each other.

Compared with the traditional MOPSO algorithm, the hierarchical optimization strategy is adopted when updating the particle optimal position and the global optimal position. According to the priority order of the optimization targets, the optimization of the highest priority target is completed first, and then the optimization of other targets is completed based on the corresponding priority order. Corresponding, the particle swarm solution set satisfying the highest priority target is firstly found. In the particle swarm solution set of highest priority target, solution sets satisfying other targets are found successively. The pseudocode of H-MOPSO algorithm is shown as follows, where A represents the total iteration number.

Algorithm 1: H-MOPSO algorithm
1: Parameter initialization
2: Initialization $\boldsymbol{x_0}$ (position) and $\boldsymbol{v_0}$ (velocity)
3: $[Me_1{}^0, Me_2{}^0]$ = **function**($\boldsymbol{x^0}$)
4: **Record** P_i (particle optimal position) and G (global optimal position)
5: **for** $a = 1{:}A$ (The a-th iteration)
6:     **for** $i = 1{:}N$ (The i-th particle)
7:         **Update** $\boldsymbol{v_a}$ and $\boldsymbol{x_a}$
8:         $[Me_1{}^a, Me_2{}^a]$ = **function**($\boldsymbol{x_a}$)
9:         **if** $Me_1(i_{a+1}) < Me_1(i_a)$, **update** P_{ibest}
10:         **if** $Me_1(i_{a+1}) = Me_1(i_a)$ and $Me_2(i_{a+1}) < Me_2(i_a)$, **update** P_{ibest}
11:         **if** $Me_1(P_i) <= Me_1(G)$ and $Me_2(P_i) <= Me_2(G)$, **update** G_{best}
12:     **end**
13: **end**
14: **Output** G_{best}

18.3 Case of Subsea Production Control System

18.3.1 Introduction of the Subsea Production Control System

The subsea production system can be divided into subsea production control system and subsea production equipment. According to the actual failures and maintenance records, subsea equipment in the production control system is prone to failures, such as breakage of the umbilical, insulation drop of the communication cable,

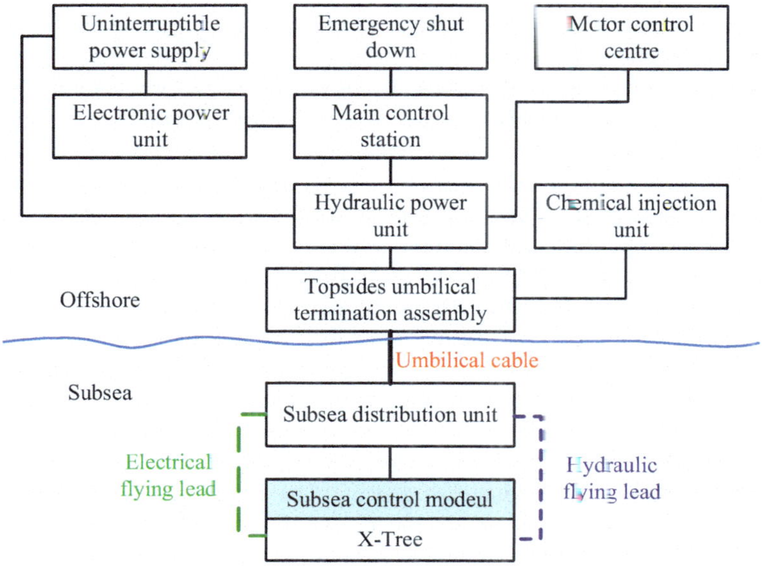

Fig. 18.3 Structure of the subsea production control system

communication module failure, and downhole communication interruption. Therefore, some important subsea equipment in the control system, such as subsea control module (SCM), hydraulic flying lead (HFL), electric flying lead (EFL) and umbilical cable (UM), are selected as research objects. The simplified structure of the subsea production control system is shown is Fig. 18.3.

SCM is an electro-hydraulic control module mounted on the subsea X-tree. SCM can guide the hydraulic fluid to open or close subsea valves on command, collect real-time sensor information from subsea wellheads and manifold, and transmit data to topside control station. UM is used to transmit control signals, power and hydraulic fluid between the topside umbilical termination assembly and subsea umbilical termination assembly. EFL connects from the subsea distribution unit to SCM and transmits power and signals to the SCM. HFL is used to transfer hydraulic power and filling chemical agents.

18.3.2 Performance Degradation Prediction of Subsea Control System

The unique working environment of the subsea production control system poses a challenge in terms of data availability, resulting in limited monitoring data. Consequently, this insufficient data hinders the effective implementation of degradation prediction processes. The performance prediction model is established based on simulation data, which is generated based on failure parameters derived from OREDA database (DNV 2015). Based on the failure parameters obtained from the OREDA database, a total of 11 sets of degradation data are generated. 10 sets of degradation data are set as historical data, which are used for estimation of fixed parameters. A set of data is set as monitoring data for updating drift parameters. The failure parameters for the subsea control system are shown in Table 18.1. Based on the historical data sets, the fixed parameters are estimated according to the maximum likelihood algorithm. The prior drift parameters $\mu_{\lambda 0}$ and ${\sigma_{\lambda 0}}^2$ are estimated by combining the simulated monitoring data and the estimated fixed parameters. The estimated fixed parameters and prior drift parameters are shown in Table 18.2.

Kalman filter algorithm is used to update drift parameters $\mu_{\lambda}(t)$ and ${\sigma_{\lambda}}^2(t)$, and the obtained results is shown in Fig. 18.4. The system performance degradation process follows the nonlinear law, so the mean degradation rate is not a constant value, and its change reflects the degradation rate of the components. Degradation rates of umbilical cable and HFL are gradually increasing over time, and degradation rates of SCM and EFL are gradually decreasing. The variance tends to be stable after a certain number of iterations, which means that the randomness in the prediction process is gradually eliminated, and the estimated value of degradation is gradually approaching to the real value.

The cumulative degradation values of SCM, UM, EFL and HFL in the normal degradation process are shown in Fig. 18.5. It can be seen that the cumulative degradation values increase rapidly during the initial phase and the end phase. In the mid-term of the whole service life cycle, cumulative degradation experiences a relatively slow growth. During the initial phase, a steep rise in cumulative degradation values can often be traced back to issues in system design or manufacturing defects. Conversely, in the later stages, the rapid accumulation of degradation is more likely to be attributed to the natural aging process of the equipment.

Table 18.1 Failure parameters for the subsea control system

Parameters	SCM	Umbilical cable	EFL	HFL
Failure rate/per month	0.018	0.0044	0.0127	0.00745
SD	0.03	0.0093	0.0173	0.014

Table 18.2 Estimated fixed parameters and prior drift parameters for the nonlinear Wiener model

Parameters	SCM	Umbilical cable	EFL	HFL
θ	6.25×10^{-5}	1.18×10^{-4}	1.56×10^{-5}	3.13×10^{-5}
${\sigma_B}^2$	1.1×10^{-3}	1.15×10^{-4}	4.28×10^{-4}	2.88×10^{-4}
${\sigma_\varepsilon}^2$	1.6×10^{-7}	1.4×10^{-8}	1.00×10^{-7}	6.27×10^{-8}
$\mu_{\lambda 0}$	0.0407	0.0075	0.015	0.0028
${\sigma_{\lambda 0}}^2$	1.2×10^{-3}	1.12×10^{-4}	4.3×10^{-4}	2.89×10^{-4}

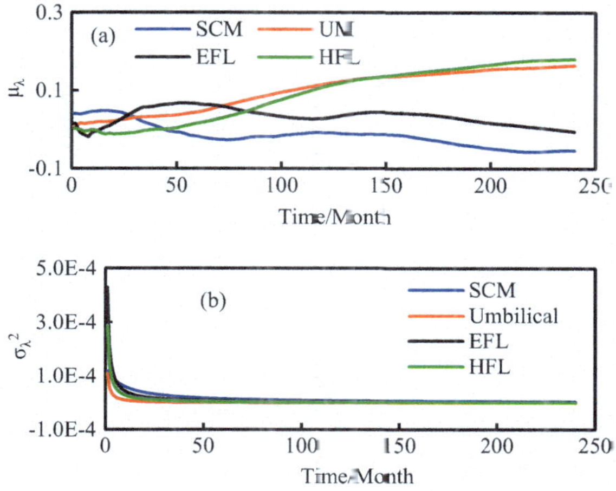

Fig. 18.4 Updating of drift parameters

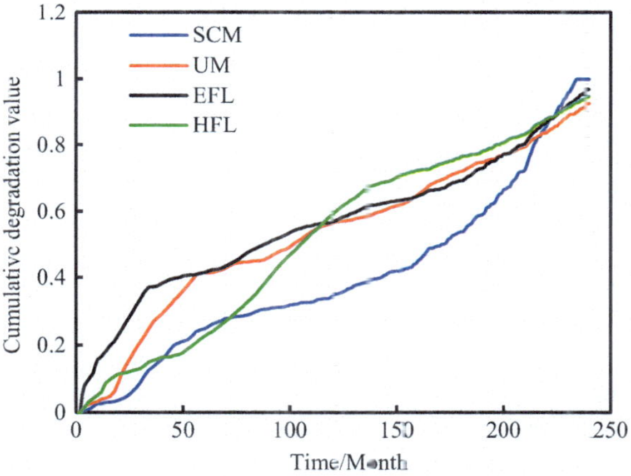

Fig. 18.5 The normal degradation process for each component

In the subsea control system, components are interconnected in a series. Utilizing the cumulative degradation of each component, the integrated importance degrees and weighting coefficients for each component throughout the whole service life are calculated, which are shown in Fig. 18.6. UM is the most important component in the control system, which is consistent with the actual engineering records. In practical engineering applications, with the increasing of service time, the failures of the umbilical cable occur more frequently.

The subsea environment of the control system is complex, and the system performance may be affected by internal wave, typhoons, etc. Cumulative impact model is used to describe the impact of natural disasters on subsea facilities. To reflect the actual impact process, parameter estimation of the cumulative impact model is carried out according to statistical data such as water level data, wind wave current main

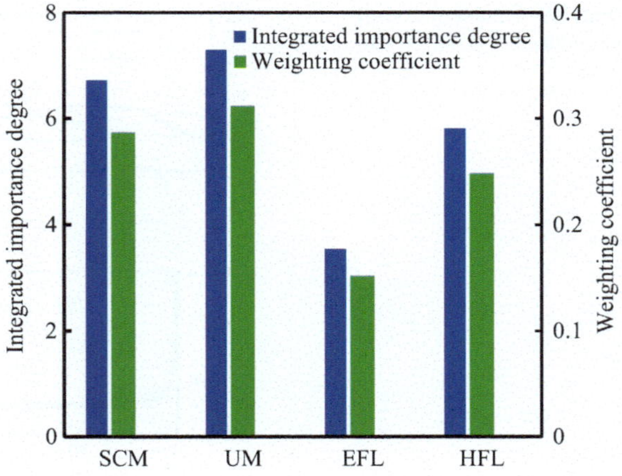

Fig. 18.6 Integrated importance degrees and weighting coefficients for each component

Table 18.3 Parameters of the cumulative impact model

Parameters	v_s	μ_W	${\sigma_W}^2$
Values	0.06	0.145	0.012

extreme value data, wind wave current condition extreme value data and internal wave current extreme value data in the actual oilfield sea area. The estimated parameters are shown in Table 18.3.

The actual degradation process of the system is consisted of two parts, the normal degradation effect and the effect of experienced external shocks, as shown in Fig. 18.7. In Fig. 18.7a, the normal degradation process over the whole service life cycle is calculated according to the cumulative degradation values of each component and their corresponding weighting coefficients. The degradation value of the system under external shocks is shown in Fig. 18.7b. The actual degradation process of the system is shown in Fig. 18.7c. The system will be completely failure when the cumulative degradation value reaches 1. Due to external random shocks, the system will be completely failure at the 105th month without the intervention of any maintenance activity, which is well below the expected design life. So, EM activity and CBM activity in the whole service life are necessary.

18.3.3 Joint Optimization for the Subsea Control System

To minimize maintenance downtime, a replacement strategy is generally adopted for the subsea equipment. The subsea production system is usually designed with a modular approach. The failed or damaged equipment can be replaced directly in case of an accident. Corresponding spare parts are stored in the platform warehouse or base warehouse, and the replaced equipment will be checked repaired, which can be

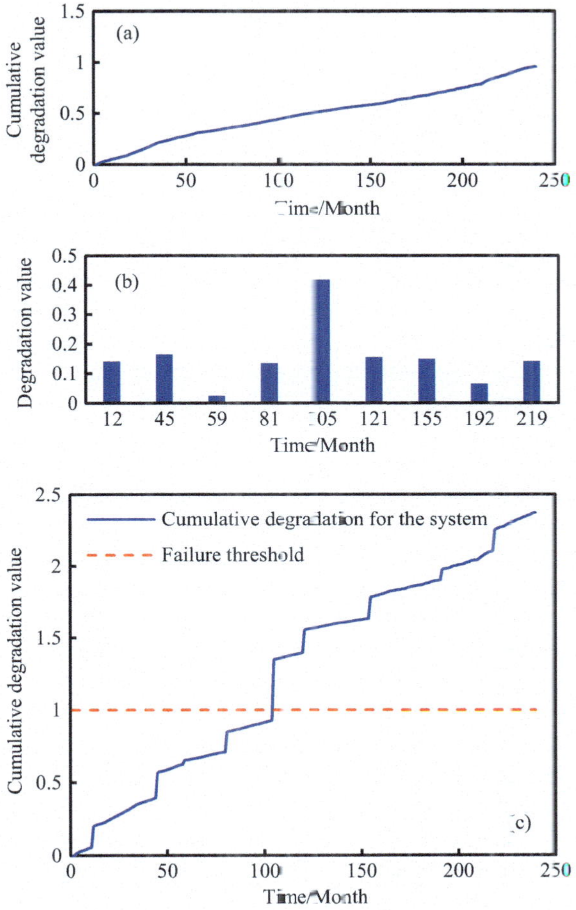

Fig. 18.7 Actual degradation process of the control system, **a** the normal degradation process, **b** effects of the random shocks, **c** the actual degradation process

further used as the spare part. CBM is specific to each component and can be regarded as a perfect maintenance method. In this way, the component performance can be restored to the initial state. The imperfect maintenance effect of EM is judged based on practical engineering applications. Due to limitations of maintenance time and maintenance resources, a more effective method will be taken to restore production as soon as possible in the EM process for the subsea production system. Taking the communication cable failure as an example, a standby power cable was transformed into the communication cable, instead of building a new all-electric umbilical control cable. The regression of cumulative degradation is a stepwise curve (Liu et al. 2023).

To effectively optimize the maintenance process of the subsea control system, the primary objectives are to minimize both performance loss and maintenance costs. According to practical engineering requirements, the priority of minimizing the performance loss is higher than minimizing maintenance cost. In the H-MOPSO

Table 18.4 Average CBM time and EM time for each component

Parameters	SCM	Umbilical	EFL	HFL
Average CBM time/day	7	12	5	5
Average EM time/day	4	5	2	2

algorithm, there are five decision variables in total, including one EM threshold for the system and four CBM thresholds for SCM, UM, EFL and HFL. The values of the five decision variables constitute the position of a particle. The fitness values are derived by inserting the solution into the respective fitness functions, specifically the performance loss function and the maintenance cost function. According to the priority of multiple optimization objectives, the optimization of the minimum performance loss with the highest priority is completed firstly. The position ranges of particles satisfying the minimum performance loss are found. Then, within these ranges, the position ranges of particles that meet the next optimization objective will be found, the optimization of the minimum maintenance cost is completed.

The maintenance cost of subsea equipment mainly includes maintenance resource costs and downtime loss. Maintenance resource cost includes maintenance equipment rental cost and the salary of operators. The maintenance equipment needed includes ROV equipment, basic inspection tools, and a subsea positioning system. The maintenance operators include operation leader, ROV supervisor, ROV helmsman, inspection engineer and positioning engineer. Based on the actual maintenance quote, the daily cost of maintenance resource is about 97,465 CNY. In order to quantify the production loss, the tree flow rate in the subsea production system is considered constant, and the average daily oil value of the subsea production system is about 1,023,200 CNY. According to the actual maintenance records of the subsea production system, the average EM time and CBM time of each component are estimated, as shown in Table 18.4. The total downtime of EM is calculated as the sum of the EM time of each component. Based on the daily cost of maintenance resource and downtime loss, the maintenance cost of EM and CBM can be obtained.

According to H-MOPSO algorithm and target priority, the multi-objective problem for the joint optimization process is optimized by adopting the hierarchical update strategy. The number of particles is set to 50. The maintenance thresholds that require optimization include the EM threshold for the system and the CBM thresholds of each component. The value range of each threshold is segmented into 50 intervals. A random value is then sampled from each interval and assigned to 50 particles, constituting the initial population. The inertia weighting coefficient is set as 0.9, and scale factors c_1 and c_2 are set as 1.5.

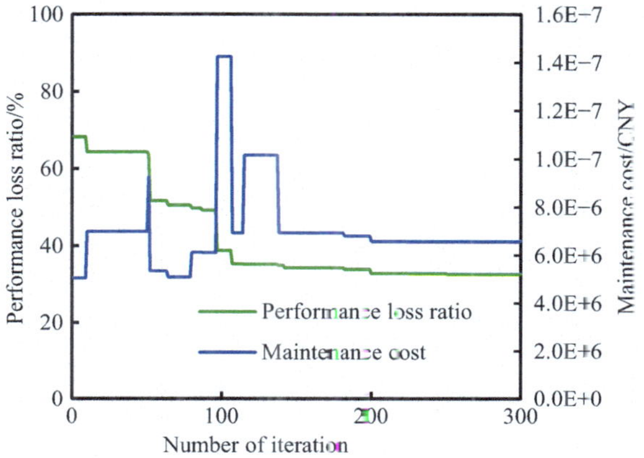

Fig. 18.8 Optimization processes for performance loss ratio and maintenance cost

18.3.4 Results and Discussions

(1) Optimization results of the maintenance thresholds

Optimization processes for performance loss ratio and maintenance cost are shown in Fig. 18.8. The prioritization for goal optimization is as follows: minimum performance loss ratio and minimum maintenance cost. The performance loss ratio of the system has been decreasing throughout the iteration. Taking the current performance loss ratio as the constraint condition, the maintenance cost is optimized synchronously. In the early stage, the maintenance cost fluctuated greatly with the decrease of performance loss ratio. However, in the later stage, with the decrease of performance loss ratio, the optimization result of maintenance cost shows a downward trend and gradually became stable.

Based on the optimization algorithm, the optimal EM threshold for the control system is 0.392. The optimal CBM thresholds for SCM, umbilical cable, EFL and HFL are 0.3216, 0.292, 0.497 and 0.507, respectively. The performance variation trend of the control system during the whole service life cycle is shown in Fig. 18.9. The system underwent a total of 11 maintenance activities over its life cycle, including 3 EM activity and 8 CBM activity. The system will experience 13 CBM activities without taking into account the EM. So, the joint maintenance optimization method can reduce the maintenance times of the system, and then reduce the maintenance cost. The cumulative degradation value of the system after EM activity is reduced to 0.237. Although the system performance cannot be restored to the initial state, the degradation value fluctuates in a more stable area. In this way, the continuous degradation of the system to a worse level is avoided, and the risk of failure is avoided.

(2) Convergence procedure comparison with the traditional MOPSO

The traditional Multi-Objective Particle Swarm Optimization (MOPSO) algorithm is used for the joint maintenance optimization under the condition of multi-objective. An external population archive is used to store all current non-dominated

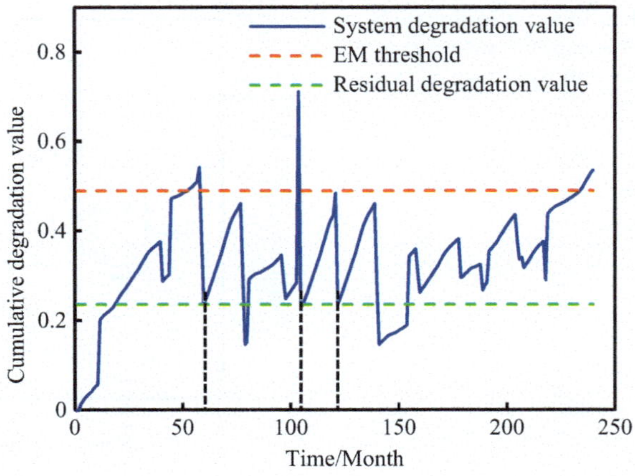

Fig. 18.9 The performance variation trend during the whole service life cycle

solutions in MOPSO. The individuals in the external archive are regarded as elite individuals, which can control the evolutionary direction of the population and guide the population to approach the real Pareto frontier. At last, the particles in the external archive are approximated as the optimal Pareto solution. The particle swarm optimization algorithm incorporates the Pareto domination relation to identify the individual historical best particle and the global best particle.

In the MOPSO optimization process, performance retention rate and maintenance cost are set as optimization objectives. The minimum maintenance cost corresponding to the maximum performance retention rate in the whole service life cycle can be obtained, meaning that the minimum maintenance cost is aligned with the minimum performance loss. In the iteration process, the positions of the global best particle are shown in Fig. 18.10. With the improvement of the performance retention rate, the maintenance cost is constantly fluctuating.

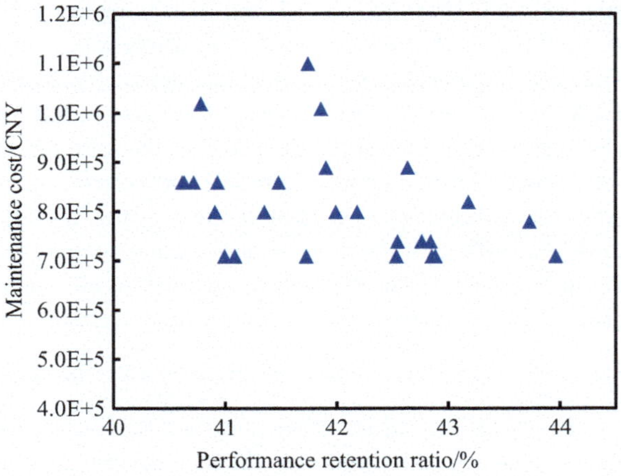

Fig. 18.10 Positions of the global best particle in the whole iteration process

Based on the traditional MOPSO algorithm, the optimal EM threshold for the control system is 0.92. The optimal CBM thresholds for SCM, umbilical cable, EFL and HFL are 0.711, 0.90, 0.44 and 0.71, respectively. It can be seen that the threshold optimization results based on the traditional MOPSO algorithm are significantly higher than the threshold results based on H-MOPSO algorithm.

The optimization iterative process based on the MOPSO algorithm and result comparison with H-MOPSO algorithm are shown in Fig. 18.11. The system performance loss and maintenance cost are both decreasing during the iteration process. However, compared with the results of H-MOPSO, it is found that the optimal performance loss ratio based on MOPSO is much greater. During its whole service lifetime, the system underwent a total of 6 maintenance activities. The purpose of saving maintenance cost is achieved by reducing the number of maintenance activities. Under the combined influence of external random shocks and internal degradation process, the cumulative degradation value of the system can reach to 92%, which means that the control system may fail at any time. Once the system fails, it will lead to a huge downtime loss and cause environmental pollution.

(3) Optimization of EM

Usually, only one maintenance team is equipped for the subsea production system. In the EM process, each component of the system is repaired in a certain maintenance sequence. For example, the system is attacked by a random shock at the 105th month, and the cumulative degradation value rapidly increases from 0.2856 to 0.7128. At the moment, an EM activity is needed to avoid further deterioration of the system state. To reasonably optimize the EM process, a microscopic time perspective is adopted, and the unit of time is set to days. The performance loss ratios under different maintenance sequences are calculated. The comparison of system performance loss ratios under different maintenance sequences is shown in Fig. 18.12. From the point of minimizing performance loss, the optimal maintenance sequence during emergency maintenance is determined. Under the 18th maintenance sequence, the component

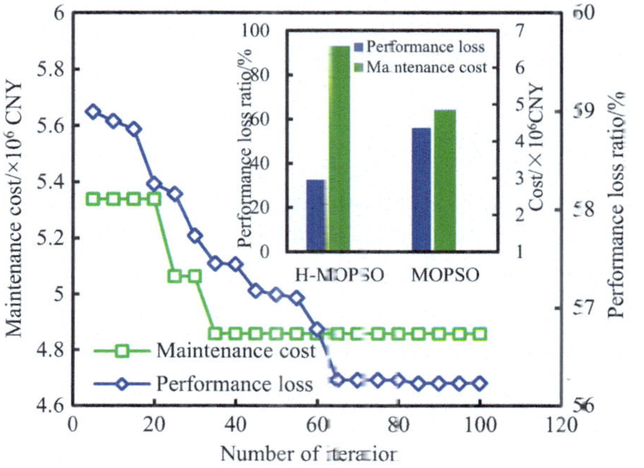

Fig. 18.11 Optimization iterative process of MOPSO and result comparison with H-MOPSO

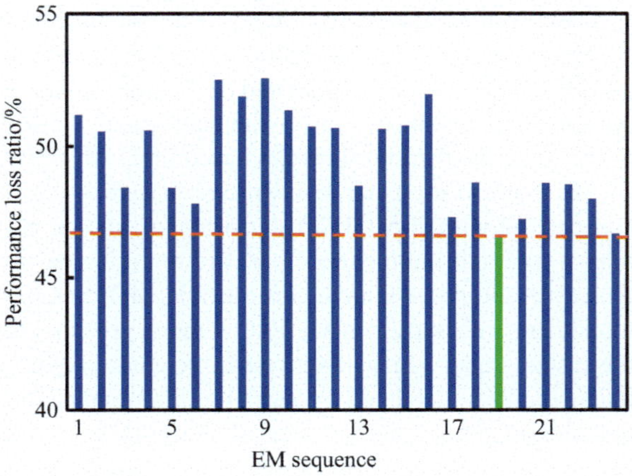

Fig. 18.12 Comparison of system performance loss ratios under different maintenance sequences

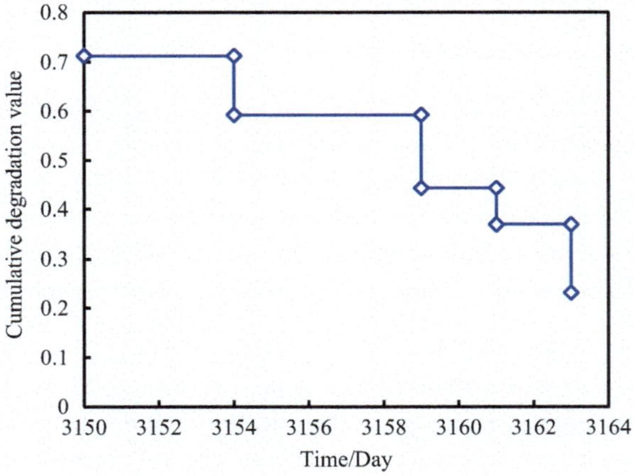

Fig. 18.13 The stepwise recovery curve in the EM progress

performance loss ratio is the smallest, which is 46.63%. The maintenance sequence is HFL, SCM, EFL and UM.

Effective EM activity after disruptions is essential to improve the recovery capability of the system. Under the EM activity, the gradual regression of the cumulative degradation value of the system is shown in Fig. 18.13. Due to limited maintenance resources, the subsea control system faces challenges in achieving simultaneous recovery for all its components. In this way, the stepwise performance recovery curve is closer to reality. The gradual recovery curve offers a more accurate portrayal of the incremental gain achieved through each component recovery, thereby furnishing vital insights that inform effective EM decisions.

Chapter 19
Maintenance Strategies for Sustainability Development

Abstract Due to the hostile subsea environments, it is challenging to conduct efficient maintenance activities for the complex systems with failure dependences, especially when try to keep the maintenance activities sustainably simultaneously. Currently, there is no standardized method to assess the coupling impact of failure dependences and the effectiveness of maintenance on the overall sustainability. This chapter introduces an innovative, integrated framework that uses Bayesian networks to comprehensively analyze the combined effects of component degradation, failure dependences, and maintenance management on the sustainability of these complex subsea systems. The proposed model explores the impact of various maintenance strategies and failure dependences on overall sustainability. A case study demonstrates the importance of considering failure dependences and the practicality of the sustainability evaluation model, offering solutions for optimizing maintenance strategies from a sustainable perspective. The research contributes to improving the maintenance and reliability of complex subsea systems at reasonable costs, providing valuable guidance for decision-makers seeking sustainable maintenance practices.

19.1 Introduction

Sustainability development strives to fulfill the requirements of the current generation while ensuring that future generations remain capable of satisfying their own needs (Development 1987). In the realm of engineering, sustainability refers to the ability of a system to sustain a continual long-term process, considering the integration of environmental, social, and economic factors. The three pillars of sustainability are holistic enough to basically encompass the requirements of the system, as presented in Fig. 19.1. The integration of sustainability evaluation in the analysis becomes essential in light of the worldwide push towards environmentally and socially responsible practices. In particular, subsea industrial engineering is expanding quickly so as to achieve economic growth, carrying along problems related to social and environmental difficulties, such as the sustainability issues mentioned above. Sustainability in the context of subsea systems involves not only operational efficiency

B. Cai et al., *Intelligent Operation and Maintenance for Subsea Production Systems*,
https://doi.org/10.1007/978-981-96-2677-9_19

but also environmental and economic impact. Due to the hostile subsea environments, it is challenging for subsea production systems to ensure their reliability while keeping sustainable during the operation and maintenance. Besides, the expansion of subsea industrial engineering leads to ever-more complex systems to satisfy human demands. Within the subsea production system, certain components are functionally or structurally interconnected, leading to two types of failure dependence, requiring an increasing concern. More complex subsea production systems are more vulnerable to failure dependences, which can result in higher risks, greater financial losses, and more severe environmental pollution—all of which are signs of decreased sustainability from a variety of perspectives. In this context, the operation and expansion of subsea production systems subject to failure dependences pose significant threats to the marine ecosystem in terms of sustainability. To improve the sustainability of the complex subsea production system, failure dependences should be mitigated, and the maintenance activities should also be conducted. Hence, there is a desire for investigation on sustainability issues of complex subsea production systems. This brings up the potential research question. How to evaluate the coupling impact of failure dependences and maintenance activities on the sustainability of complex subsea production systems?

The impact of failure dependences on overall sustainability is reflected by two ways. To start with, failure dependences influence the component states and degradation rates, and thus influence the system operation and production, which directly influence the overall sustainability. Besides, the mitigation of failure dependences calls for attention and some specific maintenance activities to decouple the components, which indirectly influence the overall sustainability. The decoupling maintenance (DM) activities refers to the behavior of reducing the degree of failure dependence between the coupled parts or components of a system. Such decoupling maintenance activities and the conventional maintenance practices for subsea components both present a set of unique challenges due to hostile subsea environments such as high pressure and low temperature, corrosive seawater, and constant exposure to the wastes. Similar as failure dependences in this context, the maintenance is placed as an essential element within the paradigm of sustainable development (Ghaleb

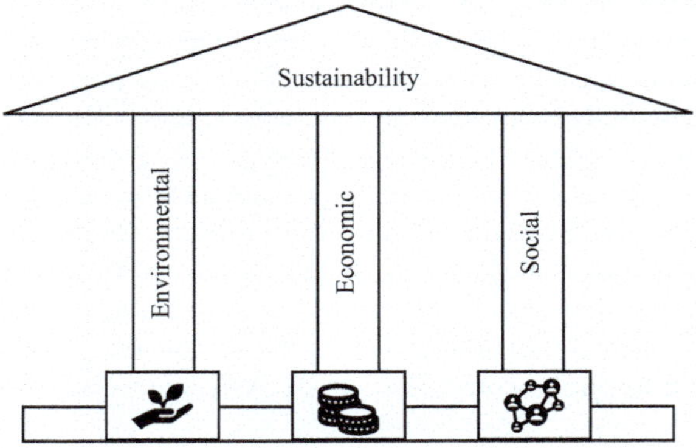

Fig. 19.1 Three pillars of the sustainability

and Taghipour 2022), and may also have an impact on the sustainability directly or indirectly. For instance, maintenance activities impact sustainability in two ways: indirectly improve sustainability by enhancing system performance, or directly cause sustainability changes by carrying out activities.

Given the fact that the failures such as leakage, collapse, erosion, the accelerated degradation due to failure dependences, as well as the challenging maintenance activities could all significantly impact sustainability, the mitigation of failure dependences and the evaluation of sustainability for subsea systems considering maintenance activities become imperative tasks. Achieving an optimal balance between the performance of the subsea system and its sustainable relationship with surroundings requires a holistic under-standing of the natural degradation of components, the complex failure dependence among them and their maintenance activities.

Directing to the research issue, this chapter proposed a novel framework to evaluate the system sustainability with consideration of failure dependence and maintenance activities, and subsequently applied this model to a subsea transmission system. In this chapter, the decoupling maintenance is delimitated and modeled to eliminate the failure dependences. In addition, a comprehensive methodology integrating the impact of component degradation, the failure dependences among components, and the maintenance activities on the overall sustainability is provided.

19.2 Integrated Framework with Failure Dependence and Maintenance

Figure 19.2 shows an innovative methodology incorporating the impact of failure dependences and maintenance activities to evaluate sustainability over time. This integrated framework includes three sub-models. Within this framework, the degradation model estimates the components states and failure dependences among components using expert experience and historical data. Additionally, drawing from expert experience and maintenance records, it is possible to determine the traditional maintenance strategies and the maintenance model, which aids in recognizing the impacts that the maintenance strategies may have on the components. It is important to highlight that the innovative DM activities are also embedded in the maintenance model to mitigate the failure dependence. In addition, a universal sustainability model is employed to evaluate changes in sustainability throughout the system operation and maintenance activities, thereby facilitating optimization of the decision-making scheme. It is worth noting that the degradation process and maintenance activities of components within systems are closely interconnected and have a reciprocal impact on each other. The degradation process determines the state of components and thus influences the maintenance policies. Conversely, the maintenance activities changed both the component states and degradation process. This prompt introducing the integration merging degradation and maintenance models, denoted as degradation-maintenance model for short.

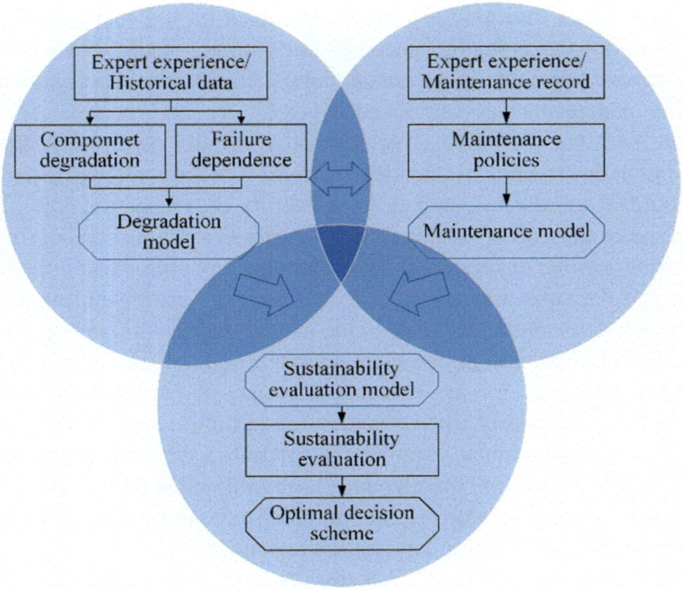

Fig. 19.2 The integrated framework of the proposed methodology

Considering the dynamics of a system, the DBN serves as a suitable tool for capturing the d evolving behaviors of the system within a constantly shifting environment. DBN can handle the uncertainty of complex systems, allowing engineers to better understand the evolution of the system over time and better formulate maintenance strategies. In this chapter, the DBN model addresses the components and the system at different time intervals. Comparisons of overall sustainability before and after required maintenance activities can therefore be studied. The DBN calculation is implemented through the software GeNIe. The GeNIe primarily serves the purpose of constructing, assessing, and visualizing probabilistic graphical models, with a particular focus on BNs and DBNs (Ma et al. 2018). The GeNIe offers a free development environment for graphical decision-theoretic models to analyze the probabilistic relationships among various variables (Cheng et al. 2021). This facilitates comprehension and management of the probabilistic events in complex systems.

19.3 Degradation-Maintenance Model

The degradation-maintenance model is illustrated in Fig. 19.3. Taking component 1 as an illustrative example, the grey nodes labeled C1 and C1′ denote its state of at time t and $t + \Delta t$ respectively. The nodes CR1 and DM1 in yellow represent the maintenance activities concerning the component and the maintenance activities to mitigate the failure dependences, respectively. The node FD1 denotes the overall failure dependence originating from other components towards component 1. The red arcs emanating from the nodes representing the other components and terminating at the node FD1 visually depict the failure dependence that exists between component 1 and each of these components.

The state of the n components system is denoted by $X = (x_1, x_2, \ldots x_n)^T$, where $x_i \in [0, 1]$ characterizes the state of component i in this system. When $x_i = 0$, the component i is in the As-Good-As-New state, i.e., the component is brand new; while when $x_i = 1$, the component i is in Failed state. As the component degrades, its state value increases. Use a saturation function $\varphi(x)$ to restrict the state of the components in [0, 1]

$$x_i = \varphi(x_i) = \begin{cases} 0, & x_i < 0 \\ x_i, & 0 \le x_i \le 1 \\ 1, & x_i > 1 \end{cases} \tag{19.1}$$

The components experience degradation and maintenance intervention over time. Let $x_i(t)$ represent the state of component i at time t. Then $x_i(t + \Delta t)$ is the state of component i at time $t + \Delta t$, where Δt is the period of component degradation and maintenance intervention, starting from the moment the component states are observed until completion of the maintenance, which could be abbreviated as *delay period*. The delay period may be affected by the following factors: maintenance type, component importance, failure impact, maintenance procedures and resource availability, etc. Overall, the delay period is generally determined based on balancing factors such as component availability, cost-effectiveness, and safety. Its specific scope varies based on specific application scenarios and component characteristics. The function to determine $x_i(t + \Delta t)$ refers to Eq. (19.2).

$$x_i(t + \Delta t) = \varphi(x_i(t) \cdot D_i + \eta) \tag{19.2}$$

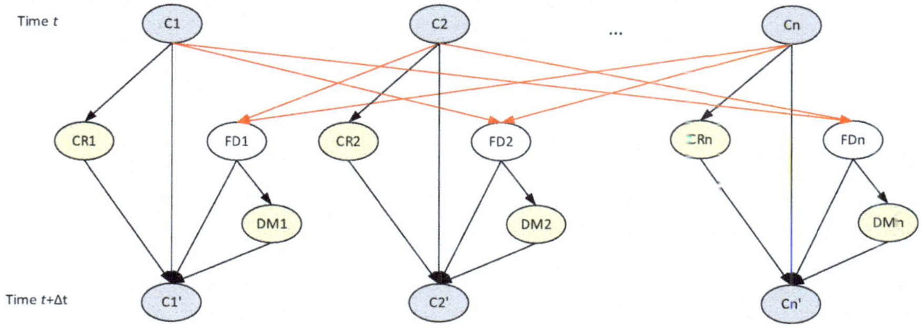

Fig. 19.3 DBN model for degradation process and maintenance activities considering CAFs

where D_i is the failure dependences from other components on component i, implying that the state of a component can be affected by the degradation or failure of other components.

$$D_i = \prod_{j=1}^{n} \left(1 + x_j(t)\right) \tag{19.3}$$

and η is the state change of the component, representing the degradation and the maintenance effectiveness.

$$\eta = \begin{cases} \eta_D \Delta t, & \text{for NM} \\ \eta_{\text{PM}}, & \text{for PM} \\ \eta_{\text{CM}}, & \text{for CM} \end{cases} \tag{19.4}$$

In Eq. (19.4), η_D is the degradation decrement of component i during time unit. When there is no maintenance intervention, the state of component i only changes due to degradation and failure dependences. A value between 0 and 1 is assigned to state change rate due to degradation. If there is maintenance intervention during Δt, the state of component i changes due to maintenance intervention and failure dependences. Let η_{PM} and η_{CM} denote the maintenance effectiveness of Preventive Maintenance (PM) activities and Corrective Maintenance (CM) activities respectively, then a value between -1 and 0 is assigned to state change caused to each PM and CM activities. In this model, the η_D is assigned a positive value since it signifies the degradation of the component. On the other hand, η_{PM} and η_{CM} are assigned negative values because they denote the improvement of the component state, leading to a decrement in its state value. This model assumes that the maintenance activities could eliminate the cumulative effect of both previous and current damage. Therefore, the degradation during the period of Δt is not considered in the equations related to the maintenance activities.

In addition, the maintenance activities aiming at diminishing the failure dependences, i.e., DM, are also examined. When such maintenance activities are implemented, the failure dependence is assumed to drop to 0, and the state of component i at time $t + \Delta t$ is

$$x_i(t + \Delta t) = \varphi(x_i(t) + \eta) \tag{19.5}$$

In summary, various maintenance activities are proposed, including the PM and CM activities aligned with the component states, as well as the implementation of DM activities aligned with the failure dependences among components.

- No Maintenance (NM) activities: the state of the component is acceptable, which does not surpass the threshold for any maintenance activity ($x \leq a$), thereby necessitating no maintenance intervention.
- Preventive Maintenance (PM) activities: the PM involves imperfect maintenance and could only improve the component performance to a certain extent.

Common PM activities for subsea systems include coating repair, cleaning, anti-rust treatment, etc.

- Corrective Maintenance (CM) activities: the CM is perfect maintenance, fully restoring the component to its normal operating condition. Typical CM activities for subsea systems involve replacements of broken pipelines or damaged critical devices, etc.
- Decoupling Maintenance (DM) activities: the DM aims to decouple dependent components, specifically addressing the failure dependences. The failure dependence is assumed to be eliminated completely after DM activities. Examples of DM activities for subsea systems encompass physical isolation between equipment, redundant design, interlocking systems, etc.

19.4 Sustainability Evaluation Model

19.4.1 Evaluation of Impacts of Component Performance on Sustainability

The system performance determines not only the output efficiency but also the pollution it generates during both regular operation and in the event of failures, i.e., emissions, waste, and noise. While the regular operation of the system constantly results in a certain amount of emission or pollution to the surrounding environment, its production will contribute positively to both society and the economy. To evaluate the impact of the component performance on the sustainability more comprehensively, the suggested procedure is outlined in Table 19.1.

Step 1. System familiarization.

Table 19.1 Suggested stepwise procedure of the sustainability evaluation concerning the impact component performance

Step	Description
Step 1	System familiarization
Step 2	Information acquisition and determination of nominal states of the components
Step 3	Scoring the effects of the component performance on each sustainability indicator (SI)
Step 4	Weighing the contribution of the sustainability indicators to the sustainability pillar (SP)
Step 5	Determination of the importance of each sustainability pillar
Step 6	Determination of the overall sustainability score (OSS) by impacts of component performance

For a comprehensive understanding of the system, it is crucial to have a well-defined framework with clear physical boundaries and specific details regarding its operational requirements and relationship with sustainability. Operational familiarization includes the system structure, components functions and dependence among components. Environmental familiarization involves identification of all potential sustainability pillars and sustainability indicators that may contribute to each sustainability pillar.

Step 2. Information acquisition and determination of nominal states of the components.

Based on the component states defined as $x_i \in [0, 1]$, if the component is normally working, and no maintenance is required, it is supposed that the damage from the component performance to the sustainability is infinitely close to 0 and could be neglected. The state of component that falls below the threshold for maintenance activities could be then considered as the nominal state. Information about the nominal values of component states could be obtained from OREDA (SINTEF and NTNU 2015), historical data, maintenance records, and other relevant sources.

Step 3. Scoring the effects of the component performance on each sustainability indicator.

Components that are functioning below the maintenance threshold are assumed to be in a nominal state and do not significantly impact the system sustainability indicators. However, when a component surpasses this nominal state, the variation between the component state and the nominal state can be utilized quantitatively to evaluate the effects of the component performance on sustainability indicators. Compared to the nominal states of the components, the effects of the component performance on each sustainability indicator could be denoted as

$$\sigma_{i \to SI_j} = \begin{cases} 0, & x_i < a \\ x_i - a, & x_i \geq a \end{cases} \tag{19.6}$$

The integrated effects of the component performance from all the components on each sustainability indicator can be then given by $\sum_{i=1}^{n} \varepsilon_{i \to SI_j} \sigma_{i \to SI_j}$, where $\varepsilon_{i \to SI_j}$ represents the relationship between the state of component i and the sustainability indicator SI_j, and

$$\sum_{i=1}^{n} \varepsilon_{i \to SI_j} = 1, \quad \text{for} \quad j = 1, 2, \ldots r \tag{19.7}$$

The scores of $\varepsilon_{i \to SI_j}$ can be derived from technical documentation, operational information, expert assessments, accident reports, and interviews with engineers. When the performance of the component declines, indicating an increase in the value of the component state, and such degradation negatively impacts the sustainability,

$\varepsilon_{i \to SI_j}$ assumes a negative value within the range of -1 to 0. Conversely, when the component degradation positively affects the sustainability, $\varepsilon_{i \to SI_j}$ assumes a positive value within the range of 0 to 1.

Step 4. Weighing the contribution of the sustainability indicators to sustainability pillars.

Sustainability pillars and sustainability indicators are a pair of concepts, but they serve distinct purposes. Sustainability pillars are the fundamental categories that represent the various aspects of sustainability, such as Environmental, Social, and Economic. On the other hand, sustainability indicators are the tools to assess how well the sustainability goals are achieved within these sustainability pillars. They are specific metrics used to evaluate the performance of a system in terms of sustainability. For instance, under the Environmental pillar, sustainability indicators might include hazardous substances, carbon emissions, wastes, and noise. The sustainability indicators contributing to a sustainability pillar may have varying weights of influence. The contributing weight of sustainability indicator SI_j for the sustainability pillar SP_k is denoted as w_{jk}. Should there be no relation between sustainability indicator SI_j and the sustainability pillar SP_k, then $w_{jk} = 0$. The weights could be scaled in

$$\sum_{j=1}^{r} w_{jk} = 1, \text{for} k = 1, 2, \ldots p \tag{19.8}$$

Step 5. Determination of the importance of each sustainability pillar.

The sustainability pillars provide a framework for conceptualizing the different sustainability indicators of systems. However, distinct systems may seek diverse sustainability objectives, leading to variations in the importance of sustainability pillars. To ensure alignment with the system distinctiveness and sustainability objectives, it is essential to assign weights to the sustainability pillars individually according to their importance to the overall sustainability when conducting sustainability evaluations. Suppose that there are p sustainability pillars in total, then

$$\sum_{k=1}^{p} I_k = 1 \tag{19.9}$$

Step 6. Determination of the overall sustainability score by impacts of component performance.

Finally, the overall sustainability can be calculated based on the obtained scores and weights, as shown in Eq. (19.10). While the influence of component performance on sustainability indicators and pillars may not perfectly align with the reality, it is still a reasonably accurate approximation for quantifying the GSS

$$OSS = \sum_{k=1}^{p} I_k \cdot \left[\sum_{j=1}^{r} w_{jk} \cdot \left(\sum_{i=1}^{n} \varepsilon_{i \to SI_j} \sigma_{i \to SI_j} \right) \right] \tag{19.10}$$

Since Eq. (19.10) solely represents the influence of component performance on sustainability, which is, the evaluation of system sustainability prior to the execution of maintenance activities, it could be used to evaluate the system sustainability at time t. This system sustainability can be confined within a range from -1 to 1. A lower value of the *OSS* closer to -1 signifies a lower degree of acceptance for the system sustainability, and a higher *OSS* indicates an increased level of acceptability for the system sustainability.

19.4.2 Evaluation of Impacts of Maintenance Activities on Sustainability

Maintenance activities significantly impact sustainability in two main ways. Firstly, by enhancing system performance, maintenance activities indirectly promote sustainability, which could be categorized to the discussion in the last subsection. However, maintenance activities themselves can directly result in sustainability changes. For example, Emissions arise from the transportation and deployment of old and new devices, as well as the maintenance crews. Additionally, the replacement of failed components also contributes to resource wastage. This subsection examines the second type of sustainability impact resulting from maintenance activities.

Assign a correlation value ranging from 0 to 1 to map and assess the impact of each maintenance activity. Let $\sigma_{MSSi \to SI_j}$ represent the impact of maintenance activities for component i on sustainability indicator SI_j, and $\varepsilon_{MSSi \to SI_j}$ denote the relationship between the maintenance activities of component i and the sustainability indicator SI_j. In this case, the cumulative impact from all the maintenance activities on sustainability indicator SI_j could be denoted as $\sum_{i=1}^{n} \varepsilon_{MSSi \to SI_j} \sigma_{MSSi \to SI_j}$. Afterwards, the assessment of relationships among sustainability indicators, sustainability pillars, and the overall sustainability could adhere to the same steps that examines how component performance influences sustainability.

Every maintenance activity and component state change lead to the overall sustainability changes. The overall sustainability evaluation at time $t + \Delta t$ contains both of their effects. Since the impact of maintenance activities on sustainability is considered independent of the impact of component state change, their impact on sustainability indicators can be simply superimposed. Therefore, the overall sustainability score at time $t + \Delta t$ (OSS^*) can be calculated as below.

$$OSS^* = \sum_{k=1}^{p} I_k \cdot \left[\sum_{j=1}^{r} w_{jk} \cdot \left(\sum_{i=1}^{n} \left(\varepsilon_{i \to SI_j} \sigma_{i \to SI_j} + \varepsilon_{MSSi \to SI_j} \sigma_{MSSi \to SI_j} \right) \right) \right] \tag{19.11}$$

Compared with the OSS at time t, the overall sustainability at time $t + \Delta t$ considers the influence of maintenance activities as an independent factor. Consequently, the combination of the effects from component performance and the maintenance activities on sustainability can lead to an OSS^* value at time $t - \Delta t$ that falls beyond the range of $[-1, 1]$. Nonetheless, similarly with OSS, a higher OSS value still indicates an increased level of acceptability for system sustainability. Besides, the value of OSS^* can comprehensively reflect the impact of component performance and maintenance activities on sustainability from three aspects (environmental, social, and economic).

19.4.3 Process of Overall Sustainability Evaluation

Figure 19.4 illustrates the whole process to evaluate the overall sustainability. The blue nodes labeled OSS and OSS^* represent the overall sustainability score at time t and $t + \Delta t$. The OSS is calculated only relying on the states of the components at time t, while OSS^* calculation involves the effects of the component states at time $t + \Delta t$, as well as the effects of maintenance activities during the period Δt. At time t, the maintenance activities are not carried out, and the overall sustainability score depends only on the component states, denoted by the grey nodes. At time $t + \Delta t$, the overall sustainability score depends both on the component states and on the impact of maintenance activities for the component on the sustainability, which encompasses the impact of maintenance activities CR and the impact of maintenance activities DM. By assessing the impact of these maintenance activities and the new component states at time $t + \Delta t$, an estimation of the new overall sustainability score can be made.

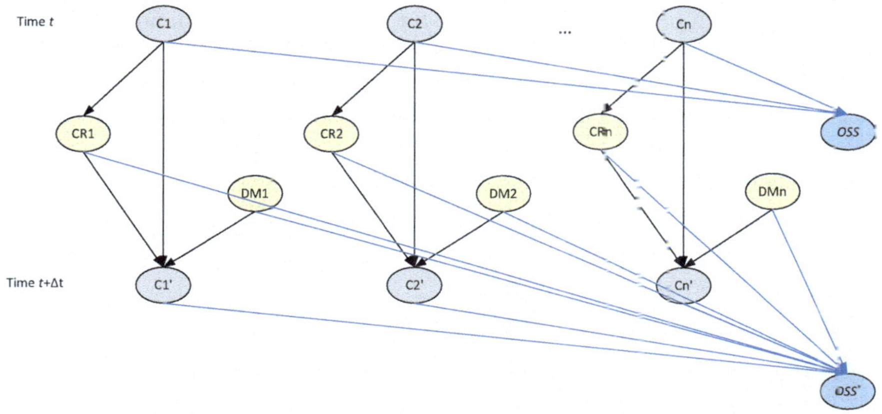

Fig. 19.4 DBN model for sustainability evaluation

19.5 Case of Subsea Transmission System

The subsea transmission system in Fig. 17.6 suffers failure dependences and faces sustainability issues, which could be selected in this case study. In this subsea transmission system, the component failure can easily trigger CAFs due to failure dependences and cause more severe accidents. Moreover, the inherent challenges in maintenance activities for subsea transmission system contribute to increased costs and resource demands. These factors collectively exert significant impacts on the subsea environment, society, and the economy, thereby influencing the overall sustainability of the system.

19.5.1 Numerical Analysis

According to the proposed model, the state of component i is characterized by $x_i \in [0, 1]$. Assume that the state of component i follows a beta distribution $X \sim \text{Beta}(\alpha, \beta)$ to restrict the state values within the range of [0, 1]. The Beta distribution is widely used in statistical applications as a prior distribution for proportions in Bayesian analysis due to its analytical tractability, versatility (Fernández and Pérez-González 2012) and flexibility for modelling (Horn and Leira 2019). In particular the beta distribution could be used to describe failure probability estimation in quantitative risk analysis (QRA)(Steijn et al. 2017) and denote failure rates of offshore components in reliability assessment (Horn and Leira 2019), which demonstrates its relevance to the research. The beta distribution parameters could be adjusted for a more realistic representation to align the distribution shape with expert experience and empirical data (SINTEF and NTNU 2015; O'Connor et al. 2016). Determine the probability distributions of the states for the compressor and two pumps as: $X_1 \sim \text{Beta}(1, 5)$, $X_2 \sim \text{Beta}\ (2, 4)$, $X_3 \sim \text{Beta}(2, 5)$. Their distributions are capable of being defined within the software GeNIe, and the screenshots depicting their representation in the software can be observed in Fig. 19.5. Figure 19.5 illustrates the probability values associated with different states of distinct components at the initial time point t.

Each component exists in one of four states: normally operating (state 0), moderately degraded (state 1), degraded (state 2), and failed (state 3). To scale the four

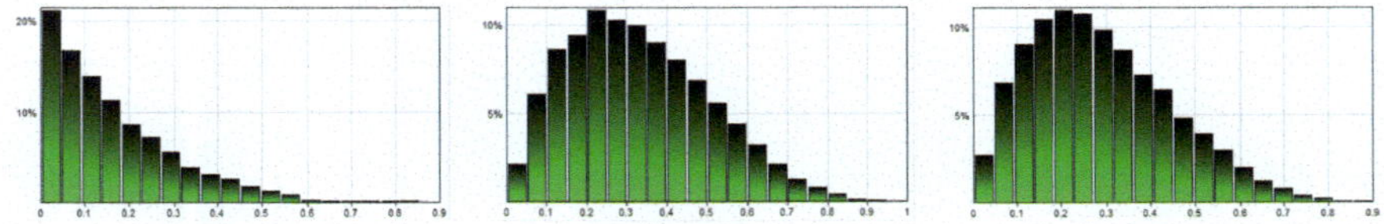

(a) Compressor (b) Pump 1 (c) Pump 2

Fig. 19.5 State distributions of the components at time t states of component, the range of the value characterizing each state is set to be 0.25: Normally Operating state falls within the interval of [0, 0.25); Moderately Degraded state lies in the interval of [0.25, 0.5) and so forth. During one time unit, the degradation decrement of the compressor is 0.01, and the degradation rates of the pumps is 0.05. If the component is degraded (state 2), PM is required to restore the component state to moderately degraded (state 1). Considering that PM solely restores the component to its prior state, and given that the gap between distinct states is 0.25, the assigned value for PM in this context is −0.25. This ensures that components undergoing PM revert to their previous state. If the component fails (state 3), CM is needed to take the component to the normally operating state (state 0). To handle the variety of CM activities, the CM activities are allocated a value of −0.75, guaranteeing that, following CM, the component falls within the Normally Operating state range [0, 0.25), regardless of its initial state. In this case, the state change rates due to degradation and the maintenance effectiveness could be valued as listed in Table 19.2.

In practice, the implementation of maintenance activities depends not solely on the components states but also on various factors, such as the urgency of the degradation or failures, the complexity of maintenance procedures, and the constraints of available maintenance resources. This means that there are often no fixed, predetermined maintenance activities for a particular state of the component. The probabilities of choosing a specific maintenance activity for a particular component state are given according to the historical data, maintenance records and expert opinions. The historical data and maintenance records are documented by the industry, while expert opinions are sourced from one engineer in the industry, one researcher in the safety institute, one professor in the university, and one PhD working on maintenance management. According to their experience, the probabilities of various maintenance activities, considering the state of components are deliberated, gathered, and assessed. These probabilities can be viewed as the prior probabilities and generalized into conditional probability tables (CPTs) when applied to DBNs, as shown in Table 19.3.

Similar to PM activities and CM activities, the prior probabilities of DM activities are also conditional. The probabilities carrying out the DM activities depend on the strength of failure dependence. Due to the limited occurrence of DM maintenance activities addressing failure dependences in practical situations, the level of failure dependence is quantified by Eq. (19.3) and simply classified into Weak and Strong. Furthermore, the probability of undertaking corresponding maintenance activities for various components, based on these two levels, is approximately equal, as suggested

Table 19.2 Parameters for the degradation decrement and maintenance effectiveness

Parameters	Degradation decrement η_D		Maintenance effectiveness for PM η_{PM}	Maintenance effectiveness for CM η_{CM}
	Compressor	Pumps		
Values	0.01	0.05	−0.25	−0.75

Table 19.3 CPT for the PM and CM maintenance policies

Maintenance activities	Component 1				Component 2				Component3			
	State 0	State 1	State 2	State 3	State 0	State 1	State 2	State 3	State 0	State 1	State 2	State 3
NM	1	0.45	0	0	1	0.6	0	0	1	0.6	0	0
PM	0	0.55	0.6	0.1	0	0.4	0.7	0.1	0	0.4	0.7	0.1
CM	0	0	0.4	0.9	0	0	0.3	0.9	0	0	0.3	0.9

by maintenance records and expert opinions. From a different perspective, reciprocal nature of failure dependence among components results in a more evenly distributed probability of performing DM maintenance activities for them. Table 19.4 shows the CPT for the DM maintenance policies.

Since the methodology to determine *OSS* is not the primary focus, three sustainability-related pillars were assigned weights based on the BWM method applied in a reference (Ghaleb and Taghipour 2022): Environmental(0.4); Social(0.2); and Economic(0.4). These sustainability pillars are specifically expressed by sustainability indicators. The sustainability indicators associated with each sustainability pillar does not have any contribution or connection on the other sustainability pillars. The contribution weights of sustainability indicators for each sustainability pillar should be determined by experts. Table 19.5 lists the results of expert estimation.

Table 19.4 CPT for the DM maintenance policies

DM activities	Component 1		Component 2		Component 3	
	Weak	Strong	Weak	Strong	Weak	Strong
NO	1	0.4	1	0.4	1	0.4
YES	0	0.6	0	0.6	0	0.6

Table 19.5 Contribution weights of SIs for each SP

SPs	SIs		
	Environmental	Social	Economic
Energy consumption	0.6	–	–
Wastes	0.25	–	–
Noise	0.15	–	–
Accidents	–	0.55	–
Injury frequency	–	0.45	–
Resource usage costs	–	–	0.3
Productivity	–	–	0.3
Downtime	–	–	0.4

The effects of the component performance on each sustainability indicator and the impacts of maintenance activities on sustainability indicators are also evaluated through expert opinions, primarily derived from the information contained within maintenance records. An example serves to highlight variations in the influence of maintenance activities on sustainability indicators. Maintenance activities such as replacement of components can significantly impact both SI-wastes and SI-downtime. An escalation in such maintenance activities leads to greater wastes and extended downtime, consequently diminishing sustainability. Consequently, the impact of these maintenance activities on sustainability indicators SI-wastes and SI-downtime can be quantitatively negative. Conversely, maintenance activities that address failure dependences may result in some wastes but do not significantly impact component downtime. Therefore, the impact of DM activities on the sustainability indicator SI-downtime can be negligible, with a value close to zero or potentially disregarded as zero when quantified.

After scoring all the values and weights, the overall sustainability score at time t and $t + \Delta t$ could be calculated using the software GeNIe. Suppose that the maintenance activities are completed in one time unit. Regarding to this case, the results of OSS and OSS^* are in distribution terms and shown in the right of Fig. 19.6.

As depicted in Fig. 19.6, prior to the implementation of the maintenance activities, the probability distribution of sustainability values lies to the left of 0, with a higher probability for values closer to 0. However, after experiencing component degradation and maintenance activities during the delay period Δt, the probability distribution of sustainability values shifts leftward, and some smaller sustainability values appear with a certain probability. This suggests that, even after the application of maintenance activities, the overall sustainability still exhibits a declining trend in general over time. Since GeNIe displays a handful of parameters of the distribution, more specific values of the results could be calculated and listed in Table 19.6. In Table 19.6, the StdDev is the abbreviation of Standard Deviation, which is used to

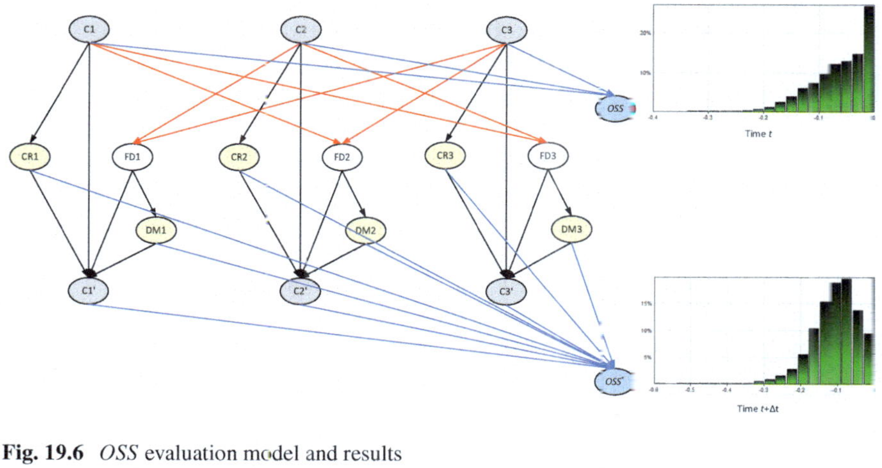

Fig. 19.6 *OSS* evaluation model and results

Table 19.6 Results for the OSS and OSS* evaluation

Values	StdDev	Maximum value	Minimum value	Mean value
OSS	0.0542	0	−0.3125	−0.0634
OSS^*	0.0765	0	−0.5687	−0.1259

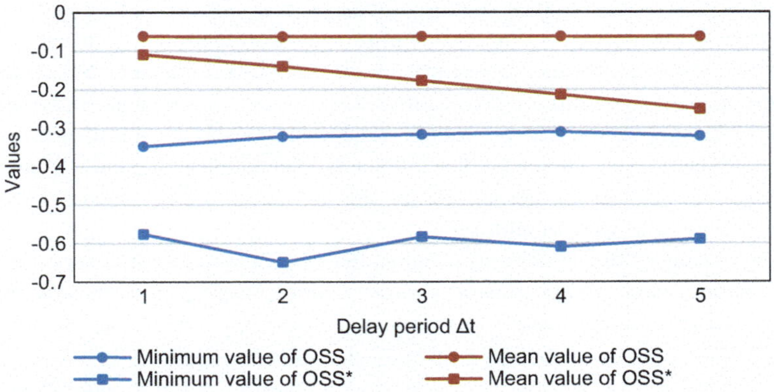

Fig. 19.7 OSS and OSS^* results with different delay period

measure the dispersion of a set of data. The maximum value indicates the peak level of overall sustainability acceptability, which is consistent for both OSS and OSS^*. Conversely, the minimum value and the mean value signify the lowest level and the average level of overall sustainability acceptability, and these two parameters may serve as estimates for overall sustainability. Therefore, we could also obtain a similar conclusion that the overall sustainability declines by comparison of the results of these two parameters.

Assum an extension in the duration of component degradation and maintenance interventions, which alters the value of delay period Δt, and observe the changes of the overall sustainability by comparing its mean values and minimum values. The results are shown in Fig. 19.7. From this figure, the mean values of OSS are almost stable on a horizontal line, while the minimum values of OSS exhibit minimal variation with changes in the delay period. This phenomenon can be easily comprehended since the value of OSS is independent of the delay period and the activities that occur during it. Hence, to assess the impact of component degradation and various maintenance activities on overall sustainability during the delay period, it is rational to exclusively consider the values of OSS^*. Regarding OSS^*, its minimum values experience fluctuations without obvious time-dependent pattern. However, the mean values of OSS^* demonstrate a consistent and steady decline as the delay period extends, making it a more suitable fundamental parameter for analysis, compared to the minimum value. Additionally, both the mean values and minimum values of OSS^* are notably lower than their counterparts of OSS, reinforcing the conclusion that the overall sustainability diminishes following the delay period of component degradation and maintenance activities.

19.5.2 Results and Discussion

In the DBN model, the CPT of nodes can be adjusted with flexibility, enabling the reassessment of overall sustainability under the hypothesis of implementing different maintenance strategies. This reassessment can then guide the determination of an optimal maintenance decision, as mentioned in Eq. (19.1). The aforementioned maintenance strategy in the case study is labeled as maintenance strategy 1. Some other maintenance strategies are also considered. The impact of these strategies on the reassessed sustainability varies compared to maintenance strategy 1, highlighting the potential for optimization through different maintenance approaches. Figure 19.8 shows the mean values of the overall sustainability after delay period with six kinds of maintenance decision. as explained below.

- Maintenance strategy 2: Decreasing the probability of DM activities from 0.6 to 0.3 for all components;
- Maintenance strategy 3: Ignoring the DM activities (the probability of DM activities will be 0);
- Maintenance strategy 4: Using PM activities instead of CM activities (the probability of NM activities remains the same and the probability of PM activities will be 1 − Pr(NM));ss

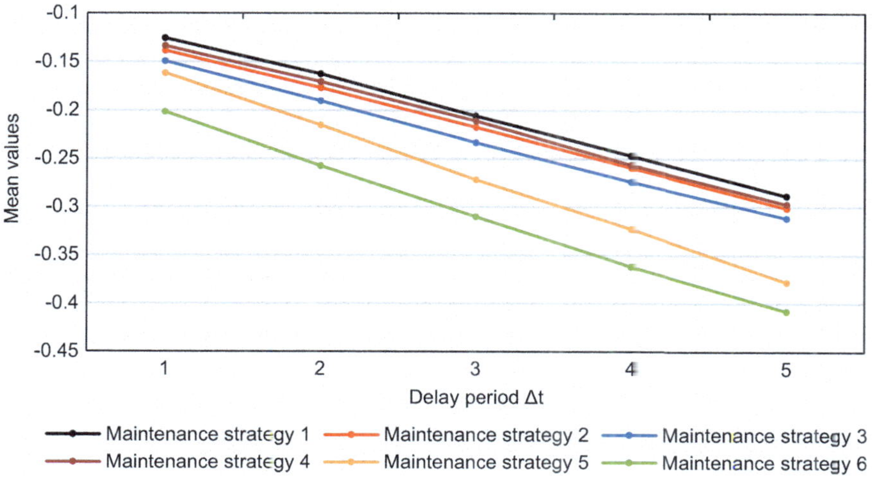

Fig. 19.8 Mean values of OSS^* with various maintenance strategies

- Maintenance strategy 5: Using NM activities instead of PM activities (the probability of CM activities remains the same and the probability of NM activities will be 1 − Pr(CM));
- Maintenance strategy 6: Ignoring all kinds of maintenance activities (the probability of NM activities will be 1, and the probability of DM activities will be 0).

As shown in Fig. 19.8, the mean values of the overall sustainability associated with other maintenance strategies are consistently lower than that of maintenance strategy 1. The mean values of OSS^* show the biggest difference between the maintenance strategy 1 and maintenance strategy 6. It implies that the worst-case scenario for overall sustainability occurs when no maintenance interventions are implemented, and the component degradation is the sole factor causing a negative impact on overall sustainability. While the maintenance strategy may inherently have a somewhat adverse impact on sustainability, their primary benefit of enhancing component performance significantly outweighs these drawbacks. In addition, a more detailed discussion of maintenance strategies can reveal the different impacts that different maintenance strategies have on overall sustainability. Among them, maintenance strategies 2 and 3 focus on the alteration of DM maintenance activities, while maintenance strategies 4 and 5 focus on the alteration of PM and CM maintenance activities. Compared the maintenance strategies 1, 2 and 3, as the probability of implementing DM activities becomes smaller (0.7 to 0.3 to 0), the mean value of the overall sustainability is lower, while the reduction is not substantial. This implies that the DM activities have a modest capacity to enhance the overall sustainability, though their impact is not particularly pronounced. Thus, in practical engineering, the DM activities could be considered when budgets and resources permit. In terms of the maintenance strategies 1, 4, the differences between their numerical results are also not obvious, denoting that if all the maintenance activities are taken with CM, the overall sustainability could remain a pretty good level. However, ignoring moderate or severe degradation through the absence of maintenance, as in strategy 5, the overall sustainability is largely decreased. Another observation is that as the delay period increases, the mean value of overall sustainability in the case of maintenance strategy 5 diminishes more rapidly than that in the case of maintenance strategy 1, in other words, the difference between them becomes larger. This could be attributed to the assumption that some components with moderate degradation or degradation are not maintained when considering only CM activities. As delay times lengthen, the degradation of these components intensifies. When components degrade severely, their hostile effect on the sustainability largely surpasses the possible negative impact of maintenance activities on the sustainability. Moreover, as component degradation escalates, its adverse effects accelerate at a faster pace, contributing to a more rapid overall decline in sustainability over longer delay periods.

Table 19.7 CPT for the optimized PM and CM maintenance policies

Maintenance activities	Component 1				Component 2				Component3			
	State 0	State 1	State 2	State 3	State 0	State 1	State 2	State 3	State 0	State 1	State 2	State 3
NM	1	0.4	0.05	0	1	0.35	0	0	1	0.35	0	0
PM	0	0.55	0.7	0.1	0	0.6	0.6	0.15	0	0.6	0.6	0.15
CM	0	0.05	0.25	0.9	0	0.05	0.4	0.85	0	0.05	0.4	0.85

Table 19.8 CPT for the optimized DM maintenance policies

DM activities	Component 1		Component 2		Component 3	
	Weak	Strong	Weak	Strong	Weak	Strong
NO	1	0.3	1	0.3	1	0.3
YES	0	0.7	0	0.7	0	0.7

The preceding discussions regarding maintenance strategies under extreme conditions can serve as reference for investigating the influence of various maintenance activities on the overall sustainability. To further consider the optimization of maintenance strategies based on the case study, altering the parameters of maintenance activities within the CPTs may yield diverse simulation results, revealing more effective maintenance strategies. As an illustration, an optimized maintenance strategy is presented, with its specific details outlined in Tables 19.7 and 19.8.

Figure 19.9 shows the comparison between the sustainability evaluation of a previously applied maintenance strategy and an optimized one. As shown in Fig. 19.9, the mean values of OSS^* under optimized maintenance strategy are always higher than that under previous maintenance strategy, which implies that the optimized maintenance strategy notably enhances the overall sustainability. This example demonstrates the practical applicability of the proposed methodology in refining maintenance strategies to enhance sustainability. In practical applications, this model can be used to simulate the enhancement of system sustainability under different maintenance strategies before decision-making, which is achieved by adjusting the types and associated probabilities of maintenance activities, thereby identifying the optimal strategy that maximizes system sustainability.

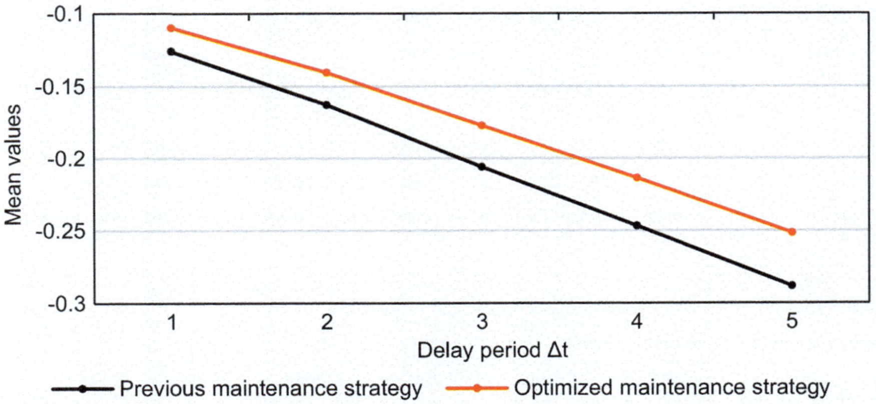

Fig. 19.9 Mean values of OSS^* under previous and optimized maintenance strategy

Part III
Concluding Remarks

In conclusion, Part III *Maintenance* has explored the complexities and advanced methodologies involved in maintenance methodologies for subsea production systems, addressing critical challenges such as imperfect maintenance, spare parts management, maintenance delays, discrete operational states, and failure dependences across system components. By presenting a series of advanced models and strategies, we have highlighted how maintenance strategies especially CBM can be optimized to meet the unique demands of subsea operations, where reliability and timely intervention are crucial for system availability and safety. The integration of EM with CBM and the emphasis on sustainability underline the need for a balanced approach that considers both technical performance and environmental impact. Through the application of real-world case studies, the part demonstrates that CBM is not just a theoretical framework but a practical and adaptable solution for modern subsea production systems. These insights pave the way for further innovations in maintenance strategies that can enhance the efficiency and sustainability of subsea operations.

References

Amin, M.T., F. Khan, S. Ahmed, and S. Imtiaz. 2021. Risk-based fault detection and diagnosis for nonlinear and non-Gaussian process systems using R-vine copula. *Process Safety and Environmental Protection* 150: 123–136.

Attaianese, C., M. D'Arpino, M. Di Monaco, and L.P. Di Noia. 2023. Model-based detection and estimation of DC offset of phase current sensors for field oriented PMSM drives. *IEEE Transactions on Industrial Electronics* 70 (6): 6316–6325.

Aydemir, G., and K. Paynabar. 2020. Image-based prognostics using deep learning approach. *IEEE Transactions on Industrial Informatics* 16 (9): 5956–5964.

Azzam, I., J. Hwang, F. Breidi, J. Lumkes, and T. Salem. 2023. Automated method for selecting optimal digital pump operating strategy. *Expert Systems with Applications* 232.

Bai, Y., and Q. Bai. 2012. *Subsea Engineering Handbook*. USA: Gulf Professional Publishing.

Bhardwaj, U., A. P. Teixeira, and C. G. Soares. 2022. Bayesian framework for reliability prediction of subsea processing systems accounting for influencing factors uncertainty. *Reliability Engineering & System Safety* 218.

Cai, B., L. Huang, and M. Xie. 2017. Bayesian networks in fault diagnosis. *IEEE Transactions on Industrial Informatics* 13 (5): 2227–2240.

Cai, B., X. Kong, Y. Liu, J. Lin, X. Yuan, H. Xu, and R. Ji. 2019. Application of Bayesian networks in reliability evaluation. *IEEE Transactions on Industrial Informatics* 15 (4): 2146–2157.

Cai, B., C. Sheng, C. Gao, Y. Liu, M. Shi, Z. Liu, Q. Feng, and G. Liu. 2021a. Artificial Intelligence enhanced reliability assessment methodology with small samples. *IEEE Transactions on Neural Networks and Learning Systems* 34 (9): 6578–6590.

Cai, B. P., K. K. Hao, Z. Wang, C. Yang, X. D. Kong, Z. K. Liu, R. J. Ji, and Y. H. Liu. 2021. Data-driven early fault diagnostic methodology of permanent magnet synchronous motor. *Expert Systems with Applications* 177.

Cai, B.P., H.L. Liu, and M. Xie. 2016. A real-time fault diagnosis methodology of complex systems using object-oriented Bayesian networks. *Mechanical Systems and Signal Processing* 80: 31–44.

Cai, B.P., Y.H. Liu, Q. Fan, Y.W. Zhang, Z.K. Liu, S.L. Yu, and R.J. Ji. 2014. Multi-source information fusion based fault diagnosis of ground-source heat pump using Bayesian network. *Applied Energy* 114: 1–9.

Cai, B.P., Y.H. Liu, Z.K. Liu, X.J. Tian, X. Dong, and S.L. Yu. 2012a. Using Bayesian networks in reliability evaluation for subsea blowout preventer control system. *Reliability Engineering & System Safety* 108: 32–41.

Cai, B.P., Y.H. Liu, Z.K. Liu, F. Wang, X.J. Tian, and Y.Z. Zhang. 2012b. Development of an automatic subsea blowout preventer stack control system using PLC based SCADA. *ISA Transactions* 51 (1): 198–207.

B. Cai et al., *Intelligent Operation and Maintenance for Subsea Production Systems*,
https://doi.org/10.1007/978-981-96-2677-9

Cai, B.P., Z.D. Wang, H.M. Zhu, Y.H. Liu, K.K. Hao, Z.Q. Yang, Y. Ren, Q. Feng, and Z.K. Liu. 2022. Artificial Intelligence enhanced two-stage hybrid fault prognosis methodology of PMSM. *IEEE Transactions on Industrial Informatics* 18 (10): 7262–7273.

Cai, B.P., M. Xie, Y.H. Liu, Y.L. Liu, and Q. Feng. 2018. Availability-based engineering resilience metric and its corresponding evaluation methodology. *Reliability Engineering & System Safety* 172: 216–224.

Cai, B. P., Y. P. Zhang, H. F. Wang, Y. H. Liu, R. J. Ji, C. T. Gao, X. D. Kong, and J. Liu. 2021. Resilience evaluation methodology of engineering systems with dynamic-Bayesian-network-based degradation and maintenance. *Reliability Engineering & System Safety* 209.

Cao, Y.S. 2021. Modeling the effects of dependence between competing failure processes on the condition-based preventive maintenance policy. *Applied Mathematical Modelling* 99: 400–417.

Carlin, H. M., A. Goodall, R. I. M. Young, and A. West. 2024. An interactive framework to support decision-making for Digital Twin design. *Journal of Industrial Information Integration* 41.

Casillas, M.V., L.E. Garza-Castañ6n, and V. Puig. 2015. Optimal sensor placement for leak location in water distribution networks using evolutionary algorithms. *Water* 7 (11): 6496–6515.

Casillas, M.V., V. Puig, L.E. Garza-Castañ6n, and A. Rosich. 2013. Optimal sensor placement for leak location in water distribution networks using genetic algorithms. *Sensors (Basel, Switzerland)* 13: 14984–15005.

Chang, Y.J., J.Y. Li, C.S. Zhang, G.M. Chen, X.H. Li, S.Y. Zhang, L.B. Xu, and L.X. Sheng. 2019. Dynamic risk assessment approach of riser recoil control failure during production test of marine natural gas hydrate. *Energy Science & Engineering* 7 (5): 1808–1822.

Chen, L.Y., E. Arzaghi, M.M. Abaei, V. Garaniya, and R. Abbassi. 2018. Condition monitoring of subsea pipelines considering stress observation and structural deterioration. *Journal of Loss Prevention in the Process Industries* 51: 178–185.

Chen, T.R., C. Wang, G. Chen, Z.Y. Dong, and D.J. Hill. 2019. Small fault detection for a class of closed-loop systems via deterministic learning. *IEEE Transactions on Cybernetics* 49 (3): 897–906.

Cheng, A.K., and N.Z. Chen. 2017. Corrosion fatigue crack growth modelling for subsea pipeline steels. *Ocean Engineering* 142: 10–19.

Cheng, T.T., Q. Wu, B. Wu, and X.P. Yan. 2021. A probabilistic decision-making system for joining traffic lanes within an inland traffic separation scheme. *Marine Technology Society Journal* 55 (5): 44–63.

Coelho, L.D., and M.A.B. Cunha. 2011. Adaptive cascade control of a hydraulic actuator with an adaptive dead-zone compensation and optimization based on evolutionary algorithms. *Expert Systems with Applications* 38 (10): 12262–12269.

Compare, M., P. Baraldi, I. Bani, E. Zio, and D. McDonnell. 2020. Industrial equipment reliability estimation: a Bayesian Weibull regression model with covariate selection. *Reliability Engineering & System Safety* 200.

Cui, Y.Q., J.Y. Shi, and Z.L. Wang. 2016. Quantum assimilation-based state-of-health assessment and remaining useful life estimation for electronic systems. *IEEE Transactions on Industrial Electronics* 63 (4): 2379–2390.

de Waard, C., U. Lotz, and D.E. Milliams. 1991. Predictive model for CO_2 corrosion engineering in wet natural gas pipelines. *Corrosion* 47 (12): 976–985.

Development, W. C. o. E. a. 1987. Our Common Future. 383. Oxford.

Di Mauro, M., G. Galatro, F. Postiglione, W. Song, and A. Liotta. 2024. Hybrid learning strategies for multivariate time series forecasting of network quality metrics. *Computer Networks* 243.

Ding, F.F., and Z.G. Tian. 2012. Opportunistic maintenance for wind farms considering multi-level imperfect maintenance thresholds. *Renewable Energy* 45: 175–182.

DNV. 2015. *OREDA Handbook: Offshore and Onshore Reliability Data*. Norway: OREDA Participants.

Dui, H., K. X. Liu, and S. M. Wu. 2023. Cascading failures and resilience optimization of hospital infrastructure systems against the COVID-19. *Computers & Industrial Engineering* 179.

Dui, H., M. Liu, J. Y. Song, and S. M. Wu. 2023. Importance measure-based resilience management: review, methodology and perspectives on maintenance. *Reliability Engineering & System Safety* 237.

Dui, H., Y. L. Zhang, X. M. Wu, and L. D. Xing. 2024. Dynamic reliability assessment model for IoT-enabled smart offshore wind farm. *IEEE Transactions on Reliability*.

Dui, H. Y., T. Z. Tian, J. B. Zhao, and S. M. Wu. 2022. Comparing with the joint importance under consideration of consecutive-k-out-of-n system structure changes. *Reliability Engineering & System Safety* 219.

Dui, H., X. Wu, S. Wu, and M. Xie. 2024. Importance measure-based maintenance strategy optimization: fundamentals, applications and future directions in AI and IoT. *Frontiers of Engineering Management* **11**: 542–567.

Fan, D. Y., H. Sun, J. Yao, K. Zhang, X. Yan, and Z. X. Sun. 2021. Well production forecasting based on ARIMA-LSTM model considering manual operations. *Energy* 220.

Feng, L.J., and C.H. Zhao. 2021. Fault description based attribute transfer for zero-sample industrial fault diagnosis. *IEEE Transactions on Industrial Informatics* 17 (3): 1852–1862.

Feng, T. T., S. C. Li, L. Guo, H. L. Gao, T. Chen, and Y. X. Yu. 2023. A degradation-shock dependent competing failure processes based method for remaining useful life prediction of drill bit considering time-shifting sudden failure threshold. *Reliability Engineering & System Safety* 230.

Fernández, A.J., and C.J. Pérez-González. 2012. Generalized beta prior models on fraction defective in reliability test planning. *Journal of Computational and Applied Mathematics* 236 (13): 3147–3159.

Fu, A.W.C., E. Keogh, L.Y.H. Lau, C.A. Ratanamahatana, and R.C.W. Wong. 2008. Scaling and time warping in time series querying. *Vldb Journal* 17 (4): 899–921.

Fu, X.W., Y.S. Yang, and O. Postolache. 2019a. Invulnerability of clustering wireless sensor networks against cascading failures. *IEEE Systems Journal* 13 (2): 1431–1442.

Fu, X.W., H.Q. Yao, and Y.S. Yang. 2019b. Modeling cascading failures for wireless sensor networks with node and link capacity. *IEEE Transactions on Vehicular Technology* 68 (8): 7828–7840.

Ghaleb, M., and S. Taghipour. 2022. Assessing the impact of maintenance practices on asset's sustainability. *Reliability Engineering & System Safety* 228.

Groden, M., and M. Collette. 2017. Fusing fleet in-service measurements using Bayesian networks. *Marine Structures* 54: 38–49.

Guo, C.M., W.B. Wang, B. Guo, and X.S. Si. 2013. A maintenance optimization model for mission-oriented systems based on Wiener degradation. *Reliability Engineering & System Safety* 111: 183–194.

Hafsa, W., B. Chebel-Morello, C. Varnier, K. Medjaher, and N. Zerhouni. 2015. Prognostics of health status of multi-component systems with degradation interactions. In *Proceedings of 2015 International Conference on Industrial Engineering and Systems Management*, Seville, Spain. IEEE.

Han, Y.Y., C.L. Ma, S.J. Tang, F.F. Wang, X.Y. Sun, and X.S. Si. 2023. Residual life estimation of lithium-ion batteries based on nonlinear Wiener process with measurement error. *Proceedings of the Institution of Mechanical Engineers Part O-Journal of Risk and Reliability* 237 (1): 133–151.

Holmes, J., and S. Hassini. 2021. Discrete-time Markov chain modelling of the Ontario air quality health index. *Water Air and Soil Pollution* 232(4).

Horn, J. T., and B. J. Leira. 2019. Fatigue reliability assessment of offshore wind turbines with stochastic availability. *Reliability Engineering & System Safety* 191.

Hu, C.H., H. Pei, Z.Q. Wang, X.S. Si, and Z.X. Zhang. 2018. A new remaining useful life estimation method for equipment subjected to intervention of imperfect maintenance activities. *Chinese Journal of Aeronautics* 31 (3): 514–528.

Hu, C. H., Y. X. Xing, D. B. Du, X. S. Si, and J. X. Zhang. 2023. Remaining useful life estimation for two-phase nonlinear degradation processes. *Reliability Engineering & System Safety* 230.

Hu, J. W., A. C. Xu, B. Li, and H. T. Liao. 2021. Condition-based maintenance planning for multi-state systems under time-varying environmental conditions. *Computers & Industrial Engineering* 158.

Huang, X.P., T. Moan, and W.C. Cui. 2008. An engineering model of fatigue crack growth under variable amplitude loading. *International Journal of Fatigue* 30 (1): 2–10.

Jardine, A.K.S., D.M. Lin, and D. Banjevic. 2006. A review on machinery diagnostics and prognostics implementing condition-based maintenance. *Mechanical Systems and Signal Processing* 20 (7): 1483–1510.

Jia, Z.Y., J.W. Ma, F.J. Wang, and W. Liu. 2010. Characteristics forecasting of hydraulic valve based on grey correlation and ANFIS. *Expert Systems with Applications* 37 (2): 1250–1255.

Jiang, H.F., S.F. Qin, J.L. Fu, J. Zhang, and G.F. Ding. 2021. How to model and implement connections between physical and virtual models for digital twin application. *Journal of Manufacturing Systems* 58: 36–51.

Jiujian, W., Y. Shaopu, Y. Liu, and W. Guilin. 2021. A method of bearing remaining useful life estimation based on convolutional long short-term memory neural network. *Journal of Mechanical Engineering*.

Karandikar, J.M., N.H. Kim, and T.L. Schmitz. 2012. Prediction of remaining useful life for fatigue-damaged structures using Bayesian inference. *Engineering Fracture Mechanics* 96: 588–605.

Karatu, C., and Y. Arslanoglu. 2022. Development of condition-based maintenance strategy for fault diagnosis for ship engine systems. *Ocean Engineering* 256.

Kelley, J., and M. Hagan. 2020. New fault diagnosis procedure and demonstration on hydraulic servo-motor for single faults. *IEEE-Asme Transactions on Mechatronics* 25 (3): 1499–1509.

Kohtz, S., J. H. Zhao, A. Renteria, A. Lalwani, Y. W. Xu, X. L. Zhang, K. S. Haran, D. Senesky, and P. F. Wang. 2024. Optimal sensor placement for permanent magnet synchronous motor condition monitoring using a digital twin-assisted fault diagnosis approach. *Reliability Engineering & System Safety* 242.

Kong, X. F., J. Yang, and L. Li. 2022. Reliability analysis for multi-component systems considering stochastic dependency based on factor analysis. *Mechanical Systems and Signal Processing* 169.

Lei, Y.G., N.P. Li, L. Guo, N.B. Li, T. Yan, and J. Lin. 2018. Machinery health prognostics: a systematic review from data acquisition to RUL prediction. *Mechanical Systems and Signal Processing* 104: 799–834.

Levitin, G., L. D. Xing, and Y. S. Dai. 2022. Mission aborting and system rescue for multi-state systems with arbitrary structure. *Reliability Engineering & System Safety* 219.

Li, F., L. Jia, and Y. Gu. 2023a. Identification of nonlinear process described by neural fuzzy Hammerstein-Wiener model using multi-signal processing. *Advances in Manufacturing* 11 (4): 694–707.

Li, W., Y. X. Huang, L. L. Ji, L. L. Ma, R. K. Agarwal, and M. Awais. 2023. Prediction model for energy conversion characteristics during transient processes in a mixed-flow pump. *Energy* 271.

Ling, M.H., H.K.T. Ng, and K.L. Tsui. 2019. Bayesian and likelihood inferences on remaining useful life in two-phase degradation models under gamma process. *Reliability Engineering & System Safety* 184: 77–85.

Liu, D. Q., S. F. Song, T. Huang, S. H. Shen, X. P. Li, and J. Gong. 2024. Application of Bayesian networks based on Sequential Monte Carlo simulation and physical model in fault diagnosis of horizontal three-phase separator system. *Ocean Engineering* 306.

Liu, M., Q. Feng, D.M. Fan, H.Y. Dui, B. Sun, Y. Ren, D.Z. Yang, and Z.L. Wang. 2023. Resilience importance measure and optimization considering the stepwise recovery of system performance. *IEEE Transactions on Reliability* 72 (3): 1064–1077.

Longo, F., A. Padovano, F. De Felice, A. Petrillo, and M. Elbasheer. 2023. From "prepare for the unknown" to "train for what's coming": a digital twin-driven and cognitive training approach for the workforce of the future in smart factories. *Journal of Industrial Information Integration* 32.

Loughney, S., and J. Wang. 2018. Bayesian network modelling of an offshore electrical generation system for applications within an asset integrity case for normally unattended offshore installations. *Proceedings of the Institution of Mechanical Engineers Part M-Journal of Engineering for the Maritime Environment* 232 (4): 402–420.

Luo, K.P. 2022. Water flow optimizer: a nature-inspired evolutionary algorithm for global optimization. *IEEE Transactions on Cybernetics* 52 (8): 7753–7764.

Luque, J., and D. Straub. 2016. Reliability analysis and updating of deteriorating systems with dynamic Bayesian networks. *Structural Safety* 62: 34–46.

Lv, Z.H., J.K. Guo, and H.B. Lv. 2023. Safety Poka Yoke in zero-defect manufacturing based on digital twins. *IEEE Transactions on Industrial Informatics* 19 (2): 1176–1184.

Ma, J., L. Cai, G. B. Liao, H. P. Yin, X. S. Si, and P. Zhang. 2023. A multi-phase Wiener process-based degradation model with imperfect maintenance activities. *Reliability Engineering & System Safety* 232.

Ma, X., Y. Xing, and J. Lu. 2018. Causation analysis of hazardous material road transportation accidents by Bayesian network using genie. *Journal of Advanced Transportation* 2018: 1–12.

Mebarki, N., S. Benmoussa, M. Djeziri, and L. H. Mouss. 2022. New approach for failure prognosis using a bond graph, gaussian mixture model and similarity techniques. *Processes* 10(3).

Motahari-Nezhad, M. and S. M. Jafari. 2021. Bearing remaining useful life prediction under starved lubricating condition using time domain acoustic emission signal processing. *Expert Systems with Applications* 168.

Nguyen, K.T.P., P. Do, K.T. Huynh, C. Bérenguer, and A. Grall. 2019. Joint optimization of monitoring quality and replacement decisions in condition-based maintenance. *Reliability Engineering & System Safety* 189: 177–195.

Ning, P. Y., Z. J. Zhou, Y. Cao, S. W. Tang, and J. Wang (2022). A concurrent fault diagnosis model via the evidential reasoning rule. *IEEE Transactions on Instrumentation and Measurement* 71.

Nisar, K. S., K. K. Ali, M. Inc, M. S. Mehanna, H. Rezazadeh, and L. Akinyemi. 2022. New solutions for the generalized resonant nonlinear Schrodinger equation. *Results in Physics* 33.

O'Connor, A. N., M. Modarres, and A. Mosleh. 2016. *Probability Distributions Used in Reliability Engineering*. University of Maryland, College Park, Maryland, USA, Center for Risk and Reliability.

Ochella, S., M. Shafiee, and C. Sansom. 2021. Adopting machine learning and condition monitoring P-F curves in determining and prioritizing high-value assets for life extension. *Expert Systems with Applications* 176.

Pillai, S., and P. Vadakkepat. 2021. Two stage deep learning for prognostics using multi-loss encoder and convolutional composite features. *Expert Systems with Applications* 171.

Qi, Q.L., F. Tao, T.L. Hu, N. Anwer, A. Liu, Y.L. Wei, L.H. Wang and A.Y.C. Nee. 2021. Enabling technologies and tools for digital twin. *Journal of Manufacturing Systems* 58: 3–21.

Roos, M.H., P.H. Nguyen, J. Morren, and J.G. Slootweg. 2020. Modeling and experimental validation of power electronic loads and DERs for microgrid islanding simulations. *IEEE Transactions on Power Systems* 35 (3): 2279–2288.

Sahin, I. 2023. Data-driven stochastic model for train delay analysis and prediction. *International Journal of Rail Transportation* 11 (2): 207–226.

Salama, M.M. 2000. An alternative to API 14E erosional velocity limits for Sand-Laden fluids. *Journal of Energy Resources Technology* 122 (2): 71–77.

Sayed, A.R., C. Wang, and T.S. Bi. 2019. Resilient operational strategies for power systems considering the interactions with natural gas systems. *Applied Energy* 241: 548–566.

Shi, J.H., Y. Zhu, F. Khan, and G.M. Chen. 2019. Application of Bayesian Regularization Artificial Neural Network in explosion risk analysis of fixed offshore platform. *Journal of Loss Prevention in the Process Industries* 57: 131–141.

Shi, Y., W. H. Zhu, Y. S. Xiang, and Q. M. Feng. 2020. Condition-based maintenance optimization for multi-component systems subject to a system reliability requirement. *Reliability Engineering & System Safety* 202.

Si, S.B., M.L. Liu, Z.Y. Jiang, T.D. Jin, and Z.Q. Cai. 2019. System reliability allocation and optimization based on generalized birnbaum importance measure. *IEEE Transactions on Reliability* 68 (3): 831–843.

SINTEF and NTNU. 2015. Offshore and Onshore Reliability Data Volume 1—Topside Equipment. Norway, OREDA Participants.

Steijn, W., J. Groeneweg, F. van, J. van and P. van (2017). An integration of human factors into quantitative risk analysis: a proof of principle. *Safety and Reliability – Theory and Applications* 53–53.

Straub, D. 2009. Stochastic modeling of deterioration processes through dynamic Bayesian networks. *Journal of Engineering Mechanics* 135 (10): 1089–1099.

Sun, C., H. Zheng, W. D. Liu, and W. T. Lu. 2020. Numerical investigation of complex fracture network creation by cyclic pumping. *Engineering Fracture Mechanics* 233.

Sun, F. Q., H. Li, Y. Y. Cheng, and H. T. Liao. 2021. Reliability analysis for a system experiencing dependent degradation processes and random shocks based on a nonlinear Wiener process model. *Reliability Engineering & System Safety* 215.

Sun, H., D. Cao, Z. Zhao, and X. Kang. 2018. A hybrid approach to cutting tool remaining useful life prediction based on the Wiener process. *IEEE Transactions on Reliability* 67 (3): 1294–1303.

Tang, S. N., Y. Zhu, and S. Q. Yuan. 2022. Intelligent fault identification of hydraulic pump using deep adaptive normalized CNN and synchrosqueezed wavelet transform. *Reliability Engineering & System Safety* 224.

Teng, W., C. Han, Y.K. Hu, X. Cheng, L. Song, and Y.B. Liu. 2020. A robust model-based approach for bearing remaining useful life prognosis in wind turbines. *IEEE Access* 8: 47133–47143.

Tlili, L., M. Radhoui, and A. Chelbi. 2015. Condition-based maintenance strategy for production systems generating environmental damage. *Mathematical Problems in Engineering* 2015.

Tobon-Mejia, D.A., K. Medjaher, and N. Zerhouni. 2012. CNC machine tool's wear diagnostic and prognostic by using dynamic Bayesian networks. *Mechanical Systems and Signal Processing* 28: 167–182.

Wang, J., Z. G. Li, G. H. Bai, and M. J. Zuo. 2020. An improved model for dependent competing risks considering continuous degradation and random shocks. *Reliability Engineering & System Safety* 193.

Wang, J., Y.T. Yang, and Y.Q. Fu. 2024a. Joint optimization of condition-based maintenance and condition-based reallocation for a system with multiple degrading components. *Quality and Reliability Engineering International* 40 (1): 339–361.

Wang, J. H., X. H. Zhang, and J. C. Zeng. 2022. Dynamic group-maintenance strategy for wind farms based on imperfect maintenance model. *Ocean Engineering* 259.

Wang, L. B., Z. B. Zhu, and X. F. Zhao. 2024. Dynamic predictive maintenance strategy for system remaining useful life prediction via deep learning ensemble method. *Reliability Engineering & System Safety* 245.

Wang, W.C., K.W. Chau, D.M. Xu, and X.Y. Chen. 2015. Improving forecasting accuracy of annual runoff time series using ARIMA based on EEMD decomposition. *Water Resources Management* 29 (8): 2655–2675.

Wang, X. Y., J. X. Wang, R. Ning, and X. Chen. 2023. Joint optimization of maintenance and spare parts inventory strategies for emergency engineering equipment considering demand priorities. Mathematics 11(17).

Woo, S. W., D. L. O'Neal, and M. Pecht. 2023. Improving the lifetime of mechanical systems during transit established on quantum/transport life-stress prototype and sample size. *Mechanical Systems and Signal Processing* 193.

Wu, F. H., J. Tang, Z. P. Jiang, Y. B. Sun, Z. Chen, and B. S. Guo. 2023. The remaining useful life prediction method of a hydraulic pump under unknown degradation model with limited data. Sensors 23(13).

Xu, Q. Q., H. Liu, Z. H. Song, S. H. Dong, L. B. Zhang, and X. L. Zhang 2023. Dynamic risk assessment for underground gas storage facilities based on Bayesian network. *Journal of Loss Prevention in the Process Industries* 82.

Xu, X. D., C. Q. Yu, S. J. Tang, X. Y. Sun, X. S. Si, and L. F. Wu. 2019. Remaining useful life prediction of lithium-ion batteries based on wiener processes with considering the relaxation effect. *Energies* 12(9).

Xu, X.J., X.P. Yan, C.X. Sheng, C.Q. Yuan, D.L. Xu, and J.B. Yang. 2020. A belief rule-based expert system for fault diagnosis of Marine Diesel Engines. *IEEE Transactions on Systems Man Cybernetics-Systems* 50 (2): 656–672.

Xu, Z.G., Y.D. Ji, and D.H. Zhou. 2009. Real-time reliability prediction for dynamic systems with both deteriorating and unreliable components. *Science in China Series F-Information Sciences* 52 (11): 2234–2246.

Yang, D.S., Y.H. Pang, B.W. Zhou, and K. Li. 2019. Fault diagnosis for energy internet using correlation processing-based convolutional neural networks. *IEEE Transactions on Systems Man Cybernetics-Systems* 49 (8): 1739–1748.

Yang, X., C. I. Oriaku, J. P. Zubelli, and M. F. Pereira. 2017. Automated numerical characterization of dilute semiconductors per comparison with luminescence. *Optical and Quantum Electronics* 49(3).

Zang, Y., S. G. Wei, B. G. Cai, H. S. Wang, and M. G. Pecht. 2021. Hybrid remaining useful life prediction method. A case study on railway D-cables. *Reliability Engineering & System Safety* 213.

Zeng, L. Q., J. H. Zheng, L. Yao, and Z. Q. Ge. 2023. Dynamic Bayesian networks for feature learning and transfer applications in remaining useful life estimation. *IEEE Transactions on Instrumentation and Measurement* 72.

Zhang, L.B., and J.Q. Hu. 2013. Safety prognostic technology in complex petroleum engineering systems: progress, challenges and emerging trends. *Petroleum Science* 10 (4): 486–493.

Zhang, N., A. Alipour, and L. Coronel. 2018. Application of novel recovery techniques to enhance the resilience of transportation networks. *Transportation Research Record* 2672 (1): 138–147.

Zhang, N., K. Q. Cai, Y. J. Deng, and J. Zhang. 2024. Joint optimization of condition-based maintenance and condition-based production of a single equipment considering random yield and maintenance delay. *Reliability Engineering & System Safety* 24[illegible].

Zhang, N., M. Fouladirad, A. Barros, and J. Zhang. 2021a. Reliability and maintenance analysis of a degradation-threshold-shock model for a system in a dynamic environment. *Applied Mathematical Modelling* 91: 549–562.

Zhang, S.Y., Y.F. Yan, L. Shi, L.F. Li, L. Zhao, R. Wang, and X.Z. Yan. 2021b. A semi-empirical model for underground gas storage injection-production string time series remaining useful life analysis in process safety operation. *Process Safety and Environmental Protection* 154: 1–17.

Zhang, Y. L., S. H. Chai, C. An, F. Lim, and M. L. Duan. 2022. A new method for optimal sensor placement considering multiple factors and its application to deepwater riser monitoring systems. *Ocean Engineering* 244.

Zhao, Y., W. Gao, and C. Smidts. 2021. Sequential Bayesian inference of transition rates in the hidden Markov model for multi-state system degradation. *Reliability Engineering & System Safety* 214.

Zhao, Y. X., B. P. Cai, H. H. S. Kang, and Y. L. Liu. 2023. Cascading failure analysis of multi-state loading dependent systems with application in an overloading piping network. *Reliability Engineering & System Safety* 231.

Zheng, L., N.S. Matin, J. Landon, G.A. Thomas, and K. Liu. 2016. CO2 loading-dependent corrosion of carbon steel and formation of corrosion products in anoxic 30 wt.% monoethanolamine-based solutions. *Corrosion Science* 102: 44–54.

Zheng, R., Y. Xing, and X. Y. Ren. 2023. Multilevel preventive replacement for a system subject to internal deterioration, external shocks, and dynamic missions. *Reliability Engineering & System Safety* 239.

Zhou, P., and P.T. Yin. 2019. An opportunistic condition-based maintenance strategy for offshore wind farm based on predictive analytics. *Renewable & Sustainable Energy Reviews* 109: 1–9.

Zhou, W., D.F. Wang, J.Y. Sheng, and B. Guo. 2012. Collaborative optimization of maintenance and spare ordering of continuously degrading systems. *Journal of Systems Engineering and Electronics* 23 (1): 63–70.

Zou, W.D., Y.Q. Xia, and H.F. Li. 2018. Fault diagnosis of Tennessee-Eastman process using orthogonal incremental extreme learning machine based on driving amount. *IEEE Transactions on Cybernetics* 48 (12): 3403–3410.

The manufacturer's authorised representative in the EU is Springer Nature Customer Service Centre GmbH, Europaplatz 3, 69115 Heidelberg, Germany. If you have any concerns regarding our products, please contact ProductSafety@springernature.com

Printed and bound by CPI Group (UK) Ltd, Croydon, CR0 4YY
01/10/2026
02225937-0017